CERAMIC MATERIALS RESEARCH TRENDS

CERAMIC MATERIALS RESEARCH TRENDS

PAUL B. LIN
EDITOR

Nova Science Publishers, Inc.
New York

For permission to use material from this book please contact us:
Telephone 631-231-7269; Fax 631-231-8175
Web Site: http://www.novapublishers.com

LIBRARY OF CONGRESS CATALOGING-IN-PUBLICATION DATA

Ceramic materials research trends / Paul B. Lin. Editor.
p. cm.
Includes index.
ISBN-13: 978-1-60021-769-2 (hardcover)
ISBN-10: 1-60021-769-9 (hardcover)
1. Ceramic materials. I. Lin, Paul B.
TA455.C43C4635 2007
620.1'4--dc22 2007019181

Published by Nova Science Publishers, Inc. ✢ New York

CONTENTS

PREFACE

Ceramics are refractory, inorganic, and nonmetallic materials. They can be divided into two classes: traditional and advanced. Traditional ceramics include clay products, silicate glass and cement; while advanced ceramics consist of carbides (SiC), pure oxides (Al2O3), nitrides (Si3N4), non-silicate glasses and many others. Ceramics offer many advantages compared to other materials. They are harder and stiffer than steel; more heat and corrosion resistant than metals or polymers; less dense than most metals and their alloys; and their raw materials are both plentiful and inexpensive. Ceramic materials display a wide range of properties which facilitate their use in many different product areas. This book presents new and important international research in this dynamic field.

Combustion synthesis (CS) is a modern, alternative fabrication method for advanced ceramics and ceramic composites that has existed since the early 1970s. CS has the advantage of being extremely time and energy efficient when compared to more conventional and commonly used processing techniques such as hot and cold isostatic pressing, hot pressing and sintering. These advantages stem from the fact that CS is dependent upon the ability of a highly exothermic chemical reaction to become self-sustaining after a short energy pulse is applied to initiate the reaction. Typically, there are two types of CS reaction modes for ceramics fabrication: (1) a propagating mode, whereby the sample is heated locally to an ignition temperature, heat generated from the chemical reaction heats adjacent layers to an ignition temperature that results in a self-sustaining combustion wave, termed *Self-propagating High-temperature Synthesis* (SHS), and (2) a simultaneous mode, whereby the sample is heated uniformly to an ignition temperature, heating finer particles faster than coarser particles, leading to a rapid, exothermic, simultaneous conversion of reactants into products, often termed *thermal explosion* (TE). These reactions have been investigated and modeled extensively in terms of thermodynamics and kinetics over this time period. Reviews of the CS process and resultant products are often concerned with discussing and documenting these details. However, a more current overview of CS processes and modifications is lacking, especially in terms of ceramic materials and how the process might be applied on a commercial and/or industrial scale and where the end products of the process may and are being used. Chapter 1 will highlight the extreme diversity of CS methods and ceramic materials that can be fabricated from these methods, so that those investigating lower cost, quicker and effective methods to produce advanced materials will be aware of the versatility, safety and potential of the CS process.

There has been growing interest in ferromagnetic manganites ($A_{1-x}T_xMnO_3$, where A is a trivalent rare-earth element such as La, Pr, Nd, Sm, Eu, etc. and T is a divalent alkaline earth ion such as Ca, Sr, Ba, Pb, etc.), owing to their unique and outstanding physical properties. In particular, the discovery of magnetoresistance (MR) and magnetoimpedance (MI) effects in these materials has provided exciting opportunities to develop them for use in magnetic sensors for magnetic recording technology, while the discovery of the magnetocaloric (MC) effect has made them promising for use in active magnetic refrigerators. In order to design and produce novel manganite materials for such applications, it is important to understand the physical mechanisms behind the MR, MI and MC phenomena. Chapter 2 aims to provide a comprehensive analysis of the complex nature of these phenomena and their correlation. Emphasis is placed on providing detailed guidance concerning the development of ferromagnetic manganites for future magnetic sensor and refrigerator technologies.

Directionally solidified eutectic oxide/oxide ceramic composites are promising for high temperature use as fiber materials for reinforcement and as bulk materials for machine parts due to the high mechanical property up to around 2000K, high oxidation-resistance, clean interface and high stability of the microstructure. In Chapter 3, focusing on the temperature dependence of mechanical properties of the directionally solidified bulk type Al_2O_3/YAG($Y_3Al_5O_{12}$) composite, the recent experimental results as well as the analytical results mainly based on the finite element analysis are presented. Following subjects are contained. (1) Temperature dependence of Young's modulus and coefficient of thermal expansion of the composite and its relation to that of the constituents (Al_2O_3 and YAG) at room temperature to 1800K. (2) The temperature- and displacement speed- dependence of the stress-strain curve, brittle-ductile transition and flow stress in the plastic deformation region of unnotched composite specimens at room temperature to 2023K. (3) The temperature- and displacement speed- dependence of the notched strength, fracture toughness, plastic zone ahead of the notch and plastic deformation of the ligament of notched specimens at room temperature to 2023K. (4) The stress- and strain rate- distributions in each constituent in plastic deformation, the influence of the morphology of the constituents on the flow stress of the composite and the relation of the microscopic behavior of the constituents to the overall flow stress of the composite at 1873K.

Perovskite-type oxides display a very large range of structural, physical and chemical properties. Due to the flexibility of the perovskite structure it is possible to make substitutions with cations of different size, formal charge and electronegativity, forming solid solutions with gradual or sudden changes in their properties. While cationic substitutions in perovskites have been intensively studied for decades comparatively little is know about substitutions in the anion lattice. Partial replacement of oxygen by nitrogen leads to oxynitrides. The lower electronegativity of nitrogen compared to oxygen results in an increase in the covalence of bonding when nitrogen is inserted in the anionic sublattice. In many cases the optical band gap is decreased to the visible region leading to intensively colored compounds. As a consequence of this optical property, oxynitride perovskites can be used as environmental-friendly pigments and visible light activated photocatalysts. In Chapter 4 the authors present different synthesis methods to produced oxynitride perovskites. Starting from the commonly used solid-state reaction, we turn to soft chemistry methods like the polyol synthesis and finally to the spray pyrolysis. The physical properties and morphologies of the compounds prepared by the various methods are discussed. Especially surface area and format aspects are compared among the samples. For this purpose, SEM micrographs are used to illustrate the different particle sizes and shapes obtained by each technique. Focusing on the oxynitrides, their optical property and photocatalytic potential were studied. A comparison between the synthesized oxynitrides and TiO_2 is presented. As known, TiO_2 is the most common photocatalyst because it has the advantages to be non-toxic, cheap, chemically/biologically inert, photo-catalytically stable and easy to produce. The big inconvenience of titanium dioxide is that it only absorbs in the ultra violet region. In contrast the oxynitride perovskites absorb in an adjustable range of the visible light and several studies have already demonstrated that some of them show considerable photocatalytic activity. Two different methods to verify the photocatalytic potential (i.e. decomposition of simple organic compound monitored by in-situ infra-red spectroscopy and the water splitting reaction) of selected samples are explained in detail.

Applications of superionic conductors in solid-oxide fuel cells (SOFCs) were considered in Chapter 5. A promising direction is the distributed power. Electrochemical current generators come to direct consumers of electric power, who operate the generators as required by their power consumption plan. As a result, the power is reasonably saved and the fuel is used efficiently. SOFCs provide a highly efficient direct conversion of the chemical energy of the fuel to electricity. Any hydrocarbons can be used as fuel in SOFCs. Natural fuels (coal, oil and gas) and products of technical activities of man (industrial and agricultural wastes) are used. Electrochemical "low-temperature" burning provides high ecological compatibility of the process. All these factors open up good prospects for the use of SOFCs to satisfy ever growing requirements of mankind in the electric power [1-6].

In the last decades, separation processes employing membranes have experienced a remarkable increase at an industrial level as an alternative to traditional techniques, mainly due to the mild operating conditions and low energy consumption.

The main problem associated with membrane filtration is the fouling caused by the adsorption of material from the feed stream on to the membrane surface or inside the pores. This phenomenon involves a significant decrease in both selectivity and permeate flux and, as a consequence, appropriate cleaning protocols are needed to regenerate the membranes. An intensive research effort has been made in order to obtain porous materials which minimize fouling effects and resist aggressive cleaning.

Since the pioneering use of cellulose acetate, other polymers such as polyamide, polysulfone, polyvinylidene fluoride have been employed for membrane manufacture. More recently, ceramic membranes have been developed based on zirconium and aluminum oxides. Ceramic membranes show excellent resistance to extreme pH, temperatures and pressures, as well as to solvents and oxidizers. They are also steam-sterilizable and resistant to microorganisms. Due to these desirable properties, ceramic membranes represent a major advance in the field of membrane technology, since new applications can be devised in the food, biotechnology and chemical industry.

The aim of Chapter 6 is to present an overview reflecting the state of the art in the applications of ceramic membranes in the food industry. The chapter will be organized as follows. As an introduction, a brief description of general aspects of the pressure driven membrane separation technology will be made, including fouling and cleaning aspects. Then, the production and characterization of ceramic membranes will be undertaken and some details of commercially available membranes will be given.

The main section will be devoted to the detailed analysis of the applications. A special attention will be paid to the dairy industry, where ceramic membranes are employed in the concentration of milk for cheese manufacture, the concentration of whey protein and for the isolation of bioactive compounds. Other sectors that which benefit from membrane technology will be exposed such as fruit juices, alcoholic beverages, sugar and, finally, vegetable and animal proteins.

High temperature X-ray and neutron diffraction combined with Rietveld analysis are powerful tools to investigate the thermal atomic shifts in the crystal structure of ceramic materials and infer the distortions responsible for their dilatometric behaviour. Chapter 7 concerns several phosphates of medium and big tetravalent cations (Zr, Hf, Th, U) with very low or negative thermal expansion, including β-$Zr_2O(PO_4)_2$, a ceramic refractory known for its high resistance to thermal shocks. These materials have revealed new mechanisms or combinations of mechanisms that widen the usual field of researches for low expansion materials. Structural features previously ignored like cavities, pairs or chains of cations at short distances, or considered as unsuitable like polyhedra connected by edges should be considered now as possible hints for contraction mechanisms.

In Chapter 8, diopside was introduced into alumina matrix ceramic materials as a sintering aid and fine structural alumina matrix ceramic materials were fabricated by the technology of pressureless sintering, during which new phases were produced. The relative density, hardness, fracture toughness and bending strength of the composites were tested. The experiment results show that the introduction of diopside into alumina matrix can enhance the densification rate due to the generation of liquid phase in the sintering process, which leads to the improvement of the mechanical properties of the composites. The best mechanical properties were achieved when 3vol.% diopside was added into the composite, the hardness, bending strength and fracture toughness of which are 15.57GPa, 417MPa and 5.2MPa•$m^{1/2}$, respectively. The effects of diopside on mechanical properties, fracture mechanism and interspaces of fine structural alumina matrix ceramic materials were analyzed together with the toughening mechanism of the composites.

The preparation of nanoporous ceramic membranes and their performance in non-aqueous nanofiltration is discussed. Starting from the typical manufacturing method for hydrophilic nanofiltration membranes, an evaluation is made of the performance of these membranes, relative to other membrane types. Fluxes for polar and non-polar solvents are studied for TiO_2 membranes with changes in the preparation procedure (e.g., the firing temperature). Rejections for some typical organic compounds are given. Suitable membranes in the nanofiltration range, i.e., with a molecular weight cut-off around 500 g/mole, could be obtained.

Possibilities for obtaining a (more) hydrophobic surface are described in Chapter 9. One of the most attractive options is the introduction of silane groups into a mesoporous membrane structure. It can be shown that after silane coupling, a hydrophobic surface is obtained suitable for filtration of non-polar solvents. Fluxes and rejections (and, correspondingly, pore sizes) depend on the chain length of the silane coupling agent. Furthermore, the hydrophobic nature and the stability of the membrane depends on the conditions used during the modification procedure. In particular, it was shown that contact with water in some cases partly reverses the modification.

The permeabilities of hydrophilic and hydrophobic membranes were analysed in terms of (non-)viscous behaviour. It was found that for hydrophilic membranes with typical nanofiltration separation characteristics, permeabilities deviate from viscous flow behaviour due to polarity effects. Mesoporous hydrophilic membranes are less influenced by polarity effects and are generally closer to viscous behaviour. Hydrophobic membranes also deviate from viscous behaviour, although the conclusions for these membranes are less obvious.

The phenomenon of non-Ohmic conduction (varistor effect) observed in tin dioxide based and zinc oxide based semiconductor oxide ceramics is discussed. Although ZnO varistors are widely known, last years significant progress in the development of highly nonlinear SnO_2-based varistors is reached since in 1986-1987 the first SnO_2 varistors without zinc oxide addition in a system SnO_2-Bi_2O_3-Co_3O_4-BaO-Nb_2O_5 with the nonlinearity coefficient $\beta = 21$ and the electric field E_1=3500 V/cm were obtained. Results on SnO_2 and ZnO varistors obtained using a wide range of techniques are discussed. Special attention to newly developed SnO_2-based ceramic varistors with quite high nonlinearity coefficient about 40-60 and higher is given. Barrier nature of electrical conduction in oxide ceramics is discussed. Experimental evidence of grain-boundary potential barriers in oxide varistor ceramics is outlined. Common features and differences between SnO_2 and ZnO varistors are analyzed. Earlier conclusion that electrical properties of SnO_2 varistors are very similar to the properties of ZnO varistors and that non-Ohmic conduction in SnO_2 and ZnO varistors have a common physical nature related to electron transport across the grain-boundary barriers is additionally confirmed. Critical review of recent results in the field of SnO_2 varistors is presented. The approach to oxide varistor ceramics as a system with different grain boundaries is discussed. From fairly general assumptions about the nature of non-Ohmic conduction in oxide ceramics (decrease of the barrier height on the electric field) the relationship $j = \sigma_0 E \exp(\alpha E)$ between the current density j and the electric field E is suggested. This expression can be used for the approximation of current-voltage characteristics of SnO_2 and ZnO varistors. The nonlinearity factor α or the normalized nonlinearity coefficient β_E can be used instead of the traditional but empirical nonlinearity coefficient β for an estimation of the nonlinearity of current-voltage characteristic of a varistor and for a comparison of the nonlinearity of varistors having different values of the electric field at fixed current density. The relation between them is given by the expression $\alpha \approx \beta_E = (\beta / E_1)$. The problems in the area of tin dioxide varistors and desirable future developments are outlined in Chapter 10.

The immobilization of chromium ion in ceramic bodies produced from ashes generated by the incineration of chromium tanned leather shavings (ACS) and using waste of soda-lime glass as raw material is discussed in Chapter 11.

The raw materials, leather shavings and soda-lime glass, were mixed in different proportions. In order to investigate the phenomena acting in the chromium ion immobilization process, pure oxides - Na_2O, TiO_2, MgO and CaO – were mixed with soda-lime glass or added to ceramic mixtures with ASC and soda-lime glass. These samples were dry-pressed and fired in an electrical furnace, at temperatures of 750, 800, 950 and 1000°C. The ceramic bodies were evaluated with respect to physical properties and mineralogical composition by x-ray diffraction and EDS mapping. The chromium immobilization was characterized by leaching tests according Brazilian Regulations (NBR 10005).

The higher the firing temperature, the more effective the chromium immobilization in the ceramic bodies investigated. Raising the firing temperature diminishes the open porosity, by glass phase formation, leading to a lower content of chromium in the leaching water.

However, when adding only glass to the ACS, the content of chromium in the leaching water increased. The lixiviation decreased when higher temperatures were used. This occurred because there are more leaching phases, such as, sodium chromates, originated from soda-lime glass and ACS reactions. All ceramic bodies containing sodium chromate investigated were inefficient in immobilizing the chromium ion.

The results confirmed the immobilization of the chromium ion within the allowed limits of Brazilian regulations, NBR 10004 (5 mg/L), by means of incorporation in glass and glazing/densification agents such as titanium and magnesium oxides.

As explained in Chapter 12, impedance spectroscopy enables the many contributions to the dielectric and resistive properties of electroceramic materials to be deconvoluted and characterised separately. The charge transport mechanism and the macroscopic dielectric constant in polycrystalline materials commonly consist of several contributions such as sample-electrode interface, grain boundary and bulk resistance/capacitance. Impedance spectroscopy is shown to be a powerful tool to account for and model such behaviour. In order to gain precise understanding of the functionality of polycrystalline electroceramic materials it is essential to deconvolute these contributions.

Basically, the formation of nanoislands can be achieved through two main mechanisms:

- Nucleation of islands, growing and sometimes coarsening; this scheme is often observed for PVD-CVD processes (refs).
- Surface roughening, islanding and growing; this scheme is observed both for sol-gel and PVD-CVD processes (refs).

Chapter 13 focuses on the second idea and the main objective of this study is to highlight the main physical parameters responsible for the islanding of thin solid films through surface roughening synthesized either by sol-gel or PVD-CVD processes. Therefore the authors present two physical models dedicated to simulate the formation of nanoislands without nucleation, either during thermal treatments of thin solid films synthesized by sol-gel dip-coating or during the deposition of matter on a substrate by PVD-CVD processes. They highlight the numerical results by experimental data concerning either the islanding process of yttria-stabilized zirconia thin films deposited on corundum substrates or the formation of germanium QD's on silicon substrates.

The models describe thin solid films surface rougheness evolution from continuous layers to nanoislands. The authors chose classical Monte Carlo approaches with Metropolis scheme to describe the evolution of the systems from mesoscopic length scale. These approach allows one to compare, at least qualitatively, the main physical parameters responsible for the formation of nanoislands between the two synthesis process described previously through simple physical concepts.

In section 2, they present experimental details and morphological analysis of yttriastabilized zirconia nanoislands on sapphire substrates. The section 3 gives a description of the physical models used to simulate the islanding mechanism for thin solid films synthesized either by sol-gel or PVD-CVD processes and section 4 is the general conclusion.

Expert Commentaries

In: Ceramic Materials Research Trends
Editor: Paul B. Lin, pp. 3-4
ISBN: 1-60021-769-9

Cementing the Future: A Review of Past, Present and Future Use of Cement and Its Combination with Piezoelectric PZT Ceramic

*Arnon Chaipanich**
Department of Physics, Faculty of Science, Chiang Mai University, Chiang Mai 50200, Thailand

The first cement developed was Portland cement by Joseph Aspdin in 1824. This is because of the resemblance to the Portland stone. Today, it is still used as one of the main construction materials, although more so in combination with by-products which are pozzolanic meaning that they can react with calcium hydroxide, a hydration product from Portland cement and give better performance. The use of such by-products replacing Portland cement help the environment by cutting down the overall usage of Portland cement thus indirectly reducing the carbon dioxide emission. Also the energy required would be lower. In the future, much work still needed to be carried out on the development of such materials those perhaps can be referred to as low-energy low-cost materials. Ideally to produce reasonably high strength materials by using minimum energy and cost. This indeed is a great challenge to tackle since the most obvious source of low-energy and low-cost are waste materials, and they do not always have the suitable properties to be used as construction materials. In some materials which were first recognized as waste due to no utilization of the materials initially but as soon as they are found to posses the right properties, they become valuable both in the desirable properties themselves and the cost that seemed to be continually rising due to demand. Physical and chemical characteristics of materials used are so critical to the performance of hardened cements. Sub-micron or nano size materials can physically improve the cement performance such as in the case of silica fume which is apporoximately 100nm in diameter. The extensive use of silica fume is found in structure where high strength is required. Recent investigation has also been on the use of carbon nanotubes in order to improve the cementitious properties. Of course, as soon as one material is blended with

* E-mail address: arnon@chiangmai.ac.th

another it either from a binary or multi-blended cements. The recognition nowadays is that each cement type is referred to in its own right.

Lets now move on to the use of cement with piezoelectric ceramic material. Lead zirconate titanate, $Pb(Zr_{0.52}Ti_{0.48})O_3$ (PZT), piezoelectric ceramic exhibits high dielectric constant and high piezoelectric coupling coefficient and is recognized as a smart material where it can be used in numerous applications such as sensors and actuators. Examples of smart systems are smart toilet that is capable of analyzing urine to identify health problems and smart transducer which can "talk" to fish. For modern structures, sensors made of piezoelectric material such as PZT can be inserted or placed outside the structure where it can detect loadings exerted on the structures, thus a sophisticated structure sometimes being referred to as smart structure. However, PZT itself does not match the host structure, concrete and there have been developments in the use of cement based composites consisting of PZT in order to provide the matching required. The use of such piezoelectric-cement composite would provide an advantage in the matching of structural concrete. Therefore the use of piezoelectric PZT with cement was found to be compatible to the host construction material, concrete. These piezoelectric PZT-cement based composites are being developed for use as sensors in smart structures. The benefit from the use of such piezoelectric-cement based composite is that these composites possessed better matching to the concrete host structure than the normal piezoelectric ceramic or other types of piezoelectric composites. There have been investigations both by the author and other researchers in this area. The main difficulty found and reported is the poling process in order to obtain piezoelectricity. This process requires both skill and patience but once achieved, the composites possess high enough piezoelectric coefficient for them to be used in sensor applications.

In: Ceramic Materials Research Trends
Editor: Paul B. Lin, pp. 5-7
ISBN: 1-60021-769-9

COOPERATION INSTEAD OF COMPETITION MAY DO THE TRICK

Rainer Schmidt*
University of Cambridge, Department of Materials Science & Metallurgy, Pembroke Street, Cambridge CB2 3QZ, United Kingdom

Ceramic materials commonly exhibit strong potential disorder. Therefore, the Hamiltonian for valence electrons contains kinetic and potential energy terms, where the potential needs to account for the strong disorder. This is often achieved by calculating the mean potential field created by such disorder. In many materials the disorder is not time invariant though, and the most advanced theory to date to account for such complex behaviour may be "dynamical mean field theory". A much simplified approach is the assumption of electron localisation due to such disorder and the introduction of a central Coulomb potential into the Hamiltonian. This model neglects the time dependency of the potential and reduces the complicated mean field problem to a simple scaling problem. The electron localisation due to disorder is averaged and described by an effective electron localisation length α, which is sufficient to treat many aspects of the electronic structure of ceramics. For example, electron transport can then be understood in a simple intuitive picture of localised electron hopping and transport data can be described precisely. It is fully established now that the low temperature ground state of valence electrons in ceramics is a localised electronic structure. Still, valence electrons can move out of the ground state and delocalise, for example in the presence of strong double-exchange interactions.

Electron localisation leads to strong interactions between electrons and electrons, and electrons and crystal lattice, which can give rise not only to double-exchange but to a rich variety of various magnetic spin interactions, ordering phenomena, and type II superconductivity. The lattice constitutes a phonon heat batch (therefore electron localisation is also referred to as electron-phonon coupling), and the degree of electron localisation and the electron phonon coupling strength determine or strongly influence ordering and interaction phenomena.

* E-mail address: rainerxschmidt@googlemail.com

Different exchange interactions between electrons, or electrons and crystal lattice, often compete with each other. In the most extreme case, this can lead to phase coexistence on different length scales within one single crystal, if the competition of different mechanisms for predominance is close. Small local variations in the crystal field potential, stoichiometry or strain can then result in phase separation.

Such phase separation or coexistence has been the object of a great boom in research activity in the last 15 years and has produced numerous outstanding publications in top ranked journals. The "manganites" have been the main study object in this context. These are compounds with perovskite or perovskite related structures of the general formula $(Re_{1-x}Ae_x)MnO_3$, where Re is a rare earth element (usually La or Pr, more exotic compounds can contain Nd or Eu), and Ae an alkaline earth metal (usually Ca or Sr, sometimes Ba). Despite an inflation of publications on effects occurring in these materials, such as the phase coexistence based Colossal Magneto-Resistance (CMR) effect, charge and orbital ordering and exotic phase transitions, industrial applications of these materials have been scarce, because application requirements have been largely unmet so far.

More recently, a new boom in research activity has unfolded concerning ceramic "multiferroics", and it is hoped that it will generate a more fruitful outcome in terms of applications.

The basis of this new trend is the fact that different electron lattice or electron electron interactions can also coexist or even occur in cooperation with each other. The multiferroics are the most heavily studied compounds in this context, because they exhibit at least two ferroic properties simultaneously (thus termed "multi-ferroics"). In most cases this is (anti-) ferromagnetism/ferrimagnetism and (anti-)ferroelectricity. The research interest into these compounds is driven by long-term technological aspirations, for example by potential application of such materials in multi-state memories or memories with dual read-write mechanism or other devices. Multi-state memories would require the magnetic and electric order to occur independently from each other, or at least that they can be switched independently. This is probably the exception in single phase multiferroic materials. Most multiferroics discovered so far usually show a strong magneto-electric coupling effect, which implies that magnetic order can be addressed by the application of an electric field and vice versa. This would allow the realisation of dual-read-write mechanisms in multiferroic data storage media.

The examples of compounds with co-existing ferromagnetic and ferroelectric order are scarce due to the obvious incompatibility: usually ferromagnets are metals and ferroelectrics are insulators. Such incompatibility can be overcome in binary oxides, a family of compounds, which contains several of such multiferroic species. Perovskite $BiFeO_3$ and $BiMnO_3$ are key examples of such binary oxides. Ferroelectricity arises from trivalent A-site Bi 6s "lone pair" electrons causing a displacement of the B-site cations from the octahedral centro-symmetric positions and the resulting dipole moment.

(Anti-) Ferromagnetism arises from electron spin alignment due to superexchange (antiferromagnetic) or double exchange (ferromagnetic) interactions between the perovskite B-site trivalent cations. $BiMnO_3$ is ferromagnetic below $T_C \sim 105K$ with a spontaneous magnetisation of $\sim 3.6\ \mu_B/Mn$ in bulk and $\sim 2.2\ \mu_B/Mn$ in epitaxial films. Due to $T_C \sim 105K$, $BiMnO_3$ is not useful for room temperature multiferroic application, for example in data storage media. However, the large magnetic moment below T_C may facilitate the study of fundamental aspects of multiferroic magnetism.

Bulk $BiFeO_3$ is anti-ferromagnetic, where the anti-ferromagnetic spins are canted and a spiral spin structure leads to a zero net magnetisation. The latent magnetisation can be released though in epitaxial $BiFeO_3$ films, because the spiral spin structure is suppressed by epitaxial constraint and/or the fact that film thickness may be of the same order of magnitude as the spiral spin structure. $BiFeO_3$ is the most extensively studied multiferroic to date, because it is the only material known to exhibit spontaneous magnetic and electric order at room temperature (in epitaxial films it is ferrimagnetism and ferroelectricity). However, the ferrimagnetic moment in epitaxial layers has been proven to be too small to be useful for applications ($\sim 0.05\ \mu_B$/Mn), and the search for the "holy grail" of multiferroics continues: a room temperature multiferroic with a useful ferromagnetic moment and either strong or non-existent coupling of ferromagnetism and ferroelectricity.

A promising approach may be the fabrication of heterogeneous multiferroics. These are two component systems made of a ferroelectric and a ferromagnetic material. Strong coupling between the two components could be realised for example in thin films. The two different phases may be distributed laterally, e.g. ferromagnetic nano-pillars in a ferroelectric matrix or vice versa. Alternatively, a vertical multilayer arrangement may be possible also. The growth of ferromagnetic epitaxial films, possibly ruthenates or manganites, on a ferroelectric substrate such as $BaTiO_3$ may result in useful coupling between magnetic and electric order. More importantly, the fabrication of composites could allow realisation of uncoupled devices for multi-state memories.

RESEARCH AND REVIEW STUDIES

In: Ceramic Materials Research Trends
Editor: Paul B. Lin, pp. 11-46
ISBN: 1-60021-769-9

Chapter 1

A Review of the Application of Combustion Synthesis in Advanced Ceramic Materials Processing

Douglas E. Burkes[1] and John J. Moore[2]

[1]Idaho National Laboratory, Nuclear Fuels and Materials Division, P. O. Box 1625, Idaho Falls, ID, U. S. A. 83415-6188

[2]Reaction Synthesis Laboratory, Colorado School of Mines, Department of Metallurgical and Materials Engineering, 1500 Illinois Street, Golden, CO, U. S. A. 80401-1887

Abstract

Combustion synthesis (CS) is a modern, alternative fabrication method for advanced ceramics and ceramic composites that has existed since the early 1970s. CS has the advantage of being extremely time and energy efficient when compared to more conventional and commonly used processing techniques such as hot and cold isostatic pressing, hot pressing and sintering. These advantages stem from the fact that CS is dependent upon the ability of a highly exothermic chemical reaction to become self-sustaining after a short energy pulse is applied to initiate the reaction. Typically, there are two types of CS reaction modes for ceramics fabrication: (1) a propagating mode, whereby the sample is heated locally to an ignition temperature, heat generated from the chemical reaction heats adjacent layers to an ignition temperature that results in a self-sustaining combustion wave, termed *Self-propagating High-temperature Synthesis* (SHS), and (2) a simultaneous mode, whereby the sample is heated uniformly to an ignition temperature, heating finer particles faster than coarser particles, leading to a rapid, exothermic, simultaneous conversion of reactants into products, often termed *thermal explosion* (TE). These reactions have been investigated and modeled extensively in terms of thermodynamics and kinetics over this time period. Reviews of the CS process and resultant products are often concerned with discussing and documenting these details. However, a more current overview of CS processes and modifications is lacking, especially in terms of ceramic materials and how the process might be applied on a commercial and/or industrial scale and where the end products of the process may and are being used. This chapter will highlight the extreme diversity of CS methods and ceramic materials that can be fabricated from these methods, so that those investigating lower cost, quicker and effective methods to produce advanced materials will be aware of the versatility, safety and potential of the CS process.

Introduction

Combustion synthesis (CS) is an alternative technique of producing advanced ceramics, ceramic composites (CERCERs) and ceramic-metallic composites (CERMETs) [1]. Modern CS methods first appeared in the early 1970s in the former Soviet Union resulting from investigations by A. G. Merzhanov and colleagues [2]. CS has the advantage of being extremely time and energy efficient when compared to more conventional and commonly used ceramic processing techniques such as press and sintering, hot and cold isostatic pressing (HIPing and CIPing), hot pressing and reaction bonding. These advantages stem from the fact that CS is dependent upon the ability of a highly exothermic chemical reaction to become self-sustaining after a short energy pulse is applied to initiate the reaction. In addition, high temperatures and extended heating durations are typically required for fabrication of ceramic components employing conventional sintering and/or hot pressing methods because of the low diffusivity of ceramics. The low diffusivity is the result of a growth layer that thickens as a function of time acting as a diffusion barrier resulting in a decreased conversion rate. Rapid synthesis of the ceramic, such as that permitted by CS reactions, overcomes these limiting diffusion kinetics [3]. Converse to most other processing methods that require fine particle sizes to overcome the diffusion limitation, CS requires non-metallic particles that are coarser and in fact for a porous green body, fine particles are actually more detrimental to the CS reaction due to the decrease in thermal conductivity between particles. This of course can be overcome by increasing the green body density prior to the CS reaction.

Typically, there are two types of CS reaction modes: (1) a propagating mode, whereby the sample is heated locally to an ignition temperature, and where the heat generated from the chemical reaction heats adjacent layers to an ignition temperature, resulting in a self-sustaining combustion wave, termed *Self-propagating High-temperature Synthesis* (SHS), and (2) a simultaneous mode, whereby the sample is heated uniformly to an ignition temperature, heating finer particles faster than coarser particles, leading to a rapid, exothermic, near-simultaneous conversion of reactants into products, often termed *thermal explosion* (TE). Use of the word 'explosion' should not be misinterpreted, as this can often be misconstrued to mean 'unsafe' or 'uncontrollable'. The term 'explosion' only implies a simultaneous reaction whereby all of the reactants are converted to products simultaneously, without the presence of a combustion wave. The CS technique is extremely safe and controllable through the initial parameters of the reaction system, e.g. reactant stoichiometry, initial sample 'green' density, reactant particle size, etc. CS, when used in the appropriate manner, contains no more risks than any other fabrication method. The CS process can also be combined with the simultaneous application of pressure, resulting in the fabrication of fully (99.9%) or near-fully dense (>95%) ceramics, often termed reaction forging. This aspect, combined with the time and energy savings of the CS process, make this an extremely viable alternative to the traditionally accepted fabrication techniques. The selection of the CS reaction mode is dependent upon a number of initial reaction parameters, such as desired product density, chemical reactivity (effective thermal conductivity and initial 'green' density of the reactant mixture), adiabatic temperature (effective heat capacity of the reactant mixture), desired grain size, impurity content, microstructure, etc.

CS has been investigated for over forty years, during which time the technology of the process has continually evolved. Commercial and governmental applications of this amazing process have only recently begun to come to fruition, within the last ten years. However, the full potential of the CS fabrication process to such applications is nowhere close to being met. The current review seeks to discuss the versatility and applications of the CS process in terms of ceramic and ceramic composite (CERCER and CERMET) fabrication, rather than focusing on the mechanistic and fundamental impacts, such as kinetic, thermodynamic and modeling aspects. Reviews and discussions of the latter can be found throughout the hundreds of journal articles available on CS processes [2,4,5]. Often times, only one or two sentences in the introduction of CS papers discuss a "proposed" or "ideal" application for the final product. Again, the intention of this review is to document some of these applications, both real and perceived, as well as providing a supporting basis for the versatility of fabrication employing CS. Specific CS fabrication techniques and modifications along with associated parameters of the selected ceramic systems are also discussed.

Combustion Synthesis Methods

In order to give an appropriate review of applications of CS reactions, a brief overview of the different CS fabrication methods and terminology is required. For the purposes of this paper and to minimize the use of specific processes, a description of each process is used in general terms so that it may be interchangeable with similar CS methods, although subtle differences between the methods may exist. The following paragraphs will provide this brief background, although some variations in terminology and processes not yet well documented may not be included.

Spark Plasma Sintering (SPS) is a technique to fabricate dense materials by pulsation of an AC or DC current through a reactant mixture while simultaneously applying a uniaxial pressure. As such, advantages over more conventional techniques such as hot-pressing and pressure-less sintering include lower sintering temperature, shorter hold time and marked comparative improvements in material properties that result in increased control of grain size, and therefore mechanical properties, and the ability to consolidate nanometric powders [6]. For the context of this paper, "synthesis" is interchangeable with "sintering" in SPS, since applications discussed involve a chemical reaction step prior to or during consolidation. The SPS technique seeks to apply Joule heating to the reactant compact allowing high heating rates and minimizing surface diffusion mechanisms (e.g. Ostwald ripening) while applying the pressure that aids in particle re-arrangement and enhanced consolidation of the product [7]. A schematic of the SPS set-up is provided in Figure 1.

Reaction Synthesis (RS) has recently been developed for use in joining and welding applications of metallic, ceramic and polymeric systems. RS represents an ideal joining and welding process since heating is localized to the interface during joining so that temperature sensitive and thermally incompatible components or materials are joined without thermal damage [8]. In addition, no external heat source except that necessary to initiate the exothermic reaction is required and the reaction can be carried out under vacuum or any type of gaseous atmosphere. Thus, the RS process can be used to fabricate near-net shape joints, for *in-situ* joining applications, allow greater freedom over joining designs and result in increased material and time savings.

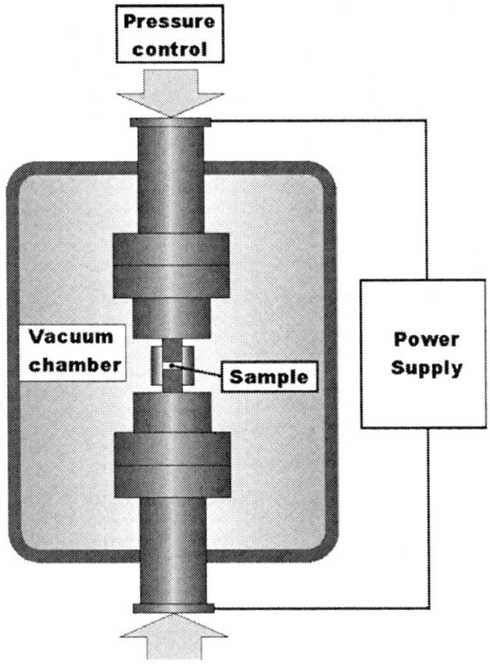

Figure 1. A schematic of typical SPS system set-up [Reprinted from Ref. 6, with permission from Springer].

Figure 2. Photograph of a typical AICS reaction to fabricate nanocrystalline yttria-doped zirconia powders [Reprinted from Ref. 9, with permission from Elsevier].

Auto-Ignition Combustion Synthesis (AICS) is a process characterized by rapidly heating an aqueous solution containing metallic, organometallic (e.g. nitrates, phosphates, acetates) and/or organic (e.g. urea, stearic acid, carbohydrazide) components that trigger an exothermic reaction resulting in high purity micro- to nano-sized powders. An example of an AICS reaction carried out on the laboratory scale in a Pyrex beaker on a hot plate is provided in Figure 2. The significant advantages of this fabrication method over traditional powder fabrication methods, such as direct nitridation, include high temperatures achieved during reaction, volatilization of any impurities and the organic components, and rapid heating and cooling times, minimizing agglomeration experienced by the powder product(s).

Biomaterial Systems

As the population of the world begins to age and older generations strive to enjoy an active life, effective biomaterial implants with minimal cost are becoming increasingly more important. It is estimated that over 500,000 joint replacement procedures are performed in the United States each year alone, while an additional 40 million Americans suffer from some type of arthritic condition [10]. Porous materials have shown promise as medical devices ranging from dental implants to large bone and joint replacement implants. Materials for bone and joint replacement applications typically require product porosity in the range of 40 to 60% and pore sizes between 200 and 500 microns, although this is dependent upon the intended anatomical location of the implant [11]. CS offers an attractive fabrication route for porous implants since porosity is an inherent attribute of the reaction providing a high volume of interconnected (open) pores accommodating vascular tissue that is required for continued mineralized bone growth [12,13]. In addition, CS products are often fabricated in a sterile condition since CS reactions occur at elevated temperatures, minimizing the need for additional cleaning before use. Ceramics are usually selected to fill the role of implant materials based on their ability to facilitate good bone integration features over other biomaterial options that tend to result in an encapsulation of the implant, e.g. metal alloys (steel or cobalt) and polymers (methyl methacrylate) [14]. The main characteristics of ceramic materials for favorable use as biomaterials are provided in Figure 3. Ceramics can compose the entire portion of the implant, such as calcium phosphates used for bone scaffolds or drug delivery systems, or they can be deposited onto a metallic substrate providing a 'bioactive'[1] layer to stimulate bone ingrowth into the more durable metallic core. Human bone, such as that found in a hip joint, differs from person to person. Commercial production of ceramic implants often assumes that an implant fits all patients. This assumption keeps fabrication costs low and requires doctors performing the implantation to make changes 'on the fly' during surgery. This can potentially lead to increased risk of infection and/or increased recovery times. Coating processes, such as chemical vapor deposition (CVD) and physical vapor deposition (PVD), are capable of coating only small implants, and are often fairly expensive to implement. Other techniques like thermal spraying are lower cost, but may not provide a coherent, durable, uniform coating to the implant. Thus, CS finds itself in an attractive position allowing for individualized, near net shape ceramic implants to be

[1] A bioactive material allows a specific biological response at its interface, enabling the formation of links between the tissues and said material [15]

fabricated before surgery begins. The concept is based on pre-surgical measurements on the surgical area to be addressed. These measurements are then fed into a rapid prototyping machine creating a green implant from the reactant powders, e.g. calcium oxide and phosphorous pentoxide for calcium phosphate, that are then ignited resulting in the desired implant in a sterile condition. An example photograph of a calcium phosphate shim fit to a specific shape before subjecting to the CS process is provided in Figure 4. Employing such a method ultimately keep costs low, since only the necessary amount of implants is fabricated, and would presumably improve patient recovery since surgical time is reduced.

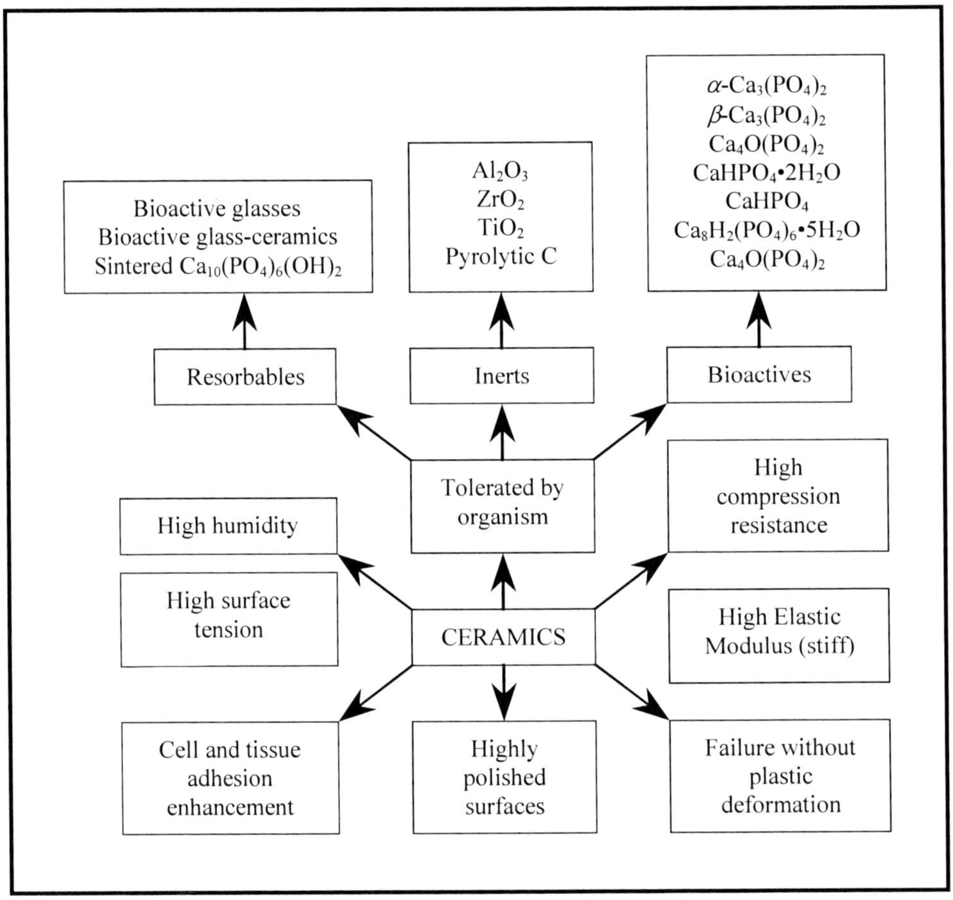

Figure 3. The main characteristics and types of ceramic materials for favorable medical implant applications (after Ref. 14).

Calcium phosphates are desirable for use as biomaterials since this compound is the mineral component of natural bone, and thus have a high degree of biocompatibility and may even stimulate bone growth. Porous calcium phosphate compounds (tri-calcium phosphate, pyrophosphate and orthophosphate) that could potentially be employed as bone replacement scaffolds have been produced employing propagating mode CS from CaO and P_2O_5 reactants. A direct correlation was observed between the amount of CaO reactant and combustion temperature, and therefore control of the phases produced. For example, CaO ratios greater than three ($CaO:P_2O_5$ ratio) reduced the combustion temperature and resulted in more apatitic phases, such as $Ca_5(PO_4)_3$, while higher combustion temperatures produced more tri-calcium phosphates, e.g. α-$Ca_3(PO_4)_2$ [16]. Correspondingly, as CaO is increased, the P_2O_5 reactant must be decreased, resulting in enhanced densification since P_2O_5 volatilizes during the CS reaction. A SEM photomicrograph showing the biocompatibility of SaOS-2 cells with a CS fabricated calcium phosphate shim is provided in Figure 4.

Figure 4. Unreacted, 'green' elongated 'C' calcium phosphate shim for bone reconstruction that is approximately 2 in. tall (left) and extracellular calcium phosphate (black arrowhead) adjacent to SaOS-2 cell trailing edge (white arrowhead) within a CS-produced tricalcium phosphate particle visible at the lower left corner of the image (right) [Reprinted from Ref. 16, with permission from John Wiley & Sons, Inc.].

Biocompatibility of calcium phosphate is not only based upon product stoichiometry, but also purity and crystallinity of the compound, while desirable potential sintering and mechanical characteristics are dependent upon particle size, shape, distribution and lack of agglomeration. Conventional powder fabrication methods, e.g. solid state synthesis, co precipitation, sol-gel synthesis, hydrothermal reaction, etc., involve expensive starting materials or processing equipment and may only be applicable to laboratory scale experiments. Auto-Ignition Combustion Synthesis (AICS) offers a large scale fabrication scheme, i.e. hot plate or furnace, with inexpensive starting materials, i.e. calcium nitrate, ammonium phosphate (di- or mono-basic) and a suitable organic fuel, with a short reaction time. Citric acid has been used as an organic fuel to produce nanocrystalline hydroxyapatite powder with enhanced sinterability characteristics after a low calcination treatment at 750°C [17]. Minor elements, such as Na^+, Mg^{2+}, K^+ and Cl^-, can be introduced with the hydroxyapatite and tri-calcium phosphate bioceramic powders when AICS is carried out in a simulated body fluid (SBF) and a short calcination treatment at 1150°C in air [18]. Incorporation of these minor elements into the fabrication product increases the biocompatibility of the powder because the overall chemical composition is closer to that of natural human bone.

Joining and Welding

As materials and composites become more complex in nature, especially those that are near net shaped, joining and welding becomes increasingly more challenging. Methods capable of joining these types of materials and different types of materials to one another, i.e. ceramics to metallics, metallics to polymers, etc., are therefore highly desirable. The greatest obstacle to overcome when using RS for joining and welding of materials is producing a high density reaction layer. Porosity is an inherent feature of all CS reactions and typically results from one or more of the following: differences in specific volume between the reactants and products, evaporation of impurities contained in the reactants (e.g. oxides, hydrides, adsorbed moisture, etc.), expansion of gas trapped in voids contained in the reactant compact during the high temperature reaction, the Kirkendall effect causing macroscopic expansion, and solidification shrinkage [19]. Although porosity can sometimes be a desirable product attribute, discussed in the previous section, it is extremely detrimental to the bond strength in welding and joining applications. A low-temperature brazing alloy containing a ceramic filler phase can be used between the RS joining composition and the material to be joined. Addition of a ceramic filler phase increases the exothermicity of the CS reaction, since enthalpies are typically greater for ceramics than metals, and increases the strength of the bond interlayer. Joining applications using RS offers the advantage of a highly graded bonding microstructure rather than a discrete interface produced from diffusion bonding under hot pressing conditions [20].

For example, titanium aluminide intermetallics are attractive for use in the aerospace industry, but are difficult to join with conventional fusion welding as a result of the high melting point [21] and diffusion bonding or brazing have low efficiencies and must be carried out under vacuum or in a protective environment to avoid oxidation [22-24]. Joining of TiAl has been accomplished using RS with an assisted electromagnetic field to form a TiAl-TiC bond layer at room temperature without the use of external furnaces [25].

The mechanical properties of nickel aluminides for high temperature structural applications such as the hot-end component of turbine blades can be significantly improved by joining the brittle intermetallic with a more ductile structural material, such as carbon or stainless steels. SPS was used to join nanocrystalline Ni_3Al and TiC/Ni_3Al composites in situ produced from elemental nickel, aluminum, titanium and carbon during a SPS reaction at temperatures below $1100^{o}C$ for only five minutes with a reasonably strong joint shear strength (~765 MPa) [26]. Thus, the SPS process was used to fabricate a Ni_3Al intermetallic and a TiC/Ni_3Al composite while near simultaneously joining the two materials to produce a functionally graded composite.

Zirconium-based bulk amorphous metallic glass alloys have been joined employing RS of Ni-Al multilayer foils with an observed increase in shear strength of the joints with increased foil thickness and applied pressure during the reaction [27]. RS is well suited for welding and joining of metallic glasses since the reaction provides an extremely rapid heating and cooling rate capable of maintaining the amorphous microstructure and thus desirable properties of the glass as well as allowing joining for a wide variety of joint geometries.

Casting

Melting and casting of intermetallics with the addition of a ceramic filler phase are desirable for fabrication of high-temperature components, such as turbine blades, rotors, heat-treatment trays, radiation tubes, catalytic converter substrates and high-temperature dies. Conventional powder metallurgy melting and casting demands large amounts of energy and time, combined with the fact that intermetallic castings often exhibit poor ductility that suggests the process is not ideal for fabrication of high quality components. Wear resistance and thermal creep resistance of iron based composites and conventional steels can be improved with the addition of a ceramic dispersive phase, i.e. Al_2O_3, ZrO_2, TiN_2, Si_3N_4, TiC, B_4C, VC, etc. These ceramic dispersive additions create a substantial and fitting opportunity for CS to be used in the fabrication and casting of such advanced iron based alloys and composites.

The principle behind using CS, particularly in the simultaneous mode so that the liquid alloy or composite product can be cast, is the heat generated by the reaction forming the dispersive phase, i.e. Ti + C → TiC, that is sufficient to melt the metallic iron and other additional elements as a result of the exothermic nature of the self-sustaining combustion reaction [28]. Iron additions ranging from 2.7% to 80% can be used to fabricate an alloy or composite without sacrificing the self-sustaining energy generation from the CS reaction [29].

Engineering of advanced materials often produces waste products that are discarded and assumed to have no further use. However, CS can be utilized to produce valuable ceramic composites from the unmarketable wastes in the form of a thermite reduction reaction that liberates a large amount of heat during reduction of an oxide by a metal. The main advantage of this method is the low cost associated with reduction of oxides, particularly those obtained as waste from an extraction process, in a time and energy efficient manner that does not require large amounts of heat input and complex melting and casting equipment. In this regard, CS can be employed as a waste-free processing technique, provide high conversion of the precursor components, and the possibility of producing near-net shape components.

For example, siliceous sand (73.1 wt% Fe_2O_3, 8.63 wt% SiO_2, 3.84 wt% SiO_2 and 2.51 wt% Na_2O) formed from aluminum extraction processes can combined with elemental aluminum and carbon powders and heated to initiate a CS reaction, whereby aluminum reduces the oxide phases to form a more stable aluminum oxide dispersion phase. Based on the heat generated from this process, titanium and carbon will react to form an additional TiC dispersion phase and melt the entire charge that is then cast. This process is represented by Equations 1 through 3 below [29].

$$3\ TiO_2 + 4\ Al + 3\ C \rightarrow 3\ TiC + 2\ Al_2O_3 \quad (1)$$

$$Fe_2O_3 + 2\ Al \rightarrow Al_2O_3 + 2\ Fe \quad (2)$$

$$SiO_2 + 2\ Al \rightarrow Si + Al_2O_3 \quad (3)$$

Similarly, conversion of 18Kh2N4MA-grade steel scale alloyed with 18% Cr, 2% Ni and 4% Mo though addition of a fused electric corundum and aluminum powder to fabricate porous, permeable materials is possible [30]. Porous, permeable Fe_2O_3-Al_2O_3 CERCERs fabricated employing the CS method from steel scale waste is ideal for filtering liquids and gases from particulates. These filters have been tested in an assembly line to clean motor oil and separate disperse copper particulates from industrial effluents with an operational efficiency comparable to that of commercially available filters, produced using raw materials rather than wastes.

As materials and composites become more advanced, materials to support structural aspects, machines and machining components must also advance. Advancement can be performed by using the low-cost CS method to cast high-quality, unique carbide steel alloys. For example, CS can be combined with a metallothermy process to cast carbide steels from charges containing a binder in the form of high-speed steel and tungsten carbide [31]. Similar Fe-based alloys with TiC ceramic filler materials for structural material applications can be fabricated from simultaneous mode CS with a densification step from elemental iron, titanium and carbon powders, with up to 75 wt% iron in the product [32].

High Temperature and Thermal Protection Materials

Gas turbines have become an attractive workhorse for electric power generating capability and as such, profit from an increase in thermal efficiency. In order to gain this improvement in thermal efficiency, thermal protection materials are often applied to the turbine base material allowing higher gas temperatures without a corresponding increase in the base material temperature. In light of the recent 2003 Columbia tragedy, new advancements in high temperature and thermal protection materials, especially those that can be produced 'on-the-fly', are highly desirable. Necessity of such advanced materials demands advanced fabrication techniques that provide high quality materials at lower overall costs and energy consumption, of which CS is well suited to fill this role.

For example, SiC ceramic and SiC-matrix composites are possible candidates for new, high temperature structural materials that could find use as low weight, next generation aircraft engine turbines. The application of SiC materials is dependent upon the ability to join the materials to one another, other ceramic material systems, or other metallic material systems, such as nickel-based superalloys, already widely used to manufacture the high temperature components of aircraft engine turbines. However, such an application involving SiC ceramic systems and Ni-based metallic alloys is hindered by significant differences in the thermal expansion coefficients (CTE) that result in a high concentration of thermal stress at the bond interface during joining and cooling. High concentrations of residual stress eventually lead to the formation of microcracks that result in low bond strength and potential interface failure. CS offers the ability to limit thermal alteration of heat sensitive components, the possibility of incorporating reinforcement filler phases in the bond layer, and the potential to create functionally graded bond layers that minimize compositional, physical and mechanical property mismatches, in addition to the established cost, energy and time saving benefits. For SiC ceramics, tungsten has been selected as the bond layer due to the closeness of the CTE, thereby minimizing stress concentrations at the interface [33]. Specifically, joining a SiC ceramic to a Ni-based metallic alloy was accomplished by using a tungsten foil as an intermediate layer for bonding, with a CS reaction involving layered [functionally graded] W, Ti, Ni and amorphous carbon (used as a filler phase) powders as the heat source to accomplish bonding. Ignition of the bond mixture was accomplished at 1190°C, significantly lower than the melt temperature of the reactant components, and joining was carried out at a pressure of 25.5 MPa for no more than ten minutes employing a relatively simple piece of equipment, a Gleeble 1500[2]. Bond strength increased with increasing thickness of the intermediate W bond layer resulting from an effective distribution of bond process-induced thermal stresses and a reduction in residual stress concentrations at the joint interface. The main reaction products comprising the bond layer, through which substantial diffusion occured during processing, were W, WC, TiC and Ni_3Ti intermetallic.

Silicon carbide (SiC) or reinforced SiC matrix (SiC/SiC) materials might also find use as tile replacement material(s) for the Space Shuttle or next generation vehicles, i.e. the crew launch vehicle (CLV) and crew exploration vehicle (CEV). CS processes allow preservation of the desired microstructural and mechanical properties of SiC materials due to the rapid reactive joining process and low temperature necessary to initiate the exothermic reaction, such as those that would be encountered on an Earth orbit. A Ti-C-Ni system has been investigated and found to join SiC and SiC/SiC materials effectively without degrading SiC whiskers that provide mechanical reinforcement [34]. Nickel additions were found to reduce porosity, inherent to CS, significantly at the joint interface. Residual porosity present at the interface was believed to result from the tape casting process used to fabricate the Ti-C-Ni reactive foils, and could be further reduced with alternate or improved fabrication methods.

Magnesium aluminate spinel ($MgAl_2O_4$) has a high melting point, low density and a high oxidation resistance. Based on these physical property traits, $MgAl_2O_4$ is also a potential material for high-temperature applications, but is somewhat limited as a structural member owing to the brittle nature of the spinel. Toughening of the brittle ceramic can be accomplished by adding a second ceramic phase into the matrix, i.e. a continuous ceramic matrix penetrated by a continuous ceramic reinforcement network, recently referred to as an interpenetrating phase composite (IPC) [35]. Traditional powder metallurgy techniques are limited for the fabrication of IPCs since the interpenetrating network is restricted to use of discontinuous fibers, whiskers or particle reinforcements. Infiltration of a liquid metal into a porous ceramic perform is possible, but limits use of IPC ceramics to those containing aluminum compounds based on the low melt temperature of aluminum, and consequently, such ceramics are not ideal for high-temperature applications [36]. Since CS is an *in situ* technique, fabrication of IPCs is well suited. Fabrication of IPCs employing a CS route has been shown using the $MgAl_2O_4$ spinel – TiAl intermetallic reaction system [37]. Titanium aluminide offers the added advantage of a low density, low cost, high ductility, and high melting temperature intermetallic as the reinforcement phase with excellent oxidation resistance, often used individually as a high temperature component in the aerospace industry. The process involves blending TiO_2, aluminum and magnesium powders followed by the ensuing CS reaction and densification (~98%) by a simultaneously applied external pressure of 80 MPa. This type of pressure-assisted, TE mode CS reaction is sometimes referred to as reactive forging (RF), an example schematic of which is provided in Figure 5.

[2] A Gleeble is commonly used in high temperature mechanical testing, such as creep determination, and employs similar principles to Joule heating and/or FACS to heat the initial reactants

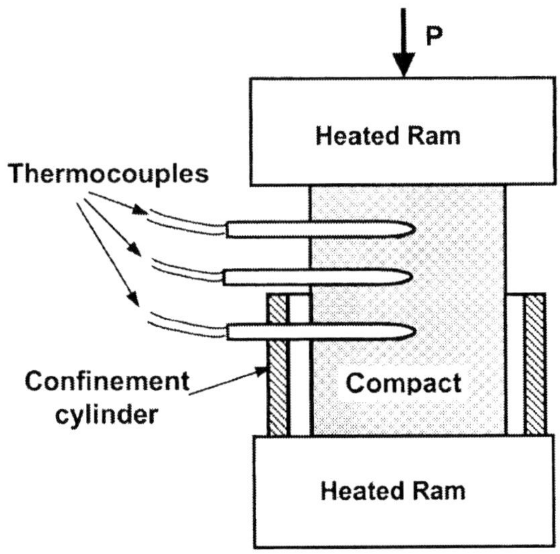

Figure 5. An example schematic of a reactive forging experimental set-up. Reactive forging combines a simultaneous mode CS reaction with an applied pressure between heated press rams [Reprinted from Ref. 37, with permission from Elsevier].

Similarly, Al_2O_3-TiAl/$TiAl_3$ IPCs have been fabricated *in situ* employing a CS reaction from TiO_2 and Al powders [38]. The ceramic Al_2O_3 matrix dominates the thermomechanical behavior of the IPC-containing intermetallic so that the desirable material properties, such as wear resistance, environmental stability and high temperature strength (creep resistance), remain the prevalent features of the IPC. The addition of the TiAl intermetallic improves overall toughness of the IPC by allowing crack bridging. A uni-axial pressure of 60 MPa applied immediately after the TE mode CS reaction was sufficient to produce a 95-98% dense Al_2O_3-TiAl/$TiAl_3$ IPC, while a pressure of 250 MPa was required to produce a similarly dense, but not completely converted IPCs after one hour of exposure employing reactive hot pressing. Employing the CS technique, up to ten samples were fabricated in a mere thirty minutes, offering extreme time and energy savings over the more conventional infiltration and reactive hot pressing techniques. Another added advantage of the CS technique is the ability to form a two-phase TiAl/Ti_3Al lamellar structure. The alternating γ (TiAl) and $\alpha 2$ ($TiAl_3$) phases, each approximately 0.2 μm thick, has higher toughness and creep resistance over the pure γ structure, therefore further enhancing the overall properties of the IPC [39].

An additional option to extend the lifetime and enhance the performance of high temperature components is to modify the working surface with a thermally protective coating. Traditionally, thermal barrier coatings comprised of a thin, porous ceramic have been employed to serve this purpose. Methods such as thermal spray and plasma or chemical vapor deposition are used to apply the coating, but encounter severe limitations since multiple bond coats must first be applied to the protected surface and are limited to small component sizes and non-complex geometries. Combining CS with a conventional powder metallurgy process would appear to optimize bonding of a thermal barrier coating to the substrate with minimal cost. Combination of these processes has demonstrated bonding of Ni_3Al intermetallic and a porous Al_2O_3-TiB_2 (~30% of theoretical) ceramic coating employing a Gleeble 1500[2] from low cost aluminum, nickel, titanium, TiO_2 and B_2O_3 powders [40]. The thermal barrier material maintained outstanding mechanical integrity during thermal cycling (300-1100 C at 200°C•min^{-1}) resulting from the continuously graded interface between the ceramic and intermetallic components. Similar thermal barrier materials consisting of a pure NiAl intermetallic graded to approximately 52 vol% Al_2O_3 have been prepared employing CS coupled with a simultaneous densification step from pre-oxidized nickel and aluminum powders [41].

Ceramics and ceramic-ceramic composites (CERCERs) composed of boride compounds possess properties desirable for high temperature applications such as high melting points, hardness, elastic modulus, electrical conductivity and ability to resist acids, molten nonferrous metals, cryolite and other non-basic slags. Since boride compounds typically have high exothermicity, *in situ* fabrication of these ceramics is easily achievable from CS reactions. For example, Al_2O_3-Ti/ZrB_2 CERCERs can be produced from zirconia, aluminum, titanium and boron oxide from propagating mode CS coupled with a simultaneous densification step [42]. The use of raw materials that are technologically more attractive and less expensive than the Ti-B system and/or Ti-C system is achievable through these CS reactions. Boron oxide is less than one hundredth the cost of elemental boron and titanium oxide is less than one tenth the cost of elemental titanium [43]. The initial starting density of the green compact must be carefully selected. On one hand, a low volume change during the conversion of reactants into the desired products should be considered, thus minimizing the amount of porosity generated during the reaction and the open volume that must be consolidated. On the other hand, the ignition of the green compact is a kinetically controlled mechanism, and not a thermodynamic mechanism. Thus, as the initial compact density increases, the ignition temperature also increases, up to a point where the CS regime is no longer dominant and instead the compact is 'reactively sintered.' Modeling studies regarding this phenomenon can be found in Ref. 44. Combustion temperature, and thus microstructure and mechanical properties, can be tailored through the amount of Al_2O_3 phase added as a diluent, or heat sink, to the reaction system.

Diborides of hafnium and zirconium, termed ultra-high temperature ceramics (UHTCs) are of interest for use as the leading edges of sharp bodied, hypersonic re-entry vehicles, nose caps, rocket nozzle inserts and air-augmented propulsion system components [45]. For example, production of 98% dense hafnium diboride (HfB_2) is possible through SPS of elemental hafnium and boron at 1700°C with a hold time of ten minutes and an applied uni-axial pressure of 95 MPa [46]. Similar to TiB_2, ZrB_2 has a high melting point, high hardness, high electrical conductivity, excellent corrosion resistance against iron, HCl, HF, other non-ferrous metals, cryolite and non-basic slags, adequate thermal shock resistance, and is wetted but not attacked by molten metals. Thus, ZrB_2 ceramics are ideal for foundry or refractory industrial applications as molten metal crucibles, thermowell tubes for steel refining and parts of electrical devices such as heaters and igniters [47]. ZrB_2 is advantageous over TiB_2 from a processing point of view since the ZrB_2 compound does not contain multiple, stable intermediate phases, as does the TiB_2 compound, i.e. TiB_{2-x}. Control over the microstructure and density of the ZrB_2 ceramic components are essential to the success in application and ensuring the desirable material properties. Sintering of ZrB_2 powders is challenging since the constituents have a high melting temperature and vapor pressure, and can only be accomplished by a pressure-assisted process at temperatures greater than 1900°C, or through addition of metal sintering aids such as Fe, Ni, Co, W [48]. A secondary reinforcing phase is desirable in the ZrB_2 ceramic to increase the toughness and minimize the stress corrosion cracking and high temperature oxidation for use as electrical or mechanical applications, such as those proposed above. The most common secondary ceramic reinforcement phases are TiB_2 or B_4C, since these boron-containing ceramics have similar properties to those of monolithic ZrB_2 [49]. Al_2O_3 can also be used as a secondary reinforcement phase to improve the mechanical properties of a ZrB_2 composite. Mixing ZrB_2 powders with the secondary reinforcement phase is somewhat undesirable, since the higher temperatures result in coarsening of the ZrB_2 grains and therefore limit the amount of densification achieved and lower the mechanical properties of the fabricated component. CS can be employed to fabricate ceramic composite powders, such as ZrB_2-Al_2O_3, that are tailored specifically for sintering dense, near net shape composites, since defects induced during the rapid heating and cooling rates associated with the process ultimately result in lower sintering temperatures [50], in addition to a more homogeneous end product. Such a fabrication sequence involves a CS reaction with ZrO_2, B_2O_3 and Al to fabricate ZrB_2-Al_2O_3 powders, milling of the resultant powders, pressing into the desired shape and sintering between 1750 and 1900 C for thirty minutes maintaining the fine grain structure [51,52].

Aluminum nitride (AlN) is a desirable ceramic for use as an electronic material, heat sink applications, cutting tool coatings and as high temperature structural components [53,54]. These desirable uses stem from the fact that AlN has high thermal and electrical conductivity, low density, adequate mechanical strength and good oxidation and thermal shock resistance. However, AlN has low toughness and lacks reliability, and as such the use of this material for load bearing applications is limited. Combining AlN with SiC forming a solid solution increases the mechanical performance of the AlN-SiC CERCER because of their similar crystal structures and high temperature properties. Small scale laboratory methods have been designed for fabrication of AlN-SiC [55]. These methods demand high temperatures in excess of 2000°C and being of small scale, are not well suited for commercial production. CS offers a viable fabrication route to produce high purity, low cost AlN-SiC composites in a matter of seconds. The solid solution can be formed via reaction of aluminum, carbon and Si_3N_4 in a gaseous nitrogen environment with the aid of an electric field. Use of an electric field to assist the CS reaction is referred to as Field Activated Combustion Synthesis (FACS), and is not entirely different from Spark Plasma Sintering (SPS) [56]. The same reaction can be carried out using elemental aluminum, silicon and carbon powders through reaction in a gaseous nitrogen environment without the use of an electric field as a result of the high exothermicity associated with the Al-N reaction [57]. This method demands the nitrogen pressure during reaction is above 12 MPa in order to avoid incomplete conversion and unreacted aluminum and silicon in the product. The nitrogen pressure must be high since the combustion temperatures range between 1800 and 2100 C, well above both the melting points of aluminum and silicon. As these reactants melt, permeation of nitrogen gas into the center of the sample is reduced, thus decreasing the formation of AlN, the driving force behind the self-sustaining CS reaction. Higher gas pressures continually force nitrogen into the molten species maintaining the propagating combustion front. However, increased gas pressure also forces a higher concentration of one of the reactants, nitrogen, leading to enhanced crystal growth kinetics. Kinetically, the reaction sequence has been determined to occur by the following: aluminum and silicon react with the nitrogen gas, forming AlN and Si_3N_4 resulting in a rapid increase in temperature, the Si_3N_4 decomposes into silicon vapor as nitrogen continually reacts with aluminum, followed by silicon and carbon atoms diffusing into the AlN crystal structure forming the AlN-SiC solid solution. Increased crystal growth kinetics lead to lower crystallinity and a less defined crystal surface than that obtained from lower nitrogen pressures. Thus, CS of AlN-SiC from purely elemental reactants requires a high pressure for full conversion, but may not provide the desired product microstructure. High combustion temperatures, thus a greater supply of aluminum in the reactants or higher nitrogen gas pressures, are the key to diffusing all of the silicon reactant into solution.

BN-AlN-TiB_2 composites can be produced from high pressure CS reactions involving boron, aluminum, BN, TiB_2 and AlN powders under a nitrogen pressure of 100 MPa [58]. The physical properties of this type of ceramic composite can be modified by the initial CS reactant mixture and the process parameters, i.e. nitrogen pressure, creating a dense ceramic composite with a low product weight with a high thermal conductivity for example. Since these types of ceramics can be machined by conventional tools, BN-AlN-TiB_2 ceramics are comparable in strength to other heat conducting non-metallic materials, such as beryllia ceramics, and significantly stronger than graphite materials.

Wear Resistant and Tooling Materials

Ti_3AlC_2 and isotypic Ti_3SiC_2 are desirable for abrasion-resistant components and elements of chemical equipment since the ternary compounds exhibit a combination of both metallic and ceramic properties. These compounds are relatively soft, elastically stiff, lightweight and easily machinable with exceptional thermal and electrical conductivity and damage tolerance [59]. However, the compounds can be challenging to fabricate since there is a narrow phase range in both ternary phase diagrams, and more conventional fabrication routes to produce dense materials typically result in undesirable kinetic and thermodynamic stability and growth mechanisms from high temperatures for extended durations of time. Bulk, polycrystalline Ti_3AlC_2 has been fabricated by SPS from elemental titanium, activated carbon or TiC and aluminum or aluminum carbide (Al_4C_3) powders at a pressure of 22 MPa and a temperature of 1250°C exhibiting excellent flexural strength and fracture toughness [60-62]. Dense, near-single-phase Ti_3AlC_2 was also fabricated employing both the self-propagating (SHS) and simultaneous (TE) CS modes coupled with a simultaneous densification resulting in a fine grain structured ternary carbide with respectable compressive strength and high machinability [63]. Similarly, dense, high purity Ti_3SiC_2 was fabricated from elemental Ti, Si or SiC and C or TiC powders employing Pulse Discharge Sintering (PDS) [64-66] or SPS [67] with relatively low sintering temperatures (1250-1300 C) and times (less than 15 minutes). The CS reaction to produce Ti_3SiC_2 using SPS is further accelerated by appropriate addition of elemental aluminum resulting in a considerably improved single phase polycrystalline, fully dense carbide [68,69]. Ti_2AlC is a variation of these ternary carbides also possessing the unique combination of properties and has been fabricated from elemental Ti, Al and carbon black using SPS with low sintering temperatures and times [70,71]. Substitution of TiAl for elemental Ti and Al followed by SHS with graphite powder effectively lowers the combustion temperature and results in formation of high purity ternary carbide phases (e.g. Ti_3AlC_2, Ti_3AlC and Ti_2AlC) [72].

Silicon nitride and carbide compounds are extremely hard, chemical and heat resistant materials that can be used for cutting tools and various types of machine and engine components [73]. Properties of monolithic silicon nitride can be significantly improved through addition of granular or fibrous silicon carbide compounds, particularly polycrystalline carbide compounds [74]. Thus, *in situ* synthesis of a silicon nitride (β-Si_3N_4)/carbide (SiC) composite with a tailored particle size and morphology is an interesting application. Such composites can be fabricated employing a CS reaction of elemental silicon and carbon soot powders of low density under a high nitrogen pressure, e.g. 7 MPa [75]. The carbide content of the composite is limited to 60-65wt%, since beyond this amount the exothermicity of the reactant mixture is no longer sufficient to drive a CS reaction. Addition of soot to the reactant mixture prevents coalescence of molten silicon particles during the reaction, allowing maximum permeability of nitrogen into the compact, resulting in formation of silicon nitride. The limitation of 60-65 wt% SiC in the composite can be overcome through the addition of fluoroplastic in the appropriate amounts (5-15 wt%). Fluoroplastic acts as an additional carburizing agent, allowing the fabrication of composites containing greater than 65 wt% SiC, all the way up to pure SiC, often difficult to fabricate using conventional CS, i.e. without the aid of pre-heating or an electrical field (FACS, SPS). Fluoroplastic carburizing agent promotes a single stage reaction regime, as opposed to a two-stage regime, i.e. low temperature regime ($p_{N2} < 1$ MPa) in which reaction only occurs between fluoroplastic and silicon and a high temperature regime ($p_{N2} > 1$ MPa) in which nitridation of silicon is possible.

Silicon, carbon and silica (SiO_2) powders can be combined in appropriate ratios and reacted under nitrogen gas to fabricate SiC and $SiON_2$ powders employing CS. Reaction pressures greater than 3 MPa result in a pure β-SiC nanopowder phase, while pressures below 3 MPa and above 3.5 MPa result in a mixture of fine β-SiC and coarser β-Si_3N_4 grained phases [76]. $SiON_2$ is conventionally fabricated by hot pressing SiN and SiO_2 in gaseous nitrogen [77] ammonolysis of hexachlorodisiloxane [78] or nitridation of Si or SiN and SiO_2 powder mixtures [79,80]. However, these techniques are not capable of fabricating finely dispersed powders with adequate control over particle characteristics. CS, on the other hand, is capable of fabricating well-dispersed silicon oxynitride nanopowder, so long as the reaction is conducted under a gaseous nitrogen pressure of 3 MPa in order to lower the contents of unrecompensed Si_3N_4 and unreacted Si [76].

Reaction bonding is a well-established ceramic technique employed to produce near net shape components from metal sponge and gaseous nitrogen or ammonia precursors [2]. This type of process has been used on a commercial scale to fabricate silicon nitride and silicon carbide ceramics. However, the technique is somewhat limited for the fabrication of other metal nitrides, such as TiN, due to the low diffusion coefficients and sintering activity of the compound, except when using vapor deposition processes, limiting use of this ceramic to thin film coatings. CS is an intermediate method to produce high purity, fully converted, high surface area TiN precursor particles under relatively mild conditions [81]. Particles fabricated from the CS method may then be conventionally reaction bonded to produce dense, near net shape TiN components for use in structural applications. The fabrication process involves magnesiothermic high temperature reduction of TiO_2 with Mg in the form of a combustion wave. Byproducts, such as MgO, unreacted Mg and TiO_2, are removed from the desired Ti product by hydrochloric acid leaching. The high temperature magnesiothermic reduction reaction produces high surface area Ti particles since a gas-phase mechanism dominates conversion, rather than a liquid intermediate mechanism. When compared to other conventional metal production methods, such as the industrial Kroll method involving reduction of $TiCl_4$ in molten magnesium to produce titanium sponge, the CS method produces

Ti particles with significantly higher specific surface areas in a significantly shorter reaction period, enabling the process to be conveniently scaled up on a commercial level. The high specific surface area Ti particles can then be compacted into desired shapes and reacted with gaseous nitrogen resulting in dense [reaction bonding], near net shape structural components with minimal shrinkage.

Advanced cutting and tooling applications require coatings that are hard and have superior wear and corrosion resistance [82,83]. Chromium coatings, in particular chromium nitride, posses these properties, but are typically fabricated by heating chromium metal for extended durations at 735-1180 C in either ammonia or nitrogen gaseous environments to reach full conversion, depending on the desired nitride phase, e.g. CrN or Cr_2N [84]. Recently, CS has been shown to produce these chromium nitrides in a matter of seconds under controlled nitrogen pressure [85]. Dense nitrides were achieved by consolidation of the CS powders by hot isostatic pressing (HIPing). The nitrogen pressure must be maintained above 0.18 MPa in order to attain ignition of the metal powder. Cr_2N powder with a particle size of ~1.5 μm and a combustion temperature of 1040°C at a nitrogen pressure of 0.18 MPa, and CrN powder with a particle size of ~2.3 μm and a combustion temperature of 1730°C at a nitrogen pressure of 2 MPa were produced from this method. The combustion temperature, and therefore the nitride phase formed, is dependent upon the nitrogen pressure used during processing. Higher pressures tend to produce coarser, more agglomerated CrN particles while the lower nitrogen pressures tend to produce finer, less agglomerated Cr_2N particles, with mixtures of both nitrides occurring for nitrogen pressures in between. Some grain growth does occur during the HIPing consolidation process, but overall the chromium nitride powders produced from direct CS have higher fracture toughness over most other ceramic systems [86].

Functionally Graded Materials

Functionally Graded Materials (FGMs) and Composites (FGCs) are desirable for a plethora of applications since desirable properties (e.g. composition, density and microstructure) of one group of materials can be combined with those of another group, in a controlled, graded structure. For example, an FGM graded from a metallic component to a ceramic component allows sufficient thermal, corrosion and oxidation resistance in the ceramic component while still maintaining a ductile, tough and strong structure in the metallic component. FGMs can overcome the brittle structure typical of ceramics, such as silicides, and toughening of ductile metallics, such as nickel, by producing a material that is significantly toughened and has improved R-curve facture behavior [87,88].

Well designed FGMs could replace nickel-based superalloys for structural applications in high temperature environments, since FGMs are capable of high temperature creep resistance and ductility as well as superior oxidation resistance at elevated temperatures, i.e. >1000°C. For example, molybdenum disilicide ($MoSi_2$) has high electrical conductivity and exceptional oxidation resistance at high temperatures, but is limited by low thermal creep resistance at elevated temperatures. Conversely, Al_2O_3 has excellent thermal creep resistance and has high electrical resistivity while maintaining exceptional oxidation resistance and a coefficient of thermal expansion similar to that of $MoSi_2$. Therefore, by functionally grading these two materials, an alternating conducting and insulating structure with acceptable mechanical properties can be fabricated for use as an electrical device. Such fabrication has been accomplished by stacking layers of MoO_3, Si or SiO_2 and Al ranging from 20 vol% $MoSi_2$ in the product to 80 vol% $MoSi_2$ (Al_2O_3 as balance) employing a tape casting process [89]. A reactive pellet initiates the ensuing CS reaction followed by application of a moderate pressure (3.4 MPa) resulting in a FGM with a strong room temperature resistivity gradient. Similarly, room temperature fracture toughness and high temperature strength and creep resistance can be improved by addition of a SiC reinforcement phase. Pressure-assisted Field-Activated Combustion Synthesis (FACS) has been employed to produce a functionally graded material from 100% $MoSi_2$ to 100% SiC in 10 mol% increments to prevent thermal cracking between layers from elemental molybdenum, silicon and carbon graphite without a significant decrease in density [90,91]. Use of FACS allows for self-propagation of a combustion wave not otherwise achievable for the $MoSi_2$-SiC reaction system as a result of low reaction exothermicity.

Ti-TiB_x FGMs can be used in a number of military applications including, but not limited to, armor, armaments and military vehicle structures resulting from the superior ballistic performance of the TiB_x ceramic. Propagating mode CS combined with a simultaneous densification step allows an energy and time efficient process to combine the desirable properties of TiB_x with a ductile metal backing structure, such as Ti, that maximizes strength and energy absorption for destroying projectiles from inexpensive, elemental starting components [92]. Substituting the metal backing structure with a SMA or pseudoelastic alloy, as described above, can enhance the FGM performance a step further, since the SMA can accommodate deformation strains up to approximately 8% without plastic damage [93]; common steels in comparison can only recover approximately 1% of induced deformation. This reaction has been demonstrated with the NiTi-TiC_x system, grading the structure from pure TiC_x to equi-atomic NiTi, although the simultaneous densification step was not employed, so that porosity was still present in the FGM [94,95]. The main advantage of the NiTi-TiC_x FGM system is that layers can be constructed so that a pure TiC_x end can be graded to an equi-atomic NiTi end employing propagating mode CS, even with the low adiabatic temperature and exothermicity of the NiTi system. Photographs of the combustion wave propagation clearly showing the layered structure are provided in Figure 6. Also shown in Figure 6 are SEM photomicrographs of the CS FGM product. TiC_x particle size decreases with increasing amounts of NiTi intermetallic as a result of decreased reaction exothermicity and Ostwald ripening. Similar reactions have been demonstrated for the TiC-Ni system resulting in dense FGMs [96,97], although the mechanical properties of this particular system are not as ideal as those demonstrated for the Ti-TiB_x and possibly the NiTi-TiC_x systems. TiC-Cu FGMs may also serve as an effective armor protection material, combining the hard, wear resistant TiC ceramic with the extremely ductile copper backing substrate. Similar to addition of Ni or Ti, addition of copper to the Ti+C reaction system limits the thermodynamics of the CS reaction so that fabrication of this FGM is constrained to between 35 and 65 wt% copper additions. The latter copper concentration can be achieved when the CS reaction is conducted under an electrothermal method, previously referred to as FACS, from elemental titanium, carbon and copper reactants with a near linearly graded structure [98].

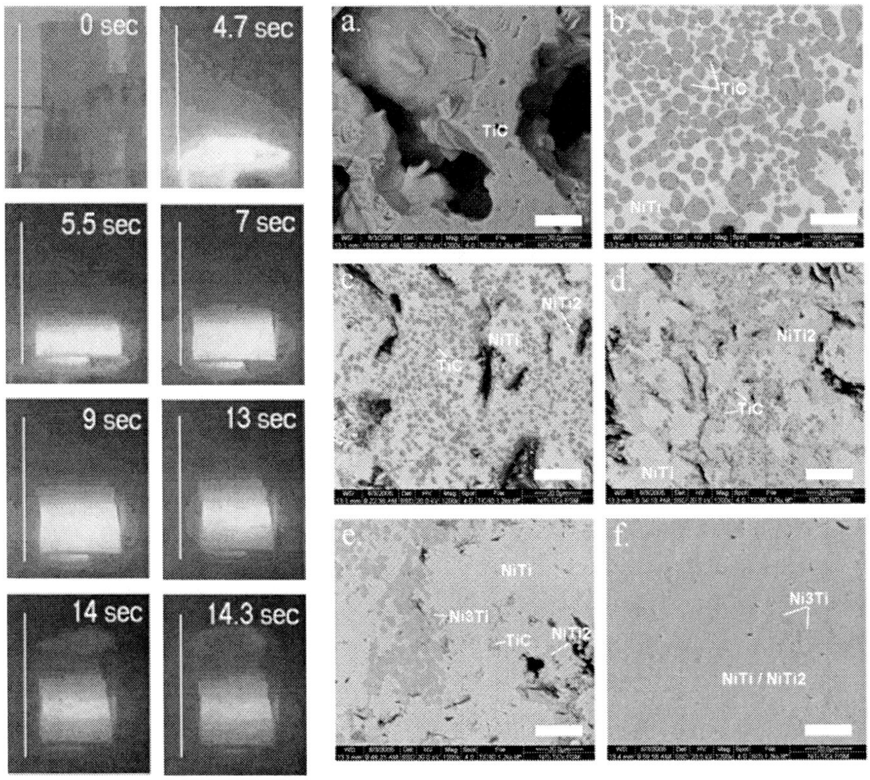

Figure 6. Left: Photographs of a NiTi-TiC_x FGM CS reaction as a function of time that clearly shows the varying layers of the FGM. The scale bar to the left of each photograph represents the original sample length of 30 mm [Reprinted from Ref. 94, with permission from ASME]. Right: SEM photomicrographs of FGM layered regions composed of (a.) TiC_x, (b.) TiC_x-20vol% NiTi, (c.) TiC_x-40vol% NiTi, (d.) TiC_x-60vol% NiTi, (e.) TiC_x-80vol% NiTi and (f.) NiTi taken at a magnification of 1.2kx with scale bars representing 20.0 μm [Reprinted from Ref. 95, with permission from Elsevier].

Armor

As the need for advanced defense and protection systems becomes more evident, the need for advanced materials capable of dissipating large amounts of energy with minimal disintegration on the backing surface and relatively low weights and fabrication costs expands [99]. Ceramic composites, or CERCERs, could be ideal candidates for armor applications due to the relatively high hardness, elastic modulus, wear resistance and fracture toughness of the composite. In most cases, such as that with TiB_2-TiC, the hardness and fracture toughness of the composite are significantly higher than those of the individual ceramic components [100]. Consolidation of CERCERs is often challenging due to the high degree of covalent bonding and low self-diffusion coefficients of the constituent elements. These poor consolidation characteristics can be overcome with CS, especially when combined with a simultaneous or near simultaneous densification step.

Near fully dense TiB_2-TiC CERCERs have been fabricated employing SHS from titanium, carbon and B_4C powders with an applied pressure of 160 MPa for five seconds [101]. Exceptional mechanical properties were obtained from the clubbed TiB_2 and equiaxed or irregular TiC grains in the resultant microstructure. SPS has been employed to fabricate near fully dense, (96 to 99% of theoretical) nanocrystalline TiB_2-TiC composites from elemental Ti, C and B or B_4C powders in less than five minutes [102,103], and explosive compaction can also be combined with the CS reaction of B_4C and elemental titanium to produce dense TiB_2/TiC CERCERs with microhardness values comparable or higher than the monolithic ceramic components [104]. Addition of an Al_2O_3 phase to the TiB_2-TiC CERCER employing dynamic compaction coupled with the simultaneous CS mode (TE) improves the oxidation resistance of the composite without a significant drop in product hardness [105].

Combination of TiB_2 with TiN would be desirable for armor applications as a result of the comparable thermochemical properties between the two ceramics. Due to the challenge associated with producing TiN, e.g. chemical or physical vapor deposition, that is significantly different than TiB_2 fabrication, e.g. hot pressing or hot isostatic pressing, these composites have not been thoroughly investigated. SPS provides an attractive fabrication route for TiB_2-TiN composites by reacting elemental titanium and boron with boron nitride producing a near fully dense (90 to 96% of theoretical) composite with nano-size crystallites and very attractive mechanical properties [106]. Similarly, TiB_2-TiN and TiB_2-TiC CERCERs with minor nickel additions have been fabricated using pressure-assisted simultaneous mode CS from B_4C, BN, titanium and nickel powders in open air with insignificant oxidation of the product [107]. Fracture toughness of the CERCERs was significantly increased by further addition of the finely dispersed, ductile nickel phase: 20.5 MPa$\sqrt{m}$ for the TiB_2-TiN-Ni and 11.9 MPa$\sqrt{m}$ for the TiB_2-TiC-Ni.

Submicrometer TiC-SiC composites with 97.9% theoretical density have been fabricated from SPS of elemental titanium and carbon with SiC at low temperatures (~1400°C), thereby minimizing Ostwald ripening kinetics and increasing sintered density in only eight minutes [108]. Solid solutions of titanium diboride-tungsten diboride (TiB_2-WB_2) were synthesized from elemental titanium, boron and tungsten reactants in a matter of minutes followed by subsequent densification to 86% of theoretical with SPS at relatively low temperatures (1800°C) [109]. Conventional fabrication methods for such a composite would require comparably higher temperatures (2000°C) for extended durations of time (>8 hours).

Light-weight, low cost high strength ceramic body armor plates can be fabricated from silicon carbide (SiC) and silicon oxynitride (Si_2ON_2) nanopowders. These ceramic compounds possess very high Young's modulus, high hardness and stiffness, constant flexural strength up to 1200°C and high fracture toughness. Traditionally, nanopowders of these compositions are not well understood or documented, mainly as a result of the significant challenges associated with fabrication. CS is an attractive technique for production of SiC and Si_2ON_2 nanopowders from silicon, carbon and silica powders in a gaseous nitrogen environment [110]. The nanopowders can then be pressed into the desired body shape and are easily sintered together employing SPS or an equivalent technique, since the nano-sized nature of the composite is of extreme importance in maintaining superior mechanical performance.

Materials used for armor applications do not necessarily need to be CERCERs, but some CERMETs that have high specific strength and elastic modulus, elevated temperature resistance, and low relative fabrication costs might also fit the bill. For example, CS can be employed to produce Al/TiC-TiB_2-αAl_2O_3 CERMETs from addition of titania and B_4C powders to molten aluminum *in situ* [111].

Acoustic and Thermal Insulators

Recently, efforts to develop ceramic foams for use as acoustic and thermal insulators in transport, building and power engineering as well as for use in electronic and biomedicine applications have been made [112]. Foams for use as thermal insulators demand high thermal stability and an appropriate degree of open and closed porosity. Fabrication of aluminum foams is well established, but use of these types of foams for thermal insulators is limited based on the low melting temperature of the component. Control over the burning rate, the amount of melt present during the reaction and the duration for which the melt is molten, difference in internal pressure in the pores and external pressure in the reaction vessel, and gravity facilitates the desired foamed ceramic properties, all controllable with the CS technique. Titanium has a much higher melting temperature and also exhibits the necessary thermal stability. An example of titanium foam for potential use as a thermal insulator is Ti-Si-Al-C compounds, examples of which are shown in Figure 7. Propagating mode CS can be employed to produce these components with a relative density of 40-60% from elemental powders and small amounts of ammonium chloride and calcium peroxide used as activation and gasifying agents [113].

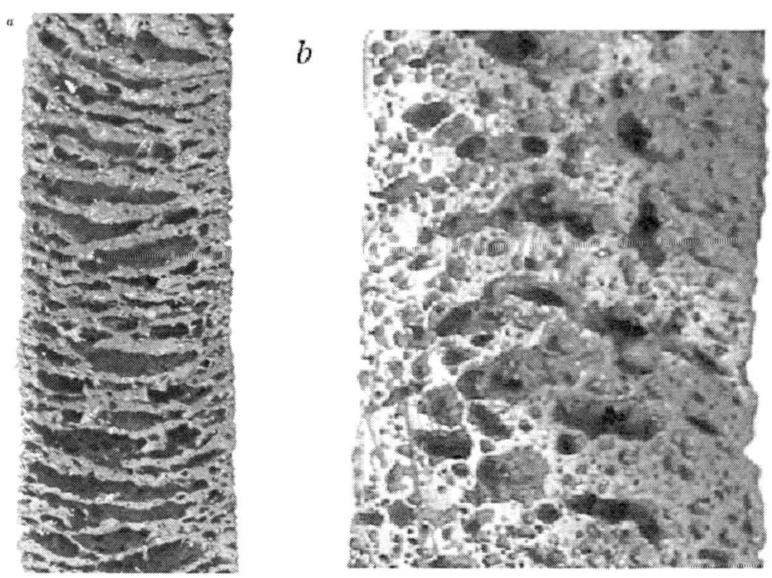

Figure 7. Photograph of the macrostructure of cylindrical Ti-Si-Al-C based foamed materials obtained using CS with an external pressure of (a.) 0.1 and (b.) 1 bar [Reprinted from Ref. 113, with permission from Springer].

Similarly, Ni_3Ti-TiC_x intermetallic-ceramic composite foams can be produced employing propagating mode CS with an appropriate gasifying agent [114]. The advantage of combining an intermetallic with a ceramic is improvement of the composite mechanical properties, especially as the porosity of the foam begins to increase. The lower melting temperature intermetallic compound can be engineered to provide a desired amount of melt, and thus control porosity and pore morphology of the foam. Calcium nitride (Ca_3N_2) and calcium carbonate ($CaCO_3$) were used as novel gasifying agents to fabricate the Ni_3Ti-TiC_x composite "foams" [115]. In addition to possible use as thermal and acoustic materials in structural applications, such foams might also find use as biomedical implant scaffolds, since residue of the gasifying agent is in the form of bioinert CaO. Porosity was found to increase anywhere from 15 – 38%, depending on the gasifying agent and reaction system, over composites fabricated with no gasifying agent.

Powder Production

Powder production of high purity nitride, oxide and carbide powders is desirable as a result of the wide range of desirable properties that can be applied towards a plethora of fabrication applications; the semi-conductor and computing industry and the medical field for example. Typical methods for oxide powder production involve calcination of starting materials at high temperatures in air for extended periods of time or wet-chemical synthesis that require complex fabrication schedules and low production rates. Metal nitride powder production include the hydride-dehydride process (used for metal powder fabrication) coupled with direct nitridation, direct synthesis, the chemical vapor deposition (CVD) process and plasma nitridation of metal oxides. The hydride-dehydride process involves the use of very pure hydrogen and direct nitridation at elevated temperatures for extended durations, not only raising safety concerns but also processing concerns. Direct synthesis is an attractive route but requires high partial pressures of nitrogen to overcome the effect of melting during the ensuing exothermic reaction, since the melt temperature of the metals is often significantly lower than the adiabatic temperature of the reaction. Carbide powder production is typically carried out employing carbothermic reduction or carburization, both of which require high temperatures for extended durations and can carry over a high percentage of impurities initially present in the sample to the final product. CS can overcome these processing limitations by offering lower processing costs, energy efficiency and high production rates using simple, easily handled metal salts and organic fuels or precursors [116]. A comparison for producing high purity powders employing different fabrication methods is provided in Table 1.

For example, titanium nitride (TiN) is desirable for high hardness, thermal stability, thermal conductivity and resistance to chemical attack for use as cutting tools, diffusion barriers in semiconductors, or other applications where a hard material or coating is warranted [118]. Crystalline titanium mononitride powders, in addition to ZrN, HfN, NbN and TaN powders, have been produced by solid state metasynthesis (SSM – a version of AICS) of metal halides with sodium azide (NaN_3), lithium nitride (Li_3N) or a combination of both. Reaction temperatures of 1370°C were reached within one second followed by rapid cooling to room temperature within thirty seconds resulting in average particle sizes of 500Å [119]. Metal azides have also been used as solid sources of nitrogen for the reaction and nitridation of titanium, zirconium and hafnium metal powders [120]. High pressure CS (HPCS) has been used to produce dense TiN compacts through ignition of titanium powder under a compaction pressure of 98 MPa and a nitrogen pressure of 200 MPa [121].

Table 1. Comparison and pertinent features of different powder production methods [after Ref. 3,117].

Production Method	*Furnace synthesis*	*Co-precipitation*	*Hydrothermal synthesis*	*Gas phase synthesis*	*Carbothermic reduction*	*Combustion synthesis*
Purity	Low	High	High	Medium to low	Medium to low	Medium to high
Small (nano) particle size	No	Yes	Yes	Yes	Yes	Yes
Compositional control	No	Good	Good	Medium	Medium	Medium
Low cost of reactants	Yes	No	No	No	Yes	No
Low cost for installation	Yes	No	No	No	No	Yes
Consumption of external energy sources	High	High	High	Medium	Medium	Low
Short duration of synthesis	No	No	No	Yes	No	Yes
Simplicity and low number of processing steps	Yes	No	No	No	No	Yes
Main applications	Low quality powder production	Simple and complex oxide production	Simple and complex oxide production	Simple covalent compound powder production	Simple covalent compound powder production	Covalent and intermetallic compound powder production

Similarly, aluminum nitride (AlN) is of interest resulting from high thermal conductivity, electrical resistivity, low thermal expansion and dielectric constant, and good corrosion and thermal-shock resistance [122]. Therefore, AlN can be substituted for material applications in place of silicon, such as integrated circuit packaging materials, heat sinks, electronic substrates and highly conductive composites. Low-pressure combustion synthesis of high purity aluminum nitride powders has been accomplished by using an ammonium additive, NH_4X where X = F, Cl, Br or I. High product yields were obtained under nitrogen pressures of around only 0.5 MPa since the ammonium additive also served as a gasifying agent, preventing molten aluminum particles from coalescing and allowing nitrogen gas to continually diffuse into the compact [123]. Furthermore, additives including urea ($CO(NH_2)_2$), stearic acid ($CH_3(CH_2)_{16}COOH$) and succinic acid ($CO_2H(CH_2)_2CO_2H$) were also found to provide high yields of aluminum nitride when the additive was combined with elemental aluminum powder, compacted and reacted under a nitrogen atmosphere of 0.5 MPa, well below reaction pressures typically required for direct synthesis of nitrides [124]. CS processing of AlN ceramics often results in desirable whisker morphology that originates from condensation of a vapor phase during the reaction [125]. These AlN whiskers can be used in polymeric systems, as an additional phase, to increase the thermal conductivity.

Boron carbonitride powders that can be used as electric insulators, high speed tools and applications requiring protection against aggressive molten salts, alloys, slags and melts can be fabricated employing filtration CS from elemental boron and carbon and or B_4C under a nitrogen atmosphere [126]. In this regard, careful control over carbon content in the reaction mixture and the pressure of the gaseous nitrogen atmosphere can result in unique B-N-C phase structures. High purity, hexagonal BN powders can be fabricated employing filtration CS from elemental boron and B_2O_3 or from B_2O_3 and Mg as a reductant, both under nitrogen atmospheres [127]. In the case of the former, the reaction takes place in the gaseous phase and is highly exothermic, while in the latter, the MgO must be removed from the product employing a chemical leaching process.

Silicon nitride (Si_3N_4) α-Fe alloy powders can be fabricated by reacting ferrosilicon in a nitrogen atmosphere employing propagating mode CS [128]. The Si_3N_4 can be isolated from the α-Fe component through acid dissolution, from which the Si_3N_4 powder can be used for advanced material applications, such as cutting and high wear resistant tools. Similarly, a silicon powder compact can be rapidly nitrided under high pressure nitrogen gas (HPCS) employing SHS (propagating mode CS) in a matter of seconds to minutes to fabricate Si_3N_4, depending on the sample size [129]. Conversely, traditional reaction-bonded Si_3N_4 typically requires more than fifty hours to complete nitridation of the silicon compact.

Tungsten carbide can be used as a coating on cutting tools and as an addition to Fe-based alloys. Carburization of tungsten metal is typically employed to produce fine, WC powders that require use of pure hydrogen and high temperatures for extended periods of time resulting in irregularly shaped particles and a low percentage of W_2C. The lower cost CS process can be employed to fabricate nano-sized WC powders from WO_3, carbon black, Mg and additional agents to control combustion temperature, and thus improved particle characteristics [130].

Silicon oxynitride (Si_2N_2O) is a unique technical ceramic of interest for cutting tools, high wear parts and heat engine components as a result of superior chemical and oxidation resistance in many harsh environments [131]. The use of this interesting material is somewhat limited due to challenges associated with fabrication of high purity powders and subsequent densification of the product, i.e. high temperatures for extended durations of time that lead to secondary phase formation or unreacted materials. However, HPCS has been employed to produce pure Si_2N_2O powder from elemental silicon and desert sand (SiO_2) synthesized in 3 MPa of nitrogen gas in a combustion furnace [132]. Near fully dense (99% theoretical) compacts were obtained employing SPS of the synthesized Si_2N_2O powder resulting in a ceramic with exceptional hardness, fracture toughness, grain structure, and oxidation and chemical resistance.

Calcium aluminate compounds are commonly used as low- and ultra-low cement castables for monolithic refractory applications or as metallurgical slags, but encounter limited use resulting from misunderstood hydration behavior. Hydration behavior is critical to engineering these compounds since the unique and desirable characteristics are dependent upon low water content and a finely dispersed matrix of calcium aluminate. More conventional processing methods, such as solid-state synthesis and wet chemical-synthesis, require multiple processing steps or expensive starting materials that hinder fabrication of a very fine crystalline, unagglomerated multicomponent oxide. Auto-Ignition Combustion Synthesis (AICS) provides an ideal alternative in the fabrication of the calcium aluminate powders employing relatively lower cost calcium and aluminum nitrates and a suitable organic fuel, such as urea or glycine. Fabrication of multi-component calcium aluminum oxides can be carried out with uncomplicated, simple equipment in a short period of time [133]. Binary compounds of the calcium aluminate system, i.e. $Ca_3Al_2O_6$, $Ca_{12}Al_{14}O_{33}$, $CaAl_2O_4$, $CaAl_4O_7$, etc., have been produced employing AICS from calcium and aluminum nitrates with carbohydrazide (CH_6N_4O) and glycine as a fuel with only a short calcination treatment of the product to promote crystallization [134].

Nuclear and Fuel Cell Materials

As the demand for energy, the decrease in potable water and the necessity of minimizing carbon released into the environment increases, alternative power generation technologies with minimal cost to the world population has become increasingly more important and urgent. Nuclear power generation is poised to effectively address a majority of these concerns, both as a near and long term solution. However, for nuclear power to continue relative competitiveness with fossil fuel power generation, the cost and performance of associated fuels and materials must be minimized. In addition, interest in fuel cells continues to grow with concerns over the diminishing world supply of fossil fuels, as a viable method to produce energy for transportation and other means.

Thoria-yttria (ThO_2-$YO_{1.5}$) solid solutions are candidate materials for electrolyte used in solid oxide fuel cells (SOFCs) [135]. A challenge associated with fabrication of ThO_2-$YO_{1.5}$ solid solutions are homogeneity and powder characteristics, since the electrolyte in the SOFC should be near fully dense to possess the desired electrical characteristics and minimize leakage of the fuel and oxidant in the form of gases [135]. AICS is a prime candidate for ThO_2-$YO_{1.5}$ nanocrystalline powder fabrication, producing powders with high sinterability and nanograin microstructure retention from thorium and yttrium nitrates with citric acid [136] or citric acid and ethylene glycol [137].

Hydrogen protonic exchange membrane fuel cells (PEMFCs) and direct methanol fuel cells (DMFCs) are limited in use due to decreases in efficiency with platinum poisoning of the anode [138]. Active metal mixed oxides with high specific surface areas can be used as substitutes for noble metals during electrocatalysis, creating regions that promote O_2^- ion transport. One such class of metal oxides are those that contain the perovskite structure, such as $LnCoO_3$ where Ln = La, Pr, Nd, Sm and Gd. Perovskites can be combined with the appropriate Pt additions resulting in a CERMET that might find use in the direct reforming of hydrocarbons in intermediate temperature solid oxide fuel cells (ITSOFCs) [139]. However, the fabrication of such oxide powders with high specific surface areas can be challenging. Typical production methods include wet chemical methods, e.g. sol-gel, chemical co-precipitation, impregnation, drip pyrolysis, etc., but are limited as described above. $SmCoO_{3-\gamma}$ and $SrCO_3$ powders have been fabricated employing the AICS process [140]. The precursors for the reaction are readily available, inert metal nitrate salts ($Sm(NO_3)_2$•$6H_2O$, $Co(NO_3)_2$•$6H_2O$, $Sr(NO_3)$, $[CH_4COCH$-$C(O)CH_3]_2Pt$, $[CH_3COCH$=$C(O)$-$CH_3]_3Ru$) and inexpensive urea fuel. These powders are then combined with Pt and Pt/Ru nano-powders (specific surface area of 12 m^2•g^{-1}) along with high specific area carbon to sinter the perovskite composites for use as an anode in a small single-cell membrane electrode assembly (MEA). The undoped samarium cobalite powder from AICS had a specific surface area of 25 $m^2 \cdot g^{-1}$, but did not possess the desired perovskite structure until after a brief calcination treatment in air at 800°C for twelve hours that also resulted in a decreased specific surface area of $9 m^2 \cdot g^{-1}$. The current density of the $SmCoO_{3-\gamma}$-Pt anode was approximately three times lower and the $SrCO_3$-Pt/Ru anode was only slightly lower than commercially available PEMFCs with an approximate methanol conversion of 65%, shown graphically in Figure 8. The behavior of the anodes can be further modified by using variable Pt and Pt/Ru particle sizes. Nonetheless, fabrication of such anodes has been shown to be relatively straightforward and simple using the AICS powder fabrication process combined with sintering. Thus, expensive processing equipment and expensive starting materials are not required, and the fabrication sequence is well suited for scale up and commercial fabrication of such anodes.

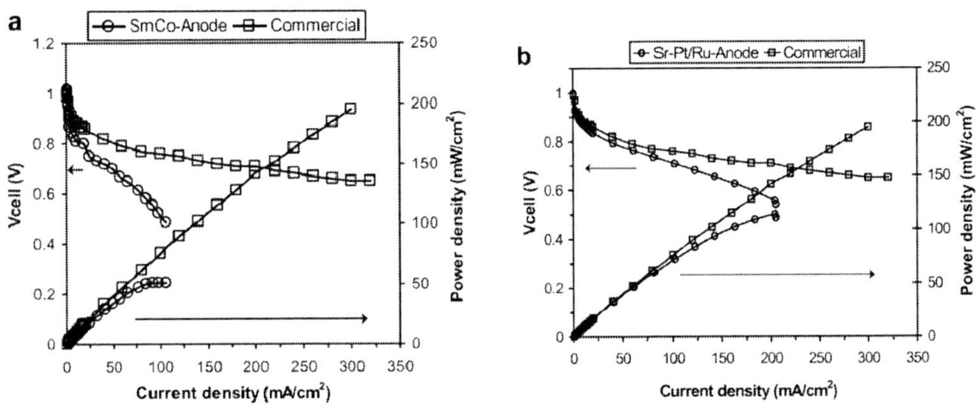

Figure 8. MEA polarization curves comparing AICS fabricated (a.) SmCo and (b.) Sr-Pt/Ru anodes against commercially available (a.) SmCo and (b.) Sr-Pt/Ru anodes at 60°C and 2 bar pressure on the electrodes [Reprinted from Ref. 140, with permission from Springer].

Yttria-stabilized-zironia (YSZ)/metallic composites (CERMETs) offer an additional candidate option for solid oxide fuel cell (SOFC) anodes based on the combination of desirable electrical and mechanical properties [135]. In order for a CERMET to be a viable SOFC anode, the composite must possess good electrical conductivity, based on careful control of composition and microstructure, have a low metallic content to minimize aging effects, and contain high porosity (50%). Conduction percolation is extremely dependent upon the metallic phase of the CERMET, so that this phase, along with the composite porosity, must have the appropriate size, shape and dispersion throughout. Typically, Ni/YSZ CERMETs are fabricated based on hydrogen reduction of poorly sintered NiO mixtures at approximately 1000°C in order to maintain a fine microstructure and high composite porosity. CS can be tailored to produce CERMETs with potential for use as SOFC anodes from a metallic oxide (nickel, copper, cobalt or iron), zirconium and yttria (Y_2O_3) powders [141]. CERMETs fabricated employing CS are characterized by adequate dispersion of the metallic phase in the ceramic matrix and an acceptable level of porosity that could be controlled by either combustion temperature, a gasifying additive or a low environmental pressure. Ni/YSZ CERMETs produced at higher combustion temperatures, through control of the initial reactant mixture, produce composites with more desirable properties, such as increased porosities and a more homogeneous YSZ matrix, but also produces the undesirable trait of coarse metallic particles. The latter can presumably be overcome by using finer metallic powders as reactants.

Tetragonal zirconia polycrystalline (TZP) ceramics can potentially be used in electrochemical cells, such as fuel cells, oxygen sensors and oxygen pumps, resulting from their high oxygen-ion conductivity at elevated temperatures [142]. The fact that these ceramics are also tough, wear resistant and possess a low thermal conductivity allows them to be used as structural materials in fuel cells [143]. The tetragonal phase of zirconia can be retained through the addition of yttria, scandia, calcia or several other rare-earth oxides as long as the grain size is between 0.2 and 1 μm [144]. AICS can be employed to produce nanocrystalline TZP ceramic powders that are then compacted and sintered at 1400-1500 C in air for 2-4 hours into the desired shape or form. The AICS process involves a red-ox reaction from zirconium oxychloride, yttrium oxide, nitric acid and ammonium hydroxide [9]. The powder product requires a calcination treatment at 600°C for 2 hours before the desired crystallinity is obtained.

Lithium ceramics, such as γ-$LiAlO_2$, Li_2TiO_3 and Li_2SiO_3, have been proposed as promising candidates for tritium-breeding materials in fusion reactor systems [145,146]. Fabrication of lithium ceramics often restrict improvement of desired properties (solid-state synthesis), require expensive starting materials and process equipment (wet-chemical methods) and/or require subsequent high temperature calcination treatments to obtain highly crystalline compounds. Fabrication of low cost, high purity γ-$LiAlO_2$ ceramics with a fine, homogeneous microstructure, high reactivity and bending strength could be achieved through AICS of glycine, urea and citric acid fuels with lithium and aluminum nitrates [147,148]. Similarly, fabrication of Li_2SiO_3 was demonstrated using a modified AICS reaction involving lithium hydroxide (LiOH) and silicic acid ($SiO_2 \bullet H_2O$) precursors with urea fuel [149]. AICS of ultra-fine crystalline Li_2TiO_3 powders was accomplished through reaction of lithium and titanium nitrates with glycine or a mixture of urea and citric acid fuel [150]. The AICS fabrication process resulted in fine particles, on the order of 30 nm, with high sinterability characteristics.

Powder metallurgy techniques, such as cold press and sinter along with hot pressing, are currently used in the nuclear industry to fabricate high density, solid-solutions of actinide oxide fuels, e.g. $(U,Pu)O_2$, $(U,Th)O_2$ and $(Th,Pu)O_2$. Powder metallurgy techniques are often meticulous in this regard and involve numerous process steps to ensure that the fabricated actinide powder is homogeneous and produced as a single phase solid solution. Fabrication can be energy, and thus cost, intensive as a result of the increased quality assurance and increased safety constraints required, and require equipment with large footprints used in batch processes for commercial production. Cerium is a common surrogate for plutonium due to the similar chemical properties the two elements possess. Uranium oxide-cerium oxide ($(U_{0.72}Ce_{0.28})O_2$) solid solutions have been fabricated from uranyl nitrate solution and cerium nitrate with citric acid employing gel-combustion synthesis in a microwave oven (referred to as microwave-assisted CS or MACS) [151]. Powders fabricated with this technique were solid solution oxides of uranium and cerium after combustion was complete at 500°C and had better sinterability characteristics than powders prepared employing a more complex and costly co-precipitation technique. Nanocrystalline, freely flowing, porous urania-thoria solid solution powders have been similarly fabricated employing MACS from uranyl and thorium nitrates with citric acid fuel that are suitable as feed material for sintering high density fuel pellets [152,153], as well as nanocrystalline ThO_2 powders with high sinterability characteristics using citric acid fuel [154] and glycine fuel [155]. Rare Earth (La, Nd and Sm) oxides that exhibit considerable solid solubility in UO_2 nuclear fuel have been produced employing AICS from metal nitrates and urea [156] while barium and strontium thorates ($BaThO_3$ and $SrThO_3$, respectively) were fabricated employing AICS from metal nitrates and citric acid that are of interest due to fission product interaction with thorium during irradiation [157].

Interest in fast reactor technology is again on the rise, as evidenced by the recent announcement of the U. S. Department of Energy Global Nuclear Energy Partnership [158], since these reactors provide adequate long-term management of plutonium and other minor actinides (transmutation and recycle), thereby minimizing proliferation risks and waste depository requirements, while still generating a respectable amount of heat for energy, hydrogen or water production. Zirconia and thoria, in particular yttria-thoria solid solutions, are prime candidates as electrolytes for oxygen monitoring in liquid sodium coolant of fast reactors [159]. As such, the product for this application must have a fine grain size, high density and acceptable mechanical properties, all of which are highly dependent upon the starting material feedstock. Nanocrystalline, high purity yttria-thoria solid solution powders were fabricated from yttrium and thorium nitrates with polyethylene glycol and citric acid solution fuels employing the AICS technique [160]. The solid-solution oxides fabricated from the AICS process resulted in sintered densities of greater than 99% while maintaining a fine grain structure, satisfying the typical requirements for oxygen monitoring electrolytes.

Thin Films

Novel composite solid thin-film coatings can be used to fulfill many industrial applications and are typically deposited on desired components by ion-plasma or magnetron sputtering physical vapor deposition (PVD) techniques. Targets for these techniques are conventionally fabricated by hot pressing, hot extrusion, HIPing or plasma sputtering of the novel components. These methods often produce targets with high residual porosity, non-uniform chemical composition and structure, and high impurity contamination contents. Examples of novel ceramic coating compounds are titanium borides, carbides and nitrides, that may also contain silicon or aluminum compounds. CS offers the ability to fabricate high density, pure, uniform, both chemically and structurally, PVD targets, and offers an additional advantage of producing functionally graded component and multi-layer targets, referred to as FGTs. FGTs are beneficial for ion-plasma sputtering since the multiple layers enhance the impact ductility and thermal resistance of the entire composite target. Use of targets containing the desired coating components without the use of a reactive gaseous environment results in uniform, nanocrystalline, high purity coatings. CS of targets 100 mm in diameter for TiB_2-TiN and TiN-Ti_5Si_3 coatings has been demonstrated starting from titanium, boron and Si_3N_4 powders, an example of which is provided in Figure 9 [161]. A TiB_2-Cu backing substrate can be fabricated *in situ*, creating a FGT that aids in maintaining high electrical and thermal conductivities necessary for magnetron sputtering, along with a TiC-TiB_2 intermediate layer to minimize thermal stress created from CTE mismatches. Good adhesion between the layers is accomplished by a distinct transition diffusional zone resulting from high temperature combustion in the presence of a [titanium, copper] liquid phase at the layer interface.

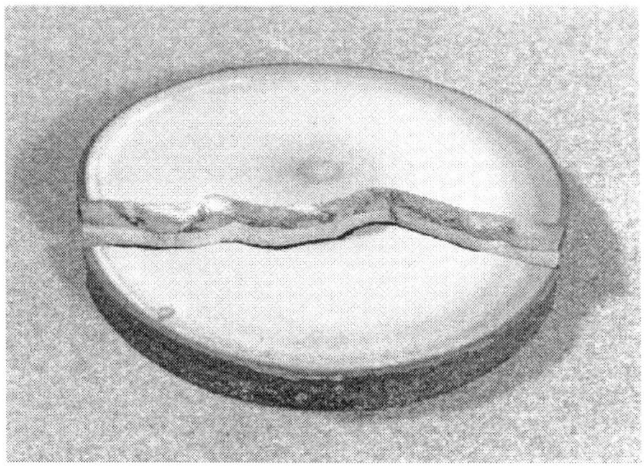

Figure 9. Photograph of a TiB_2-TiN working surface backed with Cu FGT for magnetron sputtering [Reprinted from Ref. 161, with permission from Springer].

Glass Composites

Glass-ceramic composites find use as high-strength, stable glasses with improved workability [162]. Such composites can be produced employing propagating mode CS reactions [163]. For example, a glass matrix composed of B_2O_3-Al_2O_3-MgO can be produced with a crystalline ceramic TiB_2 phase from inexpensive TiO_2, B_2O_3, Al and Mg powders. Traditional production of glasses involves fusing the oxide components in a high temperature furnace for extended durations of time, increasing energy usage, cost and time. The physical properties of the glass composite are dependent upon the duration of firing, especially the melting point of the composite. The CS technique allows *in situ* fabrication of the composites without the need for an external furnace since the heat necessary is generated from reaction of the starting components. During the CS reaction, elemental aluminum and magnesium reduce all of the TiO_2 and part of the B_2O_3 forming Al_2O_3 and MgO while liberating a large amount of heat during the process. This heat fuses the Al_2O_3, MgO and the remaining, unreduced B_2O_3 to form a glassy matrix. The by-product of TiO_2 reduction, titanium, forms with the reduced B_2O_3 resulting in a crystalline TiB_2 phase dispersed within the glassy matrix that also liberates some heat. Examples of optical micrographs of CS fabricated glass-ceramic composites are provided in Figure 10. Aluminoborate glass composites have also been produced employing the CS technique from TiO_2, B_2O_3, BaO and Al powders [164]. A comparison between some glass-ceramic composition characteristics between CS and more conventionally employed furnace melting fabrication techniques is provided in Table 2.

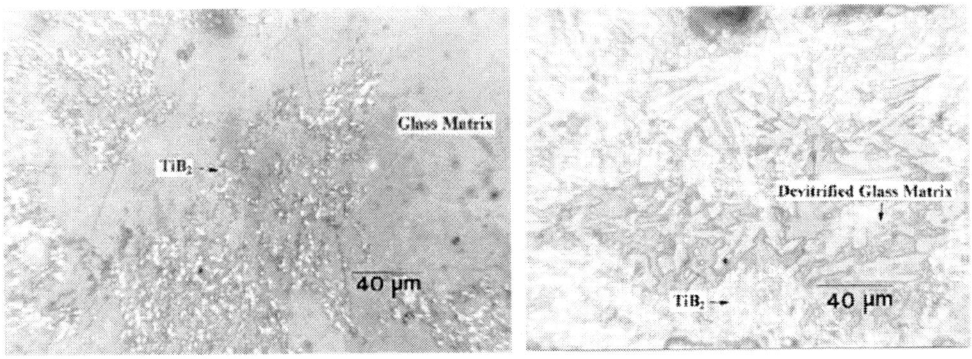

Figure 10. Optical micrographs of a sample with a matrix composition of (left) $40B_2O_3$-35MgO-$25Al_2O_3$ (mol%) showing pure glass matrix and (right) $40B_2O_3$-30MgO-$30Al_2O_3$ (mol%) showing substantial devitrification in the glass matrix [Reprinted from Ref. 163, with permission from Springer].

Table 2. Comparison of glass-ceramic composition characteristics between CS and conventional furnace melting fabrication techniques.

Compositions (mol%)	*Combustion Synthesis* [after Ref. 163,164]	*Furnace Melting* [after Ref. 162]
40B-40O-20A	Clear glass w/ TiB_2	Clear glass
40B-45O-15A	Clear glass w/ TiB_2	Clear glass
60B-20O-20A	Clear glass w/ TiB_2	Clear glass
40B-30O-30A	Clear glass w/ TiB_2	Clear glass
40B-35O-25A	Clear glass w/ TiB_2	Clear glass
40B-27O-33A	Clear glass w/ TiB_2	Devitrified
40B-22O-38A	Clear glass w/ TiB_2	Devitrified
40B-20O-40A	Devitrified w/ TiB_2	Devitrified
50B-30M-20A	Immiscible	Immiscible
50B-40M-10A	Mainly glass	Clear glass
45B-35M-20A	Pure glass	Clear glass
45B-40M-15A	Pure glass	Clear glass
40B-30M-30A	Devitrified	Devitrified
40B-35M-25A	Pure glass	Clear glass
40B-40M-20A	Mainly glass	Clear glass

B = B_2O_3, O = BaO, A = Al_2O_3, M = MgO

Conclusion

The intention of this chapter has been to demonstrate the wide capability of combustion synthesis reactions, and to provide some specific examples where the end products of these reactions can and are being used. CS is a highly versatile and dynamic material processing technology, the science of which has been studied and well-developed over 40 years by academia, government, industry and national laboratories. The process has continually been refined over these 40 years, resulting in a safe, energy and time efficient, low cost alternative to fabricate advanced materials necessary for advanced applications well into the 21st century. While every material system fabricated from CS has not and can not be referenced, the goal was to demonstrate particularly novel modifications of the CS process, and novel material systems that can either replace higher cost, more time consuming fabrication methods, or even replace current materials that are being used for specific applications.

References

[1] Munir, Z. A.; Anselmi-Tamburini, U. *Mater. Sci. Report* 1989, 3, 227-365.

[2] Merzhanov, A. G. *Combustion and Plasma Synthesis of High-Temperature Materials*; VCH: New York, NY, 1990, pp. 1-53.

[3] Pampuch, R., *J. Euro. Ceram. Soc.* 1999, 19, 2395-2404.

[4] Merzhanov, A. G. *Ceram. Int.* 1995, 21, 371-379.

[5] Moore, J. J.; Feng, H. J. *Prog. Mat. Sci.* 1995, 39, 275-316.

[6] Munir, Z. A.; Anselmi-Tamburini, U.; Ohyanagi, M. *J. Mater. Sci.* 2006, 41, 763-777.

[7] Clark, D. E.; Sutton, W. H. *Ann. Rev. Mater. Sci.* 1996, 26, 299-331.

[8] Duckham, A.; Spey, S. J; Wang, J.; Reiss, M. E.; Weihs, T. P.; Besnoin, E.; Knio, O. M. *J. Appl. Phys.* 2004, 96, 2336-2342.

[9] Juarez, R. E.; Lamas, D. G.; Lascalea, G. E.; Walsoe de Reca, N. E. *J. Euro. Ceram. Soc.* 2000, 20, 133-138.

[10] LaBerge, M. *Biomater. Forum* 2001, 23, 3.

[11] Moore, J. J.; Schowengerdt, F. D.; Ayers, R.; Zhang, X.; Castillo, M. *Sixth International Microgravity Combustion Workshop*; NASA/CP-2001 – 210826, 2001, pp. 273-276.

[12] Hulbert, S. F.; Young, F. A.; Mathews, R. S.; Klawitter, J. J.; Talbert, C. D.; Stelling, F. H. *J. Biomed. Mater. Res.* 1970, 4, 433-456.

[13] van Eeden, S. P.; Ripamonti, U. *Plastic and Reconstructive Surgery* 1994, 93, 959-966.

[14] Vallet-Regi, M. *J. Chem. Soc., Dalton Trans* 2001, 97 108.

[15] Hulbert, S. F.; Hench, L. L.; Forbers, D.; Bowman, L. S. *Ceram. Int.* 1982, 8, 131-140.

[16] Ayers, R. A.; Burkes, D. E.; Gottoli, G.; Yi, H. C.; Zhim, F.; Yahia, L.; Moore, J. J. (2007). *J. Biomed. Mater. Res. Part A*. ISSN 1552-4965.

[17] Han, Y.; Li, S.; Wang, X.; Chen, X. *Mat. Res. Bull.* 2004, 39, 25-32.

[18] Tas, A. C. *J. Euro. Ceram. Soc.* 2000, 20, 2389-2394.

[19] Rice, R. W.; McDonough, W. J. *J. Am. Ceram. Soc.* 1985, 68, C-122-C-123.

[20] Kim, H. Y.; Chung, D. S.; Enoki, M.; Hong, S. H. *J. Mater. Res.* 2006, 21, 1141-1149.

[21] Patterson, R. A.; Martin, P. L.; Damkroger, B. K.; Christodoulou, L. *Welding J.* 1990, 69, 39.

[22] Liu, H. J.; Feng, J. C. *J. Mat. Sci. Lett.* 2002, 21, 9-10.

[23] Shiue, R. K.; Wu, S. K.; Chen, S. Y. *Intermetall.* 2003, 11, 661-671.

[24] He, P.; Feng, J. C.; Zhou, H. *Mater. Sci. Eng. A* 2005, 392, 81-86.

[25] Cao, J.; Feng, J. C.; Li, Z. R. *J. Mater. Sci.* 2006, 41, 4720-4724.

[26] Liu, W.; Naka, M. *Scripta Mater.* 2003, 48, 1225-1230.

[27] Swiston Jr., A. J.; Hufnagel, T. C.; Weihs, T. P. *Scripta Mater.* 2003, 48, 1575-1580.

[28] Das, K.; Bandyopadhyay, T. K.; Das, S. *J. Mat. Sci.* 2002, 37, 3881-3892.

[29] Capaldi, M. J.; Saidi, A.; Wood, J. V. *ISIJ Int.* 1997, 37, 188-193.
[30] Tubalov, N. P.; Lebedeva, O. A.; Vereshchagin, V. I. *Refractories and Ind. Ceram.* 2003, 44, 343-345.
[31] Zhyhuts, Y. Y.; Shyrokov, V. V. *Mater. Sci.* 2005, 41, 666-672.
[32] Yang, Y.; Feng, K. Q.; Shen, K. Z.; He, H.; Lin, H. M. *Powder Metall.* 2006, 49, 135-139.
[33] Li, S.; Zhou, Y.; Duan, H.; Qiu, J.; Zhang, Y. *J. Mat. Sci.* 2003, 38, 4065-4070.
[34] Rabin, B. H. *J. Am. Ceram. Soc.* 1992, 75, 131-135.
[35] Clarke, D. R. *J. Am. Ceram. Soc.* 1992, 74, 739-759.
[36] Aghajanian, M. K.; Rocazella, M. A.; Burke, J. T.; Keck, S. D. *J. Mater. Sci.* 1991, 26, 447-454.
[37] Horvitz, D.; Gotman, I. *Acta Mater.* 2002, 50, 1961-1971.
[38] Horvitz, D.; Gotman, I.; Gutmanas, E. Y.; Claussen, N. *J. Euro. Ceram. Soc.* 2002, 22, 947-954.
[39] Chan, K. S.; Shih, D. S. *Metal. Mater. Trans.* 1997, 28*A*, 79-90.
[40] Rosario, V. M.; Chaturvedi, M. C.; Kipouros, G. J.; Caley, W. F. *Mat. Sci. Eng. A* 1999, 270, 283-290.
[41] Padmavardhani, D.; Gomez, A.; Abbaschian, R. *Intermetallics* 1998, 6, 229-241.
[42] Mishra, S. K.; Rupa, P. K. P.; Das, S. K.; Shcherbakov, V. *Met. Mat. Trans. B* 2006, 37*B*, 641-647.
[43] Meyers, M. A.; Olevsky, E. A.; Ma, J.; Jamet, M. *Mat. Sci. Eng. A* 2001, 311, 83-99.
[44] Cao, G.; Varma, A.; Morbidelli, M. *AlCHE J.* 1991, 37, 1420-1424.
[45] Upadhya, K.; Yang, J. M.; Hoffmann, W. P. *Am. Ceram. Soc. Bull.* 1997, 58, 51-56.
[46] Anselmi-Tamburini, U.; Kodera, Y.; Gasch, M.; Unuvar, C.; Munir, Z. A.; Ohyanagi, M.; Johnson, S. M. *J. Mater. Sci.* 2006, 41, 3097-3104.
[47] Morz, C. *Am. Ceram. Soc. Bull.* 1995, 74, 165-166.
[48] Woo, S.-K.; Han, I.-S.; Kang, H.-S.; Yang, J.-H.; Kim, C.-H. *J. Korean Ceram. Soc.* 1966, 33, 259-262.
[49] Monteverde, F.; Bellosi, A.; Guicciardi, S. *J. Euro. Ceram. Soc.* 2002, 22, 279-288.
[50] Bermann, O. R.; Barrington, J. *J. Am. Ceram. Soc.* 1966, 46, 502-507.
[51] Mishra, S. K.; Das, S. K.; Pathak, L. C. *Mat. Sci. Eng. A* 2006, 426, 229-234.
[52] Mishra (Pathak), S. K.; Das, S. K.; Ramachandrarao, P.; Belov, D. Yu; Mamyan, S. *Met. Mat. Trans. A* 2003, 34*A*, 1979-1983.
[53] Sheppard; Laurel, M. *Am. Ceram. Soc. Bull.* 1990, 69, 1801-1812.
[54] Prochaska, G. W.; Miller, G. R. *Mat. Res. Soc. Symp. Proc.* 1990, 167, 215-227.
[55] Landon, M.; Thevenot, F. *Ceram. Int.* 1991, 17, 97-110.
[56] Munir, Z. A. *Met. Mat. Trans. A* 1996, 27*A*, 2080-2085.
[57] Kexin, C.; Haibo, J.; Heping, Z.; Ferreira, J. M. F. *J. Euro. Ceram. Soc.* 2000, 20, 2601-2606.
[58] Kobyakov, V. P.; Bunin, V. A.; Borovinskaya, I. P. *Inorg. Mater.* 2006, 42, 1397-1403.
[59] Tzenov, N. V.; Barsoum, M. W. *J. Am. Ceram. Soc.* 2000, 83, 825-832.
[60] Zhou, A.; Wang, C.-A.; Hunag, Y. *J. Mater. Sci.* 2003, 38, 3111-3115.
[61] Zhou, A.; Wang, C.-A.; Huang, Y. *Mat. Sci. Eng. A* 2003, 352, 333-339.
[62] Zhou, W.; Mei, B.; Zhu, J.; Hong, X. *J. Mater. Sci. Lett.* 2005, 40, 2099-2100.
[63] Khoptiar, Y.; Gotman, I.; Gutmanas, E. Y. *J. Am. Ceram. Soc.* 2005, 88, 28-33.

[64] Zhang, Z. F.; Sun, Z. M.; Hashimoto, H.; Abe, T. *J. Am. Ceram. Soc.* 2003, 86, 431-436.
[65] Zhang, Z. F.; Sun, Z. M.; Hashimoto, H.; Abe, T. *J. Euro. Ceram. Soc.* 2002, 22, 2957-2961.
[66] Zhang, Z. F.; Sun, Z. M.; Hashimoto, H.; Abe, T. *Scripta Mater.* 2001, 45, 1461-1467.
[67] Gao, N. F.; Li, J. T.; Zhang, D.; Miyamoto, Y. *J. Euro. Ceram. Soc.* 2002, 22, 2365-2370.
[68] Zhu, J.-O.; Mei, B.-C.; Xu, X.-W.; Liu, J. *Scripta Mater.* 2003, 49, 693-697.
[69] Zhu, J.; Mei, B. *J. Mater. Syn. Proc.* 2002, 10, 353-358.
[70] Mei, B.; Zhou, W.; Zhu, J.; Hong, X. *J. Mater. Sci.* 2004, 39, 1471-1472.
[71] Zhou, A.; Wang, C.-A.; Ge, Z.; Wu, L. *J. Mat. Sci. Lett.* 2001, 20, 1971-1973.
[72] Lopacinski, M.; Puszynski, J.; Lis, J. *J. Am. Ceram. Soc.* 2001, 84, 3051-3053.
[73] Lange, F. *J. Am. Ceram. Soc.* 1973, 56, 445-450.
[74] Goto, Y.; Tsuge, A. *J. Am. Ceram. Soc.* 1990, 76, 1420-1424.
[75] Khachatryan, G. L.; Arutyunyan, A. B.; Kharatyan, S. L. *Comb. Expl. Shock Waves* 2006, 42, 543-548.
[76] Kata, D.; Lis, J. *Glass Phys. Chem.* 2005, 31, 364-369.
[77] Larker, R. *J. Am. Ceram. Soc.* 1992, 75, 62-66.
[78] Walter, C.; Lefever, R. A. *J. Am. Chem. Soc.* 1954, 37, 5882-5884.
[79] Jack, K. H. *J. Mater. Sci.* 1976, 11, 1135-1158.
[80] Billy, M.; Bock, P.; Dumazeau, C.; Glandus, J. C.; Goursat, P. *Ceram. Int.* 1981, 7, 13-18.
[81] Pivkina, A.; van der Put, P. J.; Frolov, Yu.; Schoonman, J. *J. Euro. Ceram. Soc.* 1996, 16, 35-42.
[82] Cholvy, G.; Derep, J. L.; Gantois, M. *Thin Solid Films* 1985, 126, 51-60.
[83] Lai, F. D.; Wu, J. K. *Surf. Coat. Tech.* 1996, 88, 183-189.
[84] McHale, A. E. (ed.) *Phase Equilibrium Diagrams in Borides, Carbides, and Nitrides*; American Ceramic Society: Columbus, OH, 1994; Vol. X.
[85] Hirota, K.; Takano, Y.; Yoshinaka, M.; Yamaguchi, O. *J. Am. Ceram. Soc.* 2001, 84, 2120-2122.
[86] Lackey, W. J.; Stinton, D. P.; Cerny, G. A.; Schaffhauser, A. C.; Fehrenbacher, L. L. *Adv. Ceram. Mater.* 1987, 2, 24-30.
[87] Heian, E. M.; Gibeling, J. C.; Munir, Z. A. *Mat. Sci. Eng. A* 2004, 368, 168-174.
[88] Ma, C. L.; Kasama, A.; Tanaka, H.; Tan, Y.; Tanaka, R.; Mishima, Y.; Hanada, S. *Mater. Trans. JIM* 2000, 41, 444-451.
[89] Dumont, A.-L.; Bonnet, J.-P.; Chartier, T.; Ferreira, J. M. F. *J. Euro. Ceram. Soc.* 2001, 21, 2353-2360.
[90] Carrillo-Heian, E. M.; Carpenter, R. D.; Paulino, G. H.; Gibeling, J. C.; Munir, Z. A. *J. Am. Ceram. Soc.* 2001, 84, 962-968.
[91] Carrillo-Heian, E. M.; Gibeling, J. C.; Munir, Z. A. *Functionally Graded Mat.* 2000, 241-248.
[92] Cirakoglu, M.; Bhaduri, S.; Bhaduri, S. B. *J. Alloy Comp.* 2002, 347, 259-265.
[93] Itin, V. I.; Gyunter, V. E.; Shabalovskaya, S. A.; Sachdeva, R. L. C. *Mater. Char.* 1994, 32, 179-187.
[94] Burkes, D. E.; Moore, J. J. *ASME J. Eng. Mat. Tech.* 2006, 128, 445-450.
[95] Burkes, D. E.; Moore, J. J. *J. Alloy Comp.* 2007, 430, 274-281.

[96] Zhang, X.-H.; Han, J.-C.; Du, S.-Y.; Wood, J. V. *J. Mat. Sci.* 2000, 35, 1925-1930.
[97] Zhang, X.-H.; Han, J.-C.; He, X.-D.; Kvanin, V. L. *J. Mat. Syn. Proc.* 2000, 8, 29-34.
[98] Shon, I.-J.; Munir, Z. A. *J. Am. Ceram. Soc.* 1998, 81, 3243-48.
[99] Chin, E. S. C. *Mat. Sci. Eng. A* 1999, 259, 155-161.
[100] Bhaumik, S. K.; Divakar, C.; Singh, A. K.; Upadhyaya, G. S. *Mat. Sci. Eng. A* 2000, 279, 275-281.
[101] Xinghong, Z.; Chuncheng, Z.; Wei, Q.; Xiaodong, H.; Kvanin, V. L. *Comp. Sci. Tech.* 2002, 62, 2037-2041.
[102] Lee, J. W.; Munir, Z. A.; Ohyanagi, M. *Mat. Sci. Eng. A* 2002, 325, 221-227.
[103] Locci, A. M.; Orru, R.; Cao, G.; Munir, Z. A. *J. Am. Ceram. Soc.* 2006, 89, 848-855.
[104] Song, I.; Wang, L.; Wixom, M.; Thompson, L. T. *J. Mat. Sci.* 2000, 35, 2611-2617.
[105] Gutmanas, E. Y.; Gotman, I. *J. Euro. Ceram. Soc.* 1999, 19, 2381-2393.
[106] Lee, J. W.; Munir, Z. A.; Shibuya, M.; Ohyanagi, M. *J. Am. Ceram. Soc.* 2001, 84, 1209-1216.
[107] Gotman, I.; Travitzky, N. A.; Gutmanas, E. Y. *Mat. Sci. Eng. A* 1998, 244, 127-137.
[108] Wang, L.; Jiang, W.; Chen, L.; Bai, S. *J. Am. Ceram. Soc.* 2004, 87, 1157-1160.
[109] Shibuya, M.; Kawata, M.; Ohyanagi, M. *J. Am. Ceram. Soc.* 2003, 85, 706-710.
[110] Kata, D.; Lis, J. *Glass Phy. Chem.* 2005, 31, 364-369.
[111] Wang, Z.; Liu, X.; Bian, X. *Adv. Eng. Mater.* 2004, 6, 977-980.
[112] Montanaro, L.; Jorand, Y.; Fantozzi, G.; Negro, A. *J. Euro. Ceram. Soc.* 1998, 18, 1339-1350.
[113] Vadchenko, S. G.; Ponomarev, V. I.; Sychev, A. E. *Comb. Expl. Shock Waves* 2006, 42, 170-176.
[114] Burkes, D. E.; Yi, H. C.; Gottoli, G.; Moore, J. J. *J. Mater. Sci.* 2006, 41, 2009-2017.
[115] Burkes, D. E.; Milwid, J.; Gottoli, G.; Moore, J. J. *J. Mater. Sci.* 2006, 41, 7944-7953.
[116] Hwang, C.-C.; Wu, T.-Y.; Wan, J. *J. Mater. Sci.* 2004, 39, 4687-4691.
[117] Pampuch, R. In *Trends in Emerging Materials and Applications*; VCH: Weinham, NY, 1995; pp. 403-414.
[118] Zhu, L.; Ohashi, M.; Yamanaka, S. *Mater. Res. Bull.* 2002, 37, 475-483.
[119] Gillan, E. G.; Kaner, R. B. *Inorg. Chem.* 1994, 33, 5693-5700.
[120] Holt J. B. U. S. Patent 4,446,242 (1984).
[121] Marin-Ayral, R. M.; Pascal, C.; Martinez, F.; Tedenac, J. C. *J. Euro. Ceram. Soc.* 2000, 20, 2679-2684.
[122] Komeya, K. *Am. Ceram. Soc. Bull.* 1984, 63, 1158-1164.
[123] Chung, S.-L.; Yu, W.-L.; Lin, C.-N. *J. Mater. Res.* 1999, 14, 1928-1933.
[124] Lin, C.-N.; Chung, S.-L. *J. Mater. Res.* 2001, 16, 2200-2208.
[125] Wang, H.; Northwood, D. O.; Han, J.; Du, S. *J. Mater. Sci.* 2006, 41, 1697-1703.
[126] Bunin, V. A.; Borovinskaya, I. P.; Senkovenko, M. Yu.; Ponomarev, V. I.; Ignat'eva, T. I. *Inorg. Mater.* 2003, 39, 357-361.
[127] Borovinskaya, I. P.; Ignat'eva, T. I.; Vershinnikov, V. I.; Khurtina, G. G.; Sachkova, N. V. *Inorg. Mater.* 2003, 39, 388-393.
[128] Chukhlomina, I. N.; Ivanov, Yu. F.; Maksimov, Yu. M.; Akhunova, Z. S.; Krivosheeva, E. N. *Inorg. Mater.* 2005, 41, 1294-1299.
[129] Ge, C.-C.; Li, J.-T.; Xia, Y.-L. *Int. J. SHS* 1996, 5, 107-117.
[130] Borovinskaya, I. P.; Ignat'eva, T. I.; Vershinnikov, V. I.; Sachkova, N. V. *Inorg. Mater.* 2004, 40, 1043-1048.

[131] Washburn, M. E. *Ceram. Bull.* 1967, 46, 667-671.
[132] Radwan, M.; Kashiwagi, T.; Miyamoto, Y. *J. Euro. Ceram. Soc.* 2003, 23, 2337-2341.
[133] Fumo, D. A.; Morelli, M. R.; Segadaes, A. M. *Mat. Res. Bull.* 1996, 31, 1243-1255.
[134] Tas, A. C. *J. Am. Ceram. Soc.* 1998, 81, 2853-2863.
[135] Minh, N. Q. *J. Am. Ceram. Soc.* 1993, 76, 563-588.
[136] Purohit, R. D.; Saha, S.; Tyagi, A. K. *J. Nucl. Mater.* 2003, 323, 36-40.
[137] Arul Antony, S.; Nagaraja, K. S.; Sreedharan, O. M. *J. Nucl. Mater.* 2001, 295, 189-192.
[138] Lago, R.; Bini, G.; Pena, M. A.; Fierro, J. L. G. *J. Catal.* 1997, 167, 198-209.
[139] Jurado, J. R.; Chinarro, E.; Colomer, M. T. *Solid State Ionics* 2000, 135, 365-372.
[140] Chinarro, E.; Segadaes, A. M.; Jurado, J. R. *J. Solid State Electrochem.* 2004, 8, 606-610.
[141] Anselmi-Tamburini, U.; Arimondi, M.; Maglia, F.; Spinolo, G.; Munir, Z. A. *J. Am. Ceram. Soc.* 1998, 81, 1765-1772.
[142] Novotny, J. (ed.) *Science of ceramic interfaces II*; Elsevier Science: Amsterdam, 1994; pp. 71-111.
[143] Garvie, R. C.; Hannink, R. H. J.; Pascoe, R. T. *Nature* 1975, 258, 703-704.
[144] Lange, F. F. *J. Mater. Sci.* 1982, 17, 240-246.
[145] Arons, R. M.; Poeppel, R. B.; Tetenbaum, M.; Johnson, C. E. *J. Nucl. Mater.* 1981, 103/104, 573-578.
[146] Johnson, C. E.; Kummerer, K. R.; Roth, E. *J. Nucl. Mater.* 1988, 155-157 , 188-201.
[147] Wen, Z.; Gu, Z.; Xu, X.; Zhu, X. *J. Nucl. Mater.* 2004, 329-333 , 1283-1286.
[148] Li, F.; Hu, K.; Li, J.; Zhang, D.; Chen, G. *J. Nucl. Mater.* 2002, 300, 82-88.
[149] Cruz, D.; Bulbulian, S. *J. Nucl. Mater.* 2003, 312, 262-265.
[150] Jung, C. H.; Park, J. Y.; Oh, S. J.; Park, H. K.; Kim, Y. S.; Kim, D. K.; Kim, J. H. *J. Nucl. Mater.* 1998, 253, 203-212.
[151] Jain, A.; Ananthasivan, K.; Anthonysamy, S.; Vasudeva Rao, P. R. *J. Nucl. Mater.* 2005, 345, 245-253.
[152] Chandramouli, V.; Anthonysamy, S.; Vasudeva Rao, P. R.; Divakar, R.; Sundararaman, D. *J. Nucl. Mater.* 1998, 254, 55-64.
[153] Anthonysamy, S.; Ananthasivan, K.; Chandramouli, V.; Kaliappan, I.; Vasudeva Rao, P. R. *J. Nucl. Mater.* 2000, 278, 346-357.
[154] Chandramouli, V.; Anthonysamy, S.; Vasudeva Rao, P. R. *J. Nucl. Mater.* 1999, 265, 255-261.
[155] Purohit, R. D.; Saha, S.; Tyagi, A. K. *J. Nucl. Mater.* 2001, 288, 7-10.
[156] Jena, H.; Asuvathraman, R.; Govindan Kutty, K. V. *J. Nucl. Mater.* 2000, 280, 312-317.
[157] Purohit, R. D.; Tyagi, A. K.; Mathews, M. D.; Saha, S. *J. Nucl. Mater.* 2000, 280, 51-55.
[158] U. S. Department of Energy, Press Release (2006)."Department of Energy Announces New Nuclear Initiative".
[159] Hobdell, M. R.; Smith, C. A. *J. Nucl. Mater.* 1982, 110, 125-139.
[160] Ganesan, R.; Vivekanandhan, S.; Gnanasekaran, T.; Periaswami, G.; Srinivasa, R. S. *J. Nucl. Mater.* 2004, 325, 134-140.
[161] Levashov, E. A.; Larikhin, D. V.; Shtansky, D. V.; Rogachev, A. S.; Grigoryan, H. E.; Moore, J. J. *J. Mater. Syn. Proc.* 2002, 10, 319-330.

[162] Hirayama, C. *J. Am. Ceram. Soc.* 1961, 44, 602-606.
[163] Yi, H. C.; Guigne, J. Y.; Moore, J. J.; Manerbino, A. R.; Robinson, L. A.; Schowengerdt, F. D. *J. Mater. Syn. Proc.* 2002, 10, 163-174.
[164] Yi, H. C.; Guigne, J. Y.; Whalen, L. A.; Moore, J. J. *J. Mater. Syn. Proc.* 2000, 8, 15-20.

In: Ceramic Materials Research Trends
Editor: Paul B. Lin, pp. 47-80

ISBN: 1-60021-769-9

Chapter 2

MAGNETORESISTANCE, MAGNETOIMPEDANCE AND MAGNETOCALORIC PHENOMENA IN FERROMAGNETIC MANGANITES

Manh-Huong Phan [a,*] ***and Seong-Cho Yu*** [b,**]

[a]Advanced Composites Centre for Innovation and Science, University of Bristol, Queens Building, Bristol BS8 1TR, England

[b]Department of Physics, Chungbuk National University, Cheongju 361-763, Korea

Abstract

There has been growing interest in ferromagnetic manganites ($A_{1-x}T_xMnO_3$, where A is a trivalent rare-earth element such as La, Pr, Nd, Sm, Eu, etc. and T is a divalent alkaline earth ion such as Ca, Sr, Ba, Pb, etc.), owing to their unique and outstanding physical properties. In particular, the discovery of magnetoresistance (MR) and magnetoimpedance (MI) effects in these materials has provided exciting opportunities to develop them for use in magnetic sensors for magnetic recording technology, while the discovery of the magnetocaloric (MC) effect has made them promising for use in active magnetic refrigerators. In order to design and produce novel manganite materials for such applications, it is important to understand the physical mechanisms behind the MR, MI and MC phenomena. This paper aims to provide a comprehensive analysis of the complex nature of these phenomena and their correlation. Emphasis is placed on providing detailed guidance concerning the development of ferromagnetic manganites for future magnetic sensor and refrigerator technologies.

PACS: 75.30.Sg; 75.30.-m; 75.50.-y

Keywords: Ferromagnetic manganites, Magneto-transport behaviors, Magnetoresistance, Magnetoimpedance, Magnetocaloric, Magnetic sensor, Magnetic refrigeration

[*] E-mail address: M.H.Phan@bristol.ac.uk

[**] E-mail address: S.C.Yu@chungbuk.ac.kr

1. Introduction

Magnetic materials play an essential role in modern technology. Recent advances in magnetic sensing technology, especially in the development of high-performance magnetic sensors, have benefited from the discovery of a new class of magnetic material with perovskite structure, known as ferromagnetic manganese oxides (or manganites) [1]. A question naturally arises concerning what a manganite is and when it was discovered? Historically, Jonker and van Santen [2], in an attempt to seek insulating and ferromagnetic materials for use in high-performance transformers, discovered in 1950 a new type of manganese oxide material (La,T)MnO_3 (T = Ca, Sr, Ba and Pb) exhibiting a metallic and ferromagnetic characteristic and, more interestingly, a relatively large magnetoresistance (MR) effect that could be exploited in magnetic sensors for use in, for example, the car, consumer electronics, and magnetic recording device industries. Since this work, considerable efforts have been devoted to improving the MR magnitude and its field sensitivity in a variety of doped manganites [3-5]. However, the term "colossal magnetoresistance – CMR" was not known until a series of new manganites were found, in 1993, to show a MR effect not by a few percent but by several orders of magnitude [6-8]. It is important to distinguish the CMR from the so-called giant magnetoresistance (GMR) - which is smaller by many orders of magnitude than the CMR and is often observed in metallic multilayers and granulated systems [9]. In principle, the GMR is related to the alignment of electron spins at the interface of different kind of magnetic materials, while the CMR observed in the vicinity of the ferromagnetic to paramagnetic transition temperature (i.e. the Curie point) is believed to be associated with the resistivity peak which can subsequently be reduced by the application of an external magnetic field. GMR technology is widely used in most modern hard drives and many read heads using GMR are commercially available. However, a search for more sensitive GMR materials that allow data to be stored even more densely and read more quickly remains in progress [9]. The reason for this is that the magnitude of GMR depends significantly on measuring temperature and is considerably reduced at high temperature (~300 K). To this extent, the development of manganite materials utilizing CMR technology is probably more desirable as it may provide a wider range of technological applications. For both cases, however, such a high field sensitivity of CMR and GMR has been achieved only under a sufficiently high magnetic field that, for practical use of a sensor device, should be as small as possible [8,9].

Fortunately, the recent discovery of a new phenomenon, i.e. a magnetoimpedance (MI) effect in soft ferromagnetic materials has opened up a new opportunity for the development of more sensitive magnetic sensors which meet the specific engineering requirements [10]. It has been shown that magnetic sensors based on the MI effect offer several advantages over those utilizing the MR effect. The most important is the ultrahigh field sensitivity of MI sensors (detectable fields up to 10^{-8} Oe). When compared with a MR sensor which has a maximum sensitivity of ~1%/Oe, the field sensitivity of a MI sensor can reach a value as high as 500%/Oe [11]. In particular, when investigating the high-frequency magneto-transport properties of $La_{1-x}Ca_xMnO_3$ manganites, Fu and co-workers [12] have demonstrated that the magnitude of MI is much larger than that of MR, indicating a tremendous advantage of the MI technology. Although MI technology is quite new and its development remains incomplete, the low prices and high flexibility of this technology will promise wide-ranging applications in the near future.

Ferromagnetic manganites have also recently been exploited for a newly discovered magnetic refrigeration technology based upon the magnetocaloric (MC) effect - which is related to the magnetic entropy change or the adiabatic temperature change of a magnetic solid substance caused by an applied external magnetic field [13]. Since the magnetic properties of manganites, Curie temperature and saturation magnetization, are strongly doping-dependent, these materials can be used for a variety of large- and small-scale magnetic refrigeration applications in a wide temperature range of 20 – 375 K. It is believed that manganite materials with superior magnetocaloric properties in addition to manufacturing flexibility and cost-effectiveness will be the option of future magnetic refrigeration technology.

With such outstanding properties, ferromagnetic manganites are attracting particular interest in the materials science community, from both fundamental and practical perspectives. It is worthy to note that application of an external magnetic field to a manganite significantly modifies its magneto-transport property, resulting in a simultaneous appearance of the interesting MR, MI and MC phenomena from which the magnetic sensing and refrigeration technologies have been developed. Despite a number of excellent review papers dealing with the magneto-transport and anomalous properties of doped manganites [8,14-19], a complete understanding of the underlying mechanisms which are responsible for the occurrence of the MR, MI and MC phenomena and their relationship has not yet been achieved.

The present paper aims to provide a systematic and comprehensive analysis of the complex nature of MR, MI and MC phenomena in ferromagnetic manganites and place a particular emphasis on giving a high level of guidance in regard to the development of these materials for the advanced magnetic sensor and refrigeration technologies. In this paper, apart from the introductory part (Section 1), we first review the most fundamental features of typical manganites involving their cryptographic and electronic structure, Jahn-Teller lattice distortion, insulator-metal transition, magnetic ordering and phase separation, and material preparation technology (Section 2). Then, we attempt to elucidate the physical origins of the MR, MI and MC phenomena and their relationship (Section 3). Finally, the criteria for selecting ferromagnetic manganites for use in high-performance magnetic sensors and active magnetic refrigerators are evaluated (Section 4). Directions for future research with respect to the development of novel manganites are also given.

2. Fundamental Features of Maganites

2.1. Perovskite Structure

An ideal perovskite structure is generally viewed in the simple form of ABO_3, where A is a large cation which occupies the center of the cub-octahedron and B is a small cation which occupies the center of the octahedron.

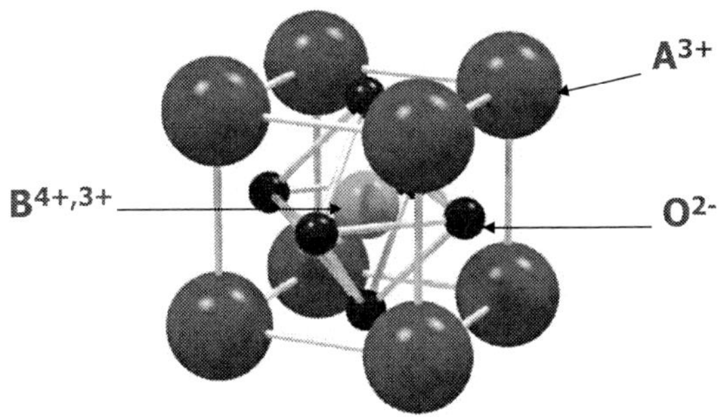

Figure 1. An ideal perovskite structure: ABO_3 (e.g. A = La, B = Mn).

To describe the perovskite structure of a doped manganite, in this paper, let us start with a well-known material of $LaMnO_3$ (A = La and B = Mn), which is insulating and antiferromagnetic. The structure of $LaMnO_3$ is illustrated in Fig. 1, where large La^{3+} ions are characterized by the twelve-fold oxygen coordination, while small Mn^{3+} cations locate at the octahedral sites and are characterized by the six-fold coordination. In reality, the structure of $LaMnO_3$ is reasonably distorted, with elongated and tilted octahedra. That is why $LaMnO_3$ is often chosen as the base compound for the synthesis of CMR materials. It is interesting to mention that doping into the A-site (A = La^{3+}) with divalent elements such as Ca^{2+}, Sr^{2+}, Ba^{2+}, Pb^{2+} leads to not only a conversion in Mn valence from Mn^{3+} to Mn^{4+}, but also considerable variations in the structure of the material. It has been shown that the structure of $LaMnO_3$ can transform from orthorhombic to rhombohedral or cubic, depending upon doping levels of divalent elements. Therefore, it is reasonable to represent a general formula for all doped manganites as $A_{1-x}T_xMnO_3$, where A stands for the trivalent rare-earth element such as La, Pr, Nd, Sm, Eu, etc. and T for the divalent alkaline earth ions such as Ca, Sr, Ba, Pb, etc. Remarkably, the replacement of a large cation (e.g., La) with a small cation (e.g., Ca) causes significant modifications in lattice sites or lattice distortions. This means that the structural change is associated with the lattice distortion. It has been shown that the stability of the perovskite structure depends on the radii of the elements, which is characterized by the Goldschmidt tolerance factor [16]:

$$t = \frac{r_A + r_O}{\sqrt{2}(r_B + r_O)}\,, \tag{1}$$

where r_A, r_B and r_O represents the averaged ionic size of each element. It should be noted that the expression (1) was incorrectly written in several references [8,14,18]. The tolerance factor t is generally considered a measure for the lattice matching of sequential AO and BO_2 planes. It is shown that $t \approx 1$ is for the occurrence of a cubic perovskite structure. As r_A or t decreases, the lattice structure can be transformed into the rhombohedral ($0.96 < t < 1$) and then into the orthorhombic structure ($t < 0.96$), where the B-O-B bond angle θ is bent and deviated from 180^0. For example, the changes of these parameters are summarized in Table 1 for the $La_{1-x}Pb_xMnO_3$ ($0.1 \leq x \leq 0.5$) system [20]. It is revealed that $LaMnO_3$ has an orthorhombic structure due to distorted, elongated and tilted octahedra [16], rather than a cubic one [1]. The doping of other elements into $LaMnO_3$ results in significant changes of the structural parameters.

Table 1. The *A*-site radius (r_A), tolerance factor (t), bond distance (Mn-O) and bond angle (Mn-O-Mn) for the $La_{1-x}Pb_xMnO_3$ ($0.1 \leq x \leq 0.5$) system [20].

x	r_A (Å)	t (Å)	Mn-O (Å)	Mn-O-Mn (^{0})	Structure
0.0	1.360	0.954	1.969	160.5	Orthorhombic
0.1	1.373	0.964	1.961	162.3	Pseudo-Rhombohedral
0.2	1.386	0.974	1.956	166.9	Pseudo-Rhombohedral
0.3	1.399	0.984	1.937	169.7	Pseudo-Rhombohedral
0.4	1.412	0.995	1.943	180.0	Rhombohedral
0.5	1.561	1.054	1.939	180.0	Cubic

These findings indicate that the difference in the radii of substituted elements causes considerable lattice distortion and hence the material structure's change. Such a distortion in the perovskite structure of a doped manganite is also illustrated in Fig. 2 [18].

Another reason for the lattice distortion in $LaMnO_3$ arises from the so-called Jahn-Teller effect, which causes the orbital degeneracy of the Mn^{3+} ions. This will further be discussed in Section 2.2. It is noticed that the Jahn-Teller effect plays a significant role in governing the magneto-transport nature of doped manganites [1,16-18].

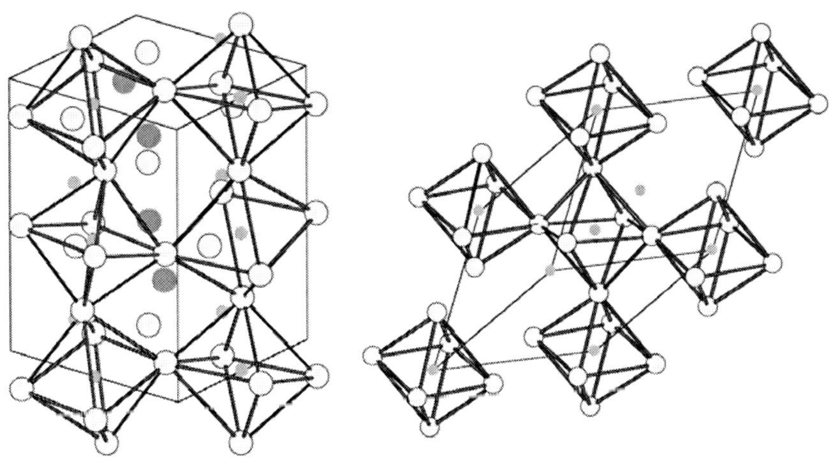

Figure 2. Structures of distorted manganites: orthorhombic (left) and rhombohedral (right) [18]

2.2. Electronic Structure

The electronic structure of a manganite can be assessed in several ways [10-19]. To provide a systematic analysis, in this paper, the electronic structure of undoped $LaMnO_3$ is first assessed, followed by that of doped $LaMnO_3$. The electronic state of $LaMnO_3$ is illustrated in Fig. 3(a) and (b). It appears that $LaMnO_3$ is an insulator with A-type antiferromagnetic ordering. In this material, Mn has d^4 electronic configuration and in 3+ formal charge state. As the 3d orbital is strongly localized, the exchange splitting (~3 eV) is larger than the crystal-field splitting (~1.7 eV) and this leads to the occurrence of high-spin sates (e_g), apart from the presence of low-spin t_{2g} states [Fig. 3(a)]. The low-spin t_{2g} states are fully occupied and the high-spin e_g states are half occupied, giving the total spin number, S = 2 [Fig. 3(b)].

The theory predicts that undistorted $LaMnO_3$ shows a metallic characteristic. However, $LaMnO_3$ itself is distorted, and its electronic structure is significantly modified due to lattice distortion. In general, there are two modes of lattice distortions. One is likely to be associated with the rotation of MnO_6 octahedra which, in turn, causes changes in the bond distance Mn-O and the bond angle Mn-O-Mn, as discussed in Section 2.1. The other is well known as the Jahn-Teller lattice distortion, which is the source of strong electron-phonon coupling. In the limit of narrow *d* band, this type of lattice distortion is reflected by the degeneracy in the high-spin e_g states [Fig. 3(b)] and the reduction in the electronic energy for the single e_g occupation [Fig. 3(c)]. However, Jahn-Teller lattice distortion can significantly be suppressed for wider bands. Therefore, the narrowing of *d* band caused by the octahedra rotation is often favorable for the Jahn-Teller effect and a band gap opens in the e_g band for a real $LaMnO_3$ [21].

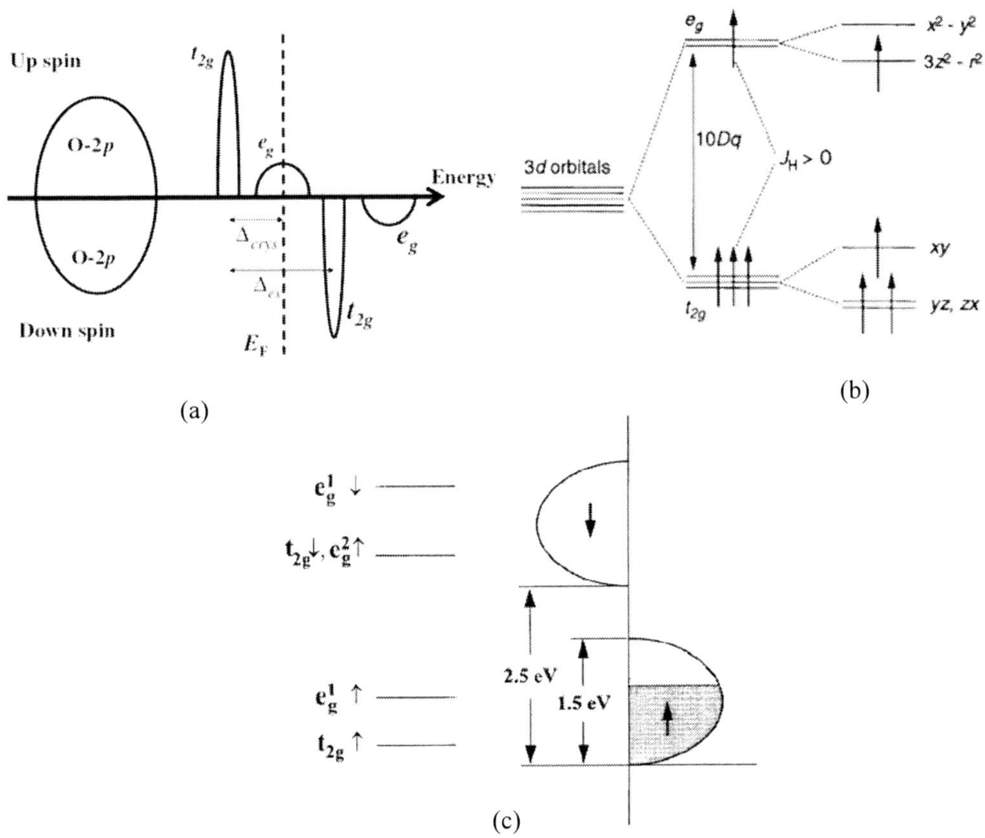

Figure 3. (a) Schematic energy diagram for $LaMnO_3$ without Jahn-Teller lattice distortion [21]; (b) the crystal field splitting of five-fold degenerate atomic 3d levels into lower t_{2g} (triply degenerate) and higher e_g (doubly degenerate) levels [14]; (c) Jahn-Teller distortion of MnO_6 octahedron further lifts each degeneracy [8].

Remarkably, the t_{2g} electrons are less hybridized with O 2p states and quite stabilized by the crystal field splitting, whereas the e_g electrons are hybridized strongly with O 2p and affected by such a correlation effect [1]. The e_g electrons tend to localize in an undoped

$LaMnO_3$, but become itinerant in a doped $LaMnO_3$ as a result of the presence of created electron vacancies or holes. The localization of e_g electrons in addition to the stability of t_{2g} electron states causes an insulating behavior (i.e., the so-called Mott insulator) in a pure $LaMnO_3$. Meanwhile, partial substitution of La^{3+} ions by divalent elements such as T = Ca^{2+}, Sr^{2+}, Ba^{2+}, Pb^{2+} leads to an appearance of Mn^{4+} ions, forming a new form of $La_{1-x}T_xMn^{3+}{}_{1-y}Mn^{4+}{}_yO_3$. This promotes the occurrence of an electron exchange process between Mn^{3+} and Mn^{4+} ions via O 2p orbital, as theoretically proposed by Zener [22] with the so-called "double-exchange" model, and, consequently, introduces a metallic and ferromagnetic behavior into the material.

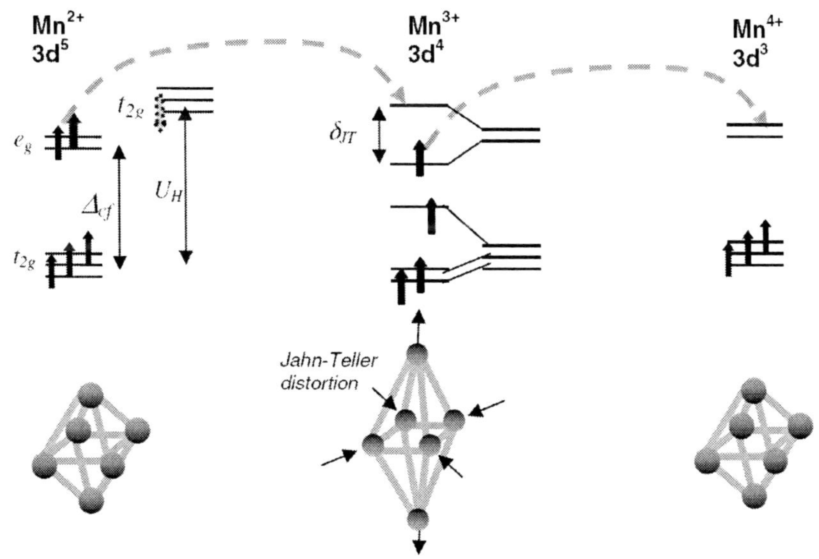

Figure 4. Single ion scheme of 3d electronic states in Mn^{2+}, Mn^{3+} and Mn^{4+} ions in perovskite manganites. Abbreviations denote Δ_{cf} octahedral crystal field splitting, U_H Hund's rule onsite coupling energy, δ_{JT} Jahn-Teller splitting energy resulting from distortion of the oxygen octahedron (as sketched below). Large dashed arrows indicate spin-conservative electron transfer between Mn ions which proceeds via intermediate O 2p orbitals [19].

It should be noted that according to the double-exchange (DE) model, the electron transfer process takes place simultaneously, i.e., one electron from Mn^{3+} jumps to a neighboring O 2p orbital and simultaneously one electron from this O 2p jumps to an adjacent Mn^{4+} ion. This is favorable for the ferromagnetic coupling, which follows Hund's rule. Since the exchange energy J_H is much larger than the intersite hoping interaction of the e_g electron between the neighboring sites, in a semiclasical approximation, the dependence of electron transfer (or the possibility of effective electron hoping - t_{ij}) between Mn ions on the angle (Θ) between their magnetic moments can be described as

$$t_{ij} = t_{ij}^0 \cos(\Theta_{ij} / 2). \tag{2}$$

The electron transfer probability varies from 1 for $\Theta = 0$ (parallel spins) to zero for $\Theta = 180^0$ (anti parallel spins). This simple relation can illustrate, to some extent, the origin of

CMR in ferromagnetic manganites, but the whole picture of CMR physics can only be understood as considering, in addition to the DE model, the role played by the Jahn-Teller lattice effect [Fig. 4].

Apart from the DE interaction, in doped manganites, other interactions such as superexchange should also be taken into consideration. In actual manganites, the presence of Mn^{2+}, Mn^{3+} and Mn^{4+} ions not only forms the ferromagnetic coupling of Mn^{3+}-Mn^{4+}, but also introduces the other couplings of Mn^{3+}-Mn^{3+}, Mn^{4+}-Mn^{4+} and Mn^{2+}-Mn^{3+} which are antiferromagnetic or weak ferromagnetic [Fig. 4]. The presence of Mn^{2+}, which leads to the coupling of Mn^{2+}-Mn^{3+}, can be negligible in a typical manganite material, because this sample is often sintered at relatively high temperatures around 1200 ^{0}C, where the only chemical reduction process of $MnO_2 \rightarrow Mn_2O_3 \rightarrow Mn_3O_4$ takes place. In other words, only the ferromagnetic interaction of Mn^{3+}-Mn^{4+} and the antiferromagnetic interactions of Mn^{3+}-Mn^{3+} and Mn^{4+}-Mn^{4+} co-exist and compete with each other, governing the magnetic nature of doped manganites.

2.3. Insulator-Metal Transition

$LaMnO_3$ is a pure insulator, but an insulator-metal transition appears to occur in it as La is properly substituted by a divalent element such as Ca, Sr, Ba, Pb. For example, the temperature dependence of electrical resistivity of a representative system of $La_{1-x}Sr_xMnO_3$ ($0 \leq x \leq 0.4$) is displayed in Fig. 5 [23].

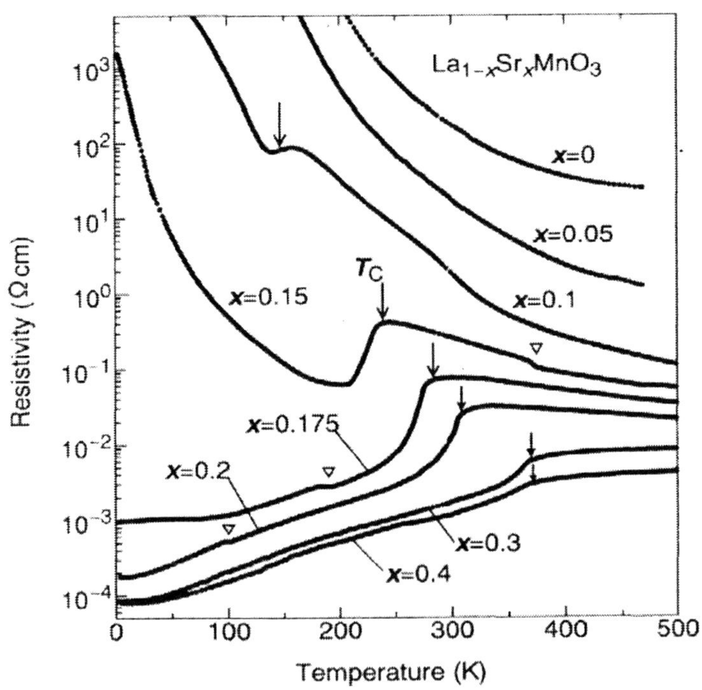

Figure 5. Resistivity against T for $La_{1-x}Sr_xMnO_3$ for various x values. The arrows denote the magnetic ordering transition as determined by magnetization measurements [23].

It can be seen that $La_{1-x}Sr_xMnO_3$ exhibits an insulating behavior for $x \geq 0.05$, but a small insulator-metal transition is induced as the Sr-doped concentration reaches $x = 0.1$. As the Sr-doped concentration exceeds $x = 0.17$, metallic conduction is observed at low temperatures below the ferromagnetic-to-paramagnetic phase transition temperature (T_c). It is worth noting that the occurrence of an insulator-metal transition is accompanied by an orthorhombic ($x \leq 0.175$) to rhombohedral ($x > 0.175$) structural transition. This structural transition reveals a considerable reduction in the Jahn-Teller lattice distortion in $LaMnO_3$ as the Sr-doped content increases. The temperature dependence of electrical resistivity (ρ) can be understood due to the carrier scattering arising from thermal spin fluctuation and the decrease of the hole-carrier effective mass. A semiconducting behavior observed above T_c for $x \leq 0.175$ can be due to the thermal spin fluctuation and the spatial randomness of the transfer interaction. Meanwhile, below T_c the remarkable increase in ρ for $x = 0.05$ and 0.10 can be attributed to the electron localization effect. It should be noted that the arrangement of ferromagnetic spins leads to a reduction of the randomness of the electron transfer, according to Eq (2). This in turn causes the occurrence of the insulator-metal transitions around T_c for $x = 0.05$ and 0.10 samples. The metallic behavior for $x \geq 0.175$ samples at temperatures below T_c is believed to be associated with the electron-electron scattering process, in which the spin wave scattering mainly contributes to the low-temperature electrical resistivity. It is noticed that the complex nature of the insulator-metal transition in manganites is significantly affected by doping and external fields [16]. Doping at the B-site (i.e. B = Mn) has also been found to cause considerable changes in the magneto-transport behavior [1,16-18].

2.4. Magnetic Ordering and Phase Separation

$LaMnO_3$ undergoes an antiferromagnetic transition at 120 K, with a spin-ordering structure of layer type in that the ferromagnetic *ab* plane is coupled antiferromagnetically along the *c* axis [8,21]. As La is substituted by Sr, the ordered spins are canted along the c axis [23]. With increasing Sr-doped content, the spin-canting angle continuously increases towards a ferromagnetic spin structure.

It can be seen from Fig. 6(a) that for the $La_{1-x}Sr_xMnO_3$ ($0 \leq x \leq 0.4$) system, the canted antiferromagnetic phase appears to persist up to $x = 0.15$. For $x > 0.15$, the ferromagnetic phase appears to occur and develops with increasing Sr-doped content up to 0.3 and then becomes saturated for higher concentrations ($x > 0.3$). Herein the change in the magnetic ordering and magnetic phase transition is likely to be related to that in the insulator-metal transition, which arises mainly from the change in the material structure. The phase diagram of $La_{1-x}Sr_xMnO_3$ is characteristic for a DE system, where the one-electron bandwidth (W) is large. It has been shown that the decrease in W due to the decrease in the A-site radii (or the tolerance factor) leads to electronic instability. This is due to the fact that the occurrence of other interactions such as charge/orbital ordering and antiferromagnetic super-exchange which compete against the existing ferromagnetic DE interaction and may become dominant in the specific doping range of x (i.e. at sufficiently small tolerance factors) and/or in the low-temperature range. For instance, as La is entirely substituted by Nd in the $R_{1-x}Sr_xMnO_3$ system (forming $Nd_{1-x}Sr_xMnO_3$), the ferromagnetic sate is only observed for $0.3 < x < 0.5$, and it changes into the charge-ordered insulating state for the $x = 0.5$ sample (i.e. the half doping) at temperatures below 160 K [see Fig. 6(b)]. The effect of the tolerance factor on the ferromagnetic and charge/orbital-ordering phases can be assessed from Fig. 6(b). Analyses on the magnetic ordering and phase separation in manganite materials have been detailed in Refs [14,16,18].

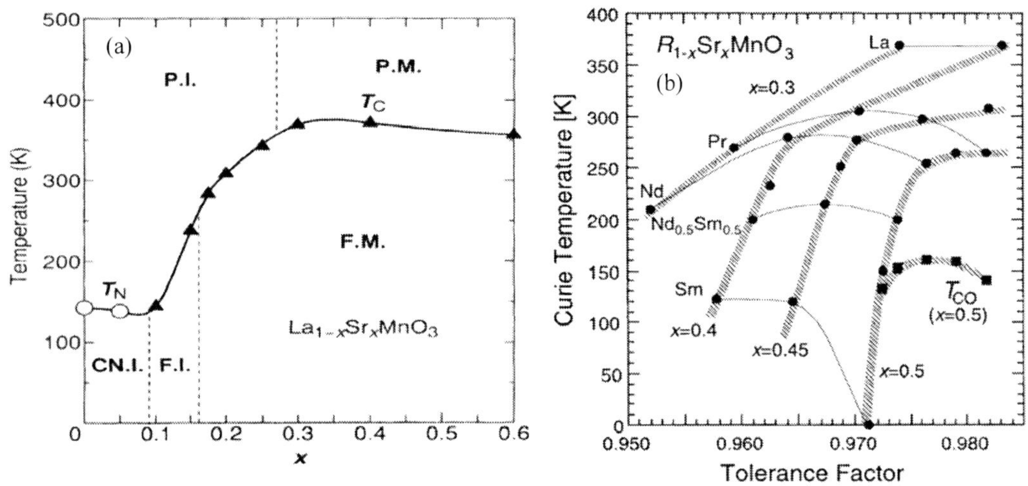

Figure 6. (a) Electronic phase diagram of $La_{1-x}Sr_xMnO_3$ ($0 \leq x \leq 0.4$). Open circles and filled triangles are the Neel (T_N) and Curie (T_c) temperatures, respectively. The abbreviations mean paramagnetic insulator (PI), paramagnetic metal (PM), spin-canted insulator (CNI), ferromagnetic insulator (FI), and ferromagnetic metal (FM) [23]. (b) The critical temperatures of the ferromagnetic transition (solid circles) and the charge-ordering transition for $x = 0.5$ (solid squares) in $RE_{1-x}Sr_xMnO_3$ (RE = La, Pr, Nd, Sm and their solid solution) crystals as a function of tolerance factor [14].

2.5. Preparation Techniques

A number of different techniques have been used for the preparation of manganite materials, including a conventional ceramic method (or a solid state reaction method), a sol-gel method and a sputtering method. The details of these techniques can be found in an excellent textbook written by Raul Valenzuela [24]. Among them, the ceramic technique has been widely employed, because of its higher processing flexibility and cost-effectiveness. Herein, we focus on describing the ceramic technique. Figure 7 illustrates the four basic operations of this technique.

In general, the ceramic technique allows the synthesis of a wide range of polycrystalline manganites. Stoichiometric samples are often pre-sintered in the temperature range of 700 - 900 ^{0}C for 10 – 20 hours, followed by grinding into compound powders. The compound powders are then pressed into pellets and sintered at 900 - 1300 ^{0}C for 10 - 30 hours, depending upon chemical compositions, to give the finished samples. It is noted that the quality of the final product, which directly affect the magnetic and electrical properties, is sensitive to the processing routes (e.g. annealing temperature and time). Therefore, an appropriate selection of annealing temperature and time is important in tailoring the novel properties of a manganite material. It is also important to note, in polycrystalline manganites, that the presence of grains and grain boundaries significantly affects the magnetic, magnetoresistive and magnetocaloric properties, other than single-crystalline ones. The grain size and boundary effect are doping dependent [see Fig. 7(a) and (b), for example]. Remarkably, the surface of a manganite sample prepared by the ceramic method is often more rough than that prepared by the sol-gel or sputtering method. It has been shown that the surface roughness becomes important in high-frequency MI studies, since the high-frequency magnetoimpedance effect is well known as a surface-related phenomenon. This is not only because the skin depth may become smaller than the surface irregularities, but also due to stray fields, which appear on the rough surface and cause a considerable reduction in the MI magnitude. For this study, the sample surface must be polished carefully, and manganite thin films - which have been prepared by the sputtering method - are often more favorable.

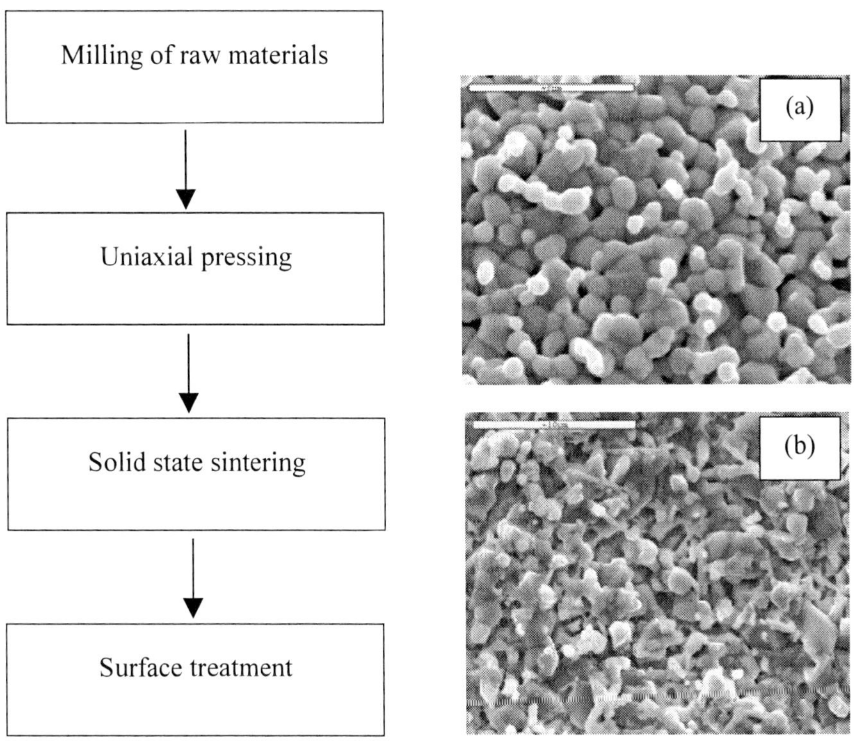

Figure 7. The four basic operations of the ceramic method for the preparation of polycrystalline manganite materials [24]. SEM images of polycrystalline $Pr_{1-x}Pb_xMnO_3$ samples with (a) $x = 0.3$ and (b) $x = 0.5$.

3. Magnetoresistance, Magnetoimpedance and Magnetocaloric Phenomena

3.1. Magnetoresistance

3.1.1. What is Magnetoresistance Effect?

When a magnetic material is subjected to a dc magnetic field, its electrical resistance increases or decreases significantly. This relative change in electrical resistance of a magnetic material upon the application of a magnetic field is known as the magnetoresistive (MR) effect. In actual manganites, the electrical resistance decreases as a magnetic field is applied, resulting in a "negative" MR effect (see Fig. 8, for example).

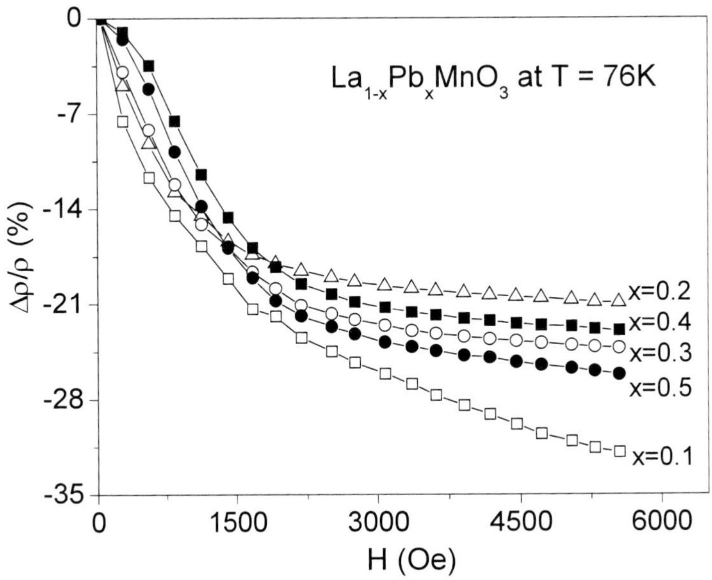

Figure 8. The MR ratio measured at T = 76 K for $La_{1-x}Pb_xMnO_3$ (x = 0.1 - 0.5) samples.

It has been shown that the MR effect of conventional materials is quite small, but manganites are found to show a much larger MR effect. Depending on the magnitude of MR ratio, it is called either as giant magnetoresistance (GMR) or colossal magnetoresistance (CMR).

The magnitude of MR effect (or the MR ratio) can be evaluated by

$$MR(\%) = 100\% \times \frac{\rho(H) - \rho(H_{\max})}{\rho(H_{\max})}, \tag{3}$$

where H_{max} is the external magnetic field sufficient to saturate the electrical resistance. In practice, the value of H_{max} is available for a given experimental equipment. Some researchers use $H_{max} = 0$ in Eq. (3), although this definition may not be appropriate, because the value of $\rho(0)$ depends on the remanent magnetic state of the material.

3.1.2. Theoretical and Experimental Aspects

The theory of MR manganites has dealt mainly with a complex problem of the correlation between magnetism and conductivity. A number of models have been proposed, but each shows a close agreement with experiment only for a certain material system. On the other hand, research has been focused mainly on elucidating the physical mechanism responsible for the occurrence of a resistive peak and a MR peak - which is observed around T_c.

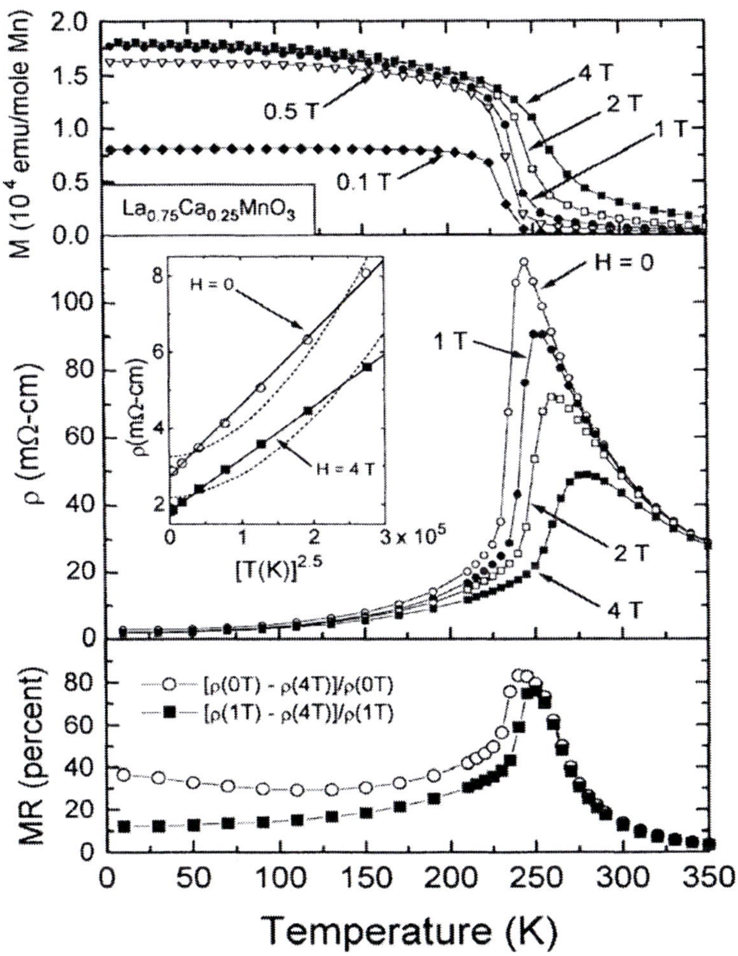

Figure 9. The temperature dependences of magnetization, resistivity and MR ratio for $La_{0.75}Ca_{0.25}MnO_3$. The inset shows ρ at low temperature at various fields, the lines are fits to the data [25].

A representative example displayed in Fig. 9 is in regard to the relationship between the magneto-transport and magnetoresistive properties [25]. Obviously, T_c shifts towards a higher temperature, together with a broadening of the ferromagnetic to paramagnetic phase transition, as the applied magnetic field is increased [see Fig. 9(a)]. This is attributed to the increase of ferromagnetic coupling in the material as the applied magnetic field is increased. It is worth noting that the resistive peak and its steep decrease with respect to an applied magnetic field are observed in the vicinity of T_c [see Fig. 9(b)]. This suggests that the relatively high resistivity above T_c is likely to be related to the scattering of conduction electrons by thermally disordered spins. Also worthy of note is that, along with the decrease in ρ, the resistive peak shifts towards a higher temperature as the applied magnetic field is increased. This is probably attributed to the suppression of the spin scattering of the e_g-state carriers.

On the other hand, it can be seen from Fig. 9 that there is a distinct correlation between the magnetization, electrical resistance and MR effect. The application of a magnetic field causes a broadening of the ferromagnetic to paramagnetic phase transition [see Fig. 9(a)] as well as the insulator-metal transition [see Fig. 9(b)]. Therefore, the shift of a resistive peak is logically linked to that of T_c as seen in Fig. 9(a) and (b). It is believed that the occurrence of MR effect at low temperatures below T_c is due to strain effects on the domains. The domain wall contribution to MR probably originates from the strong correlation between local ferromagnetic order and metallic behavior. At low temperatures, the spins in the domains are not required to be ferromagnetically coupled, which in turn causes resistance. But, this model alone cannot account for the $T^{2.5}$ temperature dependence of electrical resistivity as evidenced from the inset of Fig. 9. The role of Jahn-Teller lattice distortions may be important in this case. However, it still remains unclear if the Jahn-Teller lattice distortion has a realistic origin of the insulator-metal transition. In addition, it is quite difficult to establish the correlation between the magnetism and conductivity in La-based manganites with high doping levels ($x \geq 0.3$). The origin of charge ordering in slightly doped La-based manganites is also not well understood.

3.2. Magnetoimpedance

3.2.1. What is Magnetoimpedance Effect?

The magnetoimpedance (MI) effect is known as a large change in the ac complex impedance of a ferromagnetic conductor with an ac current, when it is subjected to a dc magnetic field. Here, it is important to distinguish the MI from the MR. It has been revealed that the MR and MI phenomena are similar, but there is a divergence in the origin of these phenomena. The MR effect reflecting a change in resistance of a magnetic material subjected to a magnetic field is connected to spin-dependent electron scattering in magnetically non-uniform systems for dc or low-frequency current. In contrast, the MI effect consists in the change of the total impedance (including both resistance and inductance) of a magnetic conductor under a dc applied magnetic field, when a high-frequency ac follows through it.

The magnitude of MI effect (or the MI ratio) can be evaluated by

$$\Delta Z / Z(\%) = 100\% \times \frac{Z(H) - Z(H_{\max})}{Z(H_{\max})}, \tag{4}$$

where H_{max} is usually the external magnetic field sufficient to saturate the impedance.

Using the magnetic field dependence of MI, it is possible to determine whether a magnetic material is suitable for sensor applications, by calculating the magnetic field sensitivity of MI (i.e. the magnetic MI response) via the following expression:

$$\xi = 2(\Delta Z / Z)_{\max} / \Delta H, \tag{5}$$

where ΔH equals to the full width at half maximum of MI curve. This has also been used for the evaluation of the magnetic field sensitivity of MR (i.e. the magnetic MR response).

3.2.2. Theoretical and Experimental Aspects

3.2.2.1. Impedance of a Magnetic Conductor

According to the definition, the complex impedance, $Z = R + j\omega L$ (R and L are resistance and inductance, respectively) of a magnetic conductor is given by the ratio V_{ac}/I_{ac}, where I_{ac} is the amplitude of a sinusoidal current $I = I_{ac}\exp(-j\omega t)$ passing through the conductor and V_{ac} the voltage measured between the ends of the conductor [26]. Figure 10 shows a schematic illustration of the impedance definition.

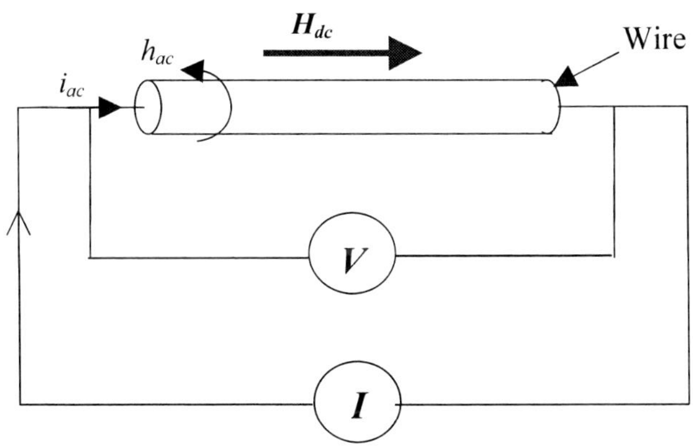

Figure 10. The definition of impedance.

Note that this definition is valid only for a uniform magnetic conductor. Nonetheless, for a metallic ferromagnet having the length L and the cross-section area q, its impedance can still be, in linear approximation, expressed as follows:

$$Z = \frac{V_{ac}}{I_{ac}} = \frac{LE_z(S)}{q\langle j_z \rangle_q} = R_{dc}\frac{j_z(S)}{\langle j_z \rangle_q}, \tag{6}$$

where E_z and j_z are the longitudinal components of an electric field and current density, respectively, and R_{dc} is the dc electrical resistance. S is the value at the surface, and $\langle\ \rangle_q$ is the average value over the cross section q. Alternatively, the expression of Z can be given in terms of the surface impedance tensor $\hat{\xi}$:

$$Z = R_{dc}\frac{q}{\rho l}\left(\xi_{zz} - \xi_{z\phi}\frac{h_z(S)}{h_\phi(S)}\right) \tag{7}$$

where ρ is the resistivity, l the length of the conductor and h_z and h_ϕ are the axial and circumferential components of ac magnetic field, respectively.

The current density *j(r)* in Eq. (6) or the magnetic field h in Eq. (7) of the conductor can be generally obtained within the framework of classical electrodynamics of continuous media, by solving simultaneously the reduced Maxwell equation

$$\nabla^2 \boldsymbol{H} - \frac{\mu_0}{\rho}\dot{\boldsymbol{H}} = \frac{\mu}{\rho}\dot{\boldsymbol{M}} - \text{grad div}\,\boldsymbol{M} \tag{8}$$

and the Landau-Lifshitz equation for the motion of the magnetisation vector

$$\boldsymbol{M} = \gamma \boldsymbol{M} \times \boldsymbol{H}_{\text{eff}} - \frac{\alpha}{M_s}\boldsymbol{M} \times \dot{\boldsymbol{M}} - \frac{1}{\tau}(\boldsymbol{M} - \boldsymbol{M}_0), \tag{9}$$

where γ is the gyromagnetic ratio, M_s the saturation magnetisation, M_0 the static magnetisation, H_{eff} the effective magnetic field, and α the damping parameter [26]. In this context, the effective field can be calculated from the free energy density of the system, which depends on the particular magnetic domain structure of the sample. According to this method, the current density depends on the material parameters and the conductor geometry, as well as the actual magnetic state of the sample [27]. It is therefore difficult to obtain the exact solution of the problem. However, a number of theoretical models have been proposed, using the simplified assumptions, to approach the solution of Eqs (8) and (9) with appropriate boundary conditions [26,27].

In order to avoid such complicated calculations, we often assume the material relationship between the induction and magnetic field is linear ($B = \mu H$ and μ is constant). Using this relationship, it means that we only solve the Maxwell equation while ignoring the Landau-Lifshitz equation of motion Eq. (9). For example, as the classical skin effect solution of Eq. (8) is obtained [28], the calculated impedance Z of a cylindrical magnetic conductor and an infinite planar film are, respectively,

$$Z = R_{DC}\, ktJ_0(kt)/2J_1(kt) \tag{10}$$

and

$$Z = R_{dc} \cdot jka \coth(jka), \tag{11}$$

where J_0 and J_1 are the Bessel functions of first kind, t is the radius of the wire, $2a$ is the thickness of the ribbon, R_{dc} is the electrical resistance for a dc, $k = (1+j)/\delta_m$ with imaginary unit j; δ_m is the penetration depth in a magnetic medium, with the circumferential permeability (μ_ϕ) for the case of the wire [26],

$$\delta_m = \frac{c}{\sqrt{4\pi^2 f \sigma \mu_\phi}}, \tag{12}$$

and with the transverse permeability (μ_T) for the case of the film [28],

$$\delta_m = \frac{c}{\sqrt{4\pi^2 f\sigma\mu_T}}, \tag{13}$$

where c is the speed of light, σ is the electrical conductivity and $f = \omega/2\pi$ is the frequency of the ac (I_{ac}) flowing along the sample.

According to Eqs (10) and (12), or Eqs (11) and (13), MI can be understood as a consequence of the increase of the skin depth until it reaches the radius of the wire (t) through the decrease of the circumferential permeability in Eq. (12) or it reaches the half thickness of the ribbon (a) through the decrease of the transverse permeability in Eq. (13) under an applied dc magnetic field. In order to obtain large MI values, the necessity is to reduce the skin depth by choosing magnetic materials that have large μ_ϕ (or μ_T) and small δ_m and R_{dc}. It is clear that the large permeability reduces the skin depth that is later increased by the applied field as shown in Figure 11. In fact, the real and imaginary components of Z change with the dc applied field, H_{dc}. In first order approximation, the in-plane component or resistance, R, can be expressed as [29]:

$$R = (\rho l)/2\pi(a - \delta_m)\delta_m. \tag{14}$$

This means that such changes in δ_m caused by H_{dc} via μ_ϕ (or μ_T) will modify R and hence Z. Therefore, the skin depth can be evaluated as a function of H_{dc} through the measurement of R. The inductance L can be expressed as [30]:

$$L = 0.175\mu_0 lf\langle\mu_r\rangle/\omega, \tag{15}$$

where μ_0 and $<\mu_r>$ are the vacuum permeability and the average relative circumferential permeability, respectively. It is pointed out that changes in $<\mu_r>$ caused by H_{dc} will alter L and hence Z. Consequently, changes in both R and L contribute to those in Z and therefore to MI.

Based on the frequency (f) of the driving ac, the giant magnetoimpedance can generally be classified into the following frequency regimes:

Low-frequency regime (up to a few kHz), where the changes in voltage at the sample's ends are mainly due the so-called magnetoinductive effect [31]. The skin effect is very weak in this case. The change in the impedance of the sample upon application of the applied field (H_{dc}) results mainly from the contribution of inductance (L), which is proportional to the circumferential permeability (μ_ϕ) for a cylindrical magnetic conductor (i.e., a magnetic wire) or the transverse permeability (μ_T) for a planar magnetic film (i.e., a magnetic ribbon) [32]. In this frequency range, the MI feature has been explained by the quasistatic model, which is based on an assumption that the measuring frequency is small enough for an equilibrium state of the system to be reached at every moment. In general, this model can describe basic features of MI at relatively low frequencies, but cannot interpret the frequency dependence of

MI in the intermediate and high frequency ranges [26]. The reason for this lies in the original assumptions made in the models.

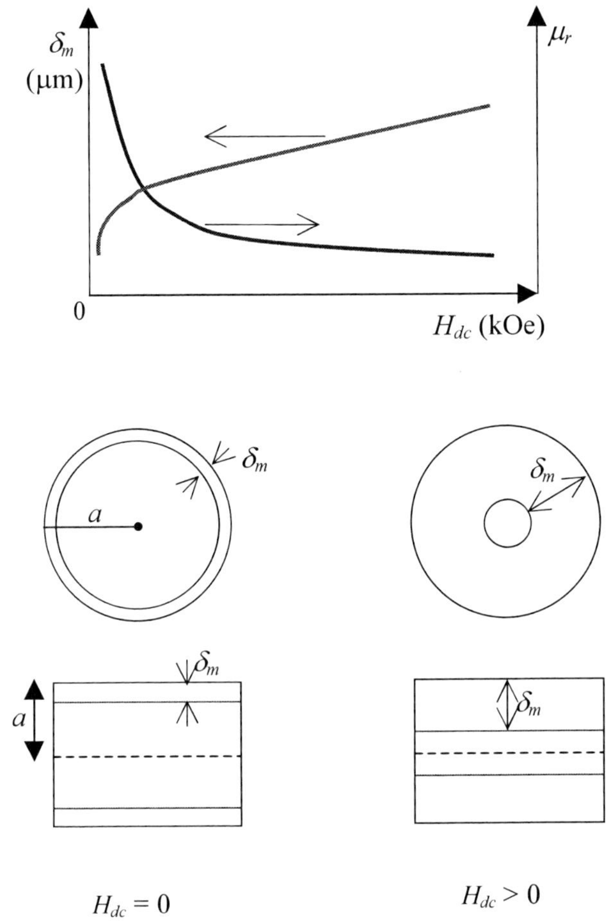

Figure 11. Upper panel: dc applied magnetic field (H_{dc}) dependences of skin depth (δ_m) and reversible permeability (μ_r). Lower panel: a schematic view of the change of δ_m with H_{dc}.

Intermediate frequency regime (between ~100 kHz and a few MHz), where MI originates mainly from the variation of the skin depth due to strong changes of the effective magnetic permeability caused by the applied dc magnetic field. It is noted here that depending on sample geometry, the MI profile can reach its peak in the intermediate frequency range (e.g., 100 kHz ~10 MHz), as a consequence of the contribution of the permeability from both domain wall motion and magnetization rotation to MI. Reduction in MI at higher frequencies is related to the domain walls becoming strongly damped by eddy currents and only magnetization rotation contributes to MI. In this frequency range, the eddy-current and domain models have been proposed and they allow explaining successfully several basic features of MI at frequencies below 100 MHz [28,32,33-35]. When a high frequency leads to a skin depth comparable with the exchange length, however, both models are inaccurate [14].

This is because ferromagnetic resonance occurs at high frequencies $f \sim 1$ GHz and becomes the main effect responsible for MI behavior.

High frequency regime (several MHz up to GHz), where the origin of MI is believed to be related to the gyromagnetic effect and ferromagnetic relaxation. The maxima in MI profiles are shifted towards higher fields, where the samples are already saturated magnetically [36,37]. Strong changes in skin depth are caused by the same mechanism as in the ferromagnetic resonance. In this context, the MI features have been explained by the exchange-conductivity and electromagnetic models. In general, the exchange-conductivity model qualitatively interprets the frequency and magnetic field dependences of MI in a wider range of frequencies than the electromagnetic models. However, neither model deal satisfactorily with the problem of MI as a whole, due to the complex domain structures of actual magnetic materials [27] and the given approximate assumptions [26].

Although the theory of MI may be understood to a reasonable degree, more research is needed to fully understand the theory. The finding of a new model that can explain all MI features in the whole frequency range is still challenging. It is believed that, to assess the underlying mechanism of MI for a variety of materials, further theoretical models must be developed on the basis of the identified domain structures of the materials. It should be noted that the formation of characteristic domain structures of MI materials is mainly related to the preparation process.

3.2.2.2. Alternating Current Dependence of MI

It is generally known that the circular (or transverse) magnetization process takes place due to the ac circular magnetic field created by an alternating current (ac). Meanwhile, the change in impedance is directly related to this magnetization process [26,27]. Therefore, a dependence

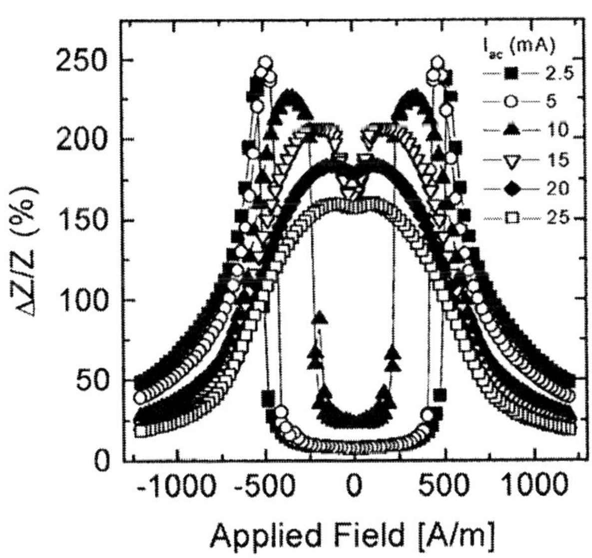

Figure 12. MI for a Co-base amorphous wire with induced circular anisotropy. Evolution from double-peak (DP) to single-peak (SP) feature with increasing current is observed [29].

of the impedance on the ac exists, according to Eq. (2.2). The higher the amplitude of the ac (I_m), the larger the impedance (Z) obtained. It has been theoretically predicted that for a double-peak (DP) feature of MI in an amorphous wire, the magnetic field (H_m) at which the value of Z reaches a maximum should decrease with increasing amplitude of the ac [33]. An example is displayed in Fig. 12 for a Co-based amorphous wire [29].

In practice, MI curves exhibit a DP behaviour for low currents where only reversible wall motion takes place, but this changes to a single-peak (SP) at high currents related to circular coercivity [27].

3.2.2.3. Magnetic Field Dependence of MI

To measure MI effect, a dc magnetic field is usually applied parallel to the ac along the longitudinal direction of the sample. This is known as the longitudinal MI effect [10]. As discussed previously, at a given frequency, the application of a dc magnetic field (H_{dc}) changes the circular or transverse permeability μ_T and hence the magnetic penetration depth δ_m that in turn alters the magnetoimpedance until the value of δ_m reaches the radius of the wire (t) or the half thickness of the sample (a). Consequently, the MI effect increases to a maximum value when the applied dc field is increased as typically displayed in Fig. 13. In reality, the applied dc magnetic field (H_{dc}) may compensate the axial magnetic anisotropy when H_{dc} reaches the switching field of the sample [27]. At this field, the quasifree magnetisation responds quickly to the external oscillating magnetic field and gives rise to a large transverse permeability [26,27]. Since the value of the switching field is actually very small in nanocrystalline samples, the maximum value of MI is observed at nearly zero field ($H_{dc} \sim 0$) for low frequencies ($f \leq 1$ MHz). The MI profile also shows an SP behavior in this frequency range (see Fig. 13). When the dc magnetic field exceeds the switching field, the transverse permeability decreases due to the unidirectional magnetostatic anisotropy caused by H_{dc}. Therefore, with increasing dc magnetic field, the skin depth increases and consequently results in an increase of the MI effect.

When studying the effects of applied dc fields on domain wall magnetisation processes at low frequencies in Fe-based nanocrystalline ribbons, Qin et al. [38] found that a small longitudinal dc field helped the domain wall unpin from the pinning field. A large longitudinal dc field increased the pinning field and damped the domain wall motion, while a transverse dc field had little influence on the pinning field.

At high frequencies ($f > 1$ MHz), MI profiles showed a DP characteristic (see Fig. 13) due to the change in the switching and anisotropy field of the sample with increasing frequency. It is the increase of the switching and anisotropy field with frequency that in turn causes the DP behaviour in MI profiles, together with a shift of the MI peak towards a higher value of H_{dc}, when the frequency is increased. However, the magnetic field dependence of MI has been correlated to the magnetic field dependence of the circular/transverse permeability. Such SP and DP features of MI are determined by the relative contribution from the domain wall motion and magnetisation rotation processes to the circular/transverse permeability. Complex domain structures and distribution of local anisotropies in actual materials can also be additional causes of the complicated dependence of MI on dc magnetic field [27].

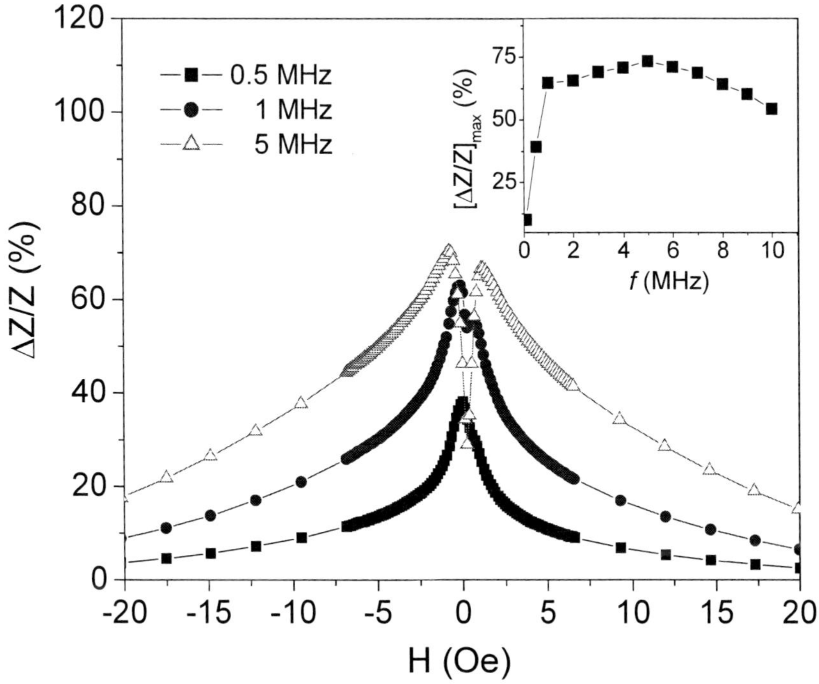

Figure 13. The dc magnetic field dependence of ΔZ/Z for a nanocrystalline Fe-based ribbon at various frequencies. The inset shows the frequency dependence of $[\Delta Z/Z]_{max}$.

3.2.2.4. Frequency Dependence of MI

According to Eqns. (10) – (13), MI and its behavior depend strongly on measuring frequency, and such frequency-dependent MI features have been experimentally studied in a wide range of frequencies up to several GHz in different materials [10,26,27].

It has been shown that, with increasing frequency, the magnetisation process via domain wall displacement relaxes at relatively low frequencies ranging between 0.1 and 1 MHz in amorphous wires [39] and ribbons [40]. For amorphous glass-covered microwires, the permeability spectra showed dispersion laws at higher frequencies [27,28]. At low frequencies $f < \sim 0.1$ MHz ($a < \delta_m$), the maximum value of MI, defined as $[\Delta Z/Z]_{max}(\%)$, is found to be relatively low due to the contribution of the induced magneto-inductive voltage to magnetoimpedance [32]. In this case, MI features can be interpreted on the basis of the quasistatic models [26]. In the frequency range of 0.1 – 10 MHz, where most MI measurements have been conducted, with increasing frequency, $[\Delta Z/Z]_{max}(\%)$ first increases up to a maximum and then decreases and even changes sign [41]. As a typical example displayed in the inset of Fig. 13, when 1 MHz $\leq f \leq$ 5 MHz ($a \approx \delta_m$), the skin effect is dominant, and a higher $[\Delta Z/Z(\%)]_{max}$ was found. Beyond $f = 5$ MHz, $[\Delta Z/Z(\%)]_{max}$ decreases with increasing frequency. It is believed that, in this frequency region ($f \geq 5$ MHz), the domain wall displacement was strongly damped owing to eddy currents, thus contributing less to the transverse permeability, i.e., a small $[\Delta Z/Z(\%)]_{max}$. It should also be noted that the magnetoimpedance increases with frequency because the impedance is proportional to

$(\omega\mu_T)^{1/2}$ even in the case of decreasing transverse permeability at high frequencies. Despite this, the frequency at which the maximum MI value is obtained depends strongly on the samples' dimension. It has also been found that MI usually reaches its maximum value at higher frequencies for amorphous microwires than for amorphous wires and ribbons [27]. At frequencies above 10 MHz, only the spin rotation mechanism occurs and the effective permeability is small. The eddy-current and domain models are valid for interpreting basic MI features in the frequency range of 0.1 – 100 MHz [26]. At higher frequencies, where ferromagnetic resonance can occur [41], MI features were successfully explained by the electromagnetic and conductivity-exchange models [36,37]. Furthermore, with increasing frequency, the low-frequency SP behaviour may become DP-like as a consequence of the frequency dependence of permeability (see the inset of Fig. 13). MI curves may also become broader with increasing frequency (see Fig. 13). When MI curves become broader, one can adapt the model developed for transverse biased permeability in thick ferromagnetic films. It is the combination of eddy current damping, ripple field and anisotropy field that gives rise to the peak in permeability as well as the broadening of permeability curves [42]. This can explain the broadening of MI curves at high frequencies.

From an experimental perspective, one should note that most MI measurements have been performed at measuring frequencies up to only 10 MHz. This is due to the fact that, in this frequency range, the voltage-current measurement technique that gives a direct measurement of the transport properties of the sample can easily be employed. At higher frequencies, both voltage and current are not always quantitatively well defined and the classic measurement technique fails. To overcome this problem, another technique based on the S-parameter measurement of a transmission line section by means of an automatic vectorial network analyzer has recently been exploited to measure the MI effect in the microwave region of soft magnetic amorphous samples.

3.3. Magnetocaloric

3.3.1. What is Magnetocaloric Effect?

The magnetic moments of the atoms in a magnetic material become reoriented when it is subjected to a sufficiently high magnetic field. The temperature of this material rises as the magnetic field is applied adiabatically, and it decreases as the magnetic field is subsequently removed. The warming and cooling in response to the application and removal of an external magnetic field is called the '*magnetocaloric effect* - MCE'. In this regard, the MCE is intrinsic to all magnetic materials and it is directly related to the magnetic entropy change [i.e. ΔS_M] and the adiabatic temperature change [i.e. ΔT_{ad}]. Schematic representation of a magnetic-refrigeration cycle is illustrated in Fig. 14 [43].

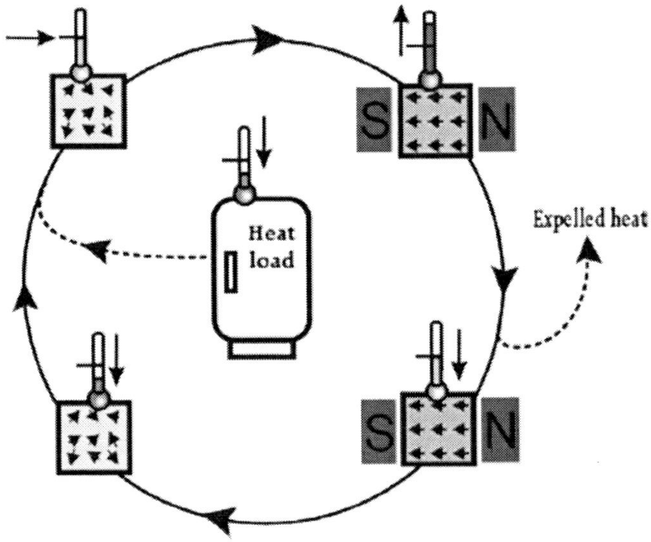

Figure 14. Schematic representation of a magnetic-refrigeration cycle, which transports heat from the heat load to its surroundings. Yellow and green depict the magnetic material in low and high magnetic fields, respectively. Initially randomly oriented magnetic moments are aligned by a magnetic field, resulting in heating of the magnetic material. This heat is removed from the material to its surroundings by a heat-transfer medium. On removing the field, the magnetic moments randomize, which leads to cooling of the magnetic material below the ambient temperature. Heat from the system to be cooled can then be extracted using a heat-transfer medium. Depending on the operating temperature, the heat-transfer medium may be water (with antifreeze) or air, and for very low temperatures, helium [43].

3.3.2. Theoretical and Experimental Aspects

3.3.2.1. Relationship between Magnetic Entropy and Adiabatic Temperature

It has been demonstrated that there is only magnetic entropy as a function of magnetic field and the change of magnetic entropy upon the application of an applied magnetic field is important in evaluating the MCE magnitude [13,44,45]. The magnetic entropy change is closely related to the adiabatic temperature change. Therefore, in this paper, we focus on the evaluation of the relationship between these two quantities.

The change of entropy (S) of a magnetic material upon the application of a magnetic field (H) is related to that of magnetization (M) with respect to temperature (T) through the thermodynamic Maxwell relation:

$$\left(\frac{\partial S}{\partial H}\right)_T = -\left(\frac{\partial M}{\partial T}\right)_H . \tag{16}$$

The magnetic entropy change, $\Delta S_M(T,H)$, is calculated by

$$\Delta S_M(T,H) = S_M(T,H) - S_M(T,0) = \int_0^H \left(\frac{\partial M(T,H)}{\partial T}\right)_H dH. \tag{17}$$

For magnetization measurements made at discrete field and temperature intervals, $\Delta S_M(T,H)$ can be approximately calculated by the following expression:

$$\Delta S_M(T,H) = \sum_i \frac{M_{i+1}(T_{i+1},H) - M_i(T_i,H)}{T_{i+1} - T_i} \Delta H \tag{18}$$

On the other hand, $\Delta S_M(T,H)$ can be obtained from calorimetric measurements of the field dependence of the heat capacity and subsequent integration:

$$\Delta S_M(T,H) = \int_0^T \frac{C(T,H) - C(T,0)}{T} dT, \tag{19}$$

where $C(T,B)$ and $C(T,0)$ are the values of the heat capacity measured in a field H and in zero field ($H = 0$), respectively. Therefore, the adiabatic temperature change (ΔT_{ad}) can be evaluated by integrating Eq. (4) over the magnetic field, which is given by

$$\Delta T_{ad} = -\int_0^H \frac{T}{C_{P,H}} \left(\frac{\partial M}{\partial T}\right)_H dH \tag{20}$$

From Eqs (18) and (20) it is easy to state that a material should have large MCE (i.e., large ΔS_M and ΔT_{ad}) when $(\partial M/\partial T)_H$ is *large* and $C(T,H)$ is *small* at the same temperature. Because $(\partial M/\partial T)_H$ peaks the magnetic ordering temperature, a large MCE is often expected close to this magnetic phase transition, and the effect may be further maximized as the change in magnetization with respect to temperature occurs in a narrow temperature interval [44]. It should be noted that although evaluating ΔS_M from magnetization measurements using Eq. (18) has been a useful tool for a rapid screening of potential magnetocaloric materials, a precise comparison of the MCE among the exiting magnetocaloric materials can only be realized by evaluating ΔT_{ad} using Eq. (20) instead of ΔS_M. This is because the magnitude of heat capacity may be much different from a magnetocaloric material family to another, for example, the heat capacity of a Gd-based alloys system is much smaller compared to that of a manganite-type materials system [10,44].

Furthermore, it is stated that the magnitude of the MCE depends not only on the magnetic moments but also on $\partial M/\partial T$. These larger these values, the higher the MCE. This is the reason why not only rare-earth elements and their compounds have large MCEs [44], but also 3d-based transition-metal compounds can have large MCEs in case of a first-order transition [45]. In the case of manganite materials, it is the rapid change of magnetization with respect to temperature in the magnetic-ordering phase transition range that causes a large magnetic entropy change, i.e., the large MCEs [10]. Indeed, Terashita et al. [46] recently revealed that a first-order structural transition in close proximity to the magnetic transition has only little influence on the MCE in doped manganites.

3.3.2.2. Theory of Molecular Field

In order to evaluate measured quantities involving the MC effect, the molecular field model has been developed by Szewczyk et al. [47]. In La-based manganites, such as $La_{1-x}Ca_xMnO_3$, it is assumed that (1-*x*) fraction of Mn ions of the sample are Mn^{3+} ions and *x* fraction of them are Mn^{4+} ions. Accordingly, spin quantum numbers for Mn^{3+} and Mn^{4+} ions are $S_3 = 2$ and $S_4 = 3/2$, respectively. The relationship between magnetic moment, *m*, of each manganese ion and the expectation value, σ, of the spin operator is given as follows: $m = -2\mu_B\sigma$, from which to determine the σ value for given *T* and *B* values by solving the entangled equation,

$$\sigma = -(1-x)S_3 B_{S_3}\left[\frac{2\mu_B}{k_B T}(B-\lambda\sigma)\right] - xS_4 B_{S_4}\left[\frac{2\mu_B}{k_B T}(B-\lambda\sigma)\right], \tag{21}$$

where μ_B and k_B denote the Bohr magneton constant and the Boltzmann constant, respectively. $B_s(x)$ is the Brillouin function,

$$B_S(x) = \frac{2S+1}{2S}\coth(\frac{2S+1}{2S}\eta) - \frac{1}{2S}\coth\frac{\eta}{2S} \tag{22}$$

with $\eta = Sg\frac{\mu_B B}{k_B T}$, and

$\lambda = 3k_B T_C / \{[(1-x)S_3(S_3+1) + xS_4(S_4+1)]2\mu_B\}$ - the molecular field constant.

In the molecular field approximation, the expression describing the magnetic entropy term of the system, in J/kg K, can be given as

$$S_M[T,B,\sigma(T,B)] = [(1-x)S_M^0(S_3,T,B,\sigma) + xS_M^0(S_4,T,B,\sigma)]k_B N_A / m_{mol}, \tag{23}$$

where N_A is the Avogadro's number, m_{mol} stands for the material molar mass, and S_M^0 is defined as

$$S_M^0[S,T,B,\sigma(T,B)] = \frac{2\mu_B}{k_B T}(B-\lambda\sigma)\sigma + \ln\left\{\frac{\sinh[(2S+1)\mu_B(B-\lambda\sigma)/(k_B T)]}{\sinh[\mu_B(B-\lambda\sigma)/(k_B T)]}\right\}. \tag{24}$$

Hence, the magnetic entropy change under influence of the external magnetic field can be written as

$$\Delta S_M(T,B) = S_M[T,B,\sigma(T,B)] - S_M[T,0,\sigma(T,0)]. \tag{25}$$

The $\Delta S_M(T,B)$ values calculated, according to the molecular field model, using Eq. (25), are verified experimentally using Eq. (18), and are depicted in Fig. 15.

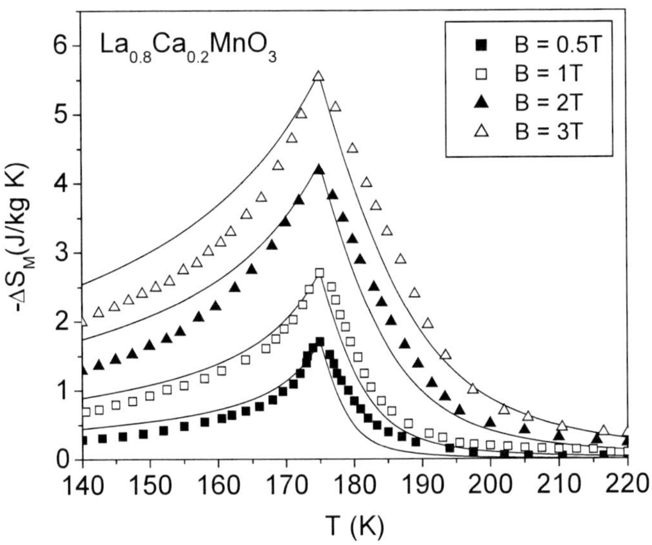

Figure 15. Change in the magnetic entropy, $-\Delta S_M$, of the $La_{0.8}Ca_{0.2}MnO_3$ sample as a function of temperature at fixed values of the external magnetic field. Solid lines - theoretical relations calculated in the molecular field approximation for the sample field intensities [48].

Obviously, the maximum magnetic entropy change, $-\Delta S_M^{max}$, reaches ~5.42 J/kg K for an applied field of 3 T. A reasonable agreement between the theoretical curves derived in the molecular field approximation and the experimental dependencies has been found. It is worth noting that in the molecular model, Eqs (23) and (25), zero field magnetic entropy reaches at T_c the value $S_M(T_C, B=0) = [(1-x)\ln 5 + x\ln 4]k_B N_A / m_{mol}$. Therefore, application of 7 T magnetic field reduces S_M by ~ 14% at T_c, or ~ 7.25 J/kg K for $-\Delta S_M^{max}$ respectively, which is practically appropriate for magnetic refrigeration applications.

In general, the theoretical calculations using the molecular field model provide a fairly good description of the magnetic entropy change above T_c, but the magnetic entropy change is overestimated below T_c. As an example displayed in Fig. 15 for the $La_{0.8}Ca_{0.2}MnO_3$ sample, the discrepancy between the theoretical calculations and the experimental data can be attributed to roughness of the model. It should be noted that this model did not take into account the roles played by the charge ordering and the Jahn-Teller effect, which actually affect the MCE value even for the case of conventional ferromagnetic manganites [47-49]. This thus warrants further study. Nevertheless, the molecular field model has been useful for describing qualitatively the magnetic entropy change and the adiabatic temperature change, particularly at temperatures close to the Curie temperature.

3.4. Relationship between Resistivity, Impedance and Entropy

3.4.1. Relation between Electrical Resistivity and Impedance

According to the MI definition, the change in impedance (Z) is directly related to that in both electrical resistance (R) and reactance (X). Note that $X = \omega L$. Therefore, there exists a contribution of MR to MI effect. In this extent, a simple model has been proposed by Barandiaran et al. [50] to explain the MR contribution to the MI effect in thin films. The theoretical calculations show that the contribution of MR to the MI effect is dominant in the low frequency regime, but can be neglected for the high-frequency range. This is because the MR only occurs at relatively low frequencies. This has been experimentally verified by Kurlyandskaya et al. [51] when investigating the MR and MI effects in Fe-Ni thin films.

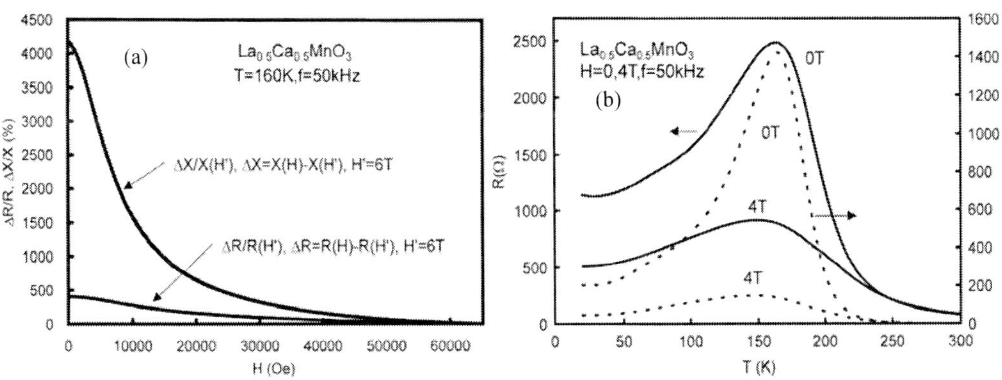

Figure 16. (a) Relative changes in resistance (R) and reactance (X) as a function of magnetic field; (b) Temperature dependence of resistance and reactance [52].

For a manganite, the relative change in reactance is much larger than that in resistance under the same applied magnetic field, as shown in Fig. 16(a) [52]. It indicates that at low frequencies, the MI effect is dominated by the reactance change ($\Delta X/X$), rather than the resistance change ($\Delta R/R$). However, at high frequencies, the contribution of $\Delta R/R$ to the MI effect is dominant [53-55]. The temperature dependence of MI is similar to that of MR, as shown in Fig. 16(b) [52]. This means that the temperature dependence of MI is related to the ratio of Mn^{3+}/Mn^{4+} and the magnetic phase [52,56]. Therefore, it is reasonable to claim that the application of a magnetic field not only decreases the transverse permeability, but also reduces the electrical resistivity by aligning the local spins and varying the transfer integral t_{ij}. Due to the co-existence of ac MI and dc MR, the observed MI effect in a perovskite manganite can originate from the combined effects of double-exchange interaction, electro-phonon coupling and skin effect [52-56]. This reflects the complex nature of MI phenomena in manganite materials. Recently, Fu et al. [12] revealed that the specific features of impedance could provide an alternative way of sensing the complex phase states of manganite materials. In particular, Patanjali et al. [57] have demonstrated the existence of secondary transitions at high temperatures in a double perovskite $La_{1.2}Sr_{1.8}Mn_2O_7$ through MI measurements; this feature was not observed using static resistivity and magnetization measurements.

3.4.2. Relation between Magnetic Entropy and Electrical Resistivity

In manganites, both MR and MC effects are often observed around the magnetic-ordering phase transition temperature (i.e. the Curie temperature) and this evidently suggests a definite relation between magnetic entropy and resistivity. In this context, Xiong et al. [58] proposed a new method for evaluating the relation between the magnetic entropy and resistivity (ρ) of a ferromagnetic manganite by means of

$$\Delta S_M(T,H) = -\alpha \int_0^H \left[\frac{\partial \ln(\rho)}{\partial T} \right]_H dH \tag{26}$$

with $\alpha = 21.72$ emu/g.

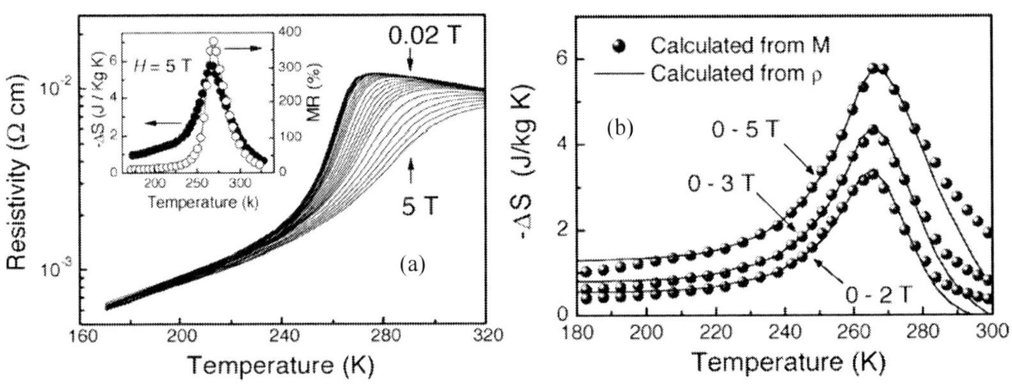

Figure 17. (a) Temperature-dependent resistivity measured under different magnetic fields. The inset shows the magnetic entropy change (ΔS) and magnetoresistance ratio (MR); (b) Comparison of $\Delta S - T$ and ρ - T curves [58].

Figure 17 shows the relationship between magnetic entropy change (ΔS) and magnetoresistance ratio (MR). The peaks of ΔS-T and MR-T curves are observed at the Curie temperature [see the inset of Fig. 17(a)]. There is a shift of these peaks to a higher temperature when the applied magnetic field increases [see Fig. 17(b)]. According to Eq. (26), a larger α leads to a more sensitive dependence of ΔS - ρ. Magnetic disorder, which is characterized by ΔS, affects magnetic polarons that influence the electric transport property thereby leading to the relation of ΔS - ρ [58,59]. This relation is only valid in a narrow temperature range, where the magnetic-ordering phase transition occurs [58]. The deviation of Eq. (26) occurring in the low temperature range is because the magnetic polarons are significantly depressed when the system is at a relative magnetic ordered state [58]. In general, the relation (26) provides an alternative method for determining the magnetic entropy change in perovskite manganites from resistive measurements.

4. Technological Applications

4.1. Magnetic Sensors

4.1.1. Criteria for Selecting Sensor Materials

The main requirements for a magnetic material to be employed for the MR/MI sensor technology include (i) the high MR/MI ratio (e.g. the large MR/MI effect) and (ii) high field sensitivity of MR/MI (e.g. the high magnetic MR/MI response). For practical applications, the manganite's Curie temperature should be in the room temperature range, since the MR/MI effect reaches a maximum at the Curie temperature. The MI is more sensitive to an applied magnetic field than the MR, so the MI technology may be more promising.

In view of existing MI materials, it is pointed out that a large MI effect should exist in magnetic materials having: (i) low resistivity, ρ; (ii) high magnetic permeability, μ; (iii) high saturation magnetisation, M_s, and (iv) small ferromagnetic relaxation parameter (or low damping parameter), α. In this context, crystalline ferromagnetic materials have the advantage of lower resistivity (ρ), but amorphous ones have better soft magnetic behavior (e.g., higher magnetic permeability, μ, and saturation magnetisation, M_s) because they lack magnetocrystalline anisotropy. The damping parameter (α) is often considered in the high frequency range where ferromagnetic resonance takes place, while it can be neglected in the intermediate frequency range (i.e. f = 0.1 – 10 MHz) where most MI-based sensing applications have been identified. In addition, for practical sensor applications, the manufacturability and cost of materials are also important factors.

4.1.2. Manganites for Magnetic Sensors

Since the Curie temperature of a manganite is doping dependent, it can easily be tuned from low to high temperature. As a result, a variety of manganites are available for the production of magnetic sensors, which operate, in optimal regime, at various temperatures. However, it is a challenging task to the MR/MI technology as a large MR/MI effect is only obtained in materials with relatively low T_c, and doping can increase T_c to room temperature but significantly reduce the MR/MI magnitude. For example, the $La_{1-x}Ca_xMnO_3$ (x = 0.1-0.5) system has been found to show the highest MR effect, but the highest T_c obtained is about 280 K, which is well below room temperature. The substitution of T (T = Sr, Ba, Pb) for Ca results in an increase in T_c, but in turn reduces the MR effect. Nevertheless, these materials are still suitable for magnetic sensor technology [1,14,21].

4.1.3. Magnetic Sensor Applications

Magnetic sensors play an essential role in modern technology. They are widely used in nearly all engineering and industrial sectors, such as high-density magnetic recording, navigation, military and security, target detection and tracking, antitheft systems, non-destructive testing, magnetic marking and labeling, geomagnetic measurements, space research, measurements of magnetic fields onboard spacecrafts, biomagnetic measurements in human body [10].

For instance, in modern computers, MR magnetic sensors play an essential part of magnetic storage disk and tape drives. In these devices, the reading module usually comprises a MR sensor which uses a multilayer film structure consisting of magnetic and nonmagnetic layers. The writing module operation induces local magnetic moments in the bit areas of the hard disk magnetic layer. Bits with remanent magnetisation cause measurable change in the resistance of the MR sensor of the reading module, which enables the two levels of digital signal to be distinguished. Since the sensitivity of a MI sensor is much higher than that of a MR sensor, thin film MI sensors might be of much more interest for future high-density magnetic recording technology.

4.2. Magnetic Refrigeration

4.2.1. Criteria for Selecting Magnetic Refrigerants

The criteria for selecting magnetic refrigerants for active magnetic refrigerators include: (i) the large magnetic entropy change and the large adiabatic temperature change (i.e., the large MCE); (ii) the large density of magnetic entropy (it is an important factor contributing to the working efficiency of materials); ferromagnets with large values of effective magneton number $P = \{J(J+1)\}^{1/2}$ are selected; (iii) the small lattice entropy (i.e., the high Debye temperature); much attention should be paid to magnetic refrigerants for room-temperature magnetic refrigerators; (iv) the MCE in the temperature range of 10 – 80 K or > 250 K, where the Curie temperature of a material is located in and, the large magnetic entropy change can be obtained in the whole temperature range of the cycle; (v) nearly zero magnetic hysteresis (it is related to the working efficiency of a magnetic refrigerant material); (vi) very small thermal hysteresis (this is related to the reversibility of the MCE of a magnetic refrigerant material); (vii) small specific heat and large thermal conductivity (these ensure remarkable temperature change and rapid heat exchange); (viii) large electric resistance (i.e., the lowering eddy current heating or the small eddy current loss); (ix) high chemical stability and simple sample synthesis route are also required for magnetic refrigerant materials. For practical applications, magnetic refrigerant materials should also be low cost.

4.2.2. Manganites for Magnetic Refrigerators

A wide range of manganites is nominated for magnetic refrigeration at various temperatures [13]. Regarding room-temperature magnetic refrigeration, it has been shown that the magnitude of MC effect of several manganites is comparable to that of gadolinium and, more interestingly, the MC peak temperature can easily be tuned in the temperature range of 290 - 360 K by selecting suitable manganites. This indicates that these manganites are attractive for room temperature magnetic refrigeration technology. However, one disadvantage of this typical material is that the adiabatic temperature change is not very large due to the relatively high heat capacity. This probably limits such manganites for active magnetic refrigeration. This task will be overcome because of the rapid development of the magnetic cooling technology as of today. Compared with Gd and other candidate materials, ferromagnetic manganites are more convenient to prepare and exhibit higher chemical stability as well as higher resistivity that is favorable for lowering eddy current heating. In addition, the manganites possess much smaller thermal and field hysteresis than any rare-earth and 3d-transition metal-based alloys. The MCE peak temperature can easily be tuned in the wide temperature range of 20 - 370 K, which is beneficial for AMR at various temperatures. In addition, the manganite materials are the cheapest (~$10 /kg) among existing magnetic refrigerants. These superior features make them very promising for future magnetic refrigeration technology.

4.2.3. Magnetic Refrigeration Applications

Magnetic refrigeration technology has wide applications in energy intensive industrial and commercial refrigerators such as large-scale air conditioners, heat pumps, supermarket refrigeration units, waste separation, chemical processing, gas liquification, liquor distilling, sugar refining, grain drying, and so forth [44].

A first superconducting-magnet-based room-temperature magnetic refrigeration device was designed and produced by the Aeronautics Corporation of America. Recently, this Corporation has successfully produced a permanent-magnet-based rotary room-temperature magnetic refrigerator. In Japan, engineers and scientists at the Chubu Electric Company, in cooperation with Toshiba Corporation, have also succeeded in developing a rotating magnetic refrigerator with permanent magnets. The Chubu's design is similar to that of the Aeronautics, with an increase in cooling capacity by a factor of 1.5 and a 1/3 decrease in driving power. The Chubu's device is also about a twentieth the size of earlier prototype refrigerators employing superconducting magnets. Potential commercial applications of such refrigerators include air conditioning, food preservation, air dehumidification, and beverage dispensing. Room temperature magnetic refrigeration and its achievements may offer several advantages in the search for overcoming the limitations of the conventional gas compression refrigeration, each with specific advantages and likely target market sectors [13,44,45].

5. Concluding Remarks and Future Outlook

This paper provides physical insights into the complex nature of magnetoresistive, magnetoimpedance and magnetocaloric phenomena in ferromagnetic manganites. It has been demonstrated that there is a coherent correlation between these phenomena.

From a technological perspective, ferromagnetic manganites are attractive candidate materials for use in magnetic sensors and magnetic refrigerators. The MI sensor technology is more promising than the MR one. Therefore, future research will be focused on seeking manganites with large MI effects and Curie temperatures in the room-temperature range.

For a variety of large- and small-scale magnetic refrigeration, ferromagnetic manganites with their Curie temperatures ranging between 20 and 375 K are available. However, most of existing manganites have large heat capacity, leading to a reduction in magnetic cooling efficiency and hence limiting them for use in active magnetic refrigerators. Future research should, therefore, be focused on tailoring the magnetocaloric properties of manganites with low heat capacity, but more reliably, on the design and fabrication of nanostructured manganites with desired MC features.

It is no doubt that ferromagnetic manganite materials with superior magnetoresistive, magnetoimpedance and magnetocaloric properties, in addition to their high flexibility and cost-effectiveness, will be the option of future magnetic sensor and magnetic refrigeration technologies.

Aknowledgements

The authors wish to acknowledge Prof. Nam Hwi Hur (Korea Research Institute of Standards and Science) and Prof. Nguyen Chau (Center for Materials Science, Vietnam) for their long-standing cooperation and supports. Research at the Chungbuk National University was supported by the Brain Korea 21 (BK21) Programme and the Korean Science and Engineering Foundation through the Research Center for Advanced Magnetic Materials at Chungnam National University. Special thank is due to Paul Harper for the aid in reading and editing the manuscript.

References

[1] Tokura Y. *Colossal magnetoresistive oxides*; Gordon & Breach Science Publishers; 2000.
[2] Jonker G. H.; van Santen J. *Physica* 1950, 16, 337-340.
[3] Searle C. W.; Wang S. T. *Can. J. Phys.* 1969, 47, 2023-2027.
[4] Kusters R. M.; Singleton D. A.; Keen D. A.; McGreevy R.; Hayes W. *Physica B* 1989, 155, 362-3625.
[5] Chabara K.; Ohno T.; Kasai M.; Kozono Y. *Appl. Phys. Lett.* 1993, 63, 1990-1992.
[6] von Helmolt R.; Wocker J.; Holzapfel B.; Schultz M.; Samwer K. *Phys. Rev. Lett.* 1993, 71, 2331-2333.
[7] Tokura Y.; Urushibara A.; Moritomo Y.; Arima T.; Asamitsu A.; Kido G.; Furukawa N. *J. Phys. Soc. Jpn.* 1994, 63, 3931-3935.
[8] Ramirez A. P. *J. Phys.: Condens. Matter.* 1997, 9, 8171-8199.
[9] Berkowitz A. E.; Mitchell J. R.; Carey M. J.; Young A. P.; Zhang S.; Spada F. E.; Parker F. T.; Hutten A.; Thomas G. *Phys. Rev. Lett.* 1992, 68, 3745 – 3748.
[10] Knobel M.; Pirota K. R. *J. Magn. Magn. Mater.* 2002, 242-245, 33-40.
[11] Vazquez M. Physica B 2001, 226-230, 693-699.
[12] Fu C. M.; Hsu K. S.; Lin M. L.; Wen Z. H. *J. Magn. Magn. Mater.* 2000, 209, 151-153.
[13] Phan M. H.; Yu S. C. *J. Magn. Magn. Mater.* 2007, 305, 325-340.
[14] Tokura Y.; Tomioka Y. *J. Magn. Magn. Mater.* 1999, 200, 1-23.
[15] Kimura T.; Tokura Y. *Annu. Rev. Mater. Sci.* 2000, 30, 451-474.
[16] Nagaev E. L. *Colossal Magnetoreistance and Phase Separation in Magnetic Semiconductors*; ISBN 1-86094-295-4, World Scientific Publisher, London, 2002.
[17] Van Tendeloo; Lebedev O. I.; Hervieu M; Raveau B. *Rep. Prog. Phys.* 2004, 67, 1315-1365.
[18] Tokura Y. *Rep. Prog. Phys.* 2006, 69, 797-851.
[19] Dorr K. *J. Phys. D: Appl. Phys.* 2006, 39, R125-R150.
[20] Chau N.; Nhat H. N.; Luong N. H.; Minh D. L.; Tho N. D.; Chau N. N. *Physica B*. 2003, 327, 270-278.
[21] Terakura K. *Prog. Magn. Sci.* 2007, 52, 388-400.

[22] Zener C. *Phys. Rev.* 1951, 82, 403-405.
[23] Urushibara A.; Moritomo Y.; Arima T.; Asamitsu A.; Kido G.; Tokura Y. *Phys. Rev. B* 1995, 51, 14103-14109.
[24] Valenzuela R.; *Magnetic ceramics*, ISBN 0-521-36485-X, Cambridge University Presss, 1994, pp. 44-92.
[25] Shiffer P.; Ramirez A.P.; Bao W.; *Phys. Rev. B* 1995, 75, 3336-3339.
[26] Kraus L. *Sen. Act. A* 2003, 106, 187-194.
[27] Knobel M; Vazquez M; Kraus L. *Giant magnetoimpedance*, in: K.H. Buschow (Ed.), Handbook of Magnetic Materials, Vol. 15, Elsevier Science B.V., Amsterdam, 2003, pp. 1–69 (Chapter 5).
[28] Panina L. V.; Mohri K. *Appl. Phys. Lett.* 1994, 65, 1189-1191.
[29] Vazquez M. *J Magn Magn Mater* 2001, 226-230, 693-699.
[30] Beach R. S.; Berkowitz A. E. *Appl. Phys. Lett.* 1994, 64, 3652-3654.
[31] Mohri K.; Kohsawa T.; Kawashima K.; Yoshida H.; Panina L. V. *IEEE Trans. Magn.* 1992, 28, 3150-3152.
[32] Panina L. V.; Mohri K.; Uchiyama T.; Noda M. *IEEE Trans. Magn.* 1995, 31, 1249-1260.
[33] Chen D. X.; Munoz J. L.; Hernando A.; Vazquez M. *Phys. Rev. B* 1998, 57, 10699-10704.
[34] Chen D. X., Munoz J. L. *IEEE Trans. Magn.* 1999, 35, 1906-1923.
[35] Betancourt I.; Valenzuela R.; Vazquez M. *J. Appl. Phys.* 2003, 93, 8110-8112.
[36] Menard D.; Britel M.; Ciureanu P.; Yelon A. *J. Appl. Phys.* 1998, 84, 2805-2814.
[37] Yelon A.; Menard D.; Brittel M.; Ciureanu P. *Appl. Phys. Lett.* 1996, 69, 3084-3085.
[38] Qin W.; Xu F.; Peng K.; Gao W.; Du Y. *J. Non-Cryst Sol.* 2003, 316, 398-402.
[39] Kim C. G.; Yoon S. S.; Yu S. C. *Appl. Phys. Lett.* 2000, 76, 3463-3465.
[40] Yoon S. S.; Kim C. G. *Appl. Phys. Lett.* 2001, 78, 3280-3282.
[41] de Araujo AEP, Machado FLA, de Aguiar FM, Rezende SM. *J. Magn. Magn. Mater.* 2001, 226-230, 724-726.
[42] Doyle W. D., He X., Tang P., Jagielinski T., Smith N. *J. Appl. Phys.* 1993, 73, 5995-5997.
[43] Tegus Q.; Bruck E.; Buschow K. H.; de Boer F. R. *Nature* 2002, 415, 150-152.
[44] Pecharsky V. K.; Gschneidner K. A.; Tsokol A. O. *Rep. Prog. Phys.* 2005, 68, 1479-1539.
[45] Bruck E. *J. Phys. D: Appl. Phys.* 2005, 38, R381-R391.
[46] Terashita H.; Myer B.; Neumeier J. J. *Phys. Rev. B* 2005, 72, 132415.
[47] Szewczyk A.; Szymczak H.; Wisniewski A.; Piotrowski K.; Kartaszynski R.; Dabrowski B.; Kolesnik S.; Bukowski Z. *Appl. Phys. Lett.* 2000, 77, 1026-1028.
[48] Phan M. H.; Yu S. C.; Hur N. H. *J. Magn. Magn. Mater.* 2005, 290-291, 665-668.
[49] Phan M. H.; Yu S. C.; Hur N. H.; Yeong Y. H. *J. Appl. Phys.* 2004, 96, 1154-1158.
[50] Barandiaran J. M.; Kurlyandskaya G. V.; Vazquez M.; Gutierrez J.; Garcia D.; Munoz J. L. *Phys. Stat. Sol. A* 1999, 171, R3-R4.
[51] Kurlyandskaya G. V.; Barandiaran J. M.; Vazquez M.; Garcia D.; Gutierrez J.; Vaskovskiy V. O.; Lepalovskij V. N. *J. Magn. Magn. Mater.* 2000, 215-216, 516-518.
[52] Hu J.; Qin H. W. *Mater. Sci. Eng. B* 2003, 100, 304-306.
[53] Hu J.; Qin H. W. *Mater. Sci. Eng. B* 2001, 79, 186-189.

[54] Hu J.; Qin H. W.; Niu H. D.; Zhu L.; Chen J.; Xiao W.; Pei Y. *J. Magn. Magn. Mater.* 2003, 261, 105-111.
[55] Qin H. W.; Hu J.; Chen J.; Wang Y.; Wang Z. X. *J. Appl. Phys.* 2002, 91, 10003-10006.
[56] Hu J.; Qin H. W. *Sol. Stat. Commun.* 2000, 116, 159-162.
[57] Patanjali P. V.; Theule P.; Zhai Z.; Hakim N.; Sridhar S.; Suryanarayanan R.; Apostu M.; Dhalenne G.; Revcolevschi A. *Phys. Rev. B* 1999, 60, 9268-9271.
[58] Xiong C. M.; Sun J. R.; Chen Y. F.; Shen B.G.; Du J., Li Y.X. *IEEE Trans. Magn.* 2005, 41, 122-124.
[59] Sun Y.; Salamon M. B.; Chun S. H.; *J. Appl. Phys.* 2002, 92, 3235-3238.

In: Ceramic Materials Research Trends
Editor: Paul B. Lin, pp. 81-110
ISBN: 1-60021-769-9

Chapter 3

Mechanical Properties of Directionally Solidified Eutectic Al_2O_3/YAG Composite at Room to Ultrahigh Temperatures

S. Ochiai[a], H. Okuda[a], Y. Waku[b] and N. Nakagawa[b]

[a] International Innovation Center, Kyoto University, Yoshida, Sakyo-ku, Kyoto 606-8501, Japan.

[b] Ube Research Laboratory, Corporate Research & Development, UBE Industries Ltd., 1978-5 Kogushi, Ube, Yamaguchi 755-8633, Japan

Abstract

Directionally solidified eutectic oxide/oxide ceramic composites are promising for high temperature use as fiber materials for reinforcement and as bulk materials for machine parts due to the high mechanical property up to around 2000K, high oxidation-resistance, clean interface and high stability of the microstructure. In the present paper, focusing on the temperature dependence of mechanical properties of the directionally solidified bulk type Al_2O_3/YAG($Y_3Al_5O_{12}$) composite, the recent experimental results as well as the analytical results mainly based on the finite element analysis are presented. Following subjects are contained. (1) Temperature dependence of Young's modulus and coefficient of thermal expansion of the composite and its relation to that of the constituents (Al_2O_3 and YAG) at room temperature to 1800K. (2) The temperature- and displacement speed- dependence of the stress-strain curve, brittle-ductile transition and flow stress in the plastic deformation region of unnotched composite specimens at room temperature to 2023K. (3) The temperature- and displacement speed- dependence of the notched strength, fracture toughness, plastic zone ahead of the notch and plastic deformation of the ligament of notched specimens at room temperature to 2023K. (4) The stress- and strain rate- distributions in each constituent in plastic deformation, the influence of the morphology of the constituents on the flow stress of the composite and the relation of the microscopic behavior of the constituents to the overall flow stress of the composite at 1873K.

Introduction

There is a strong need to develop ceramic composites that remain stable at very high temperatures to improve thermal efficiency in jet aircraft engines and high-efficiency power-generation gas turbines. Directionally solidified eutectic ceramic composites have clean interface, high thermo dynamical-compatibility between/among the constituting phases, and high static and creep strengths, as has been reported by many researchers based on the microstructural, crystallographic and mechanical studies [1-34]. Among many eutectic composites, oxide/oxide systems such as $Al_2O_3/ZrO_2(Y_2O_3)$ [3-5,18-20,24], $Al_2O_3/Y_3Al_{15}O_{12}$(YAG) [6-9,11-15,21,23,28-34], $Al_2O_3/Er_3Al_5O_{12}$ [16, 27] and $Al_2O_3/GdAlO_3$ [10, 26] are promising for high temperature use as fiber materials for reinforcement and as bulk materials for machine parts.

Concerning the Al_2O_3/YAG eutectic composite, bulk type with rather coarse microstructure and fiber type with fine structure have been developed. The bulk type Al_2O_3/YAG composite has the following features.

(a) The composite consists of continuous networks of single crystal Al_2O_3 and YAG [6-8, 11, 12, 16].
(b) The interface is clean without glassy phase [6, 8, 10, 11, 16].
(c) Figure 1 shows an example of the morphology of this composite. The Al_2O_3 and YAG phases correspond to the black and gray parts, respectively. The widths of Al_2O_3 and YAG phases are around 10 to 30 μm. The Al_2O_3 and YAG in the cross-section parallel to the direction of solidification are elongated in comparison with those in the cross-section perpendicular to it. The volume fraction of each phase is 0.5 [6, 8, 11].
(d) The microstructure is very stable. For instance, even after the exposure in air at 1973K for 1000h, it is same as before [6].
(e) The bending strength at room temperature, around 400 MPa, can be maintained almost up to the melting point of 2093K when the displacement speed is high. [6-8, 15, 21].
(f) Compressive creep strength at 1873K is about 13 times higher than that of the sintered composite with the same composition [8, 12].
(g) Due to the excellent stability of microstructure and mechanical property at ultra high temperatures mentioned above, the maximum operating temperature of these composites can be 1773~1973K, being much higher than that (1323~1373K) of Ni-based single crystal cast super alloys and that (~1300K) of monolithic Al_2O_3 [6, 10]. Accordingly, several useful applications can be considered, for instance, gas turbines and power-generation systems with non-cooled turbine blades at very high temperatures [10,31,33].

As stated above, the excellent properties of the bulk type Al_2O_3/YAG composite have been demonstrated experimentally. For practical application, it is required to reveal the property-controlling factors to provide information for further improvement of the properties and for machine design. In the present paper, focusing on the factors affecting on the mechanical property of the bulk type Al_2O_3/YAG composite, recent experimental and analytical results in the authors' laboratory are presented. The main subjects are; (1) temperature dependence of Young's modulus and coefficient of thermal expansion of the composite and its relation to that of the constituents (Al_2O_3 and YAG) at room temperature to 1800K, (2) the temperature- and displacement speed- dependence of the stress-strain curve, brittle-ductile transition and flow stress in the plastic deformation region of unnotched composite specimens at room temperature to 2023K, (3) the temperature- and displacement speed- dependence of the notched strength, fracture toughness, plastic zone ahead of the notch and plastic deformation of the ligament of notched specimens at room temperature to 2023K, and (4) the stress- and strain rate- distributions in each constituent in plastic deformation, the influence of the morphology of the constituents on the flow stress of the composite and the relation of the microscopic behavior of the constituents to the overall flow stress of the composite at 1873K.

Specimens

The bulk type Al_2O_3/YAG composite was fabricated by the following procedure [6, 7, 11, 30]. The commercially available α-Al_2O_3 powders with the purity higher than 99.99 mass % (AKP-30, produced by Sumitomo Chemical Co. Ltd.) and Y_2O_3 powders with the purity higher than 99.9 mass % (Y_2O_3-RU, submicron-type, produced by Shin-etsu Chemical Co. Ltd.) were mixed in the eutectic molar ratio of 82:18. The mixed powders were ball-milled in ethanol to achieve high homogeneity. Thus obtained slurry was dried and pre-melted by arc melting to obtain initial ingots. The ingots were crushed and the resultant powders were molten in a Mo crucible by high frequency induction heating at a pressure of 1.3×10^{-3} Pa in an argon atmosphere. After the melt was sustained at 2223K for 1.8 ks, it was solidified unidirectionally by descending the Mo crucible at a speed of 1.39×10^{-6} m/s under the atmosphere stated above. The temperature gradient at the liquid-solid interface was approximately 8 K/cm. The morphology of the fabricated composite is shown in Fig.1.

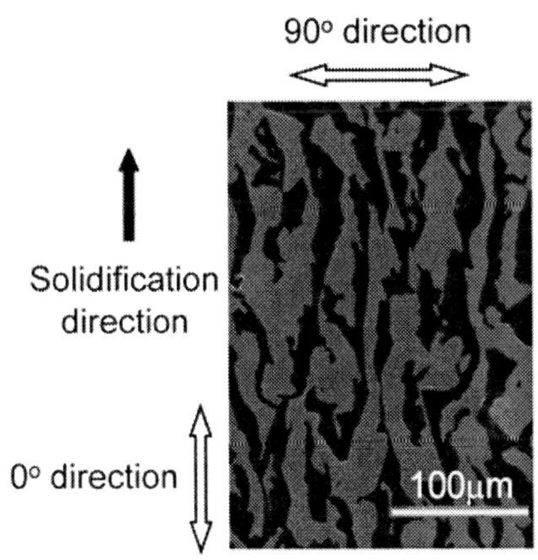

Figure 1. An example of microstructure of the Al_2O_3/YAG composite , together with the definition of 0° and 90° directions. Alumina and YAG correspond to the black and gray parts, respectively.

Temperature Dependence of Young's Modulus and Coefficient of Thermal Expansion

In this section, the experimentally observed temperature dependences of Young's modulus and coefficient of thermal expansion of the composite, and the relation of the temperature dependences of the composite to those of the constituents (Al_2O_3 and YAG), analyzed with the finite element method and the models proposed until now, are presented.

Measured Young's Modulus (E_c) and Coefficient of Thermal Expansion (α_c) of the Composite as a Function of Temperature

The Young's modulus and thermal expansion were measured by the following procedure [22]. The Young's modulus E_c of the Al_2O_3/YAG at room temperature was measured with the supersonic wave pulse echo method for the three directions in Fig.2(a); longitudinal (0°), transverse (90°) and 45° directions, corresponding to parallel, perpendicular and inclined by 45° to the solidification direction, respectively. The measured values of E_c for 0°-, 90°- and 45°- directions at room temperature were 333, 329 and 342 GPa, respectively. As the difference in Young's modulus among the directions was small although the composite consists of single crystals Al_2O_3 and YAG., the Young's modulus only of 0°-direction was measured at the temperature range from room temperature to 1800 K with the bending resonance method. The result is shown in Fig.2(b).

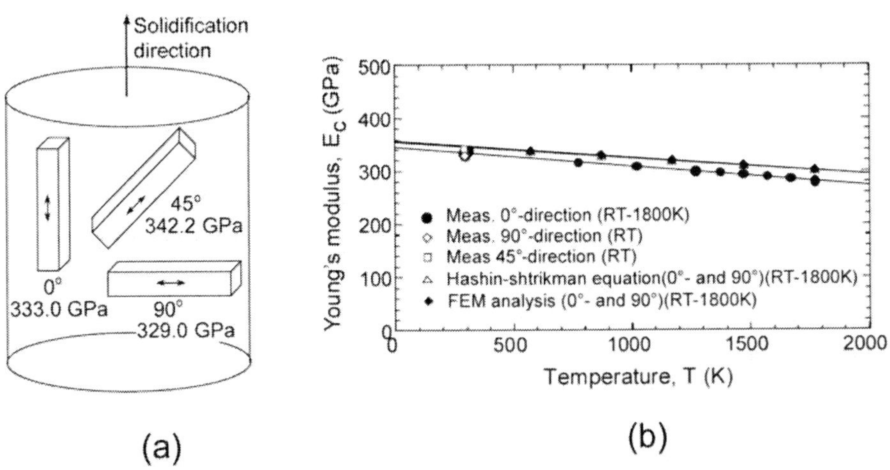

Figure 2. Temperature-dependence of Young's modulus. (a) Measured Young's modulus at room temperature for 0o, 45o and 90o specimens. (b) Measured and analyzed values at room temperature to 1800K.

The coefficient of thermal expansion α_c of the composite for 0°-, 90°- and 45°- directions was measured in the temperature range from room temperature to 1800K with the laser scanning method [22]. The measured coefficients of thermal expansions for these directions are shown in Fig.3. Again the difference in the coefficient α_c among the directions was small as well as that in Young's modulus E_c. From the experimental results stated above, the Young's modulus and coefficient of thermal expansion of the composite could be regarded to be quasi-isotropic as a first approximation. Noting the temperature as T, the averages measured for the three directions were expressed as

$$E_c = 356 - 0.0310T \text{ (GPa)} \quad (1),$$

$$\alpha_c = 6.25 + 0.00134T \text{ (x } 10^{-6}\text{/K)} \quad (2).$$

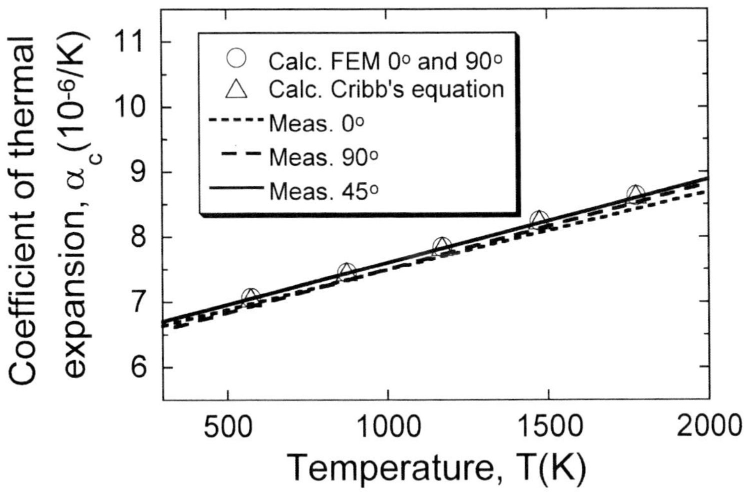

Figure 3. Measured and calculated temperature dependence of the coefficient of thermal expansion for parallel (0°), perpendicular (90°) and 45° directions.

Analysis of Young's Modulus and Coefficient of Thermal Expansion of Composite

The measured Young's modulus and coefficient of thermal expansion of the composite stated above were accounted for by the numerical analysis in a following manner [22].

For analysis of Eqs.(1) and (2), the Young's moduli of Al_2O_3 (E_A) and YAG (E_Y) and the coefficients of Al_2O_3(α_A) and YAG(α_Y) as a function of T are needed. From the experimentally observed quasi-isotropic properties of E_c and α_c stated above, the reported data [35-38] for polycrystalline Al_2O_3 and YAG were used as a first approximation. In the calculation based on the finite element method (FEM), the Young's modulus and coefficient of thermal expansion as a function of temperature T were approximated by the linear equations of the first degree ;

$$E_A=423-0.0474T \text{ (GPa)} \quad (3)$$

$$E_Y=299-0.0180T \text{ (GPa)} \quad (4)$$

$$\alpha_A=6.50+0.00146T \text{ (x}10^{-6}\text{/K)} \quad (5)$$

$$\alpha_Y = 6.09 + 0.00117T\ (\times 10^{-6}/K) \quad (6)$$

The calculated values of E_c and α_c by the finite element method are superimposed in Figs.2(b) and 3, respectively, showing a fairly good agreement with the experimental values.

It was examined whether any difference in calculation result arises or not between (a) the present FEM analysis in which discrete microstructure is used and (b) the analytical models proposed until now (Hashin-Shtrikman's [39] and Cribb's [40] models for Young's modulus and coefficient of thermal expansion, respectively) in which only volume fraction without discrete microstructure is used. The following equations and procedure were used for calculation [22]. Among bulk (K), Young's (E) and shear (G) moduli, the relation

$$E = 9KG/(3K+G) \quad (7)$$

was assumed as well as the relations of $G=E/\{2(1+\nu)\}$ and $K=E/\{3(1-2\nu)\}$. Noting the K and G-values of the composite as K_c and G_c, those of Al_2O_3 as K_A and G_A and those of YAG as K_Y and G_Y, respectively, and the volume fractions of Al_2O_3 and YAG as V_A and V_Y, respectively, the lower bounds of K and G of the composite, $K_{c,min}$ and $G_{c,min}$, and the upper bounds, $K_{c,max}$ and $G_{c,max}$, in the Hashin-Shtrikman'model [39], are given by

$$K_{c,min} = K_Y + (K_A - K_Y)V_A/\{1+3(K_A-K_Y)V_Y/(3K_Y+4G_Y)\} \quad (8)$$

$$G_{c,min} = G_Y + (G_A-G_Y)/[1+6(G_A-G_Y)(K_Y+2G_Y)V_Y/\{5G_Y(3K_Y+4G_Y)\}] \quad (9)$$

$$K_{c,max} = K_A + (K_Y-K_A)V_Y/\{1+3(K_Y-K_A)V_A/(3K_A+4G_A)\} \quad (10)$$

$$G_{c,max} = G_A + (G_Y-G_A)/[1+6(G_Y-G_A)(K_A+2G_A)V_A/\{5G_A(3K_A+4G_A)\}] \quad (11)$$

The lower and upper bounds of E_c were calculated by substituting Eqs.(8) and (9), and Eqs.(10) and (11), into Eq.(7), respectively. The difference between the upper and lower bounds of E_c was less than 1 GPa for the Young's modulus value 300-350 GPa in the temperature range investigated. As the difference between the lower and upper bounds was small, the average of the lower and upper bounds was taken as the Young's modulus value. The calculated values were in good agreement with those by FEM as shown in Fig.2(b), although the microstructure (size, shape, spatial distribution and amount of the constituents) is given discretely in the FEM models but it is given only through the volume fraction of constituents (V_A and V_Y) in Hashin-Strikman's model (Eqs.(8) to (11)). This indicates that the Young's modulus is not sensitive to microstructure of the constituents but primarily dependent on volume fraction of constituents.

According to Cribb [40], α_c is expressed by

$$\alpha_c = \{\alpha_M K_M(K_C-K_R) + \alpha_R K_R(K_M-K_C)\}/\{K_C(K_M-K_R)\} \quad (12)$$

where the subscripts M and R refer to the matrix and reinforcement, respectively. In the present composite, as the volume fraction of Al_2O_3 and YAG are the same ($V_A=V_Y=0.5$), Eq.(12) holds whichever of Al_2O_3 and YAG is matrix and reinforcement. The calculation results are presented in Fig.3. As well as for Young's modulus, the values of the coefficient of thermal expansion calculated by Cribb's equation and finite element method were almost the same. This suggests that the coefficient of thermal expansion, as well as Young's modulus, is not sensitive to the geometry of microstructure but is determined primarily by the volume fraction of constituents.

Temperature- and Displacement Speed-Dependence of Macroscopic Deformation and Fracture Behavior of Unnotched Specimens

In this section, first the bending and tensile test methods are briefly summarized. Then, the experimentally observed temperature- and displacement speed- dependences of load-displacement curve, bending strength and brittle-ductile transition of the bent unnocthed 0° composite specimens at room temperature to 2023K are presented. Following the bending test results, the experimentally observed temperature- and strain rate- dependences of the flow stress of the tensile unnotched 0° specimens tested at 1873 to 2023 K under various strain rates, and the quantitatively analyzed temperature- and strain rate- dependence of the flow stress are presented. The estimated flow stress dependence on temperature and strain rate in this section will be used in the next section to account for the temperature- and displacement speed- dependence of the notched strength and fracture toughness measured for notched specimens.

Bending and Tensile test Methods for Unnocthed Specimens

The bending and tensile tests of the unnocthed 0° specimens were carried out as follows [21, 30]. The configurations of the unnotched specimens for bending and tensile tests are shown in Fig.4. The bending test was carried out at the displacement speed of $8.3x10^{-6}$ to $8.3x10^{-8}$ m/s at room temperature to 2023K. The tensile test was carried out at the displacement speed of $1.6x10^{-8}$ to 10^{-6} m/s (corresponding to the strain rate of $1.6x10^{-6}$ to 10^{-4}/s, respectively) under a given temperature 1973K, and at the temperatures 1873 to 2023 K under a given displacement speed of 10^{-7} m/s, corresponding to the strain rate of 10^{-5}/s. In both bending and tensile tests, the test specimens were heated to the prescribed temperatures under an argon atmosphere, held for 5 min and then tested.

The size of the bending specimens was 4 mm in thickness t, 3 mm in width w, and 40 mm in length (Fig.4(a)). The span length L was 30 mm. Using the maximum load P_{max} in the load - displacement curve shown later, the bending strength σ_B was calculated by

$$\sigma_B = 3LP_{max}/(2wt^2) \tag{13}$$

The length and thickness of the gage portion of the tensile specimens were 10 and 2 mm, respectively. The flow stress in the plastic deformation region was nearly constant as shown later.

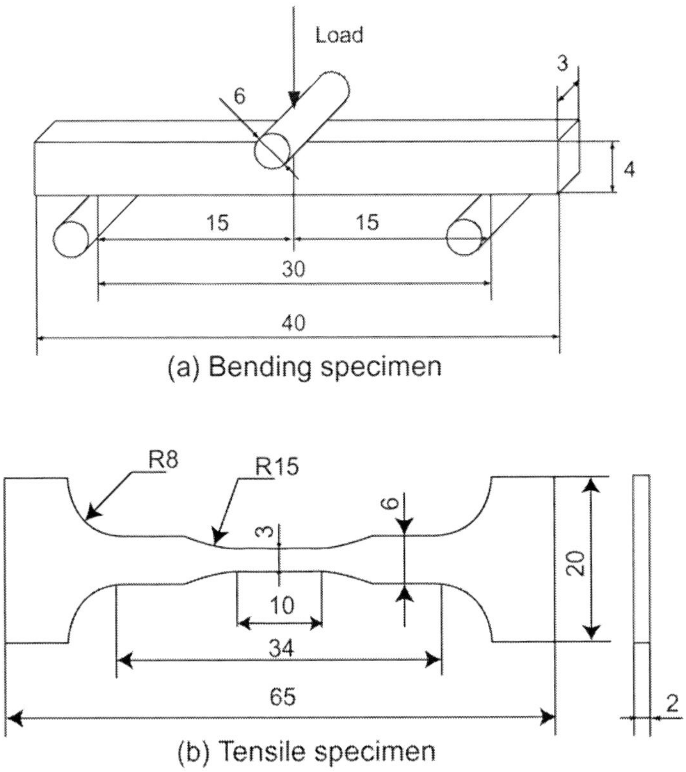

Figure 4. Specimens for bending and tensile tests.

Temperature- and Displacement Speed- Dependence of Load-Displacement Curves, Strength and Brittle-Ductile Transition, Measured by the Bending Test

The load-displacement curves and bending strength of the 0° specimens, obtained by the bending test at room temperature to 2023K under various displacement speeds, are shown in Fig.5 [30]. The load-displacement curve of unnotched specimens was dependent on the test temperature and displacement speed ((a), (b) and (c)). Below 1773 K, the specimens were brittle at all displacement speeds investigated. Beyond 1823 K, the ductility was dependent on the displacement speed. At the low displacement speed of $8.3\text{x}10^{-8}$ m/s (a), the specimens were ductile in the temperature range of 1823 to 2023 K. At 10 times higher displacement $8.3\text{x}10^{-7}$ m/s (b), the ductility was lost at 1823K while it was kept in the temperature range from 1873K to 2023K. At the highest speed of $8.3\text{x}10^{-6}$(c) in the present work, the specimens were brittle at all temperatures investigated. In this way, the unnotched specimens fractured in a brittle manner at low temperatures and at high displacement speeds but in a ductile manner accompanying plastic deformation at high temperatures and at low displacement speed. Correspondingly, the brittle-ductile transition temperature increased with increasing displacement speed [21,30].

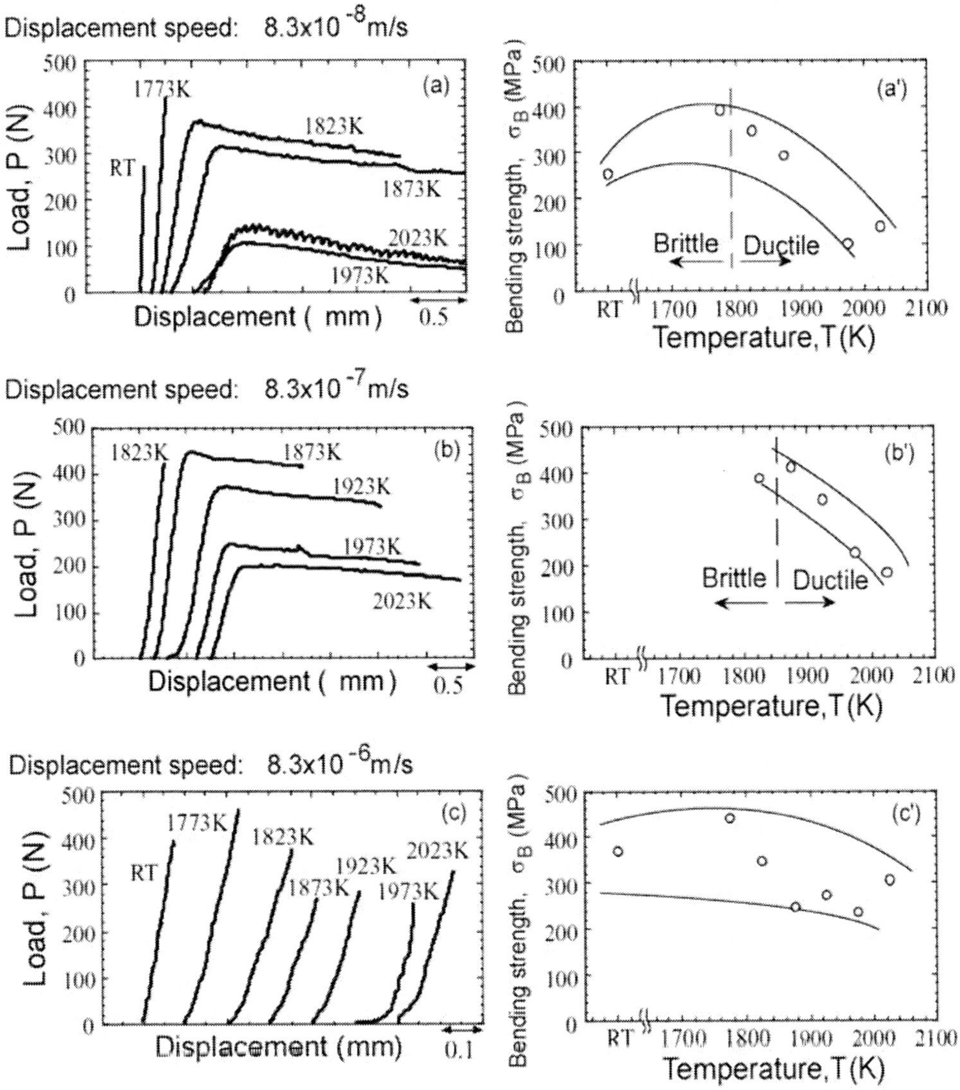

Figure 5. (a, b, c) Load-displacement curve and (a', b', c') strength of unnotched specimens measured by the three point bending test at room temperature to 2023K at the displacement speeds of (a, a') $8.3x10^{-8}$, (b, b') $8.3x10^{-7}$ and (c, c') $8.3x10^{-6}$ m/s. (From Ochiai et al [30], by permission of Elsevier.)

At the lower displacement speeds of $8.3x10^{-7}$ and $8.3x10^{-8}$ m/s above the temperature around 1800K, the unnotched strength of the specimens, which fractured in a ductile manner, decreased with increasing temperature (Fig.5(a') and (b')). At the higher displacement speed of $8.3x10^{-6}$ m/s, at which all specimens fractured in a brittle manner, the strength value at room temperature was almost maintained up to 2023K (Fig.5(c')).

In this way, the deformation and fracture behavior of the Al_2O_3/YAG composite was dependent on temperature and displacement speed. Then, the behavior in the range of ductile deformation and fracture was studied based on the tensile test results for the 0° specimens, and the temperature- and strain rate- dependence of the flow stress in the plastic deformation region was estimated, as follows.

Stress-Strain Curves And Flow Stress of the Composite, Measured by the Tensile Test At high temPeratures, and Formulation of the Temperature- and Strain Rate- Dependence of the Tensile Flow Stress

Figure 6 shows the measured tensile stress-strain curves at 1873-2023K, where all specimens showed plastic deformation. The flow stress in the plastic region is nearly constant. Such a flow stress is evidently sensitive to temperature and strain rate.

It has been shown that the relation of the strain rate ε' to the flow stress σ in plastic deformation of the present Al_2O_3/YAG composite at around 1700-2023K under compressive stress is expressed by [6, 7, 12, 15]

$$\varepsilon' = A\sigma^n \exp(-Q/RT) \tag{14}$$

where A is the constant, n the stress exponent, Q the activation energy for plastic deformation and R the gas constant. Waku et al [12] found n=5~6 and Q=670-905 kJ/mol from the compressive creep test and Yoshida et al [15] found n=6 and Q=730 kJ/mol from the static compressive test of the present composite.

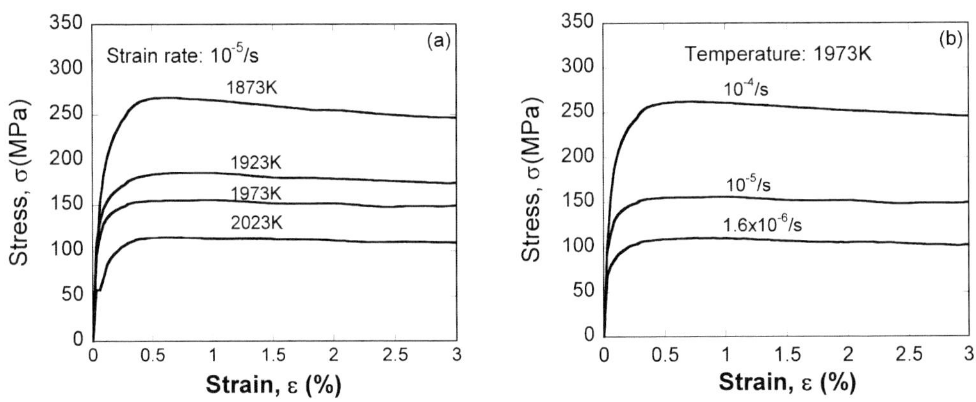

Figure 6. Tensile stress-strain curves of the composite (a) at 1873 to 2023K under a given strain rate of 10^{-5}/s and (b) at the strain rate of $1.6x10^{-6}$ to 10^{-4}/s under a given temperature 1973K.

The plastic and creep deformation of ceramics at high temperatures is controlled by dislocation mechanisms (dislocation glide and climb, dissolution of dislocation loops and so on) or boundary mechanisms (sliding accommodated by diffusion, sliding with glassy phase at grain boundary, and so on), as has been reviewed by Cannon and Langdon [41] and Davidge [42]. The boundary mechanisms are not the present case since the present composite has very clean interface between Al_2O_3 and YAG and has no grain boundary and no glassy phase. [6, 8, 11]. Waku et al [12] and Yoshida et al [15] observed the dislocation structures in both Al_2O_3 and YAG in the plastically deformed specimens with the transmission electron microscope. Their observation result, together with the measured value of n=5~6, suggests that the plastic deformation of this composite can be attributed to the dislocation glide and clime, controlled by climb [41, 42].

Concerning the tensile behavior of this composite, the temperature- and displacement speed- dependence of flow stress was not fully known. Then, from the results shown in Fig.6, the A-, n- and Q-values were determined as follows. As the variation of the flow stress with strain in the plastic region was small (Fig.6), the stress-strain curve was approximated to be elastic-perfect plastic. Then, from the fitting of the flow stress-values at 1923-2023K to Eq.(14), n and Q were estimated to be 5 and 750 kJ/mol, respectively. Thus estimated values were nearly the same as those estimated from the compressive creep test and compressive static one mentioned above. Thus we had the following flow stress of the composite σ_c (MPa) – strain rate ε' relation:

$$\sigma_c=\{\varepsilon'/(1.23x10^5)\}^{1/5}\exp\{1.81x10^4\}/T\} \quad (15)$$

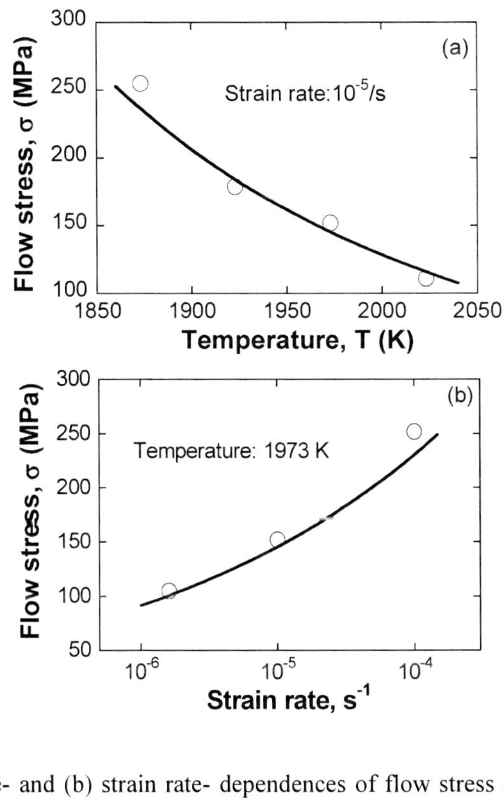

Figure 7. (a) Temperature- and (b) strain rate- dependences of flow stress of the composite in plastic deformation region.

The experimentally measured temperature- and strain rate dependence of the flow stress is well described by Eq.(15), as shown in Fig.7. As shown below, the fracture strength and fracture toughness measured by using notched specimens was also dependent on temperature and displacement speed. Such a dependence could be accounted for by using Eq.(15) as shown below.

Temperature- and Displacement Speeed- Dependence of Notched Strength and Fracture Toughness

In this section, following the test method, the temperature- and displacement speed-dependence of the fracture strength and fracture toughness of the notched specimens, measured by bending and tensile tests at room to 2023K, is presented. Then, the results analyzed by the finite element analysis, which account for the experimentally observed temperature- and displacement speed- dependence of the tensile notched strength from the view point of temperature- and displacement-dependence of the plastic zone ahead of the notch and plastic deformation of the ligament, are presented.

Bending and Tensile Test Methods for Notched Specimens

The notched strength and fracture toughness were measured by the three point bending- [21] and tensile- [30] tests by introducing the notch into the specimens shown in Fig.4. In the bending test, a notch was introduced on the tensile side. The relative notch length a/t (a: notch length, t: specimen thickness) was taken to be 0.3. The tensile test was carried out for the double edge notched specimens with a relative notch length 2a/w =0.3, where a is the notch length at each edge and w is the specimen width. The notch-tip radius was made less than 50 μm with a special saw. The notched specimens were heated to the prescribed temperatures under an argon atmosphere, held for 5 min and then tested, as similarly as the unnotched specimens. The bending test for notched specimens was carried out at the temperatures of room temperature, 1773, 1923 and 2023K under the displacement speeds of 8.3×10^{-8} and 8.3×10^{-6} m/s [21]. The tensile test was carried out at the temperatures of 1873, 1923, 1973 and 2023K under the displacement speed of 10^{-7} m/s and at the displacement speeds of 10^{-8}, 10^{-7}, 10^{-6} and 10^{-5} m/s under the temperature of 2023K [30].

The notched specimens showed brittle and ductile fracture modes depending on test temperature and displacement speed as will be shown later. For the specimens that fractured in a brittle mode, the apparent fracture toughness K_Q was estimated by the following formula.

Bending test [43]

$$K_Q=\{LP_f/(wt^{3/2})\}\{(3/2)\,\xi^{1/2}F(\xi)\} \tag{16}$$

$$F(\xi)=(1.964-2.837\xi+13.711\xi^2-23.250\xi^3+24.129\xi^4 \quad (\xi=a/t)$$

where P_f is the load at fracture.

Tensile test [44]

$$K_Q=\sigma_N(\pi a)^{1/2}F(\xi) \tag{17}$$

$$F(\xi)=(1.122-0.56\xi-0.205\xi^2+0.471\xi^3-0.190\xi^4)/(1-\xi)^{1/2} \quad (\xi=2a/w)$$

where σ_N is the notched strength, estimated by dividing the ultimate load by the gross cross-sectional area.

Temperature- and Displacement Speed- Dependence of Fracture Behavior and Fracture Toughness, Measured by the Bending Test

The measured dependence of load-displacement curve and fracture toughness of notched specimens on temperature and displacement speed, obtained by the bending test, is shown in Fig.8 [30]. When the displacement speed was low (8.3×10^{-8} m/s), the notched strength

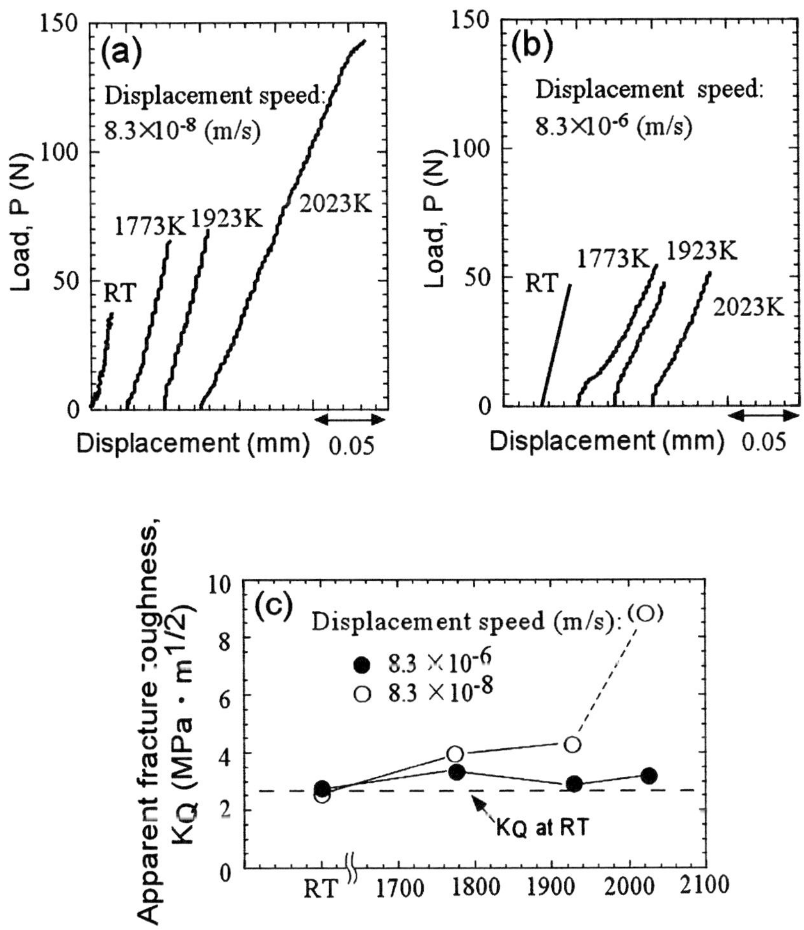

Figure 8. Load-displacement curves of the notched specimens measured by the three point bending test at room temperature to 2023K at the displacement speeds of (a) 8.3×10^{-8} and (b) 8.3×10^{-6} m/s, and the change of apparent fracture toughness value with increasing temperature (c). (From Ochiai et al [30], by permission of Elsevier.)

increased with increasing temperature (Fig.8(a)). On the contrary, when the displacement speed was high (8.3×10^{-6} m/s), the strength was nearly the same at any temperature (Fig.8(b)).

Accordingly, the apparent fracture toughness value K_Q, estimated by Eq.(16) based on the linear elastic fracture mechanics, was nearly constant for $8.3x10^{-6}$ m/s but it increased with increasing temperature for $8.3x10^{-8}$ m/s, as shown in Fig.8(c). The K_Q –value of the specimen tested at the displacement speed of $8.3x10^{-8}$m/s at 2023K is shown in the parentheses for reference since the application of the linear elastic fracture mechanics might be limited due to the enhanced plastic zone ahead of the notch (the influence of the plastic zone on the fracture strength will be discussed later). The result of the bending test indicates that the fracture behavior of the notched specimens is also dependent on temperature and displacement speed. In order to detect such a feature in more detail and to analyze numerically, tensile test was carried out as shown below.

Temperature- and Displacement Speed- Dependence of Fracture Behavior, Notched Strength and Fracture Toughness, Measured by the Tensile Test

Figure 9 shows the tensile stress-displacement curves, measured (a) at the temperatures of 1873, 1923, 1973 and 2023K under the given displacement speed of 10^{-7} m/s and (b) at the displacement speed of 10^{-8}, 10^{-6}, 10^{-6} and 10^{-5} m/s under the given temperature of 2023K. The stress is given by the gross stress, estimated by dividing the load by the cross-sectional area. In Fig.9(a), the specimens at the lower temperatures of 1873(indicated by (a)) and 1923K (b) fractured in a brittle manner in the elastic range, while those at the higher temperatures of 1973 (c) and 2023K (d) fractured in a ductile manner, accompanied by the plastic deformation in spite of the existence of the notch. In Fig.9(b), the specimens fractured in a ductile manner when deformed at the lower displacement-speeds of 10^{-8} (a) and 10^{-7} m/s (b) while they fractured in a brittle manner when deformed at the higher displacement speeds of 10^{-6} (c) and 10^{-5} m/s (d). In this way, the fracture mode turned from brittle to ductile with increasing temperature under the given displacement speed (Fig.9(a)) and with decreasing displacement speed under the given temperature (Fig.9(b)).

Figure 10 shows (a) the temperature-dependence of the measured notched strength under the displacement speed of 10^{-7} m/s and (b) displacement speed- dependence under the temperature of 2023K. (a) to (d) in Figs.10(a) and (b) refer to the specimens whose stress-displacement curves are shown in (a) to (d) in Figs.9(a) and (b), respectively. The closed and open circles refer to the specimens that fractured in the brittle and ductile manners, respectively. The apparent fracture toughness K_Q estimated by Eq.(17) is also presented. The fracture toughness values based on the elastic fracture mechanics are valid only for the specimens indicated by the closed circle, since the specimens indicated by the open circles showed overall yielding in the ligament portion, as shown later. Following features are read from Figs.8 to 10.

(a) The notched strength at a given displacement speed of 10^{-7} m/s increases, reaching maximum at 1923K, and then decreases with increasing temperature (Fig.10(a)). Also, the notched strength at a given temperature 2023K increases, reaching maximum at the displacement-speed of 10^{-6} m/s and then decreases with increasing displacement speed (Fig.10(b)).

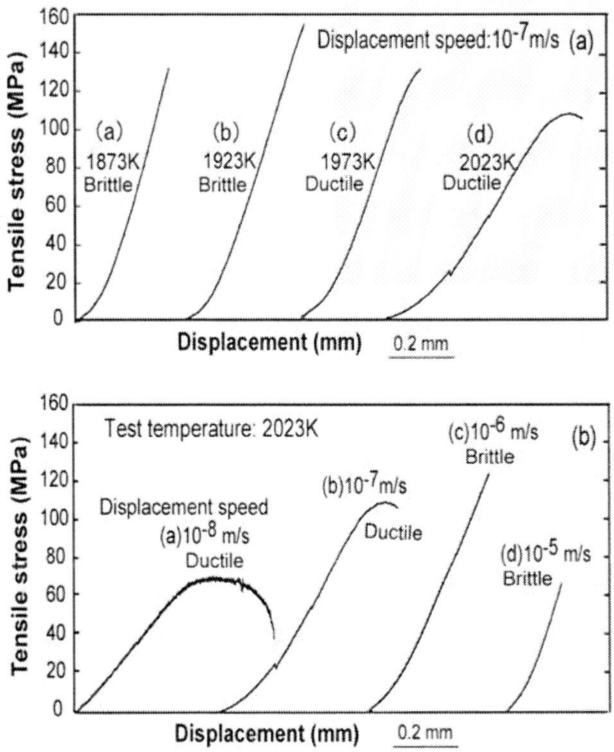

Figure 9. Gross tensile stress- displacement curves of the notched specimens measured by the tensile test (a) at 1873 to 2023K under a given displacement speed of 10^{-7} m/s and (b) at the displacement speeds of 10^{-8} to 10^{-5} m/s under a given temperature 2023 K.

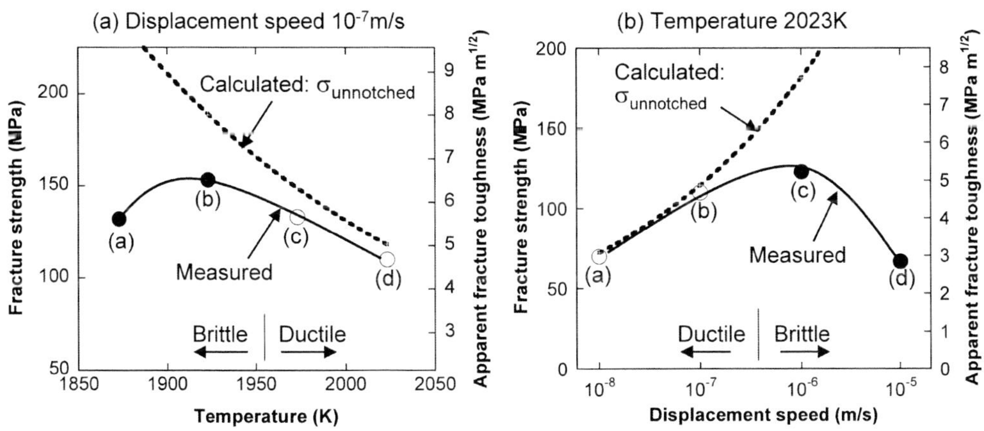

Figure 10. Fracture strength of notched specimens (a) at 1873 to 2023K under a given displacement speed of 10^{-7} m/s and (b) at the displacement speeds of 10^{-8} to 10^{-5} m/s under a given temperature 2023 K. The unnotched strengths calculated by Eq.(15) are superimposed for comparison. The open (○) and closed (●) circles refer to the ductile and brittle fracture modes, respectively. The right axis shows the apparent fracture toughness K_Q, which was estimated for the specimens fractured in brittle mode.

(b) In Fig.8(c), the apparent fracture toughness, K_Q, measured by the bending test was around 2.5~4.2 MPa·(m)$^{1/2}$ in the temperature range of room temperature to 1923K, depending on the test temperature and displacement speed. (The K_Q value at 2023K was higher than 8 MPa·(m)$^{1/2}$, but it is not certain whether the elastic fracture mechanics can be applied or not due to the possible large scale yielding in the ligament.) The measured values of K_Q obtained by the tensile test (Fig.10) were 2.8~6.2 MPa·(m)$^{1/2}$, being nearly the same as the values obtained by the bending test.

(c) The fracture mode of the notched tensile specimens turns from brittle to ductile when the temperature becomes high under the given displacement speed (Figs.9(a) and 10(a)) and it turns from ductile to brittle when the displacement speed becomes high under the given temperature (Figs.9(b) and 10(b)). In spite of the existence of the notch, the temperature- and displacement speed-dependence of the brittle-ductile transition shows the similar tendency to that of the unnotched specimens.

(d) The unnotched strength $\sigma_{unnotched}$ was calculated by Eq.(15) for reference. The results are superimposed in Fig.10. The notched strengths of the specimens that fractured in a brittle manner((a) and (b) in Fig.10(a) and (c) and (d) in Fig.10(b)) were far lower than $\sigma_{unnotched}$, indicating that the fracture was governed by the elastic fracture mechanics. On the other hand, the notched strengths of the specimens that fractured in a ductile manner ((c) and (d) in Fig.10(a) and (a) and (b) in Fig.10(b)) were lower than $\sigma_{unnotched}$, but close to it. This suggests that the ligament portion deformed plastically but at higher strain rate than the far distant portion.

Suggested Mechanism of Temperature- and Displacement Speed-Dependence of the Notched Strength and Fracture Toughness

The results shown above indicate that the plastic zone formation ahead of the notch, which plays a dominant role for determination of the K_Q-value, is dependent on displacement speed and temperature. When the temperature- and displacement speed- dependence of the growth of plastic zone is taken into consideration, the following observed behavior can be accounted for.

When the plastic flow takes place ahead of the notch, it acts to raise the notched strength. As the plastic flow stress decreases and therefore the plastic zone size increases with increasing temperature, the notched strength (and fracture toughness) is expected to increase with increasing temperature. On the other hand, as the flow stress decreases with further increasing temperature, the plastic zone ahead of the notch can be large enough to cover the ligament in the notched cross-section (overall yielding). At such high temperatures, the notch extension is suppressed due to the enhanced plastic deformation but the flow stress decreases with increasing temperature, resulting in decrease in the stress carrying capacity of the ligament. As a result, the strength of notched specimens decreases with increasing temperature. Thus, for a given displacement speed, the notched strength increases and then decreases with increasing temperature. These features are realized in the experimental results (Figs.9(a) and 10(a)).

At very high temperatures, the composite is ductile at low displacement speeds, while it is brittle at high speeds. At low displacement speed range, the plastic zone grows to cover the ligament, and the notch extension is suppressed. In such a situation, the stress carrying capacity of the ligament and therefore the strength of notched specimens increase with increasing displacement speed due to the increasing flow stress. On the other hand, at high displacement speed range where overall yielding of ligament does not take place, the size of the plastic zone ahead of the notch is limited and the specimen fractures in a brittle manner. In such a situation, as the flow stress increases with increasing displacement speed, the size of the plastic zone and therefore the notched strength decrease with increasing displacement speed. Due to such a displacement speed- dependence of the plastic zone size, the notched strength increases but then decreases with increasing displacement speed. Such features are realized in the experimental results (Figs.9(b) and 10(b)).

Based on the discussion above, the suggested mechanism of the displacement speed-dependence of the notched strength under a given temperature is summarized, as shown in Fig.11. The temperature-dependence of the notched strength under a given displacement speed is accounted for in a similar manner as the increase in temperature corresponds to decrease in displacement speed, as stated already.

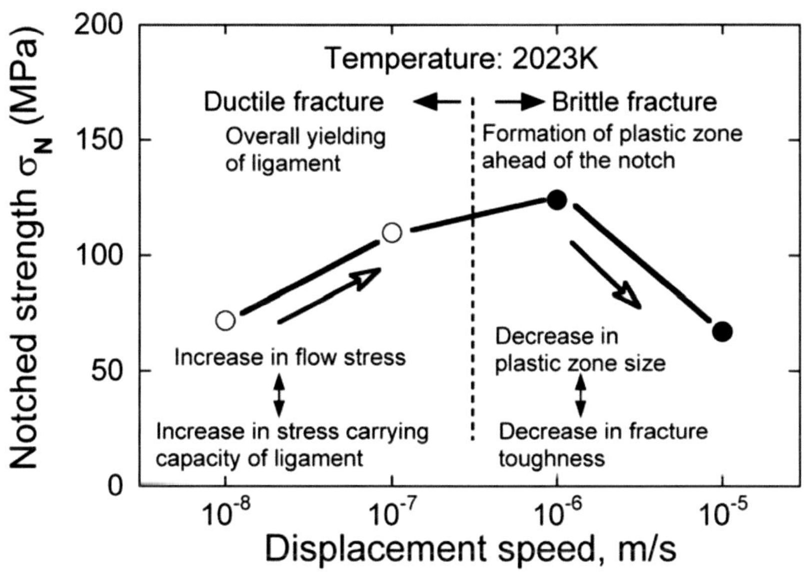

Figure 11. Mechanism for the dependence of notched strength on displacement speed under a given temperature. The temperature-dependence of the notched strength under a given displacement speed is described in a similar manner by replacing " the increase in displacement speed" by " the decrease in test temperature.

In this way, the experimentally observed temperature- and displacement speed-dependence of the notched strength (Figs.9 and 10) is accounted for qualitatively from the view point of the growth of plastic zone. In order to examine the temperature- and displacement speed- dependence of the plastic zone quantitatively, the stress- distribution and the plastic zone size were calculated by means of the finite element analysis, as follows [30].

Verification of the Suggested Mechanism by the Analysis of the Influence of Temperature and Displacement Speed on the Plastic Zone Size and Stress Distribution in the Notched Tensile Specimens by Means of the Finite Element Method

The finite element analysis for notched tensile specimens was carried out with the commercial finite element code MARC/Mentat^{TM}. The calculation was carried for the region surrounded by the broken curve (1/4 of the body concerned) shown in the top in Fig.12. The temperature (T)- dependence of the Young's modulus of the composite (E_c) was taken from Eq.(1). The Poisson's ratio, ν_c, of the composite was taken to be 0.24, since the Poisson's ratios of Al_2O_3 and YAG are 0.23 and 0.25, respectively [45] and the volume fraction of each phase is 0.5 [6-8, 11,12, 15, 21]. Assuming that the Poisson's ratio is constant over the temperature range concerned and using the relation among the shear modulus G, Young's modulus E and Poisson's ratio ν, given by Eq.(7), the shear modulus of the composite was given by G_c=144-0.0129T.

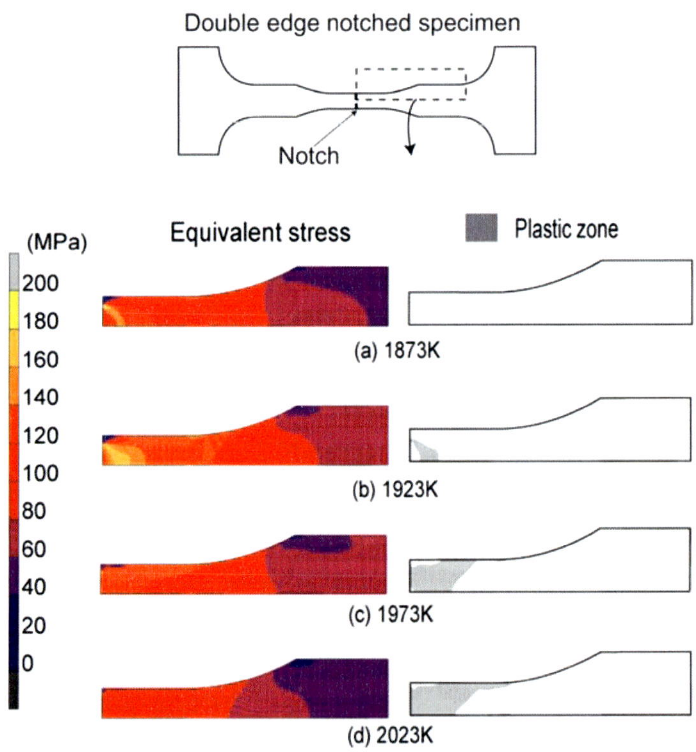

Figure 12. Distributions of equivalent stress and plastic zone in the notched specimens at the experimentally measured fracture stress, calculated by the finite element analysis. The calculation result for the specimens deformed at the temperatures of (a) 1873, (b) 1923, (c) 1973 and (d) 2023K under the displacement speed of 10^{-7} m/s is representatively presented in this figure. (From Ochiai et al [30], by permission of Elsevier.)

The calculated distribution of the equivalent stress and the shape/size of the plastic zone at the measured fracture stress of the composite at the test temperatures 1873, 1923, 1973 and 2023K under a displacement speed 10^{-7} m/s are representatively presented in Fig.12 [30]. In the specimens that fractured in a brittle manner at the temperatures of 1873 and 1923K and at the displacement speeds of 10^{-6} and 10^{-5} m/s, the plastic zone was small and the stress concentration at the notch tip was high, while in the specimens that fractured in a ductile manner at the temperatures of 1973 and 2023K and at the displacement speeds of 10^{-8} and 10^{-7} m/s, the plastic zone covered the ligament portion and the stress was nearly uniform in such a portion. The experimentally observed feature, "the notched strength increases with increasing temperature (decreasing displacement speed) in the region of brittle fracture (closed circles in Fig.10), was accounted for well by the increase in plastic zone size with increasing temperature (with decreasing displacement speed). Also the feature, "the notched strength decreases with increasing temperature (decreasing displacement speed) in the region of ductile fracture (open circles in Fig.10)", was well accounted for by the overall plastic deformation of the ligament portion, as shown in Fig.12. In this way, the present ceramic composite behave like a metal once plastic deformation takes place at ultra high temperatures.

Temperature- and Strain Rate- Dependence of the Composite, Its Relation to That of the Consituents, and Stress- and Strain Rate- Distributions in the Consituents at Ultra High Temperatures

In this section, first the experimentally observed temperature- and strain rate- dependence of the compressive stress-strain curve and flow stress of the 0° and 90° composite specimens at 1873K are presented. Then the microscopic stress- and strain rate- distributions in each constituent (Al_2O_3, YAG) under plastic deformation and the relation of the microscopic behavior of the constituents to the overall flow stress of the composite, analyzed by the finite element-, isostrain- and isostress- models, are presented.

Measured Compressive Stress-Strain Curves

The compressive stress was applied parallel (0° specimen) and perpendicular (90° specimen) to the solidification direction. The compression test of the 0° and 90° composite specimens was carried out at 1873 K under strain rates of 2.8×10^{-6}, 10^{-5} and 10^{-4} /s in argon flow. The size of the test specimens was 4x4 mm^2 in cross-section and 6 mm in height.

Figure 13 (a) and (b) shows the as-measured stress (σ_c) - strain (ε_E) curves of the 0° and 90° specimens. In the experiment, the strain ε_E was measured from the moved distance of the crossheads. Such a strain is the sum of the strain of the specimen itself, ε_c, and the additional strain, ε_A, arising from the deformation of the system for measurement. The measured σ_c- ε_E curve was converted to the σ_c- ε_c curve of the specimen itself in the following procedure (Fig.14).

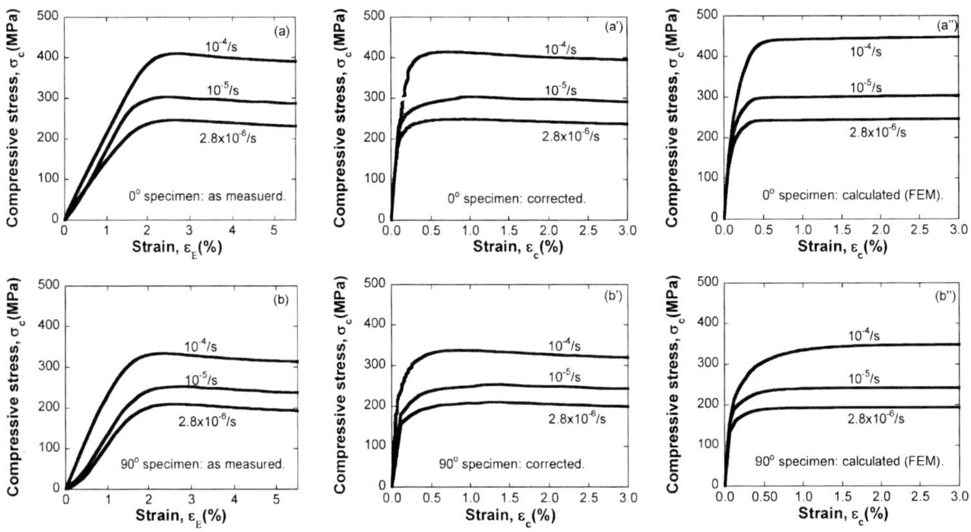

Figure 13. Compressive stress-strain curves at 1873K. (a) 0° specimen: as measured. (a') 0° specimen: corrected by the procedure shown in Figure 14. (a") 0° specimen: calculated by FEM. (b) 90° specimen: as measured. (b') 90° specimen: corrected by the procedure shown in Figure 14. (b") 90° specimen: calculated by FEM.

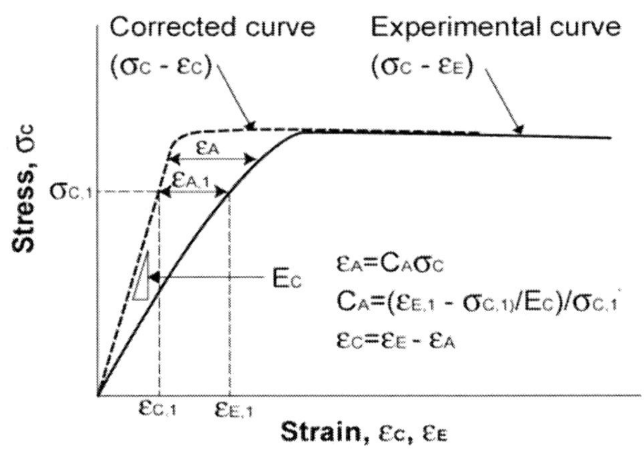

Figure 14. Schematic representation of correction procedure of the experimentally measured stress (σ_c) - strain (ε_E) curve of the composite to obtain the stress (σ_c) - strain (ε_c) curve of the specimen itself.

It was assumed that the system, which causes additional displacement, behaves like an elastic spring with a compliance C_A; namely ε_A was given by $\sigma_c C_A$. Under such an assumption, ε_E was given by $\varepsilon_E = \varepsilon_c + \varepsilon_A = \varepsilon_c + \sigma_c C_A$. In the elastic deformation stage of the specimen, at the stress $\sigma_{c,1}$ for instance (Fig.14), the C_A was estimated by $(\varepsilon_{E,1}-\sigma_{c,1}/E_c)/\sigma_{c,1}$ where E_c is the Young's modulus of the composite. The temperature (T: K) dependence of the Young's modulus (E_c) of the composite has been estimated experimentally to be E_c =356-0.0310T (GPa) (Eq.(1)) from which the E_c-value at 1873 K was taken to be 298 GPa. Using the estimated C_A–value, which is common in both elastic and plastic deformation ranges of the specimen, the strain ε_c of the composite specimen itself at any stress σ_c is calculated by $\varepsilon_E - \sigma_c C_A$. By repeating such a procedure for the whole range of ε_E, the curve (σ_c- ε_c) of the specimen itself can be obtained. Applying such a procedure to the as measured curves shown in Fig.13(a) and (b), the corrected stress (σ_c) – strain (ε_c) curves for the 0° and 90° specimens were obtained as shown in Fig.13(a') and (b'), respectively.

Modeling for Description of the Stress-Strain Curves and Flow Stress of the Composite

(A) Finite element model

Figure 15(a) shows the boundary conditions for finite element analysis. The composite specimen, in which the finite element meshes were given, was placed in ABCD. The compressive displacement was given on AB. The stress analysis was carried out using the commercial finite element code MARC/Mentat^{TM}. In the calculation, the applied strain was raised in step of 0.0001.

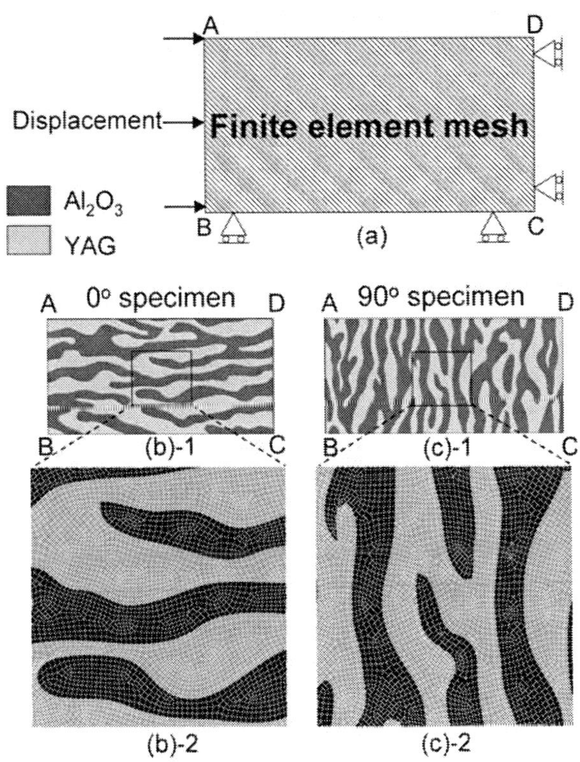

Figure 15. Finite element model for stress analysis under compressive stress. (a) Boundary condition. (b)-1 Configuration of 0 ° model. (b)-2 Mesh of 0 ° model. (c)-1 Configuration of 90 ° model. (c)-2 Mesh of 90 ° model. (From Ochiai et al [34], by permission of Elsevier.)

As has been shown in Fig.1, the alumina and YAG in the cross-section parallel to the direction of solidification are elongated in comparison with those in the cross-section perpendicular to it. For the finite element analysis, the model was taken from the observed microstructure, as shown in Fig.15(b)-1 and (c)-1 for 0° and 90° specimens, respectively. The size of the model composite was 0.3x0.15mm. The number of elements in ABCD was approximately 27000 both for 0° and 90° specimens. The meshes in the portion picked up from Fig.15(b)-1 and (c-1) are shown in (b)-2 and (c)-2, respectively.

The yield stress σ_y (and flow stress after yielding) of monolithic Al_2O_3 and YAG is dependent on strain rate and temperature. As the microstructure is complicated as shown in Fig.1, the strain rate in each constituent (Al_2O_3, YAG) in the composite is different from position to position, resulting in different yield stress from position to position. In this work, at each position (element), the equivalent strain rate was calculated and it was converted to the yield stress (σ_y) through the stress-strain rate relation in the plastic state, shown later. When the equivalent stress in the elastic state reached thus calculated yield stress σ_y, the yielding of the corresponding element was judged to occur.

(B) Isostrain and isostress models

The microstructure of the present Al_2O_3/YAG composite (Fig.1) is intermediate between the isostrain and isostress models (Fig.16). Therefore, the flow stress in plastic deformation region of the composite is expected to fall in the range between the higher and lower bounds determined by the rule of mixtures for the isostrain- and isostress models, respectively.

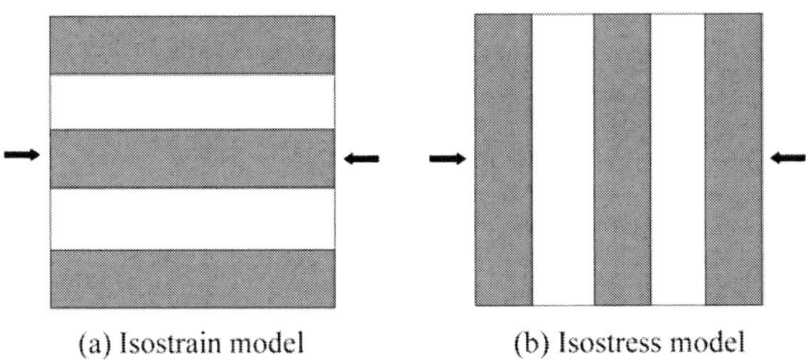

Figure 16. Isostrain and isostress models.

According to the isostrain and isostress models, in the plastic deformation range, the relation of the stress and strain rate of the composite to those of Al_2O_3 and YAG is expressed as

$$\sigma_c = \sigma_A V_A + \sigma_Y V_Y, \quad \varepsilon'_c = \varepsilon'_A = \varepsilon'_Y \text{ (Isostrain model)} \tag{18}$$

$$\sigma_c = \sigma_A = \sigma_Y, \quad \varepsilon'_c = \varepsilon'_A V_A + \varepsilon'_Y V_Y \text{ (Isostress model)} \tag{19}$$

where V is the volume fraction (=0.5 for both alumina and YAG) and the subscripts c, A and Y refer to the composite, Al_2O_3 and YAG, respectively. Substituting the relation of strain rate to flow stress in the steady state plastic deformation stage of Al_2O_3 and YAG, given later, into Eqs.(18) and (19), we had the correlation between σ_c and ε'_c at 1873K for both models.

In practical composite (Fig.1), the geometry of the components (Al_2O_3 and YAG) is complicated. In the isostrain and isostress models, a simple layered structure, in which Al_2O_3 and YAG exist alternatively in a regular form, is assumed, as has been shown in Fig.16. Thus from the comparison of the calculation result based on these models with that based on the finite element model taken from the practical microstructure, the influence of the geometrical deviation of the shape of the constituents from the isostrain and isostress models on the overall flow stress, and its difference between the 0 ° and 90 ° specimens can be detected.

(c) Input values

The temperature (T)- dependence of Young's modulus (E) and shear modulus (G) of alumina and YAG were taken from Eqs.(3) and (4) and that of shear modulus (G) were, as a first approximation, taken to be G_A=172-0.0193T and G_Y =120-0.0072T GPa by combining Eqs.(3) and (4) with $G=E/\{2(1+\nu)\}$ and the Poisson's ratios (0.23(Al_2O_3) [45] and 0.25(YAG) [45]), where A and Y refer to the alumina and YAG, respectively. These values describe satisfactorily the measured temperature- dependence of the Young's modulus of the composite specimen, as has been shown already.

It has been shown that, for both 0 ° and 90 ° composite specimens, the relation of the strain rate ε' to flow stress σ in plastic deformation at 1700-2000K is expressed by Eq.(14) [6, 7, 12, 15], for the composite and the constituents (Al_2O_3 and YAG). Yoshida et al [15] have found that n=6 and Q=637 kJ/mol for Al_2O_3, taken from the data of Kotchick and Tressler [46], and n=6, Q=789 kJ/mol for YAG, taken from the data of Corman [47], can account for the experimentally observed flow stress-strain rate relation. In the present work, these values, together with $A=2.7\times10^{-2}$ and 1.3x10 ($s^{-1}MPa^{-6}$) for Al_2O_3 and YAG, estimated from the reported data of Refs.[46] and [47], respectively, were employed.

Analyzed Overall Stress-Strain Cures and Flow Stress Based on the Microscopic Behavior of the Constituents Al_2O_3 and YAG

(A) Strain rate-dependence of the stress-strain curve of the 0 ° and 90 ° composite specimens at 1873K, calculated by the finite element model

The following features are found in the corrected stress-strain curves shown in Fig.13(a') and (b'). (i) In both 0 ° and 90 ° specimens, the flow stress in 0.5-3.0 % strain range is nearly constant, indicating that the steady state deformation occurs in the plastic region. (ii) The flow stress increases with increasing strain rate for both 0 ° and 90 ° specimens. (iii) The flow stress of 0 ° specimen is higher than that of 90 ° one at any strain rate. These features, together with the shape of the measured stress-strain curve and the flow stress level in the steady state plastic region, are satisfactorily reproduced by the finite element model for both 0 ° and 90 ° specimens as shown in Fig.13(a'') and (b'').

(B)Distributions of strain rate in Al_2O_3 and YAG in the plastic region

In the isostrain and isostress models (Fig.16), the yielding of YAG for instance occurs simultaneously, but in the practical complicated geometry of the composite microstructure (Fig.1), the stress state and strain rate are heterogeneous in each constituent. Figure 17 [34] shows (a, a') the model composites used for calculation, and the calculated spatial distribution of equivalent strain rate in the (b, c) 0 ° and (b', c') 90 ° specimens in the plastic deformation region (ε_c=1.5%) at the applied strain rate of 10^{-5}/s as an example. The blue color in (b,b',c and c') corresponds to around equal to or slightly less than the applied strain rate on composite (10^{-5}/s). The red, yellow and gray colors correspond to around 2~3, 5 and more than 10 times higher than the applied strain rate 10^{-5}/s, respectively.

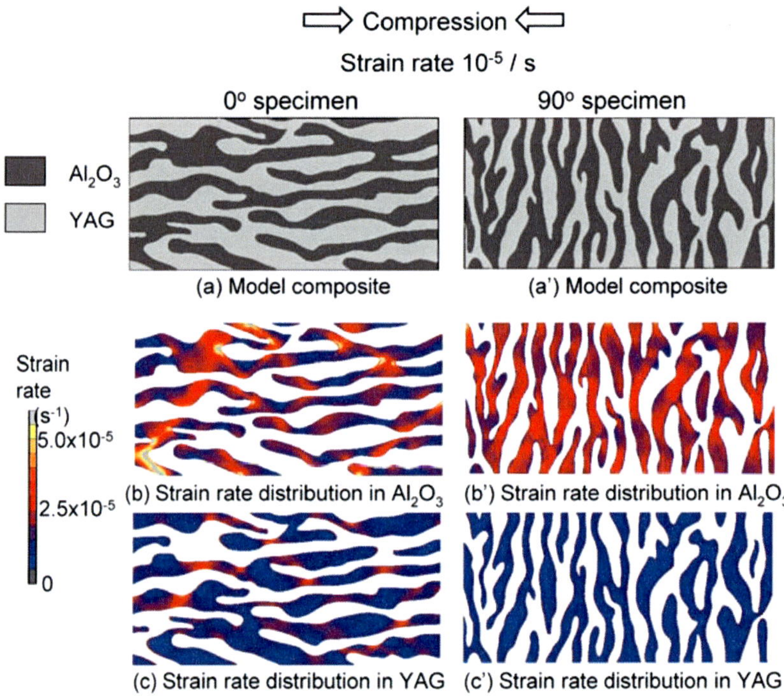

Figure 17. Model composite and calculated equivalent strain rate distribution for (a to c) 0 ° and (a' to c') 90 ° specimens deformed at a strain rate of 10^{-5}/s. (a,a') Model composite employed for calculation. (b,b') Distribution of equivalent strain rate in Al_2O_3 in the plastic region. (c,c') Distribution of equivalent strain rate in YAG in the plastic region. (From Ochiai et al [34], by permission of Elsevier.)

In the 0 ° specimen, if the structure is regular as in the isostrain model, the strain rate distribution shall be uniform to a first approximation. Due to the microstructural deviation of the finite element model (Fig.15(b)-1) from the isostrain model (Fig.16(a)), the strain rate (stress) is distributed as shown in Fig.17(b) and (c). However, still a considerable amount of both Al_2O_3 and YAG deform at similar strain rate, as expected from the isostrain model. In the 90 ° specimen, the strain rate is also distributed in both phases. However, as the Al_2O_3 is softer than YAG, the Al_2O_3 (Fig.17(b')) deforms totally at higher strain rate than YAG. (Fig.17(c')), as expected from the isostress model. In this way, the feature of the isostrain and isostress models remains in the 0 ° and 90 ° specimens, respectively.

It is noted that the strain rate (and therefore the stress) is concentrated in thick parts of Al_2O_3 and in the neighboring thin parts of YAG in the 0 ° specimen (Fig.17 (b) and (c)). In the present composite, the volume fraction of Al_2O_3 and YAG is 0.5 based on the phase diagram [6,8,11]. However, there exists a volume fraction fluctuation within the composite. Namely the regions with low and high volume fraction of YAG coexist. As YAG is harder than Al_2O_3 in the plastic state, both Al_2O_3 and YAG in the higher YAG fraction-region deform at lower strain rate. On the contrary, those in the lower YAG fraction-region deform at higher strain rate, being the stress concentrated region. Such a lower YAG fraction-region region might be a fracture initiation site in the 0 ° specimen. On the other hand, in the 90 ° specimen, the strain rate of YAG (Fig.17(c')) is totally lower than that of Al_2O_3 (Fig.17(b')). The difference in strain rate from position to position in each of Al_2O_3 and YAG (Fig.17(b') and (c')) in the 90 ° specimen is smaller than that in the 0 ° specimen (Fig.17(b) and (c)). Such a difference in spatial distribution of strain rate and therefore the stress between the 0° and 90° composite specimens leads to the difference in flow stress as follows.

(C) Comparison of the calculated flow stress by isostrain, isostress and finite element models with the measured one

As the microstructure of the present composite (Fig.1) is intermediate between the structure of isostrain and isostress models, the flow stress of the present composite is expected to fall in the range between the higher and lower bounds determined by the isostrain- and isostress models, respectively. Also it is expected that the flow stresses of the 0 ° and 90 ° specimens at any strain rate are near to those of the isostrain and isostress models, respectively, due to the morphology shown in Fig.1, where the constituents are elongated in the solidification direction. If such expectations are demonstrated to be valid for the present composite, the simple isostrain and isostress models can be a convenient tool to predict approximate upper and lower bounds of the flow stress of the composite from the behavior of the constituting phases.

Figure 18 shows the comparison of the strain rate – flow stress relation among (○) experimental result of the 0 ° specimen, (□)experimental result of the 90 ° specimen, (●) calculation result of the 0 ° specimen based on the finite element model, (■) calculation result of the 90 ° specimen based on the finite element model, (▼) calculation result based on the isostrain model, (▲) calculation result based on the isostress model, (×) monolithic YAG and (+) monolithic Al_2O_3. The calculation result of the finite element model has a good agreement with the experimental result in the whole range of strain rate (2.8×10^{-6} to 10^{-4} /s) for both 0 ° and 90 ° specimens. It is noted that the experimental values of composite are in the range between the lower bound given by the isostress model and the upper one given by the isostrain model. This result suggests that the isostrain and isostress models are useful to predict the lower and upper bounds as a first approximation.

Concerning the influence of the morphology in the 0 ° and 9 ° specimens, the following features are found in Fig.18.

(a) The measured and calculated flow stress of the 0 ° specimen is close to but slightly lower than the prediction based on the isostrain model. This suggests that the elongated microstructure of the constituting phases along the solidification direction (Fig.1) induces a behavior similar to that observed for unidirectional fiber-composites [48,49], with the slight disagreement due to the microstructural deviation (Fig.1) from the theoretical isostrain model.
(b) The measured and calculated flow stress of the 90 ° specimen is in the middle between the predictions based on the isostrain and isostress models. In this specimen,

the microstructural deviation from regularity acts to raise the overall flow stress of the composite from the prediction based on the isostress model.

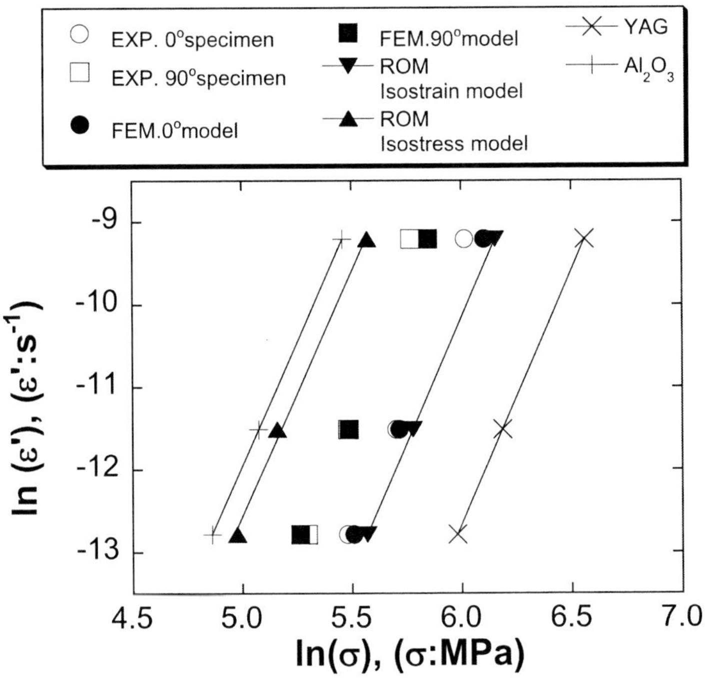

Figure 18. Comparison of the strain rate-stress relation among (○) experimental result of the 0° specimen, (□)experimental result of the 90° specimen, (●) calculation result of the 0° specimen based on the finite element model, (■) calculation result of the 90° specimen based on the finite element model, (▼) calculation result based on the isostrain model, (▲) calculation result based on the isostress model, (×) monolithic YAG and (+) monolithic Al_2O_3. (From Ochiai et al [34], by permission of Elsevier.)(c) Under the same strain rate and temperature, the flow stress of YAG is higher than that of Al_2O_3. Therefore, in the case of 0 ° specimen, YAG carries high stress since both YAG and Al_2O_3 deform nearly at the same strain rate as in the isostrain model. On the contrary, in the case of 90 ° specimen, the flow stress of YAG is forced to be nearly same as that of Al_2O_3 as in the 90 ° model. Thus high stress carrying capacity of YAG is not fully available in the 90 ° specimen, resulting in lower composite flow stress in comparison with the stress of 0 ° specimen.

Conclusion

The temperature dependence of Young's modulus and coefficient of thermal expansion, and the temperature- and displacement speed-dependence of tensile and compressive flow stress, fracture strength and fracture toughness of the directionally solidified bulk type

Al_2O_3/YAG($Y_3Al_5O_{12}$) composite were studied experimentally and analytically. The main results are summarized as follows.

(1) The measured temperature-dependences of Young's modulus and coefficient of thermal expansion in the temperature range of room temperature to 1800K were accounted for by the analytical results based on the finite element analysis, Hashin-Strikman's model and Cribb's model in which the temperature-dependences of the constituents (Al_2O_3 and YAG) were input for calculation. These properties are not sensitive to the microstructure (shape, size and spatial distribution of the constituents of Al_2O_3 and YAG) but are primarily dependent on volume fraction of the constituents.

(2) The temperature- and displacement speed- dependence of the ultimate strength, flow stress in the plastic deformation region, notched strength and fracture toughness in the temperature range of room temperature to 2023K was revealed experimentally by the bending and tensile tests for unnotched and notched specimens. The experimentally observed increase in the notched strength with increasing temperature and decreasing displacement speed up to the maximum value was accounted for by the increase in plastic zone size ahead of the notch and the observed decrease in notched strength with further increasing temperature and decreasing displacement speed was accounted for by the decrease in the stress carrying capacity of the yielded ligament based on the finite element analysis.

(3) The strain rate-dependence of the stress-strain curve and flow stress at 1873 K of the composite stressed in compression parallel ($0\ ^{o}$) and perpendicular ($90\ ^{o}$) to the solidification direction were analyzed by the finite element-, isostrain- and isostress-models. The experimentally measured stress-strain curve and flow stress in the steady state plastic deformation region of the composite was described by a finite element model. The influence of the microstructural deviation of Al_2O_3 and YAG from the theoretical isostrain and isostress models on the overall flow stress was revealed based on the calculated spatial distribution of the equivalent strain rate by the finite element analysis

Rererences

[1] Sorrell, C. C.; Beratan, H. R.; Brad, R. C.; Stubican V. S. Directional solidification of (Ti, Zr) carbide-(Ti, Zr) diboride eutectics, *J. Amer. Ceram. Soc.* 1984, vol.67, pp.190-194.

[2] Sorrell, C. C.; Stubican V. S.; Bradt, R. C. Mechanical-properties of ZrC-ZrB_2 and ZrC-TiB_2 directionally solidified eutectics. *J. Amer. Ceram. Soc.* 1986, vol.69, pp.317-321.

[3] Mazerolles, L.; Michel, D.; Portier, R. Microstructure and mechanical-behavior of Al_2O_3-ZrO_2(Y_2O_3). *J. Physique*, 1986, vol.47 (C-1), pp.335-339.

[4] Mazerolles, L.; Michel, D.; Portier R. Interfaces in oriented Al_2O_3-ZrO_2 (Y_2O_3) eutectics. *J. Amer. Ceram. Soc.* 1986, vol.69, pp.252-255.

[5] Echigoya J.; Takabayashi, Y.; Suto H.; Ishigame M. Structure and crystallography of directionally solidified Al_2O_3-ZrO_2-Y_2O_3 eutectic by the floating zone-melting method. *J. Mater. Sci. Lett.* 1986, vol.5, pp.150-152.

[6] Waku, Y.; Nakagawa, N.; Otsubo, H.; Ohsora, Y.; Kohtoku, Y. High temperature properties of unidirectionally solidified Al_2O_3/YAG composites. *J. Japan Inst. Metals*, 1995, vol.59, 71-78.

[7] Waku, Y.; Nakagawa, N.; Wakamoto, T.; Otsubo, H.; Shimizu, K.; Kohtoku, Y. Excellent high-temperature properties of YAG matrix composites reinforced with sapphire phases. In *Processing and Fabrication of Advanced Materials IV*, eds. T. S. Srivatsan & J. J. Moore, The Minerals, Metals and Materials Society, Warrendale, PA, 1996, pp.323-339.

[8] Waku, Y.; Otsubo, H.; Nakagawa, N.; Kohtoku, Y. Sapphire matrix composites reinforced with single crystal YAG phases. *J. Mater. Sci.* 1996, vol.31, pp.4663-4670.

[9] Yang, J. M.; Jeng, S. M.; Chang, S. Y. Fracture behavior of directionally solidified $Y_3Al_{15}O_{12}/Al_2O_3$ eutectic fiber. *J. Amer. Ceram. Soc.* 1996, vol.79, pp.1218-1222.

[10] Waku, Y.; Nakagawa, N.; Otsubo, H.; Shimizu, K.; Kohtoku, Y. A ductile ceramic eutectic composite with high strength at 1873K. *Nature,* 1997, vol.389, pp.49-52.

[11] Waku. Y.; Nakagawa, N.; Wakamoto, T.; Otsubo, H.; Shimizu, K.; Kohtoku, Y. High temperature strength and thermal stability of unidirectionally solidified Al_2O_3/YAG eutectic composite. *J. Mater. Sci.* 1998, vol.33, pp.1217-1224.

[12] Waku, Y.; Nakagawa, N.; Wakamoto, T.; Otsubo, H.; Shimizu, K.; Kohtoku, Y. The creep and thermal stability characteristics of a unidirectionally solidified Al_2O_3/YAG eutectic composite. *J. Mater. Sci.* 1998, vol.33, pp.4843-4951.

[13] Matson, L. E.; Hecht, N. Microstructural stability and mechanical properties of directionally solidified alumina/YAG eutectic monofilaments. *J. Europ. Ceram. Soc.* 1999, vol.19, pp.2487-2501.

[14] Brewer, L. N.; Endler, D. P.; Austin, S.; Dravid, V. P.; Collins, J. M. Interface modification for increased fracture toughness in reaction-formed yttrium aluminum garnet/alumina eutectic composites. J. Mater. Res. 1999, vol.14, pp.3907-3912.

[15] Yoshida, H.; Shimura, K.; Suginohara, S.; Ikuhara, Y.; Sakuma, T.; Nakagawa, N.; Waku, Y. High-temperature deformation in unidirectionally solidified eutectic Al_2O_3-YAG single crystal. *Key Engng. Mater.* 2000, vol.171-174, pp.855-862.

[16] Nakagawa, N.; Waku, Y.; Wakamoto, T.; Otsubo, H.; Shimizu, K.; Kohtoku, Y. High temperature properties and thermal stability of a unidirectionally solidified $Al_2O_3/Er_3Al_5O_{12}$ eutectic composite. *J. Japan Inst. Metals.* 2000, vol.64, pp.101-107.

[17] Echigoya, J.; Hayashi, S.; Obi, Y. Directional solidification and interface structure of $BaTiO_3$-$CoFe_2O_4$ eutectic. *J. Mate.r Sci.* 2000, vol.35, pp.5587-5591.

[18] Orera, V. M.; Merino, R. I.; Pardo, J. A.; Larrea, A.; Pena, J. I.; Gonalez, C.; Poza, P.; Pastor, J. Y.; Llorca, J. Microstructure and physical properties of some oxide eutectic composites processed by directional solidification. Acta Mater. 2000, vol.48, pp.4683-4689.

[19] Pastor, J. Y.; Poza, P.; Llorca, J.; Pena, J. I.; Merino, R. I.; Orera, V. M. Mechanical properties of directionally solidified Al_2O_3-$ZrO_2(Y_2O_3)$ eutectics, Mater. Sci; Engn. A. 2001, vol.308, pp.241-249.

[20] Sayir, A.; Farmer, S. C. The effect of the microstructure on mechanical properties of directionally solidified $Al_2O_3/ZrO_2(Y_2O_3)$ eutectic. Acta Mater. 2000, vol.48, pp.4691-4697.

[21] Ochiai, S.; Ueda, T.; Sato, K.; Hojo, M.; Waku, Y.; Nakagawa, N.; Sakata, S.; Mitani, A.; Takahashi, T. Deformation and fracture behavior of Al_2O_3/YAG composite from room temperature to 2023 K. *Comp. Sci. Technol.* 2001, vol.61, pp.2117-2128.

[22] Ochiai, S.; Ueda, T.; Sato, K.; Hojo, M.; Waku, Y.; Sakata, S.; Mitani, A.; Takahashi, T.; Nakagawa, N. Elastic modulus and coefficient of thermal expansion of Al_2O_3/YAG composite at ultra high temperatures. *Mater. Sci. Res. Int. Special Technical Publication-2*, Materials Science Research Institute, Kyoto, Japan, 2001, pp. 281-285.

[23] Frazer, C. S.; Dickey E. C. ; Sayir, A. Crystallographic texture and orientation variants in Al_2O_3-$Y_3Al_5O_{12}$ directionally solidified eutectic crystals. *J. Crys. Growth.* 2001, vol.233, pp.187-195.

[24] Harlan, N. R.; Merino, R. I.; Pena, J. I.; Larrea, A.; Orera, V. M.; Gonzalez, C.; Poza, P.; Llorca, J. Phase distribution and residual stresses in melt-grown Al_2O_3-$ZrO_2(Y_2O_3)$ eutectics. *J. Amer. Ceram. Soc.* 2002, vol.85, pp.2025-2032.

[25] Brewer, L. N.; Dravid, V. P.; Dhalenne, G..; Velazquez, M. Solid-solution directionally solidified oxide composites: Part I. Eutectic growth and characterization. *J. Mater. Res.* 2002, vol.17, pp.760-767.

[26] Andreeta, E. R. M.; Andreeta, M. R. B.; Hernades, A. C. Laser heated pedestal growth of Al_2O_3/$GdAlO_3$ eutectic fibers. *J. Crys. Growth.* 2002, vol.234, pp.782-785.

[27] Fernandez, J. M.; Sayir, A.; Farmer, S. C. High temperature creep deformation of directionally solidified Al_2O_3/$Er_3Al_5O_{12}$. *Acta Mater.* 2003, vol.51, pp.1705-1720.

[28] Harada, H.; Suzuki, T.; Hirano, K.; Waku Y. Ultra-high temperature compressive creep behavior of an in-situ Al_2O_3 single-crystal/YAG eutectic composite. *J. Europ. Ceram. Soc.* 2004, vol.24, pp. 2215-2222.

[29] Maton, L. E.; Hecht, N. Creep of directionally solidified alumina/YAG eutectic monofilaments. *J. Europ. Ceram. Soc.* 2005, vol.25, pp.1225-1239.

[30] Ochiai, S.; Sakai, Y.; Sato, K.; Tanaka, M.; Hojo, M.; Okuda, H.; Waku, Y.; Nakagawa, N.; Sakata, S.; Mitani, A.; Takahashi, T. Fracture Characteristics of Al_2O_3/YAG Composite at Room Temperature to 2023K. *J. Europ. Ceram. Soc.* 2005, vol.25, pp.1241-1249.

[31] Hirano, K. Application of eutectic composites to gas turbine system and fundamental fracture properties up to 1700 °C. *J. Europ. Ceram. Soc.* 2005, vol.25, pp.1191-1199.

[32] Nakagawa, N.; Ohtsubo, H.; Mitani, A.; Shimizu, K.; Waku, Y. High temperature strength and thermal stability for melt growth composite. *J. Europ. Ceram. Soc.* 2005, vol.25, pp.1251-1257.

[33] LLorca, J.; Orera, V. M. Directionally solidified eutectic ceramic oxides, *Prog. Mater. Sci.*, 2006, vol.51, pp.711-808.

[34] Ochiai, S.; Sakai, Y.; Kuhara, K.; Iwamoto, S.; Sha, J. J.; Okuda, H.; Tanaka, M.; Hojo, M.; Waku, Y.; Nakagawa, N.; Sakata, S.; Mitani, A.; Sato, M.; Ishikawa, T. Analytical modeling of stress-strain behavior at 1873K of alumina/YAG composite compressed parallel and perpendicular to the solidification direction. *Comp. Sci. Tecnol.* 2007, vol.67, pp. 270-277.

[35] Simmons, G.; H. Wang, H. *Single Crystal elastic constants and calculated aggregate properties; A Handbook*. M.I.T. Press, Massachusetts, USA, 1971, p.328.

[36] Alton, W. J.; Barlow, A. J. Temperature dependence of the elastic constants of yttrium aluminum garnet. *J. Appl. Phys.* 1967, vol.38, pp.3023-3024.

[37] Touloukian, Y. S.; Kirby, R. K.; Taylor, R. E.; Lee, T. Y. R. *Thermophysical properties of matter Vol.13, Thermal expansion, Nonmetallic solids,* Plenum, New York, USA, 1977, pp.176-193.

[38] Gupta, T. K.; Valentich, J. Thermal expansion of yttrium aluminum garnet. *J. Amer. Ceram. Soc.* 1971, vol.54, pp.335-336.

[39] Hashin, Z.; Shtrikman, S. A variational approach to the theory of the elastic behavior of multiphase materials. *J. Mech. Phys. Solids.* 1963, vol.11, pp.127-140.

[40] Cribb, J. L. Shrinkage and thermal expansion of a two phase material. *Nature.* 1968, vol.220, pp.576-577.

[41] Cannon, W. R.; Lagdon, T. G.. Review: Creep of Ceramics. *J Mater Sci.* 1983, vol.18, pp.1-50.

[42] Davidge, R. W. *Mechanical Behavior of Ceramics*. Cambridge University Press, UK, 1979, pp. 64-74.

[43] Japan Society for industrial standard method to estimate fracture toughness of fine ceramics at high temperature, JIS R 1617, 1994 (in Japanese)

[44] Okamura, H. *Introduction to Linear Elastic Fracture Mechanics* (in Japanese), Baifu-kan, Tokyo, 1975, pp.216-222.

[45] Waynant, R.; Ediger, M. *Electro-optics Handbook.* Mcgraw-Hill Inc. New York, 1994, pp.11.13-11.23.

[46] Kotchick, D. M.; Tressler, R. E. Deformation behavior of sapphire via the prismatic slip system. *J. Amer. Ceram. Soc.* 1980, vol.63, pp.429-434.

[47] Corman, G., S. Creep of yttrium aluminium garnet single crystals. *J. Mater. Sci. Lett.* 1993, vol.12, pp. 379-382.

[48] Kelly, A.; Tyson, W. R. Tensile properties of fibre-reinforced metals: copper/tungsten and copper/molybdenum. *J. Mech. Phys. Solids.* 1965, vol.13, pp.329-350.

[49] Ochiai, S. *Mechanical Properties of Metallic Composites*, Marcel Dekker, New York, USA, 1994. pp.453-510.

In: Ceramic Materials Research Trends
Editor: Paul B. Lin, pp. 111-137
ISBN: 1-60021-769-9

Chapter 4

ADVANCES IN OXYNITRIDE PEROVSKITES: SYNTHESIS AND PHOTOCATALYTIC APPLICATIONS

Rosiana Aguiar[a*], Yungi Lee[b], Kazunari Domen[b], Andreas Kalytta[a], Dmitry Logvinovich[c], Anke Weidenkaff[c], Armin Reller[a] and Stefan G. Ebbinghaus[a]

[a] Lehrstuhl für Festkörperchemie, Institut für Physik, Universität Augsburg, Universitätsstrasse 1, D-86159, Augsburg, Germany.
[b] The University of Tokyo, Department of Chemical System Engineering, School of Engineering, 7-3-1 Hongo, Bunkyo-ku, Tokyo 113-8656, Japan.
[c] Empa - Swiss Federal Laboratories for Materials Testing and Research, Ueberlandstrasse 129, CH-8600, Duebendorf, Switzerland.

Dedicated to Sir John Meurig Thomas on the Occasion of his 75th Birthday

Abstract

Perovskite-type oxides display a very large range of structural, physical and chemical properties. Due to the flexibility of the perovskite structure it is possible to make substitutions with cations of different size, formal charge and electronegativity, forming solid solutions with gradual or sudden changes in their properties. While cationic substitutions in perovskites have been intensively studied for decades comparatively little is know about substitutions in the anion lattice. Partial replacement of oxygen by nitrogen leads to oxynitrides. The lower electronegativity of nitrogen compared to oxygen results in an increase in the covalence of bonding when nitrogen is inserted in the anionic sublattice. In many cases the optical band gap is decreased to the visible region leading to intensively colored compounds. As a consequence of this optical property, oxynitride perovskites can be used as environmental-friendly pigments and visible light activated photocatalysts. In this work we present different synthesis methods to produced oxynitride perovskites. Starting from the commonly used solid-state reaction, we turn to soft chemistry methods like the polyol synthesis and finally to the spray

* E-mail address: rosiana.aguiar@physik.uni-augsburg.de. Tel: + 49 (0) 821 598 3026. Fax: + 49 (0) 821 598 3002

pyrolysis. The physical properties and morphologies of the compounds prepared by the various methods are discussed. Especially surface area and format aspects are compared among the samples. For this purpose, SEM micrographs are used to illustrate the different particle sizes and shapes obtained by each technique. Focusing on the oxynitrides, their optical property and photocatalytic potential were studied. A comparison between the synthesized oxynitrides and TiO_2 is presented. As known, TiO_2 is the most common photocatalyst because it has the advantages to be non-toxic, cheap, chemically/biologically inert, photo-catalytically stable and easy to produce. The big inconvenience of titanium dioxide is that it only absorbs in the ultra violet region. In contrast the oxynitride perovskites absorb in an adjustable range of the visible light and several studies have already demonstrated that some of them show considerable photocatalytic activity. Two different methods to verify the photocatalytic potential (i.e. decomposition of simple organic compound monitored by in-situ infra-red spectroscopy and the water splitting reaction) of selected samples are explained in detail.

1. Introduction

1.1. Perovskites and Oxynitrides

The name "perovskite" describes a structural family of compounds with the composition ABX_3. A denotes a relatively large cation of usually low charge, while the B cations have a smaller radius and in most cases a higher charge. X is an anion like O^{2-} or F^-. A typical perovskite is $SrTiO_3$, which represents the ideal cubic structure (space group $Pm\bar{3}m$) with a = 3.905 Å and can be used to describe the simple ion arrangement in these materials. The Ti atoms are located on the corners of the unit cell, and the Sr atom at the center of the cube. Oxygen ions are placed at the centers of the 12 cube edges, giving corner-sharing TiO_6 octahedra, with 90° angles. The coordination numbers of Sr, Ti and O are 12, 6 and 6, respectively [1]. For the latter, it is assumed that two Ti and four Sr, with respective distances of a/2 and √2*a/2, coordinate each oxygen atom. This atomic arrangement is illustrated in Fig. 1. The left part shows an unit cell with the large A cation in the center. On the right side of the figure the neighboring oxygen ions are included to show the three-dimensional arrangement of corner sharing octahedra.

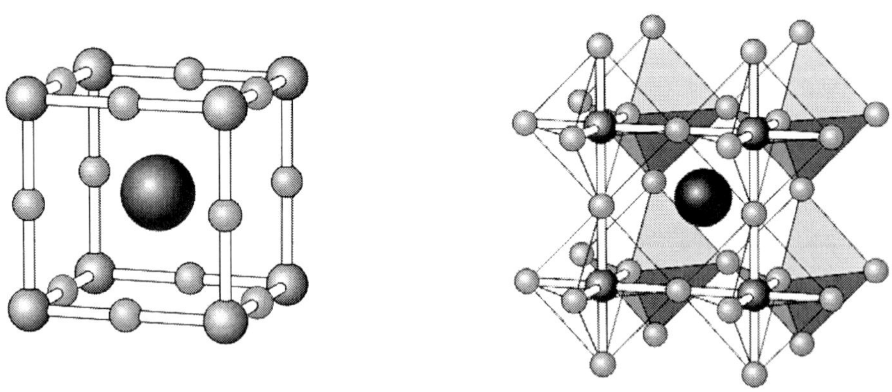

Figure 1. Ideal perovskite structure of $SrTiO_3$. Sr: blue sphere, Ti: gray spheres, O: red spheres. Left: Perovskite unit cell. Right: Perovskite structure composed of oxygen octahedra.

The perovskite structure can support a large variety of cations on the A and B positions. Around 90% of the natural metallic elements of the Periodic Table are known to be stable in a perovskite type oxide structure. This structure configuration also gives the possibility of synthesizing multicomponent perovskites by partial substitution of cations in the A and B site positions, giving rise to solid solutions with the general formula of $A_{1-x}A'_xB_{1-y}B'_yO_3$ [2]. The tolerance factor ($t = (r_A + r_O) / \sqrt{2}(r_B + r_O)$, where r_A, r_B, and r_O are the ionic radii of the A- and B-type cations and O^{2-}, respectively) can vary between 0.75 and 1. For values significantly smaller than unity, structural distortions are observed, resulting from various possible rotation schemes of the BO_6-octahedra. Because of this flexibility to incorporate so different cations in its structure, perovskites have the most variable physical/chemical properties. As example, $(Ba,Sr)TiO_3$ and $Pb(Zr,Ti)O_3$ exhibit outstanding dielectric properties and are suited for capacitor applications, particularly decoupling and tunable microwave capacitors [3]. $(La,Sr)(Ga,Mg)O_3$ is used in the electrolyte of solid oxide fuel cells (SOFC) because of its oxide ion conductivity [4]. $LaCoO_3$ and $LaMnO_{3+\delta}$ have catalytically properties and are able to destroy chlorinated volatile organic compounds. These substances are the main pollutants in the low atmosphere of the cities and responsible for the destruction of the ozone layer in the stratosphere [5]. $YBa_2Cu_3O_{7-\delta}$ (YBCO) has an oxygen defect perovskite structure and was the first material to become superconducting above 77 K, the boiling point of nitrogen [6]. This list of examples could easily be extended and a whole book would not be enough to describe all the oxide perovskites and their properties and applications.

While cationic substitutions have been thoroughly studied for many years, substitutions in the anion sublattice are by far less well examined. There are some reports on carbides [7-9], halides [10-12] and nitrides [13;14]. In this chapter we focus on a special field of anionic substitution, namely the partial replacement of oxygen by nitrogen, leading to the perovskite oxynitrides. Nitrogen is an appropriate candidate for replacing oxygen in ternary oxides because both anions have close electronic configurations and ionic radii [15]. When O^{2-} is substituted for N^{3-} in ternary oxides three possible mechanisms can compensate the additional negative charge of the nitride ion to give charge neutrality in the specific systems:

(i) Change of the oxidation state of the B-site cation e.g. $LaV^{3+}O_3 \rightarrow LaV^{3+/4+}O_{3-x}N_x$ [16].
(ii) Simultaneous A- or B- sites substitution with cations of higher valence (cross-substitution) e.g $Sr^{2+}TiO_3 \rightarrow La^{3+}{}_xSr^{2+}{}_{1-x}TiO_{3-x}N_x$ [17] or $CaZrO_3 \rightarrow CaZr_xTa_{1-x}O_{2+x}N_{1-x}$ [18].
(iii) Reduction of the number of anions as in $LaTiO_{3.5} \rightarrow LaTiO_2N$ or $SrTaO_{3.5} \rightarrow SrTaO_2N$ [19].

These possibilities of charge compensation can be accompanied by changes in the crystallographic and electronic structure of the comprising materials. For example the ammonolysis of the layered perovskite-related oxide $Sr_2Nb_2O_7$, which possesses a (pseudo-) 2D structure, leads to the 3D perovskite-type oxynitride $SrNbO_2N$ [20]. The electronic structure of the oxynitrides is influenced by the different degree of admixture of the nitrogen 2p-orbitals to the frontier molecular orbitals. In principal, the band structure of the formed oxynitrides is similar to the one of early transition metal perovskite type oxides ($KTaO_3$, $SrTiO_3$ etc.): the top of the valence band is represented by the 2p-orbitals of the anions and the bottom of the conduction band is formed by the d-orbitals of the transition metals. Despite of this general similarity, the incorporation of N^{3-} results in a pronounced change of the energy of the frontier orbitals. Since the 2p-orbitals of nitrogen have a higher energy level than the 2p-orbitals of oxygen, nitrogen insertion into the oxygen sublattice is leading to a band gap reduction. The corresponding band gap schemes for the oxide and oxynitride perovskites are illustrated in Fig. 2. With the shrinking of the band gap, the oxynitrides start to absorb in the visible light range and show bright colors. This intrinsic property of the oxynitride perovskite makes them good candidates to be employed as environmental friendly pigments [21;22] and photocatalysts [23;24].

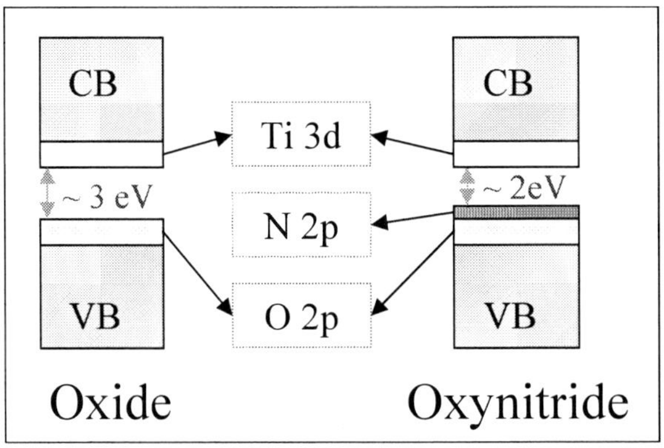

Figure 2. Scheme representing the oxide and oxynitride band structures. The oxynitride band gap decreases because of the N 2p donor level formation.

We have lately reported on possibilities to tailor the band gaps of oxynitride perovskites, and in turn adjust their colors, by conscious cationic substitution. In oxynitrides of the composition ABO_2N the band gap can be decreased by increasing the crystallographic symmetry (i.e. adjusting the size of the A cation) or by increasing the electronegativity of the B cation [22]. An example are the tantalum oxynitrides $ATaO_2N$ (A = Ca, Sr and Ba), in which the crystallographic symmetry increases with ionic radii of the A cation ($Ca^{+2} < Sr^{+2} < Ba^{+2}$). The compounds have the following crystal system, color and band gap: $CaTaO_2N$ orthorhombic, yellow, 2.4 eV; $SrTaO_2N$ tetragonal, orange, 2.1 eV and $BaTaO_2N$ cubic, red, 1.8 eV. While in these series the color is tuned by changing the A cation, it can also be modified by substitutions on the B position, in the way that the band gap is decreased by incorporation of cations with higher electronegativity. In the case of $SrBO_2N$ (B = Ta and Nb) niobium is more electronegative than tantalum. In turn $SrTaO_2N$ is orange while $SrNbO_2N$ is brown. These examples reveal that knowing the basic principles of cation substitutions and their consequences on the band gap opens the possibility to produce oxynitride with the desired color. This is illustrated in Fig. 3 and 4, which show the colors of different compounds produced as polycrystalline powders and thin films, respectively. The samples shown were prepared by thermal ammonolysis, which is the common way to insert nitrogen into ternary oxides. This method consists of the reaction between ammonia gas and solid oxides at high temperature (between 700-1100°C) and is explained in more detail in the next sections.

Figure 3. The color spectrum of different oxynitride powder compounds. Top line (from left to right): TiO_2, TiO_xN_y, $CaTaO_2N$, Sr_2TaO_3N, $SrTaO_2N$. Bottom line: $NdTaON_2$, $BaTaO_2N$, $LaTiO_2N$, $YbTaO_{2.5}N$ and $BaNbO_2N$.

Figure 4. Picture of colored $La_xSr_{(1-x)}Ti(O,N)_3$ thin films produced by spin coating deposition. From left to right: x = 1, 0.75, 0.5, 0.25 and 0 [25].

1.2. Photocatalysts

A photocatalyst is a substance that demonstrates catalytic properties under irradiation. The photocatalyst absorbs light by electronic excitations and starts a chemical reaction, based on oxidation-reduction reactions without being decomposed itself. Photosynthesis in plants is a well-known example of such a photocatalytic reaction, in which chlorophyll acts as the photocatalyst transforming water and carbon dioxide into glucose in the presence of sunlight. Photocatalysts can be extensively used in water and air purification. They have attractive applications in deodorization and sterilization. Photocatalysts can be applied indoor as well as outdoor. Some examples of their application are [26]:

- Exterior of buildings, guardrails, traffic signs, tunnels, railways, and public transportation: Elimination of toxic gases and/or contaminant molecules (as volatile organic compounds, NO_x, SO_x) coming out from vehicles exhaust gas.
- Hospitals and medical facilities: Destruction of microorganisms, as bacteria, germs, viruses and fungi, resulting in prevention of in-hospital infections.
- Hotels, restaurants and closed ambient in general: Elimination of unpleasant odor-causing organic compounds as cigarette smoke, mold, domestic smells and pets-odor. Prevention of sick-building syndrome, environmental hygienization, decomposition of allergens [27].
- Water treatment: Decontamination and purification of wastewater.

A photocatalyst has to be a semiconductor material with band-gap energy varying from circa 1.1 to 3.7 eV (e.g. TiO_2, Si, ZnO, Fe_2O_3, SiC, CdS, WO_3, ZnS, etc). A filled valence band and an empty conductor band characterize their electronic structure. When a photocatalyst is illuminated by highly energetic light (photons with energy hν > band gap) an electron jumps from the valence band to the conductor band. With this phenomenon an electron – hole pair is formed [28].

$$\text{Photocatalyst} + h\nu \rightarrow e^- + h^+ \quad (1)$$

The electrons and holes can move in the respective bands but most of them simply recombine producing thermal energy. The lifetime of an e^--h^+ pair is only a few nanoseconds, but this is enough to promote redox reactions. The electron/hole pairs near the surface can be involved in reactions.

The generated holes have strong oxidative character and produce hydroxyl radicals by reacting with water at the surface.

$$h^+ + H_2O \rightarrow OH^{\bullet} + H^+ \quad (2)$$

The generated hydroxyl radicals can cause oxidative reactions in organic compounds.

On the other hand, the generated electrons can reduce oxygen molecules producing superoxide free radicals.

$$e^- + O_2 \rightarrow O_2^- \quad (3)$$

The electrons can also be „donated“ to a metal ion deposited on the surface of the catalyst, which in turn is reduced to a lower valence state. Because organic matter (bacteria, viruses, germs, and odor molecules) is usually more oxidative than water, holes are more likely to cause oxidation of organic compounds present in the aqueous solutions. Additionally, the recombination rate of carriers (holes and electrons) decreases when the concentration of organic matter is higher. Organic compounds are thus decomposed and eventually turn into carbon dioxide and water. A scheme for the photocatalytic reaction is present in Figure 5. A general photocatalyzed oxidation of a generic chlorinated hydrocarbon is present in equation (4) [29]:

$$C_xH_yCl_z + [x+(y-z/4)]\,O_2 \xrightarrow{\text{Photocatalyst} + h\nu} x\,CO_2 + z\,HCl + (y-z/2)\,H \quad (4)$$

TiO_2 is the most popular photocatalyst employed. The possibility of photoelectrolysis was demonstrated for the first time in 1969, when A. Fujishima [30] obtained a single crystal of titanium oxide and used it as an electrode. A closed circuit using titanium oxide and platinum black as counter electrode was exposed to the light of a xenon lamp. When the surface of the TiO_2 electrodes was irradiated with light consisting of wavelengths smaller than its band gap (415 nm = 3.0 eV), gas bubbles evolved from both electrodes. These gases were found to be oxygen evolving from the TiO_2 electrode and hydrogen from the Pt side. Thus it was proved that water could be decomposed into oxygen and hydrogen under UV light irradiation without application of an external voltage. Moreover, the titanium oxide had not been consumed in the reaction even after several days of continuous photo irradiation.

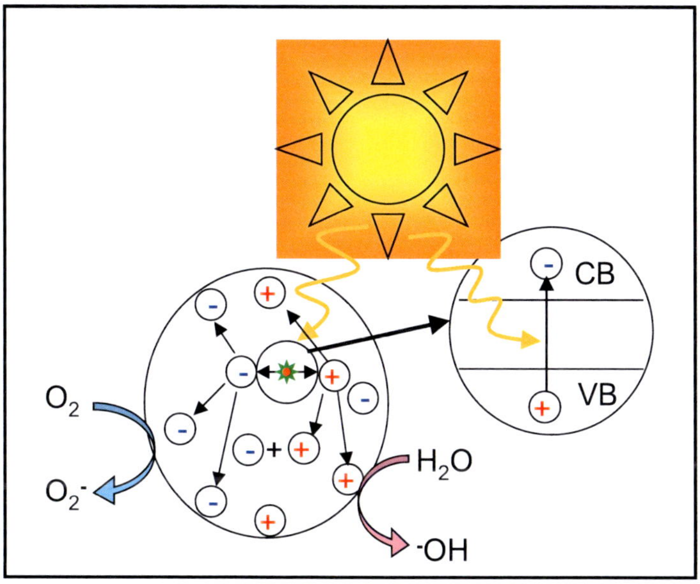

Figure 5. Scheme of photocatalytic reaction.

Generally speaking, a photocatalyst should posses the following characteristics:

- Semiconducting
- Chemically and biologically inert
- Photocatalytically stable
- Efficiently catalyze reactions
- Activated by sunlight
- Preferable it should be non-toxic, easy to produce and cheap

TiO_2 fulfill almost all of these requisites but it is only activated by a small part of the sunlight, namely the ultra violet part (UV < 400nm). Looking at the sunlight spectrum depicted in Fig. 6, one can visualize how little of the whole sunlight radiation that reaches the earth surface can actually be used by TiO_2 in its photocatalytic reactions. Only about 3-5% of the total sunlight can photoactivate TiO_2 to generate charge carries. It would be much more efficient to use more of the light that we get for free. One recent development is the introduction of nitrogen in the TiO_2 lattice. As result, the band gap reduced from 3.2 eV to 2.4 eV, and the titanium oxynitride starts to absorb parts of the visible light (λ< 517nm) what gives it a yellowish color [31;32]. With this innovative idea one door is open to tailoring the band gap of TiO_2.

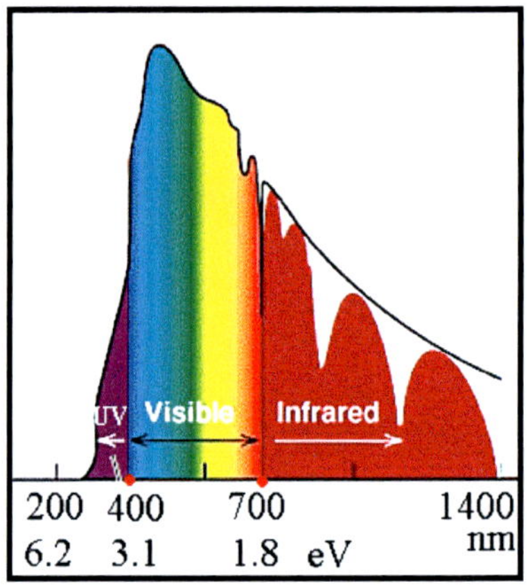

Figure 6. Solar spectrum, TiO_2 uses only the ultra violet part (λ < 400 nm) in its photocatalysis reactions.

In the last years, studies on the possibility to apply perovskites as photocatalyst have increased. For example, H. Kato and A. Kudo [33;34] have studied water splitting into H_2 and O_2 for various tantalates and niobates, such as $LiTaO_3$, $NaTaO_3$, $SrNbO_{3.5}$, $KTaO_3$, $SrTaO_{3.5}$, $CaTa_2O_6$, etc. M. Uno *et al* [35] have investigated the titanate series, as $LaTiO_{3.5}$, $SmTiO_{3.5}$, $GdTiO_{3.5}$, $YbTiO_{3.5}$ and others. S. Ikeda et al examined $K_4Nb_6O_{17}$ [36] and K. Domen et al analyzed $SrTiO_3$ [37]. The problem of these kinds of materials is the same of TiO_2, as they have high band gap energies and only absorb in the ultra violet range.

Oxynitride perovskites, on the other hand, absorb a much broader part of the sunlight as they are in the most cases colored. Since nitrogen is less electronegative the metal-nitrogen bond has a more covalent character than the corresponding metal-oxygen bond. This leads to a decrease of the electronic band gap value [38] and normally shifts their absorption edge to the visible light region. They are good candidates to be used as photocatalysts under visible light, e.g. sunlight. Some studies have been already done on this field. In 2002 Kasahara *et al* [23;39] investigated $LaTiO_2N$ as a photocatalytic active compound in the hydrogen/oxygen production from water-methanol mixtures under visible light. In 2005 Ji *et al* [24] have studied N-doped $SrNbO_{3.5}$ as photocatalyst. Rb(Ca,La)NbO:N compounds were also investigated, but because they are unstable in water, their photocatalytic activity was quite poor [40]. In this chapter we are reporting on our experiments to measure the catalytic activity of some selected oxynitride perovskites and a comparison of their activity to get the first impression of how good they could be as photocatalyst under visible light is presented.

2. Synthesis

In this section, different techniques to synthesize oxide powders as precursors for the corresponding oxynitrides are explained. A comparison between them, showing benefits and inconveniences, is developed. The obtained powers morphology, size and surface area are given for each used method.

2.1. Solid State Reaction

Solid state reactions are the most often used methods to produce ternary oxide powder because they are quite simple and easy to perform. The oxide/carbonate educts are well ground using mortar and pestle or ball mills. Normally some liquid (water, alcohol) is added to help the mixing. Afterwards, the powder is submitted to high temperature to trigger the ionic diffusion. In general, with this method the obtained particles are rather big (several μm in diameter), show a low surface area, good crystallinity and in many case a poor homogeneity, which can be improved by additional regrinding/heating cycles. Quite high temperatures and reaction times are necessary for a complete reaction. Sometimes flux materials are added to support the ion diffusion and to increase the crystallinity of the product. These fluxes can be chlorides, hydroxides or any other low melting compound that is not incorporated in the produced oxide and can be easy washed out after the reaction. Figure 7 shows as a typical example a $CaTaO_{3.5}$ powder that was produced by solid-state reaction at 1000°C within 24h, without adding flux. The average particle size is 160 μm.

Figure 7. $CaTaO_{3.5}$ produced by solid-state reaction.

2.2. Soft Chemistry Synthesis

The term Soft Chemistry (or *Chimie Douce*) summarizes various methods to prepare inorganic solids at mild reaction conditions. These reactions are nowadays also commonly used methods to produce mixed oxides. The basic approach consists of dissolving salts (e.g. carbonates, nitrates, or chlorates) of the corresponding cations in (usually) aqueous solutions together with some organic compound. The used organic molecules can be citric acid, glucose, etc. They are used to stabilize the dissolved cations in solution and to avoid their precipitation for example as hydroxides. The reaction of the salts with the organic compounds to give stable molecules is schematically shown in Figure 8. In this case citric acid and ethylene glycol were used as stabilizers. The solution has to be heated at around 70°C and kept under constant stirring until completely dissolution of the cationic precursors. Afterward, the temperature is increased up to ca. 150 °C and after evaporation of the solvent, the organic component forms a stable gel, embedding the cations. In a subsequent reaction step the organic matrix is burned out at around 600°C. Normally an amorphous powder is obtained, which contains the different cations already mixed on the atomic scale. To form the crystalline product phase heat treatments at much lower temperatures than for the solid-state reactions are required. The advantage of this technique is that a highly homogeneous powder is obtained. Normally the resulting oxides have high surface areas and small particles due to the use of organic matrix and the lower reaction temperatures, which prevent the particles from sintering. Figure 9 shows typical particle morphologies obtained by soft chemistry technique. The two compounds $SrTaO_2N$ and $SrNbO_2N$ were produced using water/citric acid solutions. The organic matrix was burned out at 600°C and the crystalline phases were obtained after heating the samples at 950°C under flowing NH_3.

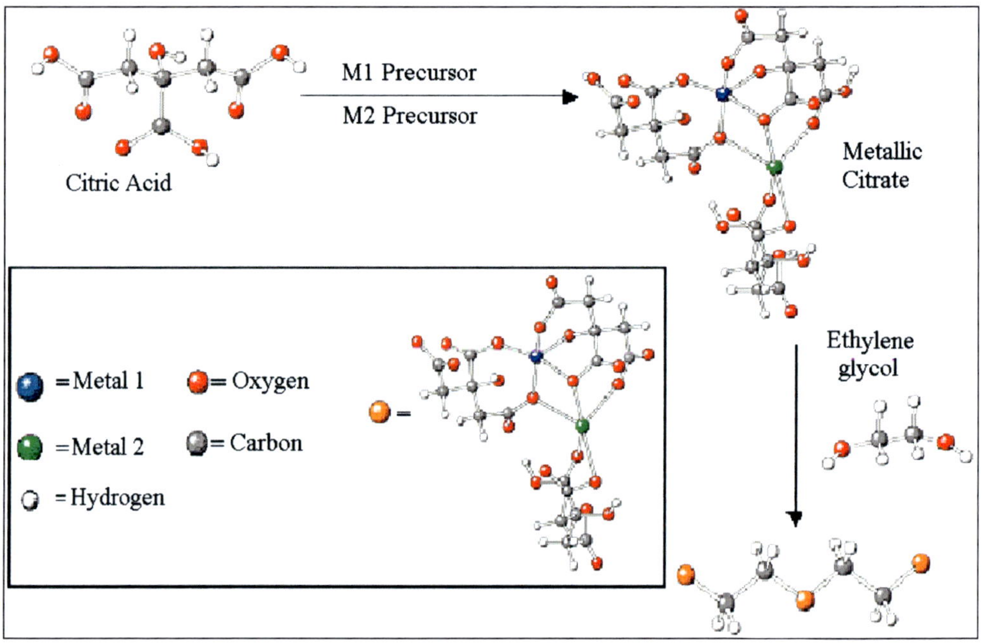

Figure 8. Scheme demonstrating how organic compounds stabilize the metallic cations in solution to avoid precipitation [41].

Figure 9. (a) $SrTaO_2N$ and (b) $SrNbO_2N$ produced by soft chemistry technique.

2.3. Polyol Method

Although the polyol synthesis has been widely used for the preparation of metals and binary oxides, it is a rather new technique to prepare ternary oxides [42;43]. The polyol method consists of dissolving metal salts in a chelating polyalcohol („polyol") at elevated temperatures. A hydrolysis agent is used to initiate the cations co-precipitation producing (amorphous) oxide nanoparticles. The experiment is carried out in a closed reaction flask attached to a reflux condenser. In some cases flowing N_2 is utilized to avoid early precipitation trigged by air humidity. Figure 10 shows a picture of the used set-up. The polyol approach is a good technique to produce nanoparticles with fine size control, because the alcohol acts as a surface-capping agent avoiding particles agglomeration, as illustrated in Figure 11. On the other hand, it is quite difficult to adjust the final composition of the precipitated oxide when complex stoichiometries are desired. The reason is that in these cases several cationic precursors have to be added. Since these precursors might differ in their hydrolysis behavior, the composition of the precipitate does not always match the one of the solution. It is also tricky to find the appropriate hydrolysis agent. In some cases the atmospheric moisture is already sufficient, while there are other cases in which not only water but also HNO_3 or KOH have to be added.

Figure 10. Picture of the polyol experiment set-up.

As an example, the synthesis of $LaFeO_3$ is described. Stoichiometric quantities of $La(NO_3)_3*6H_2O$ (p.a. - Fluka) and $Fe(NO_3)_3*9H_2O$ (p.a. - Merck) were dissolved in 30 ml of diethylene glycol (>98% - Fluka). Three different total reagents mass concentration in DEG was tested 0.96%, 1.20% and 2.84%. The solutions were heated, using a silicone oil bath, to 120°C and kept at this temperature under constant stirring for 1 hour, until clear solutions were obtained. Subsequently 0.4 mL water was added and the temperature was increased to 180°C to initiate the hydrolysis. The solutions became brownish, and after approximately 6 hours the particles precipitation was completed. The obtained suspensions were centrifuged to separate the oxide particles from the diethylene glycol. The precipitates were washed with acetone several times and dried at 90°C. Amorphous yellow-brownish powders were produced. These powders were heated at 700°C during 5h to obtain the crystalline particles.

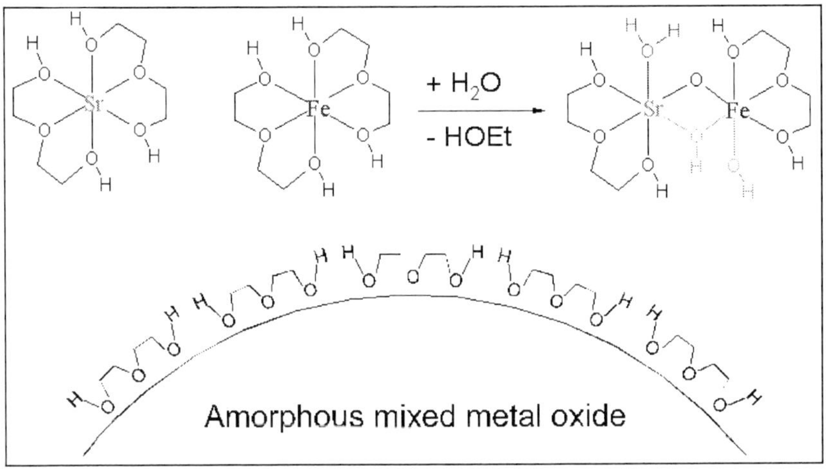

Figure 11. Polyol synthesis scheme. The metal salts are dissolved in the chelating polyalcohol. Afterwards controlled co-precipitation by hydrolysis occurs, forming the amorphous oxide nanoparticles. The alcohol molecules surround the particles, avoiding their agglomeration.

Figure 12. $LaFeO_3$ synthesized by polyol technique using two different total reagent mass concentrations: (a) 0.96% and (b) 2.84%.

The initial reagent concentration showed to have a strong influence on the particles size, as can be visualized in Figure 12. The higher the reagents concentration in DEG, the bigger particles can be obtained. The average particle sizes are 47 nm, 186 nm and 240 nm in diameter for the samples produced with reagents concentration of 0.96%, 1.20% and 2.84%, respectively. Feldmann [44] had also reported this phenomenon for binary oxides like M_2O_5 (M= V, Nb, Ta). Probably the nucleation step is much slower than particle growth. In this case a higher concentration does not yield significantly more nuclei but bigger particles are formed.

2.4. Spray Pyrolysis

The basic idea of the spray pyrolysis is to use droplets of an aerosol as microreactors. An aqueous solution containing stoichiometric ratios of the desired cations is nebulized, generally by means of an ultra-sonic vapor generator. A flowing gas transports the aerosol through a hot tube reactor [45;46]. After drying of the small droplets the corresponding solid-state reaction takes place, forming the desired phase. These reactions are quite fast. The time it takes for a droplet to pass through a vertical tube furnace kept at around 1000°C (in the order of seconds) usually is enough to complete the reaction. The formed particles are collected at the end of the quartz tube. Small and chemically homogeneous particles are usually obtained. Spray pyrolysis is a good technique to produced large quantities of powders. Figure 13 shows a spray pyrolysis apparatus in detail. There are several parameters to be controlled: gas flow rate, furnace temperature, cations concentration, type of organic additives, etc. Depending on these parameters different powder morphologies can be obtained.

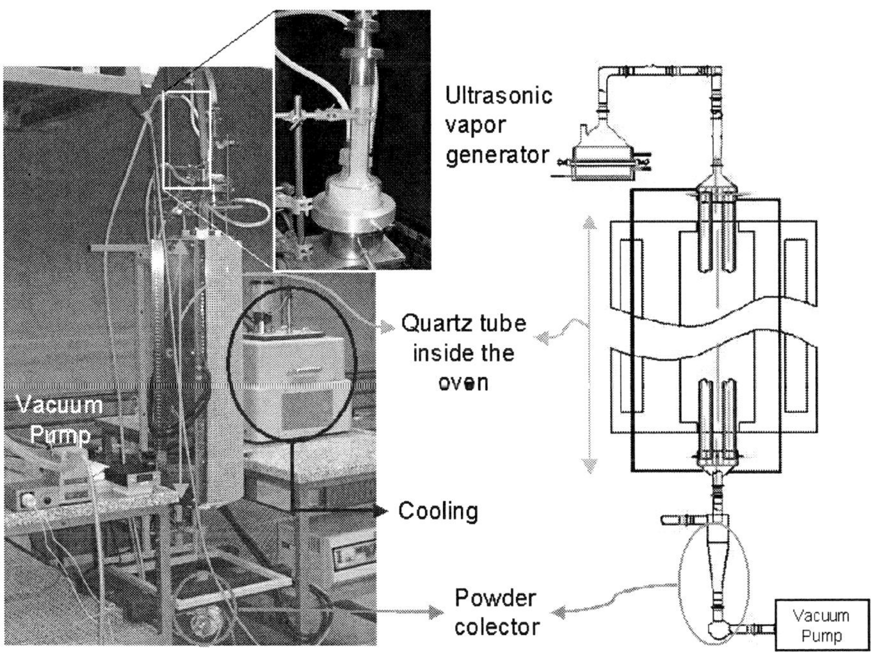

Figure 13. Picture and schematic sketch of spray pyrolysis set-up.

As an example the formation of $SrFeO_{3-x}$ by spray combustion is described. An aqueous solution containing stoichiometric amounts of $Sr(NO_3)_2$ (p.a. – Fluka) and $Fe(NO_3)_3*9H_2O$ (p.a. - Merck) was nebulized, using an ultra-sonic vapor generator, and the aerosol was passed through a tube furnace at 1000°C with a flow of 3 l/min using oxygen as carrier gas. Different alcohols were employed as additives. It was found that the type and concentration of the alcohol controls the oxygen stoichiometry of the samples $SrFeO_{3-x}$. The obtained x values varied between 0.2 and 0.33. Oxygen quantification was done by thermogravimetric analysis under H_2 atmosphere, using a Netzsch thermobalance (STA 409). The results are listed in Table 1. The utilized additives also have some influence on the morphology of the samples, as shown in Figure 14. The sample produced without alcohol has a less homogeneous morphology and bigger particles size distribution compared to the samples produced with 1% butanol.

Table 1. Oxygen quantities in $SrFeO_{3-x}$ depending on the used organic additive.

Water additive	x value (on $SrFeO_{3-x}$)
None	0.20
1% Iso-Propanol	0.20
1% Pentanol	0.28
1% Butanol	0.30
1% Methanol	0.33
1% Ethanol	0.33

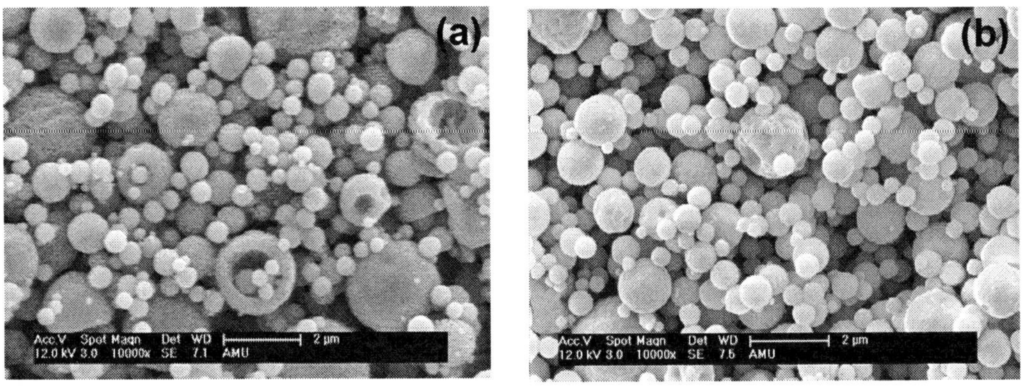

Figure 14. $SrFeO_{3-x}$ prepared (a) with water solution (x = 0.2) and (b) with 1% butanol (x = 0.3).

2.5. Ammonolysis

The oxide powders produced by the different methods were submitted to heat treatment under ammonia atmosphere to obtain the respective oxynitrides. Upon heating, the ammonia partly dissociates in H_2 and N_2, reducing the oxide. Afterwards the compound is progressively enriched by nitrogen. When the B cation is easily reduced, wet ammonia can be used to avoid the oxide decomposition and B_yN_z formation [47].

Depending on the A and B cations and on the preparation method applied to obtain the oxide precursors, different temperatures and reaction times are necessary to produce the oxynitrides. In a typical experiment, the oxide powders are placed in alumina crucibles and ammonolyzed in a tube furnace with a NH_3 flow of 20-150ml/min to result in well-crystallized oxynitrides. Normally bigger particles need higher temperature and longer time to be completely reacted. The ammonolysis parameter used to prepare the diverse oxynitride compounds and their respective colors and surface areas are listed in Table 2.

Table 2. Oxynitride samples used in the photocatalysis experiments

Sample	Synthesis Method	Ammonolysis Temperature (°C)	Ammonolysis Time (h)	Color	Surface Area (m^2/g)
$CaNbO_2N$*°	Soft Chemistry	950	60	Black	9.9
$SrNbO_2N$*	Solid state	950	54	Brown	3.4
$SrNbO_2N$°	Soft Chemistry	950	36	Brown	9.1
$BaNbO_2N$*	Solid state	950	72	Black	1.0
$SrTaO_2N$*°	Soft Chemistry	950	18	Orange	5.8
$BaTaO_2N$*	Solid state	1000	54	Red	2.6
$NdTaO_2N$*°	Soft Chemistry	1000	54	Red	6.4
$LaTiO_2N$°	Solid state	950	126	Brown	3.6
$LaTiO_2N$*°	Soft Chemistry	950	24	Brown	13.7
P25°	-	-	-	White	50

* Used to H_2 and O_2 evolution (see Section 5.2)
° Used to acetone decomposition (see Section 5.3)

Figure 15. $LaTiO_2N$ synthesized by (a) solid state reaction, (b) soft chemistry, (c) polyol and (d) spray pyrolysis.

The ammonolysis at high temperatures has the big disadvantage to cause a sintering of the particles, which results in a reduction of the surface area. Hence it is extremely difficult, not to say impossible, to kept the high surface area obtained for the oxide powders produced by the different synthesis methods described in the last sections. As example, $LaTiO_{3.5}$ produced by Pechini method (soft chemistry technique) shows a surface area of 43 m^2/g, while after ammonolysis this value is decreased to 13.7 m^2/g.

Fig. 15 shows samples of $LaTiO_2N$ prepared by ammonolysis of oxide precursors obtained by different synthesis methods. It can be seen that the oxynitride particles are rather big when produced by solid-state reaction, their average size been 2.6 μm. In contrast, the samples obtained by polyol synthesis have the smallest particles, measuring only 60 nm in average. The particles synthesized by the soft chemistry approach show an average size of 135 nm; while the ones produced by spray pyrolysis have a broad size distribution varying from 300 nm to 3.1 μm with an average of 1.28 μm. The shape of the particles changes considerably, depending on the employed synthesis method. As discussed before, each technique results in powder with typical morphology. Good examples are the particles produced by spray pyrolysis that always have a spherical shape even after ammonolysis.

3. Surface Area

The surface areas of the samples were measured by the BET method [48] using a Quanta Chrome Nova 2000 system. The measurement is based on physical adsorption of gas molecules on a solid surface. The BET equation is:

$$1/v\,[(P\,/\,P_0)-1] = (\mathrm{c}-1/\,v_m\mathrm{c})(\,P\,/\,P_0)+1/\,v_m\mathrm{c} \tag{5}$$

P and P_0 are the equilibrium and the saturation pressure of the adsorbed gas at the temperature of adsorption, v is the adsorbed gas quantity and v_m is the volume of gas adsorbed when the entire adsorbent surface is covered with a complete unimolecular layer. c is the BET constant, which is expressed by:

$$\mathrm{c} = \exp(E_1-E_L/\mathrm{RT}) \tag{6}$$

where E_1 is the heat of adsorption for the first layer, and E_L is that for the second and higher layers and is equal to the heat of liquefaction.

Equation (5) is an adsorption isotherm and can be plotted as a straight line with $1\,/\,v[(P_0\,/\,P)-1]$ on the y-axis and $P\,/\,P_0$ on the x-axis according to experimental results. The linear relationship of this equation is maintained only for low pressures, i.e. roughly in the range of $0.05 < P\,/\,P_0 < 0.35$. The values of the slope and the y-intercept of the line are used to calculate v_m and the BET constant c.

The specific surface area S is calculated using the following equation:

$$\mathrm{S} = (v_m\mathrm{Ns})/\mathrm{M}a \tag{7}$$

Where N is Avogadro's number, s adsorption cross-section, M molecular weight of the adsorbed gas and a is the weight of the solid sample.

As an example the BET adsorption isotherm curve for $LaTiO_2N$, produced by a soft chemistry technique, is plotted in Figure 16. The calculated surface area was 10.85 m^2/g. The measured surface areas for the other samples are listed in Table 2.

It is important to bear in mind that in general catalytic reactions occur only on the surface of the catalyst. As a consequence, a larger surface area of the sample provides more reactive sites and in turn increases the catalytic activity. A comparison between samples, which have the same chemical composition and different surface area values, is given in section 6.3.

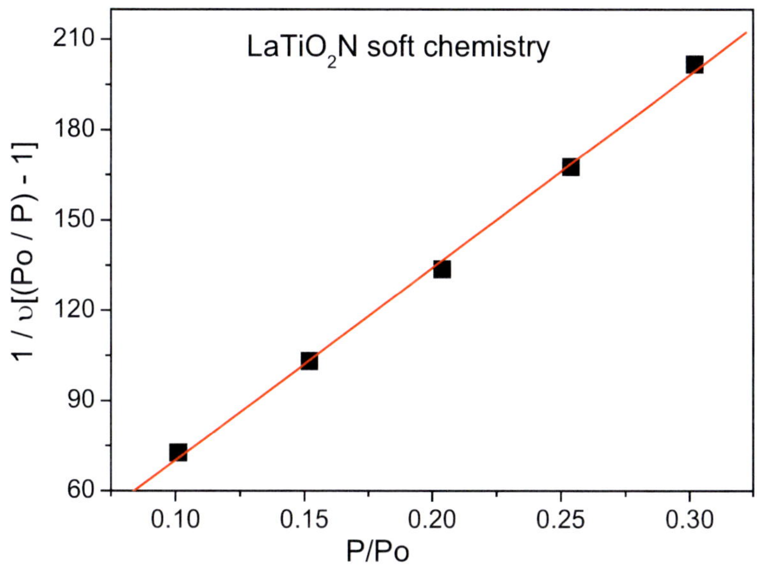

Figure 16. Adsorption isotherm curve for $LaTiO_2N$.

4. UV-Vis Spectroscopy

A Gretag Macbeth UV-Vis spectrometer model Color-Eye was used to measure the reflectance values of the samples in the range 250-750 nm. The UV-Vis comparison of TiO_2 and some oxynitrides is plotted in Figure 17. TiO_2 starts to absorb the light at wavelength lower than 400nm.

For the oxynitrides the onsets of the absorption cover a broad range of values, corresponding to the various colors of the samples. $SrTiO_3$:N is the samples with the largest band gap (considering a direct band gap transition) around 2.6 eV and presents a bluish color. It is followed by the yellow $CaTaO_2N$ that absorbs at circa 2.48 eV. $BaNbO_2N$ is the oxynitride with smallest band gap of 1.82 eV, it absorbs almost all the visible light and in turn has a black color. Depending on the A cation size (which to a large extend determines the crystallographic symmetry) and B cation electronegativity in the ABO_2N different colors can be obtained for the oxynitrides [22].

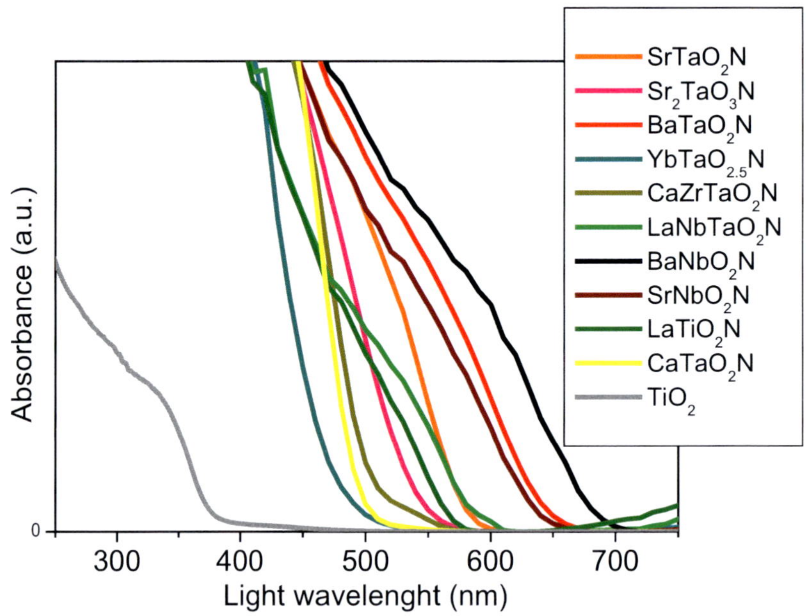

Figure 17. Absorbance *versus* light wavelength for some oxynitrides.

5. Photocatalytic Measurements

5.1. Decomposition of Organic Dyes in Solution

As mentioned above, one of the possible applications for photocatalysts is the detoxification of wastewater. To test the catalytic potential of our samples we used a simple setup consisting of a suspension of the oxynitrides in water with an organic dye dissolved. This mixture was stirred and illuminated for several hours. A halogen lamp was used to irradiate a quartz reactor vessel containing the organic aqueous solution [49]. As the halogen lamp emits a lot of heat, a water-cooling device was placed between the reaction vessel and the lamp. Running water from a cryostat through the device kept the filter at constant temperature and absorbed almost all the IR light of the halogen lamp. Fig. 18 shows a photograph of the used setup. In our measurements the investigated organic dye was methylene blue (MB). Its initial concentration in water was set to 18 mg/L. The amount of tested photocatalyst was 0.5g/L. The basic idea of the experiments is to follow the degradation (decolorization) of MB. After switching on the lamp, samples are taken at fixed time intervals and the MB concentration was measured by UV-Vis spectroscopy using a Cary 50 spectrometer. The catalyst is separated from the dye solution by centrifugation. This separation step is needed because if particles remain in suspension, the light can be absorbed and/or scattered, impeding the UV-Vis measurements. Methylene blue has its maximum light absorption at a wavelength of 600 nm, therefore this wavelength was used to determine the dye concentration.

Unfortunately, the tests did not show the expected results. Figure 19 depicts a plot using P25 and $SrTaO_2N$ as photocatalytic materials. For comparison, a pure methylene blue solution without any photocatalyst was also exposed to irradiation. Open symbols in Figure

19 are measurements performed without UV filter while the closed symbols correspond to runs with an additional UV filter (λ < 380nm) attached in front of the reaction vessel. P25 showed a better degradation rate than $SrTaO_2N$ when the measurements were done without UV filter. The decomposition rate for all experiments decreased when the UV filter was used. This finding was already expected since the most energetic part of the light was cut off. An interesting result for these tests was that P25 and $SrTaO_2N$ showed almost the same decomposition rate when the UV radiation was blocked, indicating that they possess comparable activities when activated by the visible part of the spectrum.

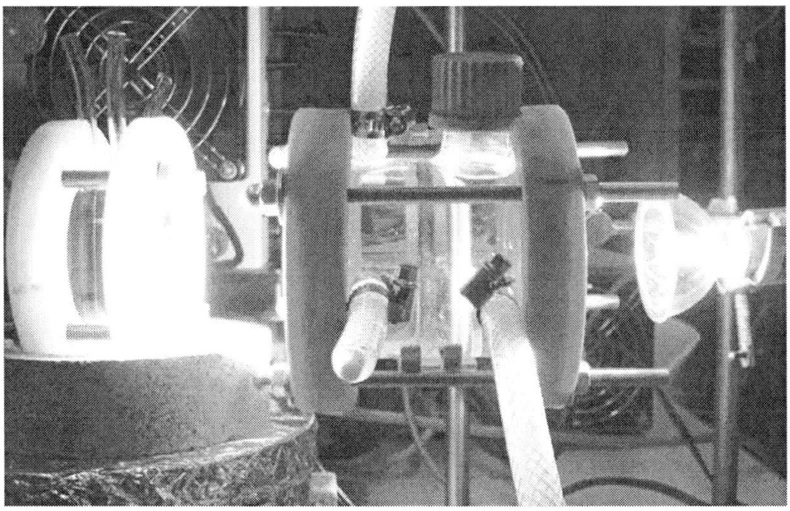

Figure 18. Setup for methylene blue photodecomposition showing from left to right the reaction vessel, the cooling device and the halogen lamp.

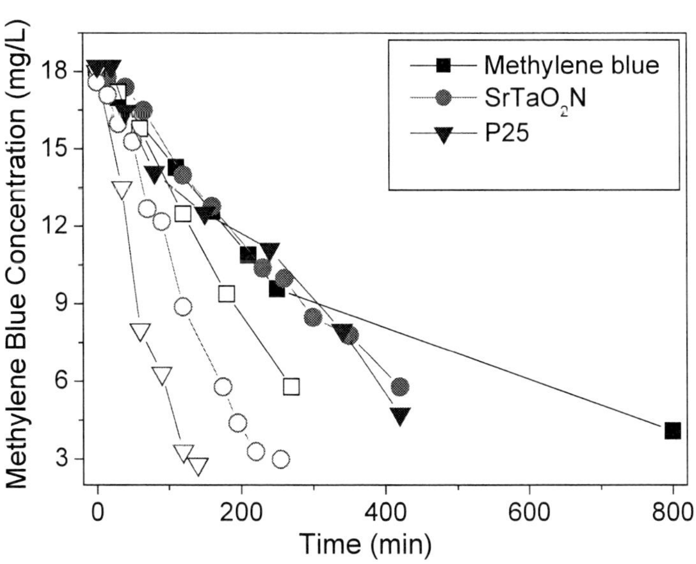

Figure 19. Methylene blue concentration *versus* time. The used photocatalytic concentration was 0.5 g/L in water solution [open symbols reaction without UV-filter, closed symbols reaction using UV-filter (λ < 380nm)].

Two problems appeared during the experiments. First it was striking that the methylene blue already decomposed under light without any photocatalytic material present. The decomposition was slower than in the presence of a photocatalyst, but it was quite high for reliable results. The second problem was that the tested photocatalytic compounds absorbed the organic dye on their surfaces. After some hours of reaction, P25 was blue and $SrTaO_2N$, which is originally orange, had changed to a green color. Because of these reasons, MB is not a good organic molecule to be used in this kind of analysis. We are currently searching for a more appropriate organic dye, which is stable under ultra violet and visible light, i.e. does not show not "self-decomposition" under irradiations and is not adsorbed on the photocatalyst surface.

5.2. H_2 and O_2 Evolution from Water

Perhaps the most interesting application for a photocatalyst is the water splitting reaction. The evolution of H_2 and O_2 from water using sunlight (which is one of the few energy sources mankind gets for free in inexhaustible quantities) would be the cheapest and most convenient way to solve energy problems.

Hydrogen is of great interest for new clean energy technologies. H_2 is used as fuel, as example in NASA's space shuttles fuel cells, to operate electrical systems. The emitted gas is water, which can even be consumed by the crew in space. Nowadays, almost all the used hydrogen (ca. 40 million tons) is produced by reforming of natural gases [50]. But the co-product is CO_2, the main responsible for greenhouse effect on the earth surface. Many research groups in the whole word are trying to develop better ways to produce hydrogen. On the other hand, oxygen is also a widely used gas. It is necessary in hospitals for patients who have difficulty in breathing or having operations. The steel industry uses oxygen to burn off impurities, such as carbon, sulphur and phosphorus from molten iron. Airplanes, which fly at high altitude, must carry oxygen. People who are climbing high mountains and divers carry gas cylinders with them.

On consideration of this background, the photocatalytic power of selected oxynitride perovskites to produce H_2 and O_2 was checked using a model reactor described in Ref [23]. In this setup, the two part reactions (formation of hydrogen and oxygen) are studied applying slightly different reaction conditions. Reduction of H^+ to H_2 was examined using 0.5 wt% Pt as co-catalyst in an aqueous solution containing 10% volume of methanol. The alcohol serves as sacrificial electron donor. Oxidation of H_2O to O_2 in the presence of Ag^+ (as electrical receptor) was measured in a separate reaction. A schematic sketch of the setup is illustrated in Figure 20. It consists of a Pyrex reactor vessel connected to a closed gas circulation and evacuation system. The system is illuminated by a 300 W Xe lamp equipped with UV cut-off filter ($\lambda > 420$ nm) and a water-cooling system. The used solution volume was 20ml with 0.2g of oxynitride for both experiments. The reaction solution was evacuated several times to remove air before the lamp was switched on. The reaction time was 5 hours. During this period the evolved gas was analyzed by gas chromatography.

The photocatalytic activities of the samples are shown in Figure 21. A significant hydrogen evolution from water-methanol solutions was observed for $NdTaON_2$, $BaTaO_2N$ and $LaTiO_2N$ with total H_2 productions of 1, 4 and 9 μmol, respectively.

Additionally, oxygen evolution was investigated using Ag^+ aqueous solution with La_2O_3 to keep the pH~8. For this reaction the compounds $SrTaO_2N$, $BaTaO_2N$, $LaTiO_2N$ and $CaNbO_2N$ were found to show a considerable activity with O_2 production of 4, 6, 22 and 37μmol, respectively. $CaNbO_2N$ has a surprisingly good oxygen photocatalytic production activity of almost twice more than $LaTiO_2N$ in the same reaction time.

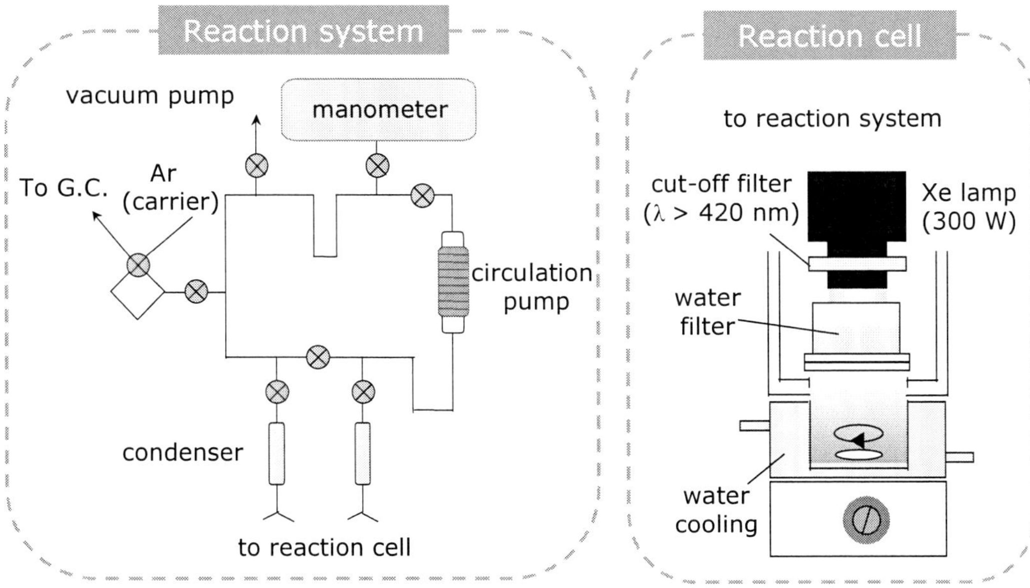

Figure 20. Scheme of the setup to produce O_2 and H_2 by photocatalysis.

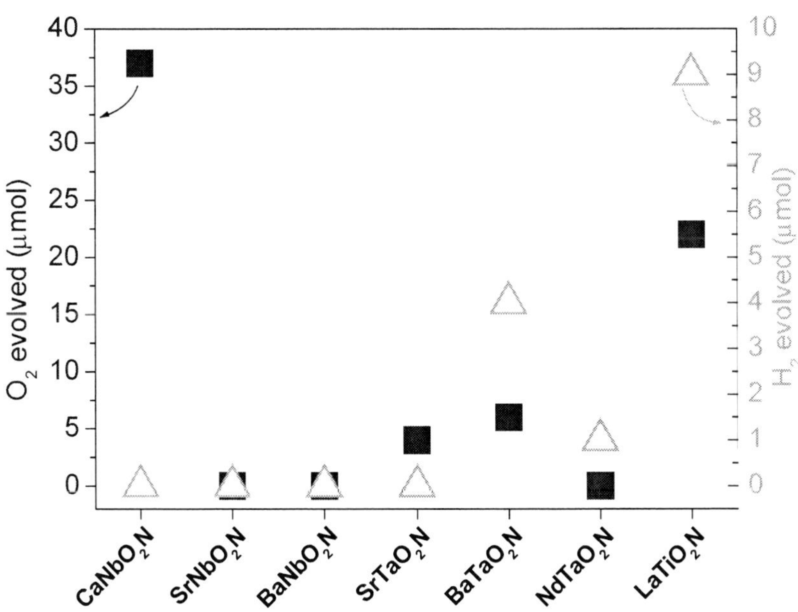

Figure 21. O_2 and H_2 evolution for selected oxynitride perovskites resulting from 5 hours of reaction.

5.3. Gas Phase Decomposition of Simple Organic Molecules

To examine whether our oxynitrides are applicable for air purification, for example to reduce disturbing odors, we studied the photocatalyzed decomposition of simple organic molecules in air. An in-situ setup was designed to measure the photocatalytic power of the oxynitride perovskites in an infrared spectrometer (Bruker Equinox 55). The basic idea of the setup is to place the oxynitride films into a quartz reactor closed by KBr windows and to illuminate them by a 150W Xenon lamp. The composition of the gas atmosphere inside the reactor is measured online with an IR laser beam passing through the windows. A scheme of the used equipment is shown in Figures 22. Acetone was chosen as organic compound because it is highly volatile and can easily be detected by its carbonyl group absorption peak. Furthermore its oxidation product CO_2 is also easy to detect. In a typical experiment 7 μl acetone was added to the reactor and after 10 minutes (the necessary time for the acetone to completely evaporate inside the vessel) the lamp was switched on. The oxynitride films were spread as thin films on glass slides. These films are made by mixing the oxynitride powder with isopropanol to form a viscous paste, which is doctor bladed onto the glass slide and dried at room temperature for several hours. Great care was taken to assure that all films have the same area and thickness.

Figure 23 and 24 show typical graphics for this kind of measurement. In Figure 23 the peaks related to acetone (1150-1280, 1600-1820, 2800-3100 cm^{-1}), CO_2 (600-765, 2250-2400 cm^{-1}) and water (1400-1750, 3550-3750 cm^{-1}) are plotted for three different reaction stages: start, middle and end. At the beginning the acetone peaks have high intensities, in the middle the intensity decreases, and in the end the peaks completely disappear. The opposite behavior is found for CO_2 and H_2O. Fig. 24 illustrates the 3D photocatalytic graphic (peaks intensity *versus* time *versus* wave number) for the photocatalyst $SrTaO_2N$.

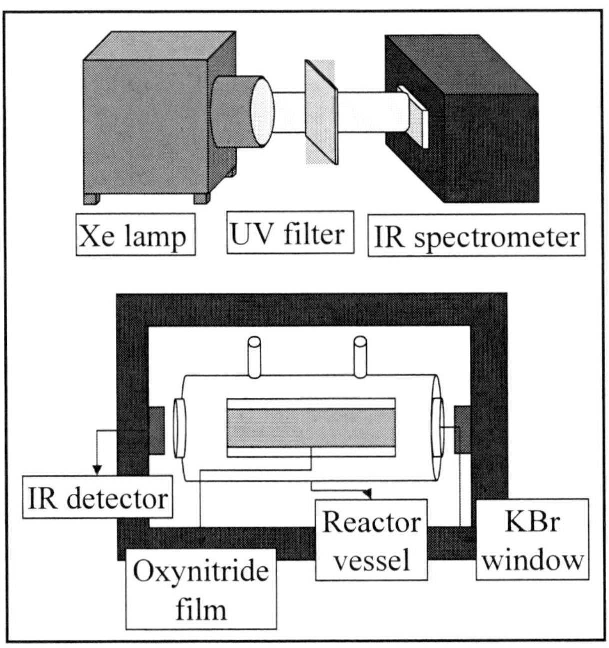

Figure 22. Schematic sketch of the infrared device and quartz reactor used on the experiment.

In closed rooms, such as kitchen, hotels and pubs the ultra violet light cannot be employed to promote the photocatalytic decomposition of odor molecules, because of its damage on the human health as skin burning and eyes injury. Therefore, it is necessary to have the photocatalysts activated only by the visible part of the light.

To test the photocatalytic power of the oxynitride samples under visible light, a UV filter (λ>420nm) was placed between the Xenon lamp and the quartz reactor. Figure 25 shows the normalized intensity of the C=O stretching mode against time for selected oxynitride perovskites and P25. For all samples the concentration of C=O decreases with time, indicating that all of them have the capability to decompose acetone under visible light radiation. Four samples showed an even better activity than P25, the best one (fastest decomposition rate) was $NdTaON_2$, followed by $LaTiO_2N$-soft chemistry product, $CaNbO_2N$ and $SrNbO_2N$, respectively. These results show that oxynitride-based compounds are in fact promising candidates for the gas phase decomposition of simple organic compounds. In this context it should be taken into account that the high activity of P25 is to a large extent due to its high surface area.

The influence of the specific surface on the photocatalytic activity of the oxynitrides was investigated using $LaTiO_2N$ as test material. Two samples, synthesized by soft chemistry and solid-state reaction with distinct surface area of 13.7 and 3.6 m^2/g, respectively, were measured. The results are also plotted in Fig. 25. As expected the sample with lower surface area, i.e. $LaTiO_2N$ by solid-state reaction, showed a much lower photocatalytic activity. After 11 hours (ca. 40000 s) of reaction, the sample produced by soft chemistry was able to decomposed 46.5% of the total acetone inside the reactor, while the $LaTiO_2N$ produced by solid-state decomposed only 3.7%. P25, for comparison, has a surface area value of approximately 50 m^2/g while the tested oxynitrides that showed a better performance than P25 only exhibited values around 10 m^2/g. It is therefore reasonable to assume that oxynitrides with a comparable specific surface would show much higher reaction rates than titanium dioxide.

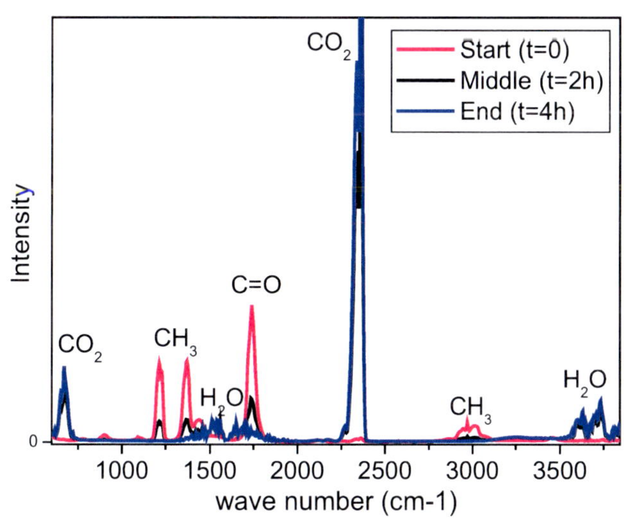

Figure 23. Peaks related to acetone, CO_2 and water for three different reaction stages.

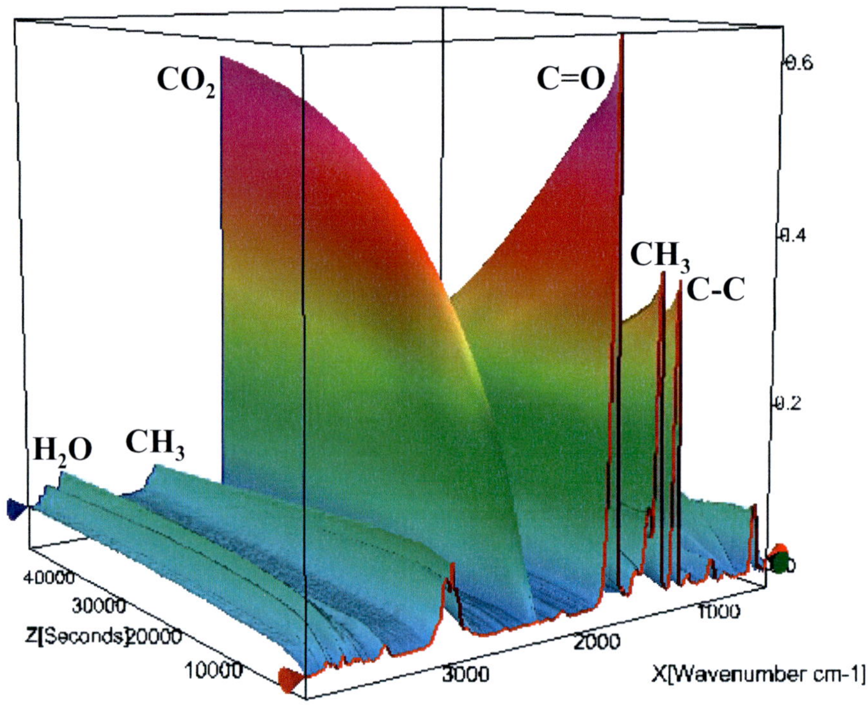

Figure 24. Acetone, CO_2 and H_2O related peaks intensity *versus* reaction time for $SrTaO_2N$ used as photocatalyst. The intensity of acetone peaks decrease with time while CO_2 and H_2O increase.

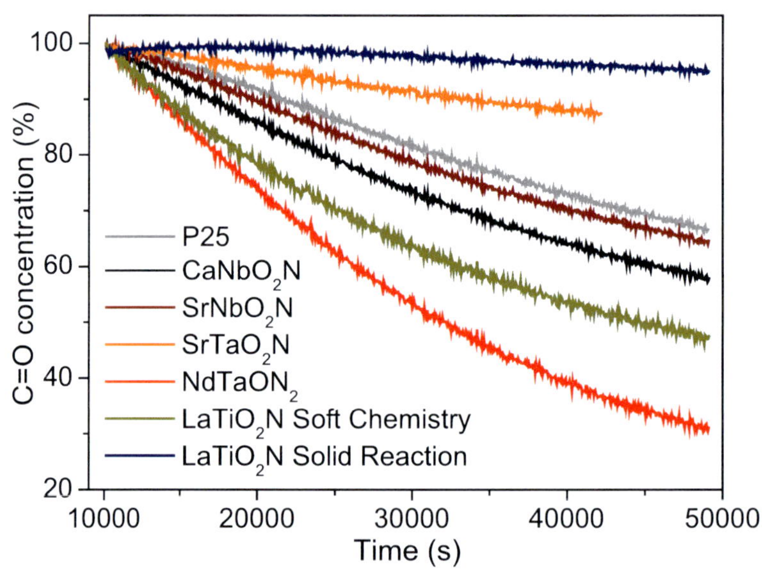

Figure 25. Integrated C=O peak intensity (e.g. acetone decomposition rate) *versus* time for different photocatalysts.

6. Conclusion

In this chapter, four synthesis techniques (solid state reaction, soft chemistry, polyol technique and spray pyrolysis) for the preparation of (amorphous) ternary oxides were described. The obtained powders exhibit different morphologies, particle sizes and surface areas and were reacted under ammonia to the respective oxynitrides of general composition $ABO_{3-x}N_x$. The introduction of nitrogen results in a pronounced reduction of the optical band gap. The samples show intense colors with the onset of absorption varying between 460 and 680 nm. The various oxynitrides were tested in two different setups concerning their photocatalytic activity. In a first experiment their capability to produce hydrogen and oxygen from aqueous solutions was investigated. Several samples were found to exhibit considerable activities. $LaTiO_2N$ was the sample that produced the largest amount of H_2 (9 μmol in 5 hours), while $CaNbO_2N$ had the highest activity for the formation of O_2 (37 μmol). The second tested setup was the decomposition of simple molecules, in this case acetone, under visible light in the gas phase. In this experiment $NdTaON_2$ was the most active sample (60.8% of the C=O bonds broken in 11h of reaction) showing an even higher reaction rate than P25 (27% of the initial acetone reacted) although the surface area of P25 is almost 8 times larger than that of $NdTaON_2$. For oxynitrides of same composition the photocatalytic activity increases with surface area as we could demonstrate for $LaTiO_2N$. The development of methods for the synthesis of oxynitrides with higher surface area to improve their catalytic activity is in progress.

In summary, it was demonstrated that oxynitride perovskites show good photocatalytic activities under visible light. They have the potential to be used for air purification in closed ambient, since they are capable to destroy organic molecules under irradiation. Furthermore, oxynitrides may also be employed for the clean H_2 and O_2 production by splitting of water.

Acknowledgment

This work was supported by the DFG within the Priority Research Program SPP 1136 and by SFB 484.

References

[1] Bhalla, A. S.; Guo, R.; Roy, R. *Mat Res Innovat* 2000, 4, 3-26.
[2] Pena, M. A.; Fierro, J. L. G. *Chem Rev* 2001, 101, 1981-2017.
[3] Dimos, D.; Mueller, C. H. *Annu Rev Mater Sci* 1998, 28, 397-419.
[4] Ishihara, T.; Shibayama, T.; Nishiguchi, H.; Takita, Y. *J Mat Sci* 2001, 36, 1125-1131.
[5] Sinquin, G.; Petit, C.; Hindermann, J. P.; Kiennemann, A. *Catalysis Today* 2001, 70, 183-196.
[6] Hyde, B. G.; Thompson, J. G.; Whitlers, R. L.; Fitzgeral, J. G.; Stewart, A. M.; Bevan, D. J. M.; Anderson, J. S.; Bitmead, J.; Paterson, M. S. *Nature* 1987, 327, 402-403.
[7] Mollah, S. J Phys: *Condens Matter* 2004, 16, P1237–P1276.
[8] Gesing, T. M.; Wachtmann, K. H.; Jeitschko, W. Z *Naturforsch B* 1997, 52(2), 176-182.

[9] Greaves, C.; Slater C.R. *Phys C* 1991, 175, 172-178.
[10] Rubio, J. O. *J Phys Chem Solids* 1991, 52(1), 101-154.
[11] Antonyak, O. T.; Voloshinovskii, A. S.; Mel'chakov, E. N.; Mikhailik, S.; Pidzyrailo, M. S.; Stefanskii, I. V.; Rodnyi, P. A. *Optics and Spectroscopy* 1998, 84(6), 861-864.
[12] Narayan, R. L.; Suryanarayana, S. V. *Mater Lett* 1991, 11(8–9), 305-308.
[13] Diao, X. G.; Takeuchi, A. Y.; Garcia, F.; Scorzelli, R. B.; Rechenberg, H. R. *J Appl Phys* 1999, 85(8), 4485-4487.
[14] Takenaka, K.; Takagi, H. *Appl Phys Let* 2005, 87(26), Art. No. 261902.
[15] Shannon, R. D.; Prewitt, C. T. *Acta Crystallogr Sect B* 1969, 25, 925.
[16] Antoine, P.; Assabaa, R.; L'Haridon, P.; Marchand, R.; Laurent, Y. *Mater Sci Eng* 1989, B5, 43-46.
[17] Cheviré, F.; Tessier, F.; Marchand, R. *Eur J Inorg Chem* 2006, 2006(6), 1223-1230.
[18] Grins, J.; Svensson, G. *Mat Res Bul* 1994, 29(7), 801-809.
[19] Marchand, R.; Laurent, Y.; Guyader, J.; L'Haridon, P.; Verdier, P. *J Eur Cer Soc* 1991, 8, 197-213.
[20] Ebbinghaus, S. G.; Weidenkaff, A.; Rachel, A.; Reller, A. *Acta Crystallographica* 2004, **60**, i91-i93.
[21] Jansen, M.; Letschert, H. P. *Nature* 2000, 404, 980-982.
[22] Aguiar, R.; Logvinovich, D.; Weidenkaff, A.; Rachel, A.; Reller, A.; Ebbinghaus, S. G. *Dyes and Pigments* 2007, in press-doi:10.1016/j.dyepig.2006.08.029.
[23] Kasahara, A.; Nukumizu, K.; Hitoki, G.; Takata, T.; Kondo, J. N.; Hara, M.; Kobayashi, H.; Domen, K. *J Phys Chem* A 2002, 106, 6750-6753.
[24] Ji, S. M.; Borse, P. H.; Hyun, G. K.; Hwang, D. W.; Jang, J. S.; Bae, S. W.; Lee, J. S. *Phys Chem Chem Phys* 2005, 7, 1315-1321.
[25] Aguiar, R.; Weidenkaff, A.; Schneider, C. W.; Reller, A.; Ebbinghaus, S. G. *Prog Solid State Chem* 2007, in press-doi: 10.1016/j.progsolidstchem.2007.01.033.
[26] http://www cru-z com/arc-e/ 2007.
[27] Peral, J.; Domènech, X.; Ollis, D. F. *J Chem Technol Biotechnol* 1997, 70, 117-140.
[28] Kabra, K.; Chaudhary, R.; Sawhney R.L. *Ind Eng Chem Res* 2004, 43, 7683-7696.
[29] Hoffmann, M. R.; Martin, S. T.; Choi, W.; Bahnemann, D. W. *Chem Rev* 1995, 95, 69-96.
[30] Hashimoto, K.; Irie, H.; Fujishima, A. *Jpn J Appl Phys* 2005, 44(12), 8269-8285.
[31] Asahi, R.; Morikawa, T.; Aoki, K.; Taga, Y. *Science* 2001, 293, 269-271.
[32] Morikawa, T.; Asahi, R.; Ohwaki, T.; Aoki, K.; Taga, Y. *Jpn J Appl Phys* 2001, 40, L561-L563.
[33] Kato, H.; Kudo, A. *Catalysis Today* 2003, 78, 561-569.
[34] Kato, H.; Kudo, A. J *Photochem Photobiol , A* 2001, 145, 129-133.
[35] Uno, M.; Kosuga, A.; Okui, M.; Horisaka, K.; Yamanaka, S. *J Alloys Compd* 2005, 400, 270-275.
[36] Ikeda, S.; Tanaka, A.; Sinohara, K.; Hara, M.; Kondo, J. N.; Maruya, K.; Domen, K. *Micropor Mater* 1997, 9, 253-258.
[37] Domen, K.; Naito, S.; Onishi, T.; Tamaru, K. Chem Phys Lett 1982, 92, 433-434.
[38] Kim, Y.-I.; Woodward, P. M.; Baba-Kishi, K. Z.; Tai, C. W. *Chem Mater* 2004, 16(7), 1267-1276.

[39] Kasahara, A.; Nukumizu, K.; Takata, T.; Kondo, J. N.; Hara, M.; Kobayashi, H.; Domen, K. *J Phys Chem B* 2003, 107(3), 791-797.

[40] Schottenfeld, J. A.; Benesi, A. J.; Stephens, P. W.; Chen, G.; Eklund, P. C.; Mallouk, T. E. J Solid State Chem 2005, 178, 2313-2321.

[41] Albarici V. C. *Synthesis and Characterizations of Li2TiSiO5 and Na2TiSiO5 Powders.* UFSCar, Sao Carlos, Brazil, 2004, 1-80.

[42] Wohlrab, S.; Weiss, M.; Du, H.; Kaskel, S. *Chem Mater* 2006, 18(18), 4227-4230.

[43] Flores-Gonzalez, M. A.; Ledoux, G.; Roux, S.; Lebbou, K.; Perriat, P.; Tillement, O. *J Solid State Chem* 2005, 178, 989-997.

[44] Feldmann, C. *Z Anorg Allg Chem* 2004, 630, 2473-2477.

[45] Sohn, J. R.; Kang, Y. C.; Park, H. D. *Jpn J Appl Phys* 2002, 41, 3006-3009.

[46] Itoh, Y.; Lenggoro, I. W.; Pratsinis, S. E.; Okuyama, K. *J Mater Res* 2002, 17(12), 3222-3229.

[47] Clarke, S. J.; Hardstone, K. A.; Michie, C. W.; Rosseinsky, M. J. Chem Mater 2002, 14(6), 2664-2669.

[48] Brunauer, S.; Emmett, P. H.; Teller, E. *J Am Chem Soc* 1938, 60(2), 309-319.

[49] Mills, A.; Wang, J. *J Photochem Photobiol A: Chem* 1999, 127, 123-134.

[50] http://www.nrel.gov/hydrogen/ National Renewable Energy Laboratory, U. S. Department of Energy. 2007.

In: Ceramic Materials Research Trends
Editor: Paul B. Lin, pp. 139-158
ISBN: 1-60021-769-9

Chapter 5

MODERN STATUS AND FUTURE OF SOFC

Lipilin Alexander Sergeyevich *

Institute of Electrophysics, Ural Branch RAS, Ekaterinburg, Russia

Abstract

Applications of superionic conductors in solid-oxide fuel cells (SOFCs) were considered. A promising direction is the distributed power. Electrochemical current generators come to direct consumers of electric power, who operate the generators as required by their power consumption plan. As a result, the power is reasonably saved and the fuel is used efficiently. SOFCs provide a highly efficient direct conversion of the chemical energy of the fuel to electricity. Any hydrocarbons can be used as fuel in SOFCs. Natural fuels (coal, oil and gas) and products of technical activities of man (industrial and agricultural wastes) are used. Electrochemical "low-temperature" burning provides high ecological compatibility of the process. All these factors open up good prospects for the use of SOFCs to satisfy ever growing requirements of mankind in the electric power [1-6].

Introduction

Solid-oxide fuel cells have a wide spectrum of applications: from portable current power supplies (10-500 W) to isolated stationary generating plants (1-10 MW). They attract interest primarily because of a highly efficient direct conversion of the chemical energy of the fuel to electricity. The fuel can be any hydrocarbon, while the oxidizer is air. The low-temperature electrochemical burning of the fuel makes the power generation process ecologically harmless.

Since all SOFC components are solid, the cells and devices made of these cells have a variety of designs. Until 1980-1990s their designs could be divided into three groups: tubular, planar and block designs. Later cells combined favorable properties of the tubular and planar designs or the tubular and block designs. The solid state of a single cell not only provides its

* E-mail address: lipilin@iep.uran.ru. Tel.: 8 (343) 267 88 27 (office); 8 908 904 53 22 (cellular). Fax: 8 (343) 267 87 94. A.S. Lipilin, Leading Researcher at Institute of Electrophysics, Ural Branch RAS, 106 Amundsen St., 620016 Ekaterinburg

diverse designs, but also allows making battery stacks and power installations in a variety of embodiments. Therefore devices can be optimized considering customer's most important requirements, with the priority given to dimensions or the mass, a high working temperature and expensive constructional materials or a low temperature and cheap materials, a high power density or the lifetime.

Operating Principle

A dense thin layer of the SOFC electrolyte is generally placed between two porous electrodes, which are the anode and the cathode. The fuel, which is fed to the anode-electrolyte contact zone, is oxidized by the flow of oxygen ions from the solid electrolyte. Electrons, which go to the external circuit, are released on the anode. They pass through the external load of the closed circuit and arrive to the cathode where oxygen is ionized on the electrolyte boundary. So, an ion current, which is equal to the electron current flowing in the external circuit, passes in the solid electrolyte. One more SOFC component – the interconnect – is added when single cells are assembled to a battery.

Materials

At present, the most popular SOFC materials are oxygen-conducting yttrium-stabilized zirconia (YSZ) for the electrolyte, strontium lanthanum manganite (LSM) for the cathode, nickel cermet (Ni+YSZ) for the anode, and strontium- or calcium-doped lanthanum chromite (LSC or LCC) for the ceramic interconnect. A metal interconnect, which is usually made of iron-chromium steel, is finding increasing use in planar designs. The working temperature of SOFCs is 700 to 1000 °C.

Design Philosophy

With respect to their mechanical strength, SOFCs are grouped to cells with the bearing electrolyte, the bearing cathode, the bearing anode, the bearing high-porous ceramic substrate, and the bearing high-porous current collector. In recent years, to make the cells more compact, they were designed with two bearing electrodes (Fig. 1) [1] or, put it more correctly, the bearing anodic and cathodic current collectors, which simultaneously serve as gas manifolds. The last remark also applies to cells with bearing cathodes and anodes since the electrodes themselves are functionally less than 10 μm thick and essentially the current collectors are responsible for the mechanical strength.

The bearing YSZ electrolyte in SOFCs is not less than 100-150 μm thick; as a result, the resistance of these cells is high at low temperatures. If the strong ScSZ electrolyte or the combination of cubic and tetragonal electrolytes is used, the electrolyte performs the bearing function (at the same mechanical strength) when it is 20-30 μm thick (Fig. 2) [1].

If the YSZ electrolyte is thinner (10-40 μm), the bearing element generally is a thicker anode, cathode, porous ceramic or metal substrate (current collectors). The use of thin electrolytes reduces the voltage loss due to the electrolyte resistance, but brings about diffusion problems related to the supply and the discharge of gas reagents on relatively thick (1-2 mm) bearing electrodes.

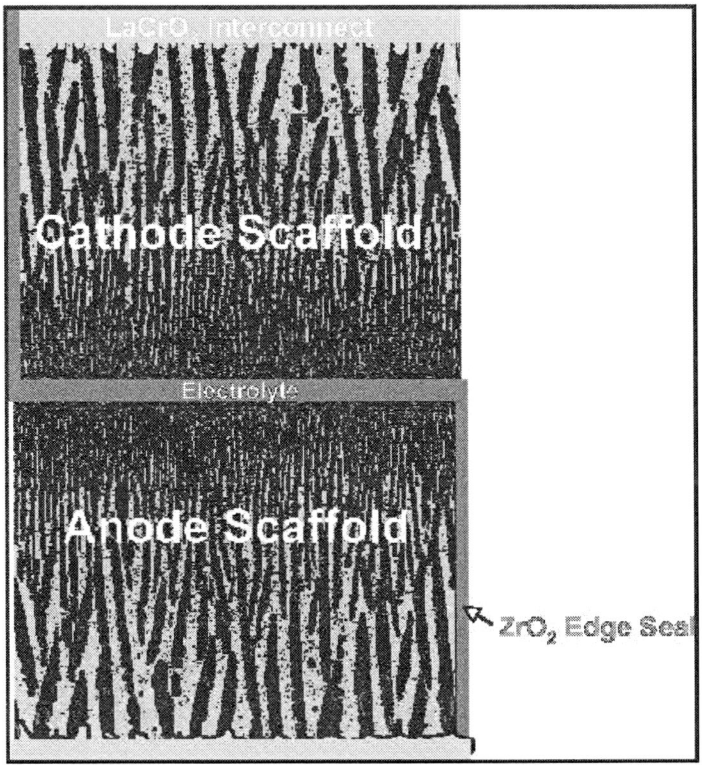

Figure 1. Cell with two bearing electrodes and a thin ceramic interconnect, but without near-electrode gas channels (1 kW/kg, NASA project) [1].

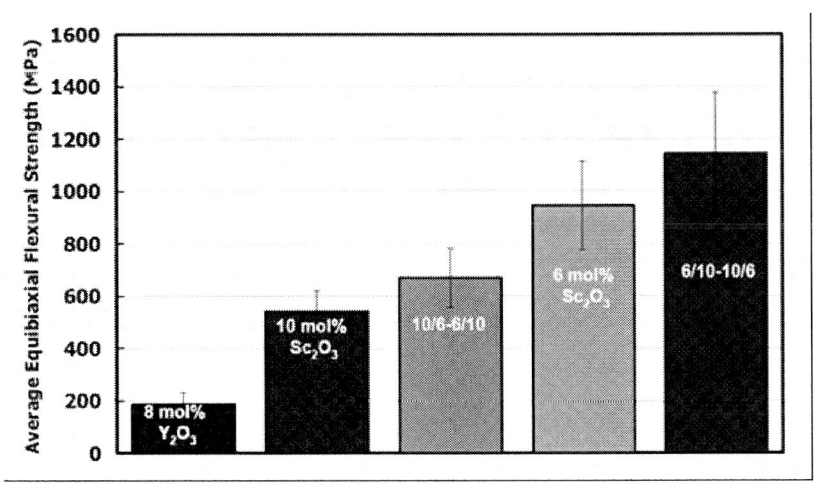

Figure 2. Mechanical strength of the ceramic as measured on discs 12 mm across and 150 μm thick [1].

In my opinion, the engineering compromise is most reasonable when the mechanical strength of SOFCs is a joint function of the electrolyte and the electrodes with current collectors reducing the diffusion problems of the supply and the discharge of reagents to a minimum.

SOFC Designs

By design, SOFCs can be conditionally divided into segmented tubular, tubular, planar, block and monolithic cells [7, 8]. The segmented tubular cells are those when several SOFCs are series-connected by the current to form a pipe or a tube and then the tubes are assembled to a battery. In tubular designs, SOFCs are usually made as tubes assembled to a battery. Planar SOFCs most frequently represent flat square plates or discs. Block SOFCs include a single cell of a complex design, for example, a rectangular parallelepiped with alternating gas ducts. In monolithic SOFCs, several green cells with a wavy surface are assembled to a battery and then are sintered during the first heating treatment.

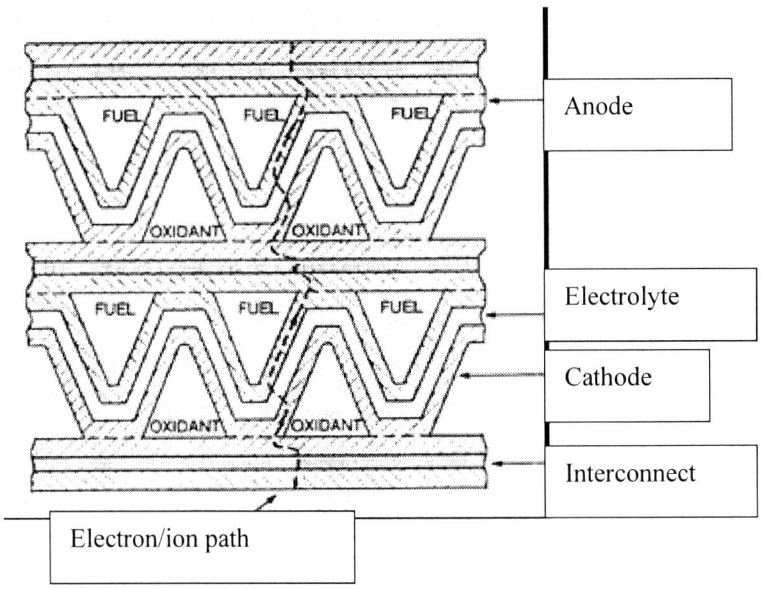

Figure 3. Electron/ion current path in monolithic SOFC [8].

With respect to the current flow in SOFCs, the cells are connected to stacks and batteries in series to increase the output voltage, in parallel for better reliability, or in series-parallel if an acceptable engineering compromise is achieved. The internal loss with the stack-connected cells is the larger, the more distant the electric connection unit is from the working zone. If this connection arrangement is used, the interconnects should have a larger cross sectional area, leading to the increase in the mass and the dimensions of the battery. The optimal flow of the current (the resistance is a minimum) is perpendicularly to all thin functional layers of SOFCs: the current collector, the anode, the electrolyte, the cathode, the connecting current collector (the interconnect), the anode of the next cell, etc. This is possible in the limit for the planar design. In real conditions, however, the near-electrode gas spaces bend and extend the current path in this design as well. Just a small portion of the current flows almost

Figure 4. Tubular cell with the current collector along the generatrix and cells comprising "oblate" pipes with 10 and 5 tubular channels [1].

perpendicularly to the electrolyte surface, whereas most of the current passes along the electrode and accumulates in the current collector as in the case of monolithic cells with a wavy electrolyte plate (Fig. 3) [8]. Therefore, to decrease the internal loss, it is necessary to reduce the layer resistance of the electrode, increase its conductivity, make it thicker, etc. The internal loss is commonly referred to constructional losses. The path of the current in the electrodes along the electrolyte is generally longer in tubular SOFCs than in planar cells. The larger the diameter of the seamless tubular cell, the longer the path of the current and the larger the loss at the same layer conductivity of the electrodes. The current flows in the electrodes circumferentially. The loss is reduced and the packing density of the cells in a battery is improved as the diameter decreases. The volume of batteries decreases, while their power is not affected. It is not by chance that Siemens Westinghouse replaced tubular cells 1.5 m long and 24 mm in diameter by cells twice as short and, essentially, combining 5 or 10 tubes of a smaller diameter (Fig. 4) [1]. Acumentric makes the interconnect in tubular cells not by the generatrix as in the seamless design by Siemens Westinghouse, but transversely (Fig. 5) [1]. They essentially resemble segmented tubular cells where the current connection in a single tube is realized not in series, but in parallel with one common bearing anode. Most

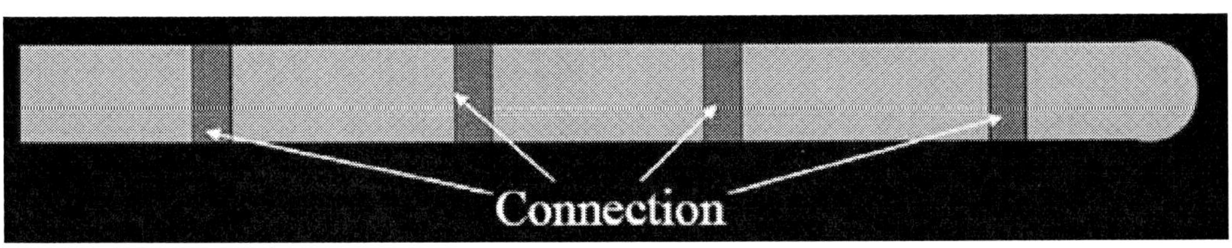

Figure 5. Tubular cell with the current collector across the tube [1].

of the current should flow along segments of a cell rather than circumferentially. In this embodiment, the loss increases with the length of the segments; the interconnect between the segments, which acts as the current collector for cells assembled to a battery, should have an adequate conductivity. It is reasonable that developers use copper and silver for this purpose.

Gas Supply to SOFCs

With respect to the flow of gas reagents in the anode and cathode spaces of SOFC stacks, they can be grouped to stacks with unidirectional parallel flows of the fuel and the oxidizer, parallel opposite flows, and perpendicular flows. The gas flows influence the potential and temperature distributions over a stack. The battery characteristics can be considerably affected by a nonuniform feed of gases to the electrode surfaces. Gas manifolds usually provide a uniform gas supply. They are made externally for tubular SOFCs and internally for planar SOFCs in combination with the electrochemical cell, distributing the gas flows over numerous channels in a single cell. The heat and mass exchange conditions in the thin flat cell should be necessarily considered for the planar designs. The heat removal function is partially imparted to the gas flows (e.g., air) and the current collector (the interconnect). In the first case, the cross-sectional area of the gas channels need be increased to avoid the pressure differential between the spaces; in the second case, the cross-sectional area of the current collector is increased to remove the heat from the center of the cell to its periphery. The anodic and cathodic spaces of the cells in a battery should be sealed reliably. All these factors complicate the design and increase the thickness of the current collector between the cells. Also, planar SOFCs become less attractive since their main advantage over tubular cells is a high packing density of the cells in a battery and its compactness.

The problem of the uniform feed of gaseous reagents to the anode and cathode spaces has been already mentioned above. But the gas flows in the near-electrode spaces are just the first part of the problem. The second part of the problem is the diffusion feed of the reagents at the right angle to the gas flow. Since electrochemical electrode reactions take place on the three-phase boundary (electrolyte – electrode – gas), the required amounts of the fuel and the oxidizer should be supplied to this reaction zone through the porous structures of the electrode manifolds and the electrodes themselves. Clearly, this task involves diffusion problems. The measure of these problems is the limiting currents. The formation of the limiting currents reduces the cell power. The diffusion problems apply to the cathode and an insufficiently porous electrode even in pure oxygen; naturally, they are aggravated when air is fed as the oxidizer and nitrogen is removed. A certain correlation of the limiting cathode currents is observed in SOFCs with air or oxygen as the oxidizer. For example, if the cathode is about 300 μm thick, the pores are 0.75 μm in size on the average and the open porosity is 26.6%, the limiting current is about 300 mA/cm^2 for air and 1750 mA/cm^2 for oxygen, i.e. almost six times larger. Of course, diffusion problems arise on the anode as well when pure hydrogen is supplied and water is removed. These problems will be worse on the anode when hydrogen is replaced by the synthetic gas or methane subject to the conversion on the electrode. The technical solution to these diffusion problems will be the increase in the porosity and the pore size, and the decrease in the thickness of the electrodes and the current collectors to obtain the limiting current density higher than the one required to produce the target specific power.

Evaluation of Technologies

Diverse engineering solutions for single cells and stacks led to a variety of manufacturing technologies. Today the cheapest technologies providing a high quality and a 5- to 10-year lifetime of SOFCs are selected for the fabrication of commercially attractive distributed power systems. The development of technologies for the formation of various components of SOFCs – a dense thin-layer solid electrolyte, a gas-tight ceramic interconnect, porous structures of the electrodes and current collectors – should be discussed in a separate report.

Organizational Arrangements for SOFC Development

The USA is an example of a wise public policy aimed at the commercialization of SOFC-based electrochemical generators. In October 2001 the Solid State Energy Conversion Alliance (SECA) was established and a program was adopted by which 5-kW (3-10 kW) SOFC-based power systems worth $400/kW will be developed and manufactured on the industrial scale by 2010. At present the SECA program covers six SOFC projects approaching the commercialization level and developed by competing industrial groups: Cummins-SOFCo, Delphi-Battelle, General Electric (GE), Siemens Westinghouse (SW), Acumentrics, and FuelCell Energy (FCE).

Figure 6. Maximum power of 5.6 kW, maximum efficiency of 49%, fuel utilization factor of 65-80%, and air utilization factor of 21-24% [9].

All the industrial groups have achieved much success by now. The first phase of the development work was complete in 2006 and different designs of stacks were subject to demonstration testing. The output specific power was increased. The developers reported their achievements at the 7th SECA Annual Meeting held in September 12-14, 2006 in Philadelphia, USA [for details see 4].

GE develops a compact household 5-kW generator based on a flat disc SOFC with the bearing anode burning the natural gas at 700-800 °C, Fig. 6 [for details see 9]. GE proposes several designs of the stack. A single stack of 21 cells provided the specific power of 307 mW/cm^2 at 800 °C. A single cell produced 400 mW/cm^2 during tests. The largest surface area of the cell developed at GE is 900 cm^2.

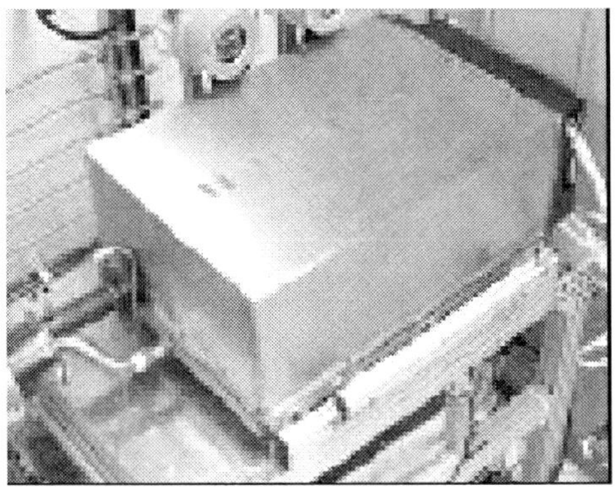

Figure 7. Power of 4.24 kW ($767/kW), efficiency of 37%, and degradation of 7.3% in 1500 hours [10].

Figure 8. Power of 3.3 kW (0.73 V per cell) at 982 °C, fuel utilization factor of 68%, efficiency of 35%, and degradation of 0% in 1500 hours [11].

Delphi works on the third generation of the project and focuses on the development of a very compact light-weight system for use in transport means, Fig. 7 [for details see 10]. Two stacks of 30 square planar cells produced 420 mW/cm^2. A stack of 5 cells provided nearly 600 mW/cm^2.

SW made tubular cells with the bearing cathode, which generated 300 mW/cm^2 at 85% utilization of the fuel and 1000 °C. The cells, which were twice as short and had an updated design in the form of an oblate tube made up of flat square LSM pipes, provided the specific power of about 400 mW/cm^2 (Fig. 4). The cells of the next generation with a set of triangular pipes produced 600 mW/cm^2. A power plant with cells of square pipes was tested under the SECA project [for details see 11].

Acumentrics tested a device made up of tubular cells on the bearing anode in the state of Alaska. The manufacturing cost of the tested stack was much lower than the cost of the previous variants. The specific power of 400 mW/cm^2 was achieved (Fig. 9). In three years the developers increased the surface area of a single cell; as a result, the power picked off a single cell increased 10 times (up to 60 W) [for details see 12].

Figure 9. Maximum power of 6.1 kW, efficiency of 36.9%, fuel utilization factor of up to 97.5%, degradation of 0.75% in 500 hours, and \$685/kW [12].

Cummins-SOFCo is the largest manufacturer of diesel generators in the market. The company correctly evaluates the market of SOFC-based generators rated at 5 kW. Their 1-kW pilot prototype developed under the SECA program operated continuously on the natural gas for 2000 hours. Their fuel processor underwent tests for 2900 hours at constant and cyclic loads in accordance with the test requirements (phase 1 of the SECA program). An evaluation unit (Fig. 10) was tested [for details see 13].

For the purpose of the SECA program, Fuel Cell Energy (FCE) incorporated Global Thermoelectric Corporation (Canada) and Versa Power Systems (VPS) acting as the developer and a subcontractor into a single legal entity. The amalgamation accelerated the development and the commercialization of SOFC systems under the SECA program. VPS has made great strides towards the global technology for SOFC production. An evaluation unit is shown in Fig. 11 [for details see 14].

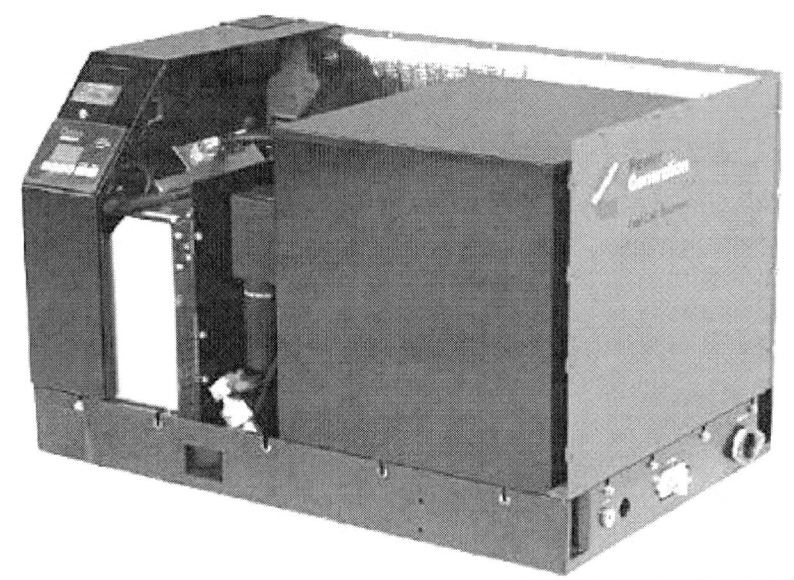

Figure 10. Power system rated at 5.4 kW and the efficiency of 50% operated for 3400 hours in three thermal cycles and provided the specific power of about 500 mW/cm^2 [13].

Figure 11. Power of up to 3.39 kW, the efficiency of up to 38% at 730 °C during 1002 hours; with the fuel utilization of 57.5% and the air utilization of 42.5%, the power density was 347 mW/cm^2, while the degradation was 1.2% in 500 hours ($774/kW) [14].

Strategic Plans

The strategy of the US Department of Energy (DOE) with respect to the fuel power engineering consists in the development of the technology of fuel cells and the demonstration of their use in large-scale devices with coal as the fuel. A new type of fuel cells is under consideration. Small-scale devices ($400/kW) burning coal gasification products will be developed at the first stage by 2010. By the year 2015 it is planned to make megawatt plants at the prime cost of US$5 000 000. Systems rated at more than 100 MW and having the efficiency of 50% or better are possible by 2020. The separation of CO_2 and its utilization by at least 90% will be provided. It is planned to design a 275-MW prototype plant, which will serve as a large-scale research laboratory-factory of carbon-hydrogen technologies. In the future such plants will produce hydrogen and CO_2 in addition to the electric power [1].

Coal as the Fuel

In November 2005 the section of the advanced research at the "Fuel Cell" seminar (Direct Carbon Conversion Workshop), which was held in the USA, considered new technologies for the direct conversion of the chemical energy of coal to the electric power by direct oxidation of coal in fuel cells (DCFCs). This method essentially brings us back, but on an advanced technical level, to the solution of the problem that was at the origin of high-temperature fuel cells late in the 18th century. We again address the problem of the more efficient burning of coal and the generation of the electric power.

On March 3, 1896 W.W. Jacques invented the first fuel cell, in which coal (one of the electrodes) was burnt on the boundary with the molten electrolyte (was electrochemically oxidized) by bubbling air. In 1897 Walter Nernst invented the mix (85% ZrO_2 + 15% Y_2O_3), which is still used as the most common YSZ electrolyte in modern SOFCs. Developers of DCFCs just borrow ideas of our ancestors and use molten carbonate and solid-oxide electrolytes or invent new technologies on their basis.

Of interest are developments by the Lawrence Livermore National Laboratories when the molten carbonate electrolyte feeds the fuel to the anode, with the fuel being nanosized coal particles.

Specialists at the Stanford Research Institute (SRI International) use a solid YSZ electrolyte and a liquid anode of molten salts supplying the fuel to the two-phase reaction zone. The biomass, oil tar, plastic scrap or powdered coal (PRB) used at thermal power plants was added as the fuel to the melt at the SRI. The fuel cells with the liquid molten anode burning coal and oil tar generated the specific power of over 100 mW/cm^2. The specific power provided by DCFCs is already comparable with the specific power of commercial power stations based on molten carbonate fuel cells burning the natural gas. Today SRI specialists work on increasing the specific power to 200 mW/cm^2, enlarging the surface area of a single cell, and developing the power systems of the kilowatt rating [for more information see 15].

Evaluation of SOFC Developing Firms

The firms are more and more definitely divided into those developing SOFC-based power systems with the view of their commercialization, producing initial materials for all SOFC components, and addressing particular problems of the optimization of individual components and units of SOFCs. The first group of the firms test systems rated at 1-5 kW and are attractive for governmental and private investments. The second group work up the market of initial materials and, knowing the specific features of the available powders, conclude contracts for the development of single cells and SOFC batteries. Firms of the third group present only the scientific interest and, most frequently, cannot interfere with SOFC production technologies elaborated by firms of the first group.

The undisputed leader in the development and the manufacture of SOFC-based power systems is Siemens Westinghouse (USA) producing 250-kW plants whose characteristics are impaired by about 0.1% in 1000 hours. The guaranteed lifetime of the plants is at least 20000 hours. The company has occupied the leading position thanks to a seamless tubular design of the cells. In recent years, while working on the design of the single cell, the company first passed to cells in the form of an oblate tube with 10 rectangular channels (this design could not but be criticized) and then turned to cells with triangular channels (for details see 11]. The company is planning to use more conductive solid electrolytes based on scandium-stabilized zirconia and gallate. They will allow decreasing the working temperature, making the lifetime longer, and reducing the kW cost of the installation. The fulfillment of these plans will guarantee the leadership to the company for the nearest 5 to 10 years.

After N.Q. Minh, who is the author of the monolithic SOFC batteries, was employed by General Electric, the company has been winning the leading position in the development of power systems with disc-shaped planar SOFCs. Research systems rated at 1-6 kW were tested. They have gained wide engineering experience in making of three-layer (anode – electrolyte – cathode) blanks by calendaring and simultaneous sintering of the layers. I think gas-tight ceramic interconnects, which can be made by this technology, will bring the world leadership to the company in the nearest 5-10 years in the development of multi-MW SOFC power plants combined with a gas turbine and a coal gasification system.

At present, the leading position in the development of power systems based on planar SOFCs is occupied by Forschungszentrum Jülich (Germany), which tested a device rated at 13 kW back in 2004. The commercialization problems made the developers to cooperate with the Saint-Gobain Concern [for details see 16]. Today the Saint-Gobain Concern, which is the world's leading company in the production of thermal insulation materials, comes to the forefront in the manufacture of materials of the new generation.

The Riso National Laboratory (Denmark) [5] is one more European center dealing with the development of SOFCs. The analysis of the SOFC developments in Europe and their results were reported at the German-Canadian seminar in 2005 [17]. The attitude of the European states to the SOFC electrochemical power engineering and, what is most significant, their resources are incomparable with the US resources. In the nearest future, this will be the decisive factor for the accelerated commercialization of SOFC power systems and the market grouping.

It is encouraging to note that at the international scientific quorum, which proved to be the world's most authoritative meeting, and the 30th annual seminar on fuel cells (Fuel Cell

Seminar) in the USA, which was held on November 13-17, 2006 in Honolulu (Hawaii), the states of the European Community claimed the development of kilowatt power systems (for more information see [2]).

SOFCs and SOFC-Based Power Systems in Russia

The majority of states aiming at the distributed electrochemical power have adopted state programs of the development of SOFC-based power systems. Unfortunately, such a program is unavailable in Russia.

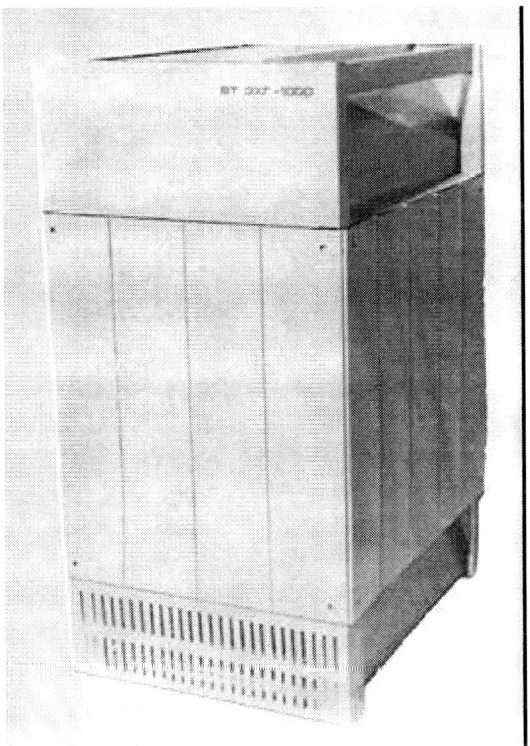

Figure 12. Power of 1.0 kW, the efficiency of 43% at 950 °C during 1500 hours; with the fuel (methane) utilization of about 90%, the specific power was about 200 mW/cm^2 [18].

Starting from the 1960s, specialists at the Institute of High-Temperature Electrochemistry, Russian Branch RAS (IHTE, Ekaterinburg) have been developing solid electrolytes with the oxygen ion conduction. Early in 1970s they began making prototype electrochemical devices based on these electrolytes. As the culmination of this work, in 1989 a temporary creative team headed by the author made a high-temperature electrochemical generator rated at 1 kW, Fig. 12 [18]. The generator consisted of six modules (Fig. 13) each comprising sixteen tubular cells in the form of tubes 210 mm long and about 10 mm in diameter. The walls of the solid electrolyte based on zirconia stabilized with a mixture of scandium and yttrium oxides were about 0.4 mm thick. With methane as the fuel and air as the oxidizer, at 950 °C the generator provided the specific power of about 200 mW/cm^2, the efficiency of 43%, and the fuel utilization factor at the maximum power of about 90%. The arrangement of the cells in the module was patented [19] and served as the prototype of all subsequent batteries and power systems with tubular SOFCs made as tubes.

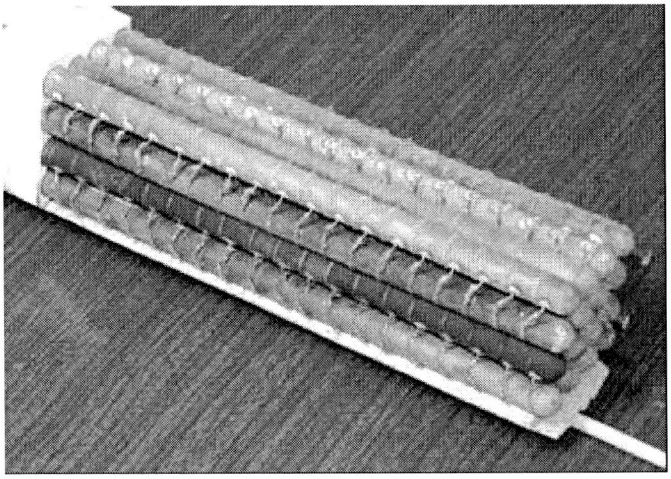

Figure 13. External appearance of the module.

Figure 14. Module of batteries rated at 2.5 kW [6, 21].

Starting from late 1980s, the development of SOFC power systems was continued, as the extension of the research performed at IHTE, almost on the enterprising basis at the Russian Federal Nuclear Center – Acad. E.I. Zababakhin's All-Russian S&R Institute of Technical Physics (ITP, Snezhinsk). Initially the developers worked on all SOFC designs (tubular, planar, block), but finally they settled on the tubular cells [20]. In recent years the work culminated in successful tests of power systems rated at 1-2.5 kW, Fig. 14 [6, 21]. The batteries, Fig. 15, comprise cells with the bearing tubular electrolyte (tubes of the YSZ electrolyte about 120-160 mm long and about 10 mm in diameter with walls less than 0.8 mm thick). All the materials and the components of the SOFCs were made at Russian enterprises. The distributed multipoint current collector from the anode and the cathode allowed decreasing the internal resistance of the cells and reaching the specific power of a single cell equal to 350 mW/cm^2 or 150 mW/cm^2 as part of a power system. The power systems are intended for use on pipelines belonging to "Gazprom" and operating in cold climatic conditions.

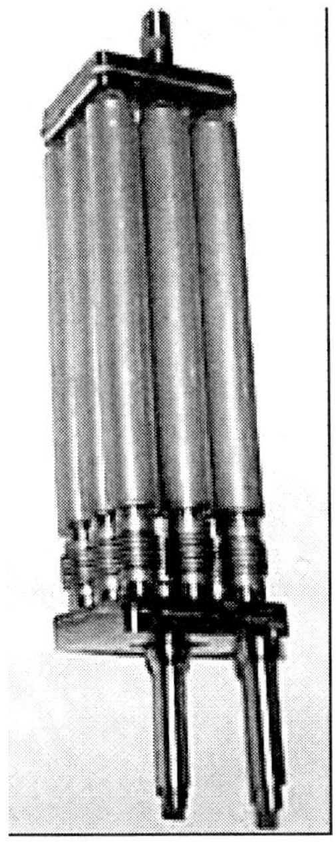

Figure 15. Battery of 8 SOFCs [6, 21].

Specialists at the Russian Federation State Scientific Center "A.I. Leipunsky's Physical and Power Engineering Institute" are developing a planar SOFC, which uses, in addition to the YSZ electrolytes, a cerium-based electrolyte for the low-temperature operation. The efficiency of up to 50% was reached at 700 °C. The maximum specific power of 700 mW/cm^2 was derived from single cells at 950 °C. The external appearance of the battery with planar cells is shown in Fig. 16. Figure 17 presents single square and disc-shaped cells [6, 22].

Many Russian scientific groups are dealing with the solid state ionics. But their activities are coordinated little if at all. The Russian Society of Solid State Ionics (RSSSI), which is part of the International Society of Solid State Ionics, was organized in 1983 (for more information see 23].

Late in 2003 Academician Yuri Osipov, who is the President of the Russian Academy of Sciences, and Mikhail Prokhorov, who is the General Director and the President of the Mining and Metallurgical Company "Norilsk Nickel", signed the general agreement on cooperation between the Russian Academy of Sciences and "Norilsk Nickel" in the field of hydrogen power engineering and fuel cells. This is the first major agreement between the Academy of Sciences and one of the leading industrial companies in Russia [for more information see 24].

Figure 16. Experimental battery of flat cells [6, 22].

Figure 17. Single cells of a planar SOFC battery [6, 22].

The cooperation is aimed primarily at the preparation, financing and realization of priority science-intensive high-tech projects whose objective is the development of competitive equipment and materials for hydrogen power engineering and fuel cells. Since the majority of scientific-engineering and fundamental problems of the development of solid-polymer, alkali and solid-oxide fuel cells have been solved by now, "Norilsk Nickel" Co. decided to finance only these technologies as they have approached most closely to the commercialization level. For this purpose, the "Norilsk Nickel" Co. has established the "New Power Projects" National Innovation Company (NPP NIC), which directly cooperates with institutes of the Russian Academy of Sciences in the development of an experimental-industrial technology for the manufacture of fuel cells and hydrogen electrochemical generators on their basis. On December 12, 2006 the Scientific and Technical Council of NPP NIC reviewed the results of the work and outlined the strategy of the development of the distributed power – electrochemical SOFC power installations rated at several kilowatts to several megawatts [for more information see 25]. STC NPP NIC decided that in 2007-2008 and for the period till 2010 one of the main priorities will be the creation of a unified series of power installations based on fuel cells and their industrial production in Russia.

Figure 18. Segments of a tubular SOFC with the bearing electrolyte 80-150 μm thick.

Figure 19. Semi-transparent ceramic of the YSZ solid electrolyte 140 μm thick.

Thanks to the financial support from the "Norilsk Nickel" Co. (last year through the NPP NIC), specialists at the Institute of Electrophysics, Ural Branch RAS (IEP, Ekaterinburg) have developed a tubular SOFC with the bearing electrolyte (150 μm) and the bearing electrode (the electrolyte 20-50 μm thick). IEP is the Russian leading institute dealing with the production of nanopowders and the development of nanotechnologies. The problem of developing SOFCs with a thin-layer electrolyte was solved at IEP through the use of the available method for synthesis of unique weakly aggregated YSZ nanopowders. Thanks to their more than 10 year experience in the production of flat and tubular components by the method of magnetic pulse compaction of nanopowders and the elaboration of the new tape casting technology, specialists at IEP have designed a semi-transparent tubular bearing YSZ solid electrolyte 80-150 μm thick (Figs. 18 and 19), a tubular bearing LSM cathode 300-500 μm thick, and a SOFC with the bearing cathode and a thin-layer (25 μm) solid electrolyte [26-29]. It was for the first time in Russia that SOFCs provided the specific characteristics of 1 W/cm^2 on the bearing electrolyte and 1.3 W/cm^2 on the bearing cathode. This work was done with the financial support from the "New Power Projects" National Innovation Company.

Conclusion

The planar and tubular designs of SOFCs will continue competing within the next few years. Then they will find their application niches. Planar SOFCs will apply most likely to situations when the most important characteristics are the volume and the mass, e.g., the use of power systems in the outer space. If these characteristics are not very important, e.g., on-land, stationary or durable applications, tubular SOFCs will be used.

If the long lifetime of SOFC power systems is a critical parameter, but the unit of the tight connection of planar cells does not satisfy this requirement, the power systems will comprise tubular cells, because they can be used in batteries and stacks that do not require gas seals in the working zone since the pipe separates the gas spaces by its design. This feature probably is the main advantage of tubular SOFCs over planar cells. If we consider in addition that the problems of the heat and mass transfer are solved easier and better by the tubular design, tubular SOFCs will occupy the place they deserve in the distributed and stationary powers.

The further development of single cells should proceed through the "reconciliation" of the two designs and the creation of more promising intermediate designs combining the pipe and the planar, bringing together their favorable properties.

I think the development of SOFCs with respect to the bearing component will also reach a reasonable technical compromise when the mechanical strength will be provided jointly by an electrolyte, which is thinner, stronger and more conductive than the YSZ electrolyte, and gradient electrodes with current collectors presenting the minimum diffusion problems as to the supply and the removal of reagents. For example, the combination of ScSZ with the tetragonal and the cubic structure will increase the strength by a factor of 5 to 6. Therefore, the proposed bearing electrolyte less than 30 μm thick will provide the same strength as YSZ-150 μm. Taking into account the higher conductivity of the former, the same power can be produced in devices at temperatures at least 20 degrees lower. This means the use of cheaper constructional materials and, of course, the increase in the reliability and the lifetime of the power systems.

I believe all the technologies currently in use for SOFCs have been elaborated differently, but all of them can be used in industry on condition they are improved to the required level. The low cost of the technologies is conditional. However, the use of simplest ceramic technologies for the synthesis of thin gas-tight electrolytes by, e.g., dipping moves technological studies to the development of polymer-powder systems, the use of nanopowders, and the synthesis of weakly agglomerated initial materials. I think these are the most promising lines of the technological development.

With respect to the fuels, SOFC power systems burning the natural gas and traditional liquid fuels (gasoline, kerosene, diesel fuel) will enter the market first. Then, as the hydrogen power engineering develops, hydrogen will serve as the fuel. However, the future of the SOFCs is the use of coal as the fuel: the traditional gasification first, and the direct oxidation of coal in fuel cells afterwards.

Hopefully, the hydrogen power engineering and fuel cells, including most promising SOFCs, will be given the deserving governmental support in Russia and the appropriate state program will be adopted envisaging the development of the economically sound, truly energy efficient and ecologically compatible electrochemical power.

References

[1] 2005 Fuel Cell Seminar, Abstracts, November 14-18, 2005, Palm Springs Convention Center, Palm Springs, California, USA.

[2] 2006 Fuel Cell Seminar, www.fuelcellseminar.com.

[3] *SECA Annual Workshop and Core Technology Program Peer Review* Tampa, FL, Jan. 28, 2005.

[4] 7thAnnual SECA WORKSHOP & PEER REVIEW, September 12-14, 2006,Philadelphia,PA,USA.<http://www.netl.doe.gov/publications/proceedings/06/seca/index.htm#Posters

[5] Proceedings of the 26th Risø International Symposium on Materials Science:*Solid State Electrochemistry* Editors: S. Linderoth, A. Smith, N. Bonanos, A. Hagen, L. Mikkelsen, K. Kammer, D. Lybye, P.V. Hendriksen, F.W. Poulsen, M. Mogensen, W.G. Wang, Risø National Laboratory, Roskilde, Denmark 2005

[6] Book of Abstracts. III All-Russian seminar with attendance by foreign participants "Fuel Cells and Power Systems on their Basis", 31.01 – 03.02.2006, Ekaterinburg, Russia.

[7] M.V. Perfiliev, A.K. Demin, B.L. Kusin, A.S. Lipilin. High-Temperature Electrolysis of Gases. Moscow, Nauka, 1988. 232 p.

[8] N.Q. Minh, T. Takahashi. *Science and technology of ceramic fuel cells*. Elsevier, 1995. 366p.

[9] http://www.netl.doe.gov/publications/proceedings/06/seca/pdf/Minh-SECA.pdf

[10] http://www.netl.doe.gov/publications/proceedings/06/seca/pdf/Shaffer.pdf

[11] http://www.netl.doe.gov/publications/proceedings/06/seca/pdf/Vora.pdf

[12] http://www.netl.doe.gov/publications/proceedings/06/seca/pdf/BessettePhilly%202006%20SECA%20Acumentrics-Pub.pdf

[13] http://www.netl.doe.gov/publications/proceedings/06/seca/pdf/Norrick.pdf

[14] http://www.netl.doe.gov/publications/proceedings/06/seca/pdf/Doyon.pdf

[15] B. Heydorn, S. Crouch-Baker// Direct carbon conversion: progressions of power, The Fuel Cell Review Institute of Physics Publishing 14.02/06 http://www.sara.com/papers/FCRJanDCFC_SARA_Reprint.pdf

[16] Fuel Cells Bulletin/ An International Newsletter, December 2006. www.fuelcellsbulletin.com

[17] f-cell 2005: Workshop "Germany meets Canada" Stuttgart, 28th September 2005 Status Quo Fuel Cell Development in Europe

[18] A.K. Demin, B.L.Kuzin, A.S. Lipilin, A.D.Neuimin, M.V.Perfiliev, S.I.Somov, "Research and Development on SOFC in the USSR", Second International Symposium of SOFC 2-5 julj 1991, Athene Greece. p. 67-73

[19] RF Patent No. 2027258 "High-temperature electrochemical generator" MKI HO1M, priority 03.07.90. S.I. Somov, A.K. Demin, B.L. Kuzin, A.S. Lipilin, M.V. Perfiliev.

[20] Solid-oxide fuel cells: *Collection of scientific-technical papers*. Snezhinsk, RFNC-ITP Publishers, 2003. 376 p. with illustrations.

[21] Chukharev V.F. // "Development of power installations based on solid-oxide fuel cells at Acad. E.I. Zababakhin's RFNC-ITP" International Forum "Hydrogen Technologies for Power Production" February 6-10, 2006, Moscow.

[22] 7th EUROPEAN SOFC FORUM, 3 – 7 July 2006, Lucerne / Switzerland http://www.efcf.com/sofc/

[23] http://roitt2007.narod.ru

[24] http://www.nornik.ru/hydrogen_energy/

[25] http://www.nic-nep.ru/default.asp?pKey=000010000700001&id=18360

[26] V.V. Ivanov, A.S. Lipilin, Yu.A. Kotov, V.R. Khrustov, S.N. Shkerin, S.N. Paranin, A.V. Spirin, A.S. Kaygorodov.// *J. of Power Sources*. 2006.V. 159. p. 605

[27] Proc. 8th International Meeting "Fundamental Problems of Solid-State Ionics". Chernogolovka, Moscow oblast. June 13-16, 2006, pp. 117-121.

[28] A Spirin, V. Ivanov, A. Lipilin, S. Paranin, V. Khrustov, A. Nikonov, Al. Rempel, S. Ivin.// *J.Advances in Science and Technology*. 2006.Vol. 45. p. 1879.

[29] V.R.Khrustov, V.V.Ivanov, A.S.Lipilin, O.F.Ivanova// Proc. of 4th int. conf. of sci., technology and applications of sintering, "Sintering'05", Grenoble, France, 2005, p155.

In: Ceramic Materials Research Trends
Editor: Paul B. Lin, pp. 159-179
ISBN: 1-60021-769-9

Chapter 6

APPLICATIONS OF CERAMIC MEMBRANES IN THE FOOD INDUSTRY

Emilia M. Guadix, Antonio Guadix and Emilio J. de la Casa
Department of Chemical Engineering, University of Granada, Spain

Abstract

In the last decades, separation processes employing membranes have experienced a remarkable increase at an industrial level as an alternative to traditional techniques, mainly due to the mild operating conditions and low energy consumption.

The main problem associated with membrane filtration is the fouling caused by the adsorption of material from the feed stream on to the membrane surface or inside the pores. This phenomenon involves a significant decrease in both selectivity and permeate flux and, as a consequence, appropriate cleaning protocols are needed to regenerate the membranes. An intensive research effort has been made in order to obtain porous materials which minimize fouling effects and resist aggressive cleaning.

Since the pioneering use of cellulose acetate, other polymers such as polyamide, polysulfone, polyvinylidene fluoride have been employed for membrane manufacture. More recently, ceramic membranes have been developed based on zirconium and aluminum oxides. Ceramic membranes show excellent resistance to extreme pH, temperatures and pressures, as well as to solvents and oxidizers. They are also steam-sterilizable and resistant to microorganisms. Due to these desirable properties, ceramic membranes represent a major advance in the field of membrane technology, since new applications can be devised in the food, biotechnology and chemical industry.

The aim of this work is to present an overview reflecting the state of the art in the applications of ceramic membranes in the food industry. The chapter will be organized as follows. As an introduction, a brief description of general aspects of the pressure driven membrane separation technology will be made, including fouling and cleaning aspects. Then, the production and characterization of ceramic membranes will be undertaken and some details of commercially available membranes will be given.

The main section will be devoted to the detailed analysis of the applications. A special attention will be paid to the dairy industry, where ceramic membranes are employed in the concentration of milk for cheese manufacture, the concentration of whey protein and for the isolation of bioactive compounds. Other sectors that which benefit from membrane technology will be exposed such as fruit juices, alcoholic beverages, sugar and, finally, vegetable and animal proteins.

Introduction

Membrane Separation Processes

In the chemical process engineering area, a membrane is defined as a selective barrier which allows the passage of certain components while retaining others. Those components which pass through the membrane determine the permeate stream, while those which are retained are called the retentate stream. If the feed flow is oriented tangentially with respect to the membrane, the operation is referred as tangential or cross-flow filtration (Zeman & Zydney, 1996).

Regarding to their morphology, membranes can be grouped into 3 categories. Homogeneous membranes are between 10 and 200 μm thick and show very low permeability. Asymmetric or anisotropic membranes are formed by a thin layer (the active layer) of 0.1 – 0.15 μm which is supported in a 50 – 150 μm porous layer. This allows the combination of selectivity from the active layer and permeability from the support layer. Finally, composite membranes consisting of a thinner active layer (10 – 100 nm), an intermediate mesoporous layer and a porous support, have been developed to deliver better permeability.

The molecular transport through the membrane can be promoted by convection or diffusion, due to a gradient of pressure, concentration, temperature or electric potential (Mulder, 1991). In Table 1, the various membrane separation processes are classified as a function of the driving force.

Table 1. Classification of membrane separation processes as a function of the driving force

Force	Pressure	Concentration	Temperature	Electric potential
Process	Microfiltration Ultrafiltration Nanofiltration Reverse osmosis	Pervaporation Gas separation Dialysis Liquid membranes	Thermoosmosis Membrane distillation	Electrodialisis Electroosmosis

Among them, pressure-driven processes highlight due to their wide range of applications and implementations at an industrial scale. Those are classified according to the size of the molecules which are fractionated, as can be seen in Figure 1.

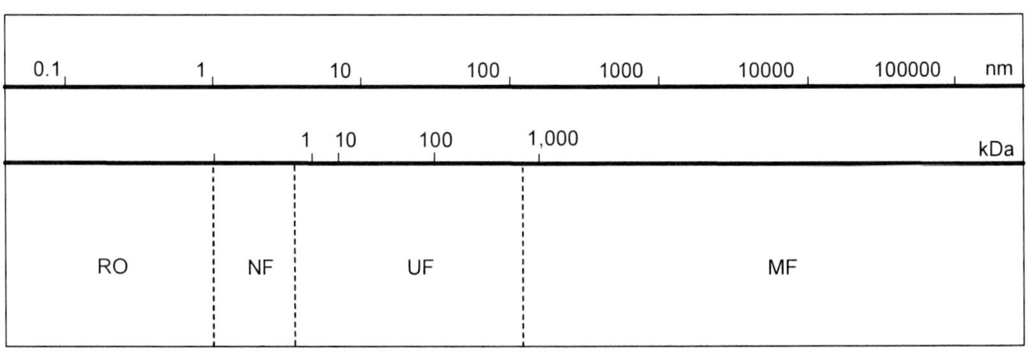

Figure 1. Membrane filtration spectrum.

It can be said that pressure-driven membrane processes find extensive uses in several industries such as pharmaceutical, food & beverage, pulp & paper, automobile and textile. Important applications can also be found for pure water production and for wastewater treatment.

Microfiltration is the most extended membrane process. It is employed to separate particles in the range 0.1 – 10 μm by applying pressures between 0.5 and 3 bar. Following ultrafiltration, species in the molecular weight interval 300 – 500000 Da are retained. To this end, membrane pore sizes between 0.001 and 0.1 μm and pressures between 2 and 10 bar are applied. By nanofiltration, low molecular weight molecules (200 – 1000 Da) and multivalent ions are retained. Typical pressures are between 5 and 30 bar. Finally, reverse osmosis employs pore sizes between 0.5 and 1.5 nm, with pressures in the 10 – 50 bar interval.

Membrane Fouling and Cleaning

The principal problem associated to membrane filtration is the progressive decrease in the flow of permeate. This process, denominated membrane fouling, involves a lower productivity and higher costs since it is necessary to implement protocols to regenerate the membranes (Song, 1998a). The severity of this phenomenon depends on the nature of the particles and the membrane and the operating conditions (concentration, temperature, pH and ionic strength) (Cheryan, 1998). As a consequence of fouling, the sieving properties of the membrane are modified and both permeability and selectivity are reduced (D´Souza & Mawson, 2005).

The most common mechanisms which originate membrane fouling are pore blocking and cake formation. Generally, in microfiltration and ultrafiltration, two different phenomena appear, namely, adsorption and deposition. Adsorption is related to the molecules in direct contact with the membrane surface. It is caused for intermolecular interactions between the particles and the membrane and takes places even in absence of filtration (Caric et al., 2000) (James et al., 2003). Adsorption occurs rapidly on the surface of the clean membrane due to electrostatic and hydrophobic interactions, disulphide bridges, van der Waals forces, among others (Hilal et al., 2005). In contrast, deposition takes place after adsorption and refers to the accumulation of material onto the membrane in a multilayer fashion.

In any case, the decrease of permeate flux at the beginning of the filtration is normally attributed to pore blocking. The maximum flux always takes place at time zero since all the pores are available. After that, pores are partially blocked depending on the relative sizes of the feed particles and the pores themselves. The more similar the particles and the pores, the more intense the blocking. Pore blocking is a very fast process: a single pass of feed solution can be sufficient to provoke a high level of this phenomenon (Song, 1998a).

After pore blocking, the decrease of permeate flux is caused by the formation and growth of a filtration cake. Given the operating conditions, the thickness of the cake is determined. This means that the filtration cake grows until an equilibrium thickness is reached and, as a consequence, permeate flux stabilizes (Song, 1998b). The accumulation of particles on the membrane surface is controlled by two opposite mass transfer mechanisms: (a) convection towards the membrane, due to permeate flow and (b) diffusion from the membrane, due to erosion promoted by the tangential flow of the retentate stream (Wang & Song, 1999).

Interestingly, it has been described the existence of a "critical flux". This is the flux value above which the deposition of particles takes place. The critical flux is determined mainly by the hydrodynamic conditions, as the shear stress, which is directly associated to the tangential flow of retentate (Grandison et al., 2000).

In order to mitigate the effects of membrane fouling, a variety of operating strategies have been developed to be carried out during the filtration process itself (Pafylias et al., 1996). One is the inclusion of turbulence promoters (rings, baffles) in the retentate channel. It is also very effective the change in the direction of flux, in which the permeate is pumped to the retentate side for a short period of time. Other possibility is to operate with uniform transmembrane pressure, which involves the presence of a permeate loop which avoids an excessive pressure gradient at the membrane inlet (Cheryan, 1998).

Although these strategies may reduce membrane fouling to some extent, it is obvious that some action should be taken in order to regenerate the membrane ideally to its original state. Here is where cleaning appears as an essential part of any filtration process. Therefore, the objective of cleaning is to recover both the permeability and selectivity of the membrane. First of all, physical cleaning is usually performed immediately after the end of the filtration. Most of times, it consists in a rinse with deionized water which is passed along the retentate channel at a high velocity and without recirculation (Nakanishi & Kessler, 1985). This is recommended by the majority of the manufacturers since the membrane should never be allowed to dry out in presence of feed solution (Cheryan, 1998). Nevertheless, the most important step in membrane cleaning is chemical cleaning. Next, the different types of chemical agents employed for membrane cleaning are enumerated.

Acids chemicals are effective for removing calcium salts and metal oxides. In both cases, the reaction involves the formation of a soluble salt. Hydrochloric and sulfuric acids are the most widely used, although the may damage the membrane modules and housings at high concentrations. Nitric acid is also employed since its oxidant properties allow to remove organic compounds. Phosphoric acid is less corrosive although its cost is higher.

Basic solutions are able to eliminate inorganic colloids and biological substances such as proteins. Their mechanisms can be related to neutralization, saponification and emulsification. Sodium hydroxide is the typical example, while sodium carbonate is also frequent.

Surfactants improve the wettability and rinsability of the membrane when included in cleaning solutions. The hydrophobic part of the surfactant remove foulants by emulsifying the fatty fractions and incorporating them in the micelles (Zeman & Zydney, 1996). Surfactants may be non-ionic such as phenol compounds and ethylene oxide, which, however, may adsorb to stainless steel involving degradation of the membrane housing. Soaps and alkyl sulphonates are anionic surfactants, while quaternary ammonium compounds are the best example of cationic surfactants. The latter are not recommended for polyamide-based membranes.

Proteases are the principal type of enzymes employed for membrane cleaning due to the importance of the fouling when proteins are present in the feed stream. The act by breaking the peptide bonds in preferred positions, resulting in shorter chains that can be removed more easily.

In some cases, for example, in the food and pharmaceutical industries, treatments leading to disinfection and sterilization are required. Sodium hypochlorite is the most popular disinfectant. It is employed at pH=10, since at pH below 9, chlorine gas is generated. This has to be avoided due to the very corrosive activity of the gas, even for stainless steel. Other oxidizing agents include hydrogen peroxide and peracetic acid. Steam sterilization for 15 minutes at 121 °C and 1 atm is usual for pharmaceutical applications. Other possibilities are sterilization with ethylene oxide or chlorofluorocarbons, although longer exposure times are to be applied. Gamma irradiation is an alternative that must be employed carefully with some materials.

Ceramic Membranes

Materials for Membrane Manufacture

Historically, the first type of material employed for the manufacture of membranes was cellulose acetate. This is based on cellulose in which one or more hydroxyl groups are replaced by acetyl ones, following an easy fabrication process that does not involve high costs. These membranes offer reduced protein adsorption so its fouling potential is lower when compared to other polymers. However, this material is not very popular nowadays because of its sensitivity to temperature, pH and chlorine, which does not allow to apply severe cleaning procedures. Moreover, the permeability and selectivity of cellulose acetate membranes are altered by compaction as a consequence of transmembrane pressure.

Taking into account the disadvantages of cellulose acetate, membranes from different polymers have been constructed. Polyamide membranes were developed for nanofiltration and reverse osmosis applications. This material presents better tolerance to high temperatures and extreme pHs, as well as good mechanical resistance. However, polyamide is very sensitive to oxidants like chlorine and is easily fouled by proteins ought to its low hydrophilicity. Polysulfone membranes are characterized by having diphenylene sulfone units in their structure. The offer average thermal stability and resistance to pH and chlorine. Acrylic polymers have also been employed for membrane manufacture, although they show low mechanical resistance and it is difficult to obtain small pore sizes. The latter also applies to polymers such as PVDF and PTFE, which are employed for microfiltration, although they are more resistant.

According to their composition, mineral or ceramic membranes are known as the third generation of membranes. Important improvements with respect to polymeric membranes must be highlighted, such as excellent thermal, chemical and mechanical resistances. Ceramic membranes can operate at high values of temperature and pressure and extreme acid or basic conditions. These membranes can be steam sterilized and back-pulsed. All these facts are crucial for many applications where polymeric membranes are not useful. As a major advantage, ceramic membranes can offer a long service lifetime, up to 10 years, which dramatically improves the period for polymeric membranes (usually 1 year) (Garmash et al., 1995) (Bobkova et al., 1997). Currently available pore sizes for ceramic membranes ranges from microfiltration to nanofiltration, covering a wide variety of potential applications.

Methods for Membrane Manufacture

Ceramic-Powder Sintering and Structural Leaching

The simplest method for the manufacture of microporous filters is sintering of ceramic powders mixed with decomposable or gas-forming agents which are added before molding to originate the pores (Komolikov, Y. I. & Blaginina, L. A., 2002). These components are of both organic (cellulose, amines, alcohols, hydrocarbons) and inorganic (carbonates, nitrates) nature. Decomposition is promoted by high temperature treatment that also controls the volume and size of the pores, which remain after the agents have been eliminated. Obviously, the granulometric composition of the starting powdered material determines the porosity and pore size of the ceramic membrane.

Methods have been described using glass-ceramic powders which involve the following steps: glass melting at around 1500 °C, ceramization at 750 and 1060 °C for 1 hour, grinding, separation according to dispersivity, molding and sintering. This technique avoids the use of binders because the residual glass phase acts as a cementing agent.

In the leaching method, a chemically unstable phase is removed with and acid or a base and, as a consequence, the porous structure is originated. Leaching is the last step of a process in which boroaluminosilicate glasses or ceramics containing a glass phase are melted, grinded, molded and sintered.

Multilayer Selective Membranes

For asymmetric and composite membranes, the selective layers are prepared using several methods (Vercauteren et al., 1998) (Komolikov, Y. I. & Blaginina, L. A., 2002). For example, in the sol-gel technology, a metal oxide sol is prepared by hydrolysis of metal alkoxides or peptization of metal hydroxides. Then, water-soluble organic binders are introduced into the sol to obtain a composite which is applied on a ceramic substrate before drying and calcination. Using this technique, the control of the membrane porosity is very accurate, i.e. nanofiltration layers with less than 3 nm of membrane pore can be prepared.

Multilayer membranes are also obtained by modifying the surface of a porous substrate. In the impregnation method, a water-soluble salt is applied. After drying, the membrane is immersed in a solution of another salt which results in the formation of an insoluble precipitate. Different porosities are obtained as a function of the salts employed.

Another technique involves the deposition of a suspension on a ceramic substrate by a slip-casting method. The aqueous suspension is prepared by dispersing the ceramic material. Then, the membrane is formed on the substrate, which is followed by drying and calcination. Usually, a low-concentration suspension (3 – 10 %) is applied first to form a separating layer and a higher one (20 – 50 %) is applied afterwards to originate the filtering layer.

Other methods include solid – phase sintering and sputtering. In the solid – phase sintering method, a porous ceramic substrate is impregnated by an ultradisperse ceramic powder, which is pressed to form a 1 μm uniform membrane layer and sintered. Membranes are also prepared by sputtering adhesive materials on a porous substrate. Here, the thickness of the sputtered layer controls the pore size.

Methods for Membrane Characterization

Structural and Functional Characterization

The characterization of ceramic membranes includes both structural and functional aspects. The most relevant structural parameters are the pore size distribution, porosity (also known as void fraction) and pore density (or number of pores per unit area of membrane surface).

With respect to functionality, the main parameters are permeability and transmission coefficients to different solutes (which determine selectivity). Permeability is defined as the volume of filtrate that passes through the unit area of membrane surface per time unit. The transmission coefficient for a given solute is defined as the ratio between the concentration of solute in the permeate and the concentration of the same solute in the retentate. Other important functional parameters are the effective diffusion coefficients, as well as mechanical, thermal and chemical resistances.

Obviously, the structural and functional characteristics of a membrane are interconnected. For example, an increase in porosity involves an increase in permeability, and an increase in the average pore size causes an increase in the transmission coefficients.

In order to determine the parameters mentioned, a number of experimental techniques have been tested including liquid penetration methods, electron and atom force microscopy, permporometry, thermporometry and others (Palacio, 1998).

Liquid Penetration Methods

These methods are based on the pressure that must be applied in a membrane to allow the flow of a fluid in which the membrane is impregnated. Bubble point, liquid – air and liquid – liquid displacement are usually employed to easily estimate the average pore size and pore size distribution in commercial membranes.

Mercury porometry is a liquid – air displacement technique in which mercury is employed to fill a dry membrane. A disadvantage of this method is that sometimes one has to use elevated pressures that may modify slightly the membrane structure. In the liquid-liquid displacement method, the pressure applied to a liquid which displaces another liquid is related to the interfacial tension between the two immiscible liquids, the contact angle and the pore radius according to the Cantor equation (Calvo et al., 2004). The permeability profile with respect to each pore size interval is obtained from the experimental curve flux – pressure by assuming laminar flow and cylindrical pores and, therefore, the Hagen – Poiseuille equation. For membranes with small pore size, liquids with low interfacial tensions are selected in order to avoid the application of high pressures. On the contrary, for large pore sizes, high interfacial tensions avoid the invasion of liquid even at low pressures.

Electron and Atom Force Microscopy

Microscopy methods allow a first visual analysis of both superficial and cross sectional images of the membranes. From these images, the pore size distribution, the porosity and the membrane thickness can be determined by employing the appropriate piece of software. The membrane is attacked with high energy electrons and the resulting interactions allow the characterization of the material.

In scanning electron microscopy (SEM), the energy ranges from 50 to 50 keV achieving magnification factors from 20 to 105. Transmission electron microscopy (TEM) utilizes electrons at 100 – 200 keV which suffer dispersion when crossing zones with different configuration or chemical composition. Field effect scanning electron microscopy (FESEM) allows very high resolution, even below 1 nm, using low energies.

The principal advantage of atom force microscopy (AFM) with respect to electron microscopy is that the former does not need a previous preparation of the sample. As explained for FESEM, AFM also works in the nanometer scale.

Permporometry and Thermporometry

Permporometry is based on the controlled blocking of the membrane pores by the condensation of the steam present in a gas mixture and in the simultaneous measure of gas flow through the pores. The pore size distribution is obtained using the Kelvin equation. Although the contact angle is usually known, the main uncertainty of this method is in the thickness of the adsorption layer, which depends on pressure.

The termporometric method was proposed in 1997 and exploits the fact that the solidification point of a gas condensed in the membrane pores is a function of the interface curvature. A differential scanning calorimeter is employed to observe the phase transition temperature and then to calculate the pore size distribution.

Solute Retention Tests

This very common and simple method is highly influenced by the operating conditions and the kind of solutes employed. The permeate flux and the rejection coefficients for a number of solutes of various molecular weights (and therefore different hydrodynamic size) is measured experimentally and correlated to the average pore size of the membrane.

Other Methods

Nuclear magnetic resonance (NMR) can be used to characterize ceramic membranes. A previous calibration with a material of known pore size is needed, as well as to assume predetermined pore geometry. X – ray diffraction allows to elucidate the degree of cristalinity of the membrane material. Finally, by performing Fourier transform infrared (FTIR) spectroscopy, one can identify the chemical composition of the material adsorpted onto membrane surfaces.

Commercial Ceramic Membranes

Membrane separation processes are more and more popular in the food industry, mainly in the dairy and beverage industries. Annual sales in this sector are estimated around 300 million euros, of which one half is devoted to microfiltration.

Currently, most of the ceramic membrane modules that are commercially available are presented in tubular form or as a multichannel monolith. This configuration avoids membrane clogging during operation and an easier cleaning procedure. The membrane support is usually made of α - alumina with a pore size between 2 and 6 μm. The active layers, with a thickness around 20 μm are formed by different proportions of $\alpha - Al_2O_3$, $\beta - Al_2O_3$, TiO_2 and ZrO_2, depending on the pore size desired. In order to incorporate the membrane elements to the whole process unit, sealing gaskets are attached at the extremes and assembled in housings, which are made usually of stainless steel.

Nevertheless, ceramic membranes have been marketed also with plane geometry. Although this configuration presents a high trend to membrane fouling, it involves significant energy savings due to the lower hydraulic diameter (Grangeon & Lescoche, 2000).

In Table 2, it can be seen a list of some manufacturers of ceramic membranes, detailing the module surface and cut-off available.

In the following sections, the different food industries in which ceramic membranes find a significant application are described. Advantages when compared to traditional procedures will be highlighted.

Table 2. List of some ceramic membrane manufacturers

Manufacturer	Surface (m^2)	Cut-off	Web site
Industrial Biotech Membranes	0.0047 - 0.63	1.04 μm - 1 kDa	www.ibmem.com
Attech Innovations	0.019 - 0.58	1.2 μm - 5 kDa	www.atech-innovations.com
CEPAration	0.001 - 5	1.4 μm - 0.3 nm	www.ceparation.com
Fairey Industrial Ceramics	0.06 - 0.33	1.4 μm - 0.2 μm	www.faireyceramics.co.uk
Inopor Membranen	0.011 - 0.43	1.0 μm - 0.9 nm	www.inopor.com
Kerafol	0.336 - 0.14	2 μ m - 7 nm	www.kerafol.com
Pall Exequia	0.24 - 0.36	5 μm - 100 nm	www.exekia.fr
Pall Schumacher	0.022 - 0.622	2 μm - 5 nm	www.schumacher-filters.de
Tami Industries	0.042 - 0.63	1.4 μm – 1kDa	www.tami-industries.com
CeraMem	0.13 - 50	0.5 μm - 5 nm	www.ceramem.com
Uniceram Advanced Materials	0.04 - 0.24	0.5 μm - 0.05 μm	www.uniceramusa.com
Novasec	0.008 - 0.34	0.8 μm – 1kDa	www.novasep.com

Applications in the Dairy Industry

Cheese Manufacture

Dairy factories are the principal industrial sector which benefit from the use of membrane technology. In particular, the ultrafiltration of milk for cheese manufacture is the largest application in terms of membrane area installed. Cheesemaking usually involves the coagulation of pasteurized milk by the action of clotting enzymes or acid, followed by separation of the solid curd and the supernatant, which is known as whey and constitutes about 90 % of the initial milk volume. Then the curd is converted into the final cheese through the ripening process thanks to specific biological reactions involving bacteria, fungi or molds.

A modification of the traditional process was introduced by Maubois, Mocquot and Vassal, inventors who gave name to the MMV process. At this point, one can see that the main step in this process is a separation in which caseins and fat are concentrated in the cheese, while mainly lactose and whey proteins remain in solution. The MMV idea is to eliminate the excess of water and lactose by means of an ultrafiltration stage in which milk is concentrated to the desired protein level (Figure 2). The final fat concentration of this "pre - cheese" can be adjusted by employing skim milk as a feed for the ultrafiltration step and adding cream afterwards. The appropriate microorganisms are inoculated to the pre – cheese, which yields the final product after ripening. Temperature must be carefully controlled during ultrafiltration, especially when operating in the batch mode, with an optimal value around 50 °C, in which a good permeate flux is obtained due to the low viscosity. In addition, microbial growth is maintained low without an excessive denaturation of the whey proteins.

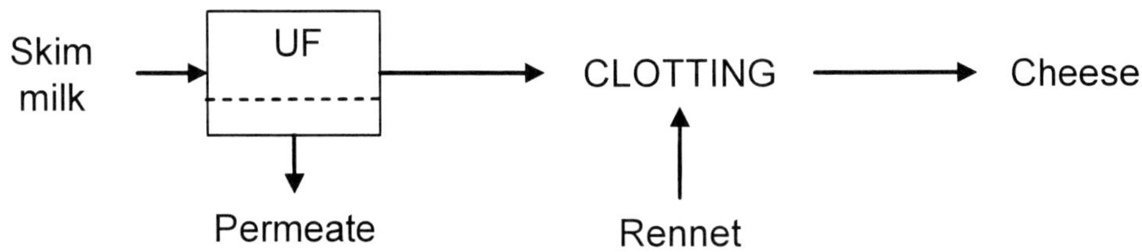

Figure 2. MMV cheese production proccess.

The main advantage of the MMW concept is an increase in the yield of the process of 10 – 30 % with respect to the conventional one, because most of the whey proteins are not lost since they are retained in the precheese during ultrafiltration. The amount of rennet enzyme needed is reduced a 80 % since ultrafiltration makes most of job of removing water.

The treatment of wastewater from cheese factories is an important environmental problem due to the high biological oxygen demand (DBO) of the whey stream. Interestingly, DBO is reduced in the MMV process, since less organic matter is contained in the ultrafiltration permeate when compared to traditional whey. MMV also eliminates the installation of tanks for curd heating (which is usually performed for an extra liberation of whey) and is more suitable for an operation in the continuous mode. As a conclusion, MMV presents significant pros in terms of process efficiency and operating costs, which makes it an attractive alternative from the economic point of view.

Whey Protein Concentration

Whey protein concentrates (WPC) are powders made by drying the retentates from ultrafiltration of whey. With a protein richness in dry matter that ranges from 35 to 85 %, WPC is an ingredient that find numerous applications in the food industry, for example, in the production of yogurt and ice cream, baked products, pasta and fruit beverages. Some possibilities of whey processing are shown in Figure 3. For example, liquid whey is concentrated by ultrafiltration to a solids content around 25 % and then diafiltered to remove more lactose and minerals. In diafiltration, water is added to the feed in order to elute low molecular components.

When whey retentate from an ultrafiltration plant is further processed by microfiltration, fat content can be reduced from around 7 % to as low as 0.4 %. In the microfiltration retentate, fat and bacteria are concentrated, while the defatted permeate is diafiltered and dried. As well as the uses mentioned above for regular WPC, defatted WPC is an interesting replacement for egg white in whipped products.

The high salt content (8 - 12 %) of whey limits its use in human foods, therefore partial or high demineralization opens new possibilities for the implementation of WPC for example in infant formula. Apart from inorganic salts, partial demineralization removes lactates and citrates. This can be achieved by cross – flow filtration using nanofiltration membranes where sodium, potassium, chloride and small molecules like lactic acid are permeated. Reduction of the chloride content can reach 70 %, while sodium and potassium reduction is around 30 – 35 %.

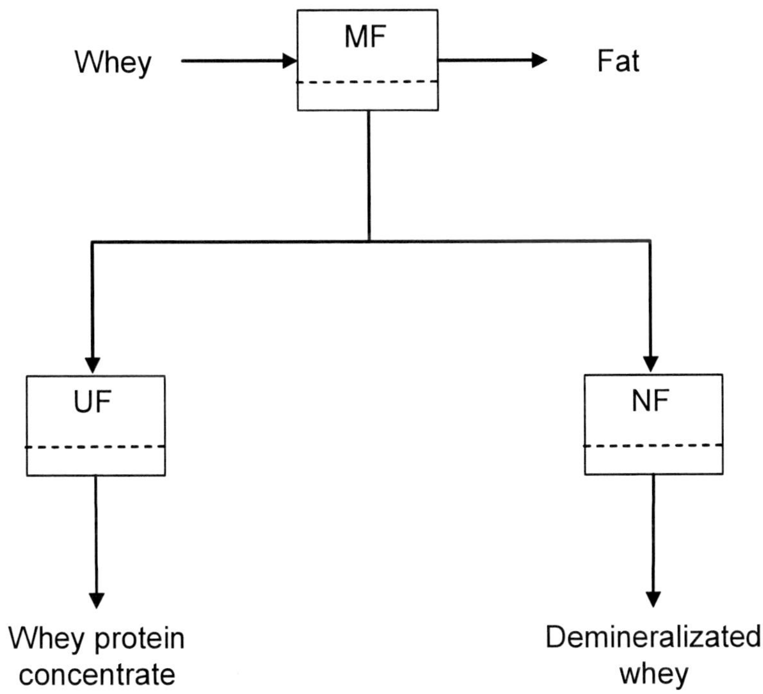

Figure 3. Whey proccessing alternatives.

Functional Ingredients

The isolation concentration of compounds with a proven beneficial biological activity is the most promising application of membrane technology in the dairy industry. These compounds can be supplemented in the formulation of functional foods and nutraceuticals. Two relevant examples of this alternative are a result of the fractionation of whey protein and the purification of oligosaccharides from caprine milk.

Individual whey proteins have their own unique nutritional, functional and biological characteristics. For instance, α-lactalbumin has been claimed as a nutraceutical and a food additive in infant formula owing to its high content in tryptophan and as a protective against ethanol and stress-induced gastric mucosal injury. β-lactoglobulin is commonly used to stabilize food emulsions because of its surface active properties. Bovine serum albumin

(BSA) has gelation properties and it is of interest in a number of food and therapeutic applications, for instance, because of its antioxidant properties. Bovine immunoglobulins can enhance the immunological properties of infant formula and they can be used therapeutically in the treatment of animal neonates and , in the form of special supplements, they can offer, in many situations, an important reduction of risk to acquire diarrhea causing infections and other illnesses. Bovine lactoferrin, which exists at relatively low concentrations, has important biological activity. Lactoferrin has been reported to have antimicrobial, immunostimulatory and anti-inflammatory activity since it affects growth and development of a wide range of infectious agents. This protein plays a fundamental role in iron metabolism as an iron-transport molecule. Furthermore, recent studies have demonstrated that lactoferrin suppresses tumor growth and metastasis in mice and rats and moreover may inhibit angiogenesis. In addition, whey proteins, partially digested, serve as a source of bioactive peptides.

Therefore, whey fractionation for the recovery and isolation of these proteins has a great scientific and commercial interest. Membrane filtration has traditionally been based solely on differences in molecular mass. Until recently, membranes were thought to achieve separation only between proteins differing in size by at least a factor of 10. The name high performance tangential flow filtration (HPTFF) has been proposed to describe a very selective process, which allows the separation of proteins with very similar molecular weight (van Reis et al., 1997). HPTFF is based on the proper choice of pH and ionic strength in order to maximize the differences in the effective hydrodynamic volume of the different proteins (de la Casa et al., 2007). Recent publications have demonstrated the feasibility of HPTFF for the fractionation of some whey proteins (Zydney, 1998) (Muller et al., 2003) (Cheang & Zydney, 2004) (Almecija et al., 2007).

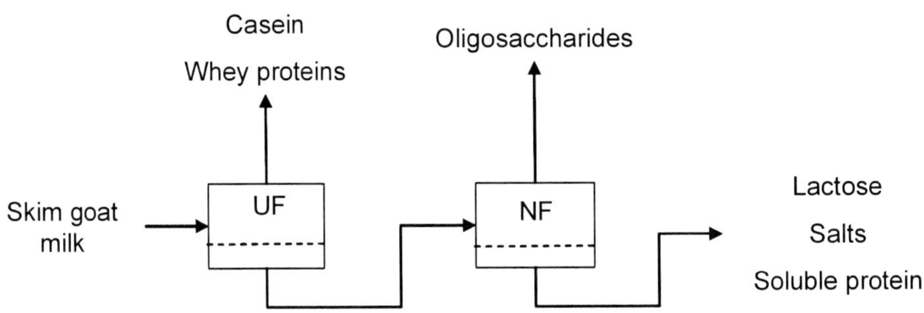

Figure 4. Scheme of the filtration process for the isolation of caprine milk oligosaccharides

Caprine milk contains a large amount of different sialylated and neutral lactose-derived oligosaccharides compared to cow or sheep milk. In addition, its oligosaccharide profile is very similar to human milk, which suggests it has similar physiological activity (Martinez-Ferez et al., 2006a). Thus, the caprine milk oligosaccharide fraction is a very promising food ingredient for human nutrition applications, especially for the supplementation of infant formulas (Daddaoua et al., 2006). A two-step cross-flow filtration process in order to recover the caprine milk oligosaccharides has been designed (Figure 4). Tubular ceramic membranes with molecular weight cut-offs of 50 and 1 kDa, respectively, were employed in two separated, consecutive continuous diafiltration steps, in which the cumulated permeate from the first step was the initial feed in the second one. A final retentate containing more than

80% of the original oligosaccharides and less than 4% of the original lactose and protein was obtained (Martinez-Ferez et al., 2006b).

Applications in Fruit Juices

The traditional process for the production of fruit juices involves several consecutive steps. After washing and peeling, the fruit if pressed and a first juice is produced which is pasteurized and treated with pectinase to reduce its turbidity. Then, a fining clarification agent (for example, gelatin) is added and held for a long time for decanting. Finally, filtration using diatomaceous earth is performed to eliminate suspended solids and colloidal particles.

The decantation and filtration steps can be replaced by membrane filtration. Ultrafiltration has effectively performed the clarification without pectinase and fining agents (Bolduan & Latz, 2000). The retention of pectin, starch, impurities and microorganisms by the membrane allows the obtention with higher yield of a clear juice obtained in the permeate, which also shows excellent flavor and color. The elimination of enzyme addition is also very positive, since some health issues about pectinase have been described.

The use of membrane involves some further advantages. First, tannin – protein complexes are reduced in the finished juice. Second, the formation of hazes during final product storage is reduces since polyphenol oxidase (which promotes enzymatic browning) is also rejected in the ultrafiltration step.

The most important fruit juices that have been assayed for membrane filtration are apple and citrus juices. Other juices include apricot (Hart et al., 1989), carrot, cherry, cranberry, grape, kiwi, melon, passion fruit, peach, pear, pineapple, plum, raspberry, tomato (Bottino et al., 2002) and strawberry (Rwabahizi & Wrolstad, 1988).

Apple juice ultrafiltration is a major example of this application (Figure 5). Pressures between 1 and 2 bar and temperatures around 50 °C are employed to minimize microbial growth. Using ceramic membranes, the permeate flux is about 100 L/(m2•h) compared to 70 for polymeric membranes. A further diafiltration stage can be implemented on the apple slurry in increase the juice yield.

Besides orange juice, other citrus juices such as lemon and lime are processed by ultrafiltration as a previous step of debittering (removal of limonin, polyphenols and other compounds) and concentration (Koseoglu et al., 1990). Citrus pulp is separated, pasteurized to deactivate enzymes and microorganisms and returned to the ultrafiltration permeate.

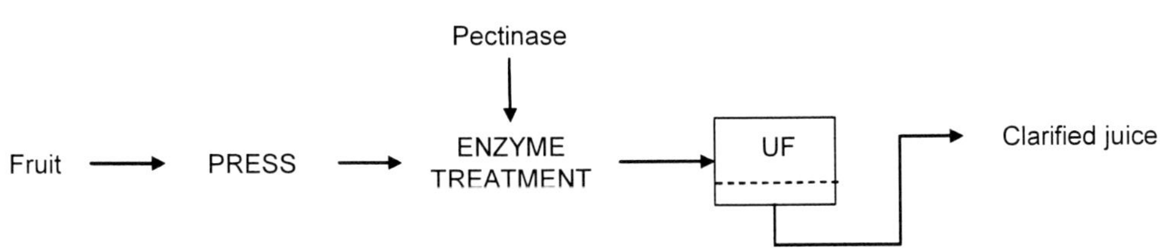

Figure 5. Manufacture of apple juice by UF processes.

Applications in Alcoholic Beverages

Wine Production

Microfiltration and ultrafiltration can be employed in the production of wine, the fermented juice of grapes (Figure 6). Before fermentation, the clarification of grape juice by ultrafiltration removes colloids and gums. Simultaneously, microorganisms are retained in the membrane which eliminates the traditional addition of bisulfite, which has been reported to cause undesirable flavor and health risk related to asthma.

The stability of the finished wine obtained after fermentation can be improved by ultrafiltration. Haze, caused by unstable (at acid pH) low molecular weight proteins (12600 – 30000 Da) and glycoproteins, is an important problem in white wines, which is traditionally treated by adding a fining agent (for example, bentonite). Unfortunately, this may involve the elimination of some flavor compounds in parallel. Following ultrafiltration through a 10000 Da membrane, the usage of fining agents is reduced up to 90 % (Hsu et al., 1987) and final wine can keeps its genuine flavor. The excess of polyphenols, which cause astringent flavor, can also be mitigated by ultrafiltration. On the other hand, polyphenol oxidase responsible of undesirable brown compounds is completely removed, thus improving wine color. Finally, wine can be microfiltered to sterility before bottling.

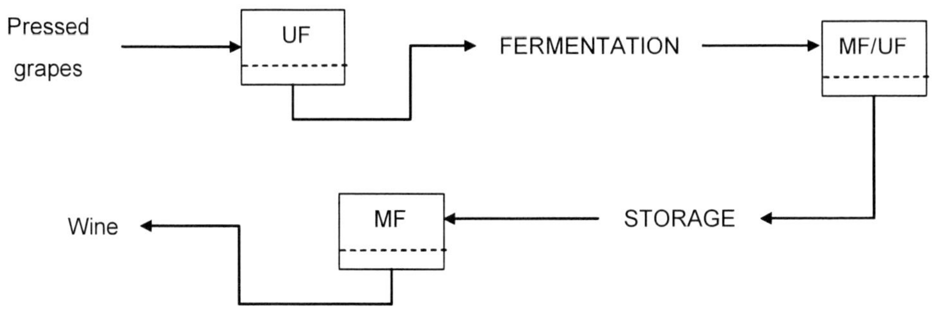

Figure 6. Membrane applications in wine production.

Beer Production

Beer is obtained by fermentation of barley or other cereals. In beer clarification, membranes can be used in order to remove the yeast cells from the beer in the fermentation process. Since other large particles formed by coagulated polyphenolic and protein compounds are removed, colloidal stability is also achieved. As a consequence, the need for diatomaceous earth filtration is eliminated.

As an alternative to pasteurization for the microbial stabilization of beer, 0.45 μm microfiltration membranes retain most of the unwanted microorganisms without exposure to heat and oxygen, thus keeping beer's natural flavor.

The productivity in beer factories can be increased by using 0.2 μm microfiltration membranes to concentrate the sediment found at the tank bottoms. This represents a 4 % of the total tank volume and 70 % of the beer can be recovered as a permeate and incorporated to the bulk production.

Unfortunately, intense membrane fouling occurs during bear filtration. Although adequate mechanical procedures such as backflushing (Gan, 2001) seems to be effective for membrane cleaning, this drawback inhibits a wide implementation at large scale of this technique.

Applications in Sugar

White sugar is obtained by extraction from cane or beets, purification and decoloration steps. Natural polysaccharides, dextrans and phenol compounds (brown color) are the main contaminants that must be removed. Following traditional operation, ion - exchange resins placed at the end of the process are fouled very quickly by traces of impurities.

Microfiltration or ultrafiltration can be used to eliminate colloidal and macromolecular species from the sugar juice to obtain a clear uncolored product without using lime or sulfite. At the same time, energy costs, yield and sugar crystallization are improved. Reduced bacterial counts and storage losses can be achieved also by employing membranes to clarify the thick juice obtained after evaporation.

A modified membrane process has been proposed in which the purified juice is consecutively processed by microfiltration, ultrafiltration and optionally nanofiltration, which is necessary if color is not acceptable after ultrafiltration (Figure 7). High quality white sugar crystals are obtained with this technique after complete decolorization by ion - exchange.

Studies on the effect of operation parameters for microfiltration and ultrafiltration with tubular ceramic membranes on the purification and decolorization of raw (Hakimzadeha et al., 2006) juice have shown that the resulting clarified juice presents better color (49 %) and turbidity (78 %) when compared to conventional lime - carbonated juices.

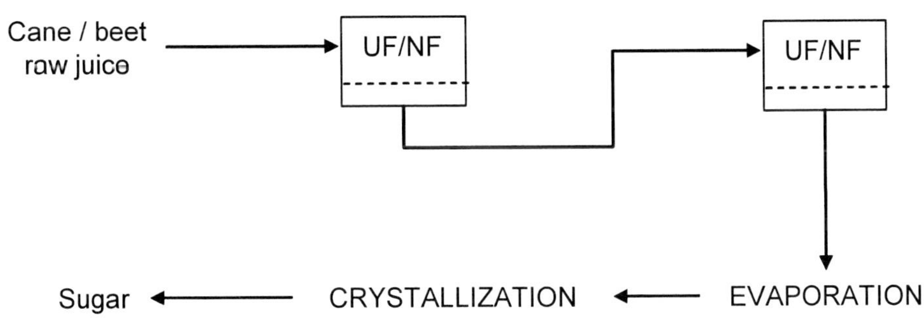

Figure 7. Possible applications of MF and UF in sugar processing.

Applications in Vegetable Proteins

In recent years, the interest for vegetable proteins has increased opposite to animal products consumption due to both health and economic issues. Soybeans, the most well established source of vegetable proteins, are more and more employed as nutritional or functional ingredients in the form of soy protein concentrates (SPC, minimum 70 % in protein) and isolates (SPI, minimum 90 %).

The traditional method for the production of SPC removes soluble sugar from defatted soy flakes employing an alcoholic or acid solution. Soy protein extract is prepared by extraction the soy flour with a basic solution and centrifugation. After acidification, precipitation, washing and neutralization, SPI is obtained by spray drying. Typical yields for SPC and SPI are 65 and 35 %, respectively.

The employment of these conventional techniques results in soy powders with low protein solubility. At the same time, functional properties are degraded due to the severe operating conditions involved in the extraction, heating, precipitation or centrifugation steps. Furthermore, considerable amount of wastewaters are generated causing an important environmental problem.

Alternatively, soy proteins can be concentrated by ultrafiltration (Alibhai et al., 2006) in which proteins are retained and washed out by diafiltration where oligosaccharides and minerals are eluted in the permeate (Figure 8). The product obtained presents better properties due to the mild process and the absence of chemicals. Again, a limitation of membrane processing is the decrease in the permeate flux due to fouling. With 50 kDa membranes, all the solubilized protein is recovered while 100 kDa membranes allow the permeation of oil molecules linked to small proteins, which results in better organoleptic properties (Samoto et al., 1998).

Regarding other proteins, it has been assayed the recovery by ultrafiltration of protein from highly polluting potato processing wastewaters. Currently, low quality protein is obtained in such process, only suitable for animal feed (Zwijnenberg et al., 2002). Other vegetable proteins studied are those from alfalfa, chickpeas, coconut, corn, cottonseed, rapeseed and sunflower seeds.

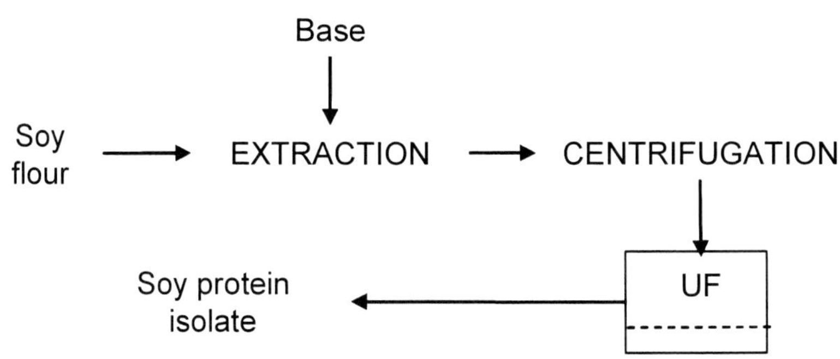

Figure 8. Production of soy isolates by ultrafiltration.

Applications in Animal Products

Meat and Poultry Processing

In terms of volume, blood is the principal waste from the meat processing industry. Its main fraction (70 %) is plasma, which is usually employed as a food supplement for animals. When obtained after centrifugation only, it is uneconomic to spray-dry plasma, which is solved if ultrafiltration is performed to preconcentrate solids up to 30 %. In order to get a better acceptance of this kind of products, techniques including chemical treatment and enzymatic hydrolysis are available for decolorization and a better palatability (Gomez-Juarez et al., 1999).

Gelatin is obtained from animal by-products, mainly from the bones of slaughtered farm animals. An aqueous extract about 3 % in protein is obtained after acid or basic treatment and is transferred into a triple effect evaporator which removes 85 % of the water. The high energy consumption of the evaporation process makes ultrafiltration (Figure 9) an alternative for the partial dewatering of gelatin liquor (Chakravorty & Singh, 1990). Due to the molecular weight range of gelatin (50000 - 70000 Da), a good separation can be achieved at low transmembrane pressures. The ultrafiltration also involves a simultaneous reduction in ash components which improves the gelling efficiency of the final product.

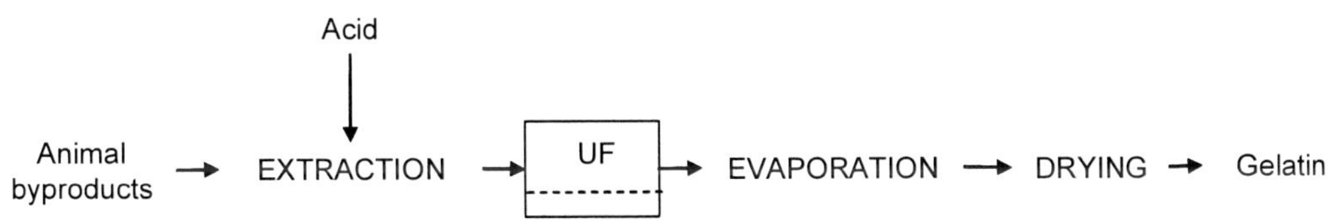

Figure 9. Membrane process in gelatin manufacture.

Large volumes of wastewaters (up to 30 L/kg) with high DBO are generated in the poultry industry in scalding and chilling operations. Conventional wastewater processing do not allow the reutilization of the valuable constituents because of the use of secondary biological treatment.

Some research (Zhang et al., 1997) has been focused on the recovery of protein from these wastewaters. Fat would be physically removed by air flotation and then protein would be retained by ultrafiltration (Lo et al., 2005).

Fish Processing

The most well established application of membrane technology in the fish industry is the recovery by ultrafiltration of enzymes from the hydrolysis of fish proteins (Figure 10). Nevertheless, a more important possibility arises for the treatment of wastewaters from the processing of fish fillets, fish meal, shellfish, crustaceans and surimi (Afonso & Bohorquez, 2002). These streams include washing, cooking and pressing waters contain a high organic load and should not be directly discharged into the sea: on the contrary, they can be treated for the recovery of proteins. Classical concentration processes such as centrifugation and evaporation may alter the proteins and are not economic for such a dilute feed. Instead, membrane separation, especially microfiltration and ultrafiltration can be employed, in which the mild operating conditions allow the simultaneous obtention of good quality proteins in the retentate and a permeate that can be recycled into the plant.

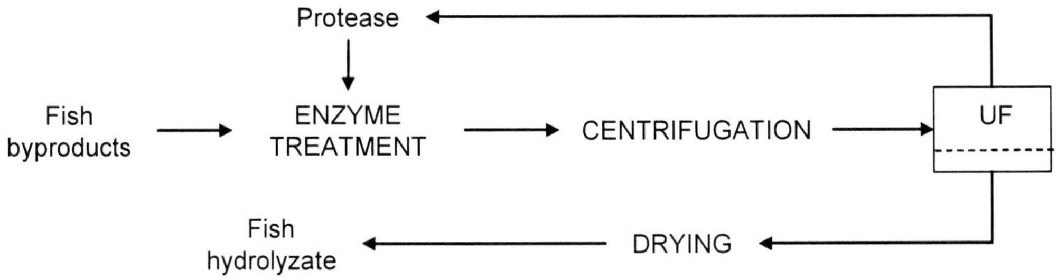

Figure 10. Process for the production of enzyme hydrolysate from fish

Egg Processing

Egg is a major ingredient in the formulation of industrial foods for its whipping and foaming functionalities. Both whole eggs and egg white, which in liquid form contain 25 % and 12 % in total solids, respectively, can be concentrated by ultrafiltration to 45 % and 24 % in total solids. In this concentrated form, egg intermediates are spray-dried without side effects and stored as a powder with an extended shelf life. Other concentration techniques, such as evaporation, are not recommended due to the sensitivity of egg protein to heat. Interestingly, glucose content in the resulting product is reduced since it is partially eliminated during ultrafiltration, which avoids the formation of brown pigments during processing.

Conclusion

This chapter has been dedicated to describe the applications of ceramic membranes in the food industry. It has been shown that membrane technology represents interesting advantages with respect to conventional separation methods, especially in terms of moderate processing temperatures and energy consumption. In addition, it has been explained how ceramic membranes are resistant to aggressive cleaning procedures, which are necessary for demanding food processes. Diverse techniques for membrane manufacture and characterization have been exposed, as well as a succinct list of membrane trademarks.

References

Afonso, M. D. & Bohorquez, R. (2002). Review of the treatment of seafood processing wastewaters and recovery of proteins therein by membrane separation processes - prospects of the ultrafiltration of wastewaters from the fish meal industry. *Desalination*, 142, 29-45.

Alibhai, Z., Mondor, M., Moresoli, C., Ippersiel, D. & Lamarche, F. (2006). Production of soy protein concentrates/isolates: traditional and membrane technologies. *Desalination*, 191, 351–358.

Almecija, M. C., Ibañez, R., Guadix, A. & Guadix, E. M. (2007). Influence of pH in the recovery of lactoferrin from whey with ceramic membranes. *Journal of Membrane Science*, 288, 28-35.

Bobkova, N. M., Barantseva, S. E. & Zalygina, O. S. (1997). Alkali-free glass ceramic rigid membranes. *Steklo i Keramika*, 6, 10-12.

Bolduan, P. & Latz, M. (2000). Ceramic membranes and their application in food and beverage processing. *Filtration and Separation*, 37, 36-38.

Bottino, A., Capannelli, G., Turchini, A., Della Valle, P. & Trevisan, M. (2002). Integrated membrane processes for the concentration of tomato juice. *Desalination*, 148, 73-77.

Calvo, J. I., Bottino, A., Capannelli, G. & Hernandez, A. (2004). Comparison of liquid-liquid displacement porosimetry and scanning electron microscopy image analysis to characterise ultrafiltration track-etched membranes. *Journal of Membrane Science*, 239, 189-197.

Caric, M. D., Milanovic, S. D., Krstic, D. M. & Tekic, M. N. (2000). Fouling of inorganic membranes by adsorption of whey proteins. *Journal of Membrane Science*, 165, 83-88.

Chakravorty, B. & Singh, D. P. (1990). Concentration and purification of gelatin liquor by ultrafiltration. *Desalination*, 78, 279-286.

Cheang, B.& Zydney, A. L. (2004). A two-stage ultrafiltration process for fractionation of whey protein isolate. *Journal of Membrane Science*, 231, 159-167.

Cheryan, M. (1998). Ultrafiltration and Microfiltration Handbook (2nd edition). Lancaster, PA: Technomic Publishing Company.

D'Souza, N. M. & Mawson, A. J. (2005). Membrane cleaning in the dairy industry: a review. *Critical Reviews in Food Science and Nutrition*, 45, 125-134.

Daddaoua, A., Puerta, V., Requena, P., Martinez-Ferez, A., Guadix, E. M., Sanchez de Medina, F., Zarzuelo, A., Suárez, M. D., Boza, J. J. & Martinez-Augustin, O. (2006). Goat milk oligosaccharides are anti-inflammatory in rats with hapten-induced colitis. *Journal of Nutrition*, 136, 672-676.

de la Casa, E. J., Guadix, A., Ibañez, R. & Guadix, E. M. (2007). Influence of pH and salt concentration on the cross-flow microfiltration of BSA through a ceramic membrane. *Biochemical Engineering Journal*, 33, 110–115.

Gan, Q. (2001). Beer clarification by cross-flow microfiltration - effect of surface hydrodynamics and reversed membrane morphology. *Chemical Engineering and Processing*, 40, 413–419.

Garmash, E. P., Kryuchkov, Y. N. & Pavlikov, V. N. (1995). Ceramic membranes for ultra- and microfiltration. *Steklo i Keramika*, 6, 19-22.

Gomez-Juarez, C., Castellanos, R., Ponce-Noyola, T., Calderón, V. & Figueroa, J. (1999). Protein recovery from slaughterhouse wastes. *Bioresource Technology*, 70, 129-133.

Grandison, A. S., Youravong, W. & Lewis, M. J. (2000). Hydrodynamic factors affecting flux and fouling during ultrafiltration of skimmed milk. *Lait*, 80, 165-174.

Grangeon, A. & Lescoche, P. (2000). Flat ceramic membranes for the treatment of dairy products: comparison with tubular ceramic membranes. *Lait*, 80, 5-14.

Hakimzadeha, V., Razavib, S. M. A., Piroozifarda, M. K. & Shahidib, M. (2006). Ultrafiltration: a means for decolorization of cane sugar solution. *Desalination*, 200, 520–522.

Hart, M. R., Ng, K. C. & Huxsoll, C. C. (1989), Microfiltration of enzyme treated apricot puree. *ACS Symposium Series*, 405, 355-367.

Hilal, N., Ogunbiyi, O., Miles, N. & Nigmatullin, R. (2005). Methods employed for control of fouling in MF and UF membranes: a comprehensive review. *Separation Science and Technology*, 40, 1957-2005.

Hsu, J. C., Heatherbell, D. A., Flores, J. H. & Watson, B. T. (1987). Heat-unstable proteins in grape juice and wine. II. Characterization and removal by ultrafiltration. *American Journal of Enology and Viticulture*, 38, 17-22.

James, B. J., Jing, Y. & Chen, X. D. (2003). Membrane fouling during filtration of milk – a microstructural study. *Journal of Food Engineering*, 60, 431-437.

Komolikov, Y. I. & Blaginina, L. A. (2002). Technology of ceramic micro and ultrafiltration membranes (review). *Refractories and Industrial Ceramics*, 43, 181-187

Koseoglu, S. S., Lawhon, J. T. & Lusas, E. W. (1990). Use of membranes in citrus juice processing. *Food Technology*, 44, 90-97.

Lo, Y. M., Cao, D., Argin-Soysal, S., Wang, J. & Hahm, T. S. (2005). Recovery of protein from poultry processing wastewater using membrane ultrafiltration. *Bioresource Technology*, 96, 687–698.

Martinez-Ferez, A., Guadix, A. & Guadix, E. M. (2006b). Recovery of caprine milk oligosaccharides with ceramic membranes. *Journal of Membrane Science*, 276, 23-30.

Martinez-Ferez, A., Rudloff, S., Guadix, A., Henkel, C. A., Pohlentz, G., Boza, J. J., Guadix, E. M. & Kunz, C. (2006a). Goats' milk as a natural source of lactose-derived oligosaccharides: isolation by membrane technology. *International Dairy Journal*, 16, 173-181.

Mulder, M. (1991). Basic Principles of Membrane Technology (1st edition). Dordrecht, the Netherlands: Kluwer Academic Publishers.

Muller, A., Chaufer, B., Merin, U. & Daufin, G. (2003). Prepurification of α-lactalbumin with ultrafiltration ceramic membranes from acid casein whey: study of operating conditions. *Lait*, 83, 111-129.

Nakanishi, K. & Kessler, K. G. (1985). Rinsing behaviour of deposited layers formed on membranes in ultrafiltration. *Journal of Food Science*, 50, 1726-1731.

Pafylias, I., Cheryan, M., Mehaia, M. & Saglam, N. (1996). Microfiltration of milk with ceramic membranes. *Food Research International*, 29, 141-146.

Palacio, L. (1999). Caracterización estructural y superficial de membranas microporosas. Ph.D. University of Valladolid.

Rwabahizi, S. & Wrolstad, R. E. (1988). Effects of mold contamination and ultrafiltration on the color stability of strawberry juice and concentrate. *Journal of Food Science*, 53, 857-861.

Samoto, M., Miyazaki, C., Kanamori, J., Akasaka, T. & Kawamura, Y. (1998). Improvement of the off-flavor of soy protein isolate by removing oil-body associated proteins and polar lipids. *Bioscience, Biotechnology and Biochemistry*, 62, 935–940.

Song, L. (1998a). Flux decline in crossflow microfiltration and ultrafiltration: mechanisms and modelling of membrane fouling. *Journal of Membrane Science*, 139, 183-200.

Song, L. (1998b). A new model for the calculation of the limiting flux in ultrafiltration. *Journal of Membrane Science*, 144, 173-185.

van Reis, R., Gadam, S., Frautschy, L. N., Orlando, S., Goodrich, E. M., Saksena, S., Kuriyel, R., Simpson, C. M., Pearl, S. & Zydney, A. L. (1997). High performance tangential flow filtration. *Biotechnology & Bioengineering*, 56, 71-82.

Vercauteren, S., Keizer, K., Vansant, E. F., Luyten, J. & Leysen, R. (1998). Porous ceramic membranes: preparation, transport properties and applications. *Journal of Porous Materials*, 5, 241-258.

Wang, L. & Song, L. (1999). Flux decline in microfiltration and ultrafiltration: experimental verification of fouling dynamycs. *Journal of Membrane Science*, 160, 41-50.

Zeman, L. J. & Zydney, A. L. (1996). Microfiltration and Ultrafiltration: Principles and Applications (1st edition). New York, NY: Marcel Dekker, Inc.

Zhang, S. Q., Kutowy, O., Kumar, A. & Malcolm, I. (1997). A laboratory study of poultry abattoir wastewater treatment by membrane technology. *Canadian Agricultural Engineering* 39, 99-105.

Zwijnenberg, H. J., Kemperman, A. J. B., Boerrigter, M. E., Lotzb, M., Dijksterhuisb, J. F., Paulsen, P. E. & Koops, G. H. (2002). Native protein recovery from potato fruit juice by ultrafiltration. *Desalination*, 144, 331-334.

Zydney, A. L. (1998). Protein separations using membrane filtration: new opportunities for whey fractionation. *International Dairy Journal*, 8, 243-250.

In: Ceramic Materials Research Trends
Editor: Paul B. Lin, pp. 181-196
ISBN: 1-60021-769-9

Chapter 7

BETA ZIRCONIUM OXIDE PHOSPHATE AND RELATED PHASES: NEW MECHANISMS OF ULTRA LOW THERMAL EXPANSION

Gilles Wallez and Michel Quarton
Université Pierre et Marie Curie-Paris 6 - CNRS UMR 7574, Laboratoire de Chimie de la Matière Condensée de Paris, 4 place Jussieu, Paris Cedex 05, F-75252 France

Abstract

High temperature X-ray and neutron diffraction combined with Rietveld analysis are powerful tools to investigate the thermal atomic shifts in the crystal structure of ceramic materials and infer the distortions responsible for their dilatometric behaviour. This study concerns several phosphates of medium and big tetravalent cations (Zr, Hf, Th, U) with very low or negative thermal expansion, including β-$Zr_2O(PO_4)_2$, a ceramic refractory known for its high resistance to thermal shocks. These materials have revealed new mechanisms or combinations of mechanisms that widen the usual field of researches for low expansion materials. Structural features previously ignored like cavities, pairs or chains of cations at short distances, or considered as unsuitable like polyhedra connected by edges should be considered now as possible hints for contraction mechanisms.

Introduction

Thermal expansion is a common topic in numerous fields of the sciences of the solid state, like the weathering of rocks, the mechanical properties of ceramics and composites, the corrosion of metals under heat treatment or the cohesiveness of glazes on substrates. Indeed, the thermal expansion determines the usefulness of most materials. Controlling this property is important for many applications like substrates and radiators of electronics devices or dental materials, but the biggest challenge is posed by ceramics and their derivates, which suffer of a high tendency to break under thermal shocks because of their poor heat conductivity.

Relieving this vulnerability has been a constant challenge during the six past decades, and researches allowed to discover oxides with very low thermal expansion coefficients like zircon ($ZrSiO_4$), spodumene ($LiAlSi_2O_6$), eucryptite ($LiAlSiO_4$), cordierite ($Mg_2Al_4Si_5O_{18}$) or amorphous silica. Refractories, catalyst supports, engine parts, lamps, brazing fixtures, induction heating liners, rocket nozzles, gauges, spark plugs and cookware are among the most common domains of application of these compounds. They are also appreciated for their dimensional stability in specific devices like large optical instruments and high precision thermometers.

Thermal Expansion and Contraction: The Basic Mechanisms

The coefficient of linear expansion in a temperature range ΔT is defined by: $\alpha_l = 1/l_o\ \Delta l/\Delta T$, where l stands for either a macroscopic length or a parameter of the crystal cell.

In a classical approach, the origin of the thermal expansion (as it is commonly known, that is, a positive definite phenomenon) in a dense material, is to be found mainly at the scale of the crystal cell. Insofar as the potential well of a non-isolated atom is dissymmetric, the length of any chemical bond increases with temperature. Actually, a covalent bond with a deep well will be less subject to thermal expansion than a metallic or an ionic one, this is why classical materials with low expansion are generally made of infinite networks of strong bonds, like those of alumina or diamond. Obviously, zero or negative expansion ($\alpha < 2\ 10^{-6}$ K^{-1}) cannot be achieved simply by this way, but only through more or less complex structural phenomena involving atomic displacements, which will oppose to the positive expansion of the bonds.

Researches concerning this important class of materials have long been ignoring the fundamental mechanisms ruling the phenomenon and remained based upon empirical approaches. In the late eighties, Roy *et al.* pointed out the lack of analytical studies dedicated to the thermal expansion of materials, and noted that "the relation of the property of thermal expansion to composition and structure eludes us, especially in the anomalously low expansion materials in which there is the greatest interest" [1].

These authors attempted to classify the heretofore known low expansion materials, but at this time, the available experimental data were not yet sufficient to allow a clear insight. Indeed, most of the progresses in this field have been made during the past twenty years, thanks to the development of adapted instruments like high temperature X-ray, synchrotron and neutron diffractometers, as well as modelization softwares.

Until nowadays, only a few tenths of materials have shown such attractive properties. From a fundamental point of view, the major milestone on this route is still the Sleight's classification, which won him the Award of the ACS in 1997 [2]. On the basis of simple specific mechanical models, Sleight defines three major modes of contraction.

i) The increase of the symmetry, which generates a thermal contraction by homogenizing the inter atom distances, like in the ferroelectric form of the perovskites. This mechanism, based upon the non-linearity of the bond strength to bond length relation, is generally effective in a limited range of temperature, typically below a transition point, and therefore of low interest regarding most applications.

ii) The bond of expansion with contraction. Among these materials, cordierite and β-eucryptite have long been predominant in the science and technology of low expansion ceramics and vitroceramics. NZP ($NaZr_2(PO_4)_3$) and its numerous derivates are other typical members of this family. In all cases, anionic polyhedra occupied by alkali or earth alkali cations share faces with small ones occupied by cations of higher oxidation state. The expansion of the inter-cations distances resulting from Coulombic repulsions is mechanically balanced by a contraction in the perpendicular direction.

iii) The rocking of polyhedra. Providing that the oxygen polyhedra are rigid enough and the bonds are not subject to strong expansion, their thermal librations can result in 2D or 3D contractions. Therefore, typical members, like fused silica, $M^{V}{}_2O_5$ and $M^{IV}P_2O_7$ compounds, are built up of small and covalent polyhedra connected by vertices. During the past years, most researches in the field have focussed on this family, in the wake of the discovery of the first isotropic contraction in cubic ZrW_2O_8 [3]. These structures can be modelized as frameworks of rigid bodies in order to predict their thermal behaviour [4]. However, although promising, these materials still suffer of a limited range of temperature, which limitates their use as refractories.

Furthermore, ceramics made of crystallites with very anisotropic expansion coefficients may exhibit microstructural defects like cracks generated during the heating and cooling cycles, that can also play a significant damping role in the expansion of the material itself, at the cost of an undesirable lowering of the mechanical strength of the material.

Even if compounds with strong negative expansion have become a very attractive topic nowadays, it should be noted that contraction mechanisms exist not only in those rare materials, but also in all those which show anomalous low expansion, insofar as their bonds also stretch with temperature. As a rule of thumb, one can expect such a mechanism in any oxide compound with a mean linear expansion lower than those of classical ceramic materials like corundum (about 8 10^{-6} K^{-1}). Furthermore, it should not be forgotten that the most interesting materials regarding important properties like resistance to thermal shocks and invariance of dimensions are those which have a near zero expansion, not necessarily a negative one.

The state of the art, nowadays, evidences significant progresses in the two latter of the three families, with the discovery of new derivates that muster strong negative expansion coefficients, but on the opposite, comprehensive studies dedicated to materials which do not belong to any of these families remain very rare in the literature. In the following, we will evoke one of these "unclassifiable" materials, β-$Zr_2O(PO_4)_2$, and its isotypes.

Beta Zirconium Oxide Phosphate, a Remarkable Material

Orthorhombic β-$Zr_2O(PO_4)_2$ is the stable variety of the compound [5,6]. It forms irreversibly between 1100 and 1200 °C from monoclinic α-$Zr_2O(PO_4)_2$ through a progressive and reconstructive transition [7]. Note that the isomeric formula $(ZrO)_2P_2O_7$, still widely in use in the technical literature, is structurally erroneous. This compound has been known since the mid eighties for the very low mean thermal expansion of the crystal cell (α_l = +1.7 10^{-6} K^{-1}

[6], +1.5 10^{-6} K^{-1} [8] from room temperature to 1000 °C) with moderate anisotropy (α_c-α_a = 7.9 10^{-6} K^{-1}) (Figure1).

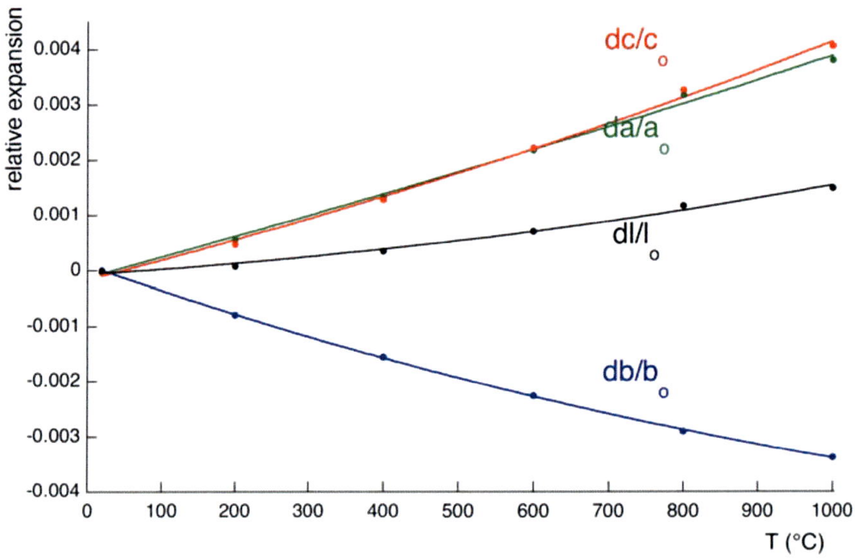

Figure 1. relative thermal evolution of the cell parameters of β-$Zr_2O(PO_4)_2$.

β-$Zr_2O(PO_4)_2$ is also stable up to 1600 °C without structural transformation and chemically inert. It can be elaborated as dense ceramics that expand at the same rate as the cell [6,9], machinable porous (28 to 30 %) materials, or composites with SiC fibers [10]. Its electrical conductivity and its dielectric permittivity are very low, thus allowing its use for the making of radomes for missiles or rockets [10]. All these remarkable performances make it a unique material, now available for industrial applications.

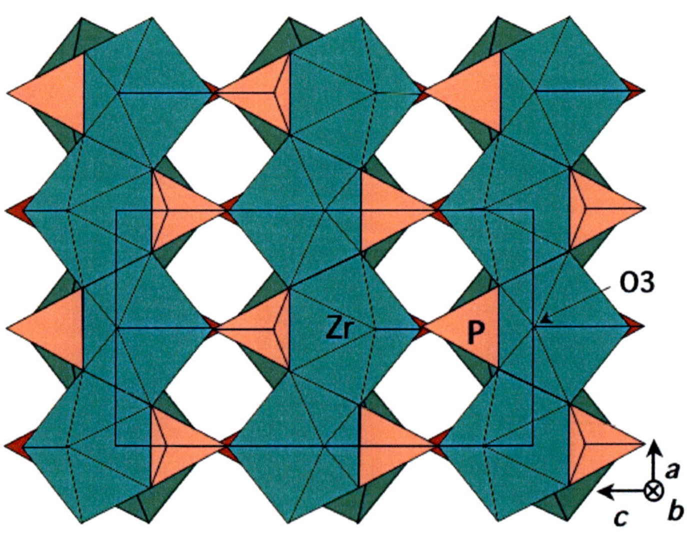

Figure 2. crystal structure of β-$Zr_2O(PO_4)_2$.

The crystal structure of β-$Zr_2O(PO_4)_2$ (Figure 2) (space group *Cmca*) [11] is made up of ZrO_7 pentagonal bipyramids forming chains along the *a* axis, and PO_4 tetrahedra. The non phosphate oxygen O(3), located on an inversion center, is strongly bonded to two zirconium atoms belonging to two parallel chains. Each zirconium atom shares two oxygen edges with two other zirconiums and one with a phosphorus. Because oxygens are in coordination 2 or 3, the structure is not fully compatible with any of the usual criteria for low expansion materials. Indeed, its behaviour, cleared up recently by Rietveld analysis of high temperature neutron diffraction patterns (Figure 3), stems from an original combination of mechanical phenomena [8].

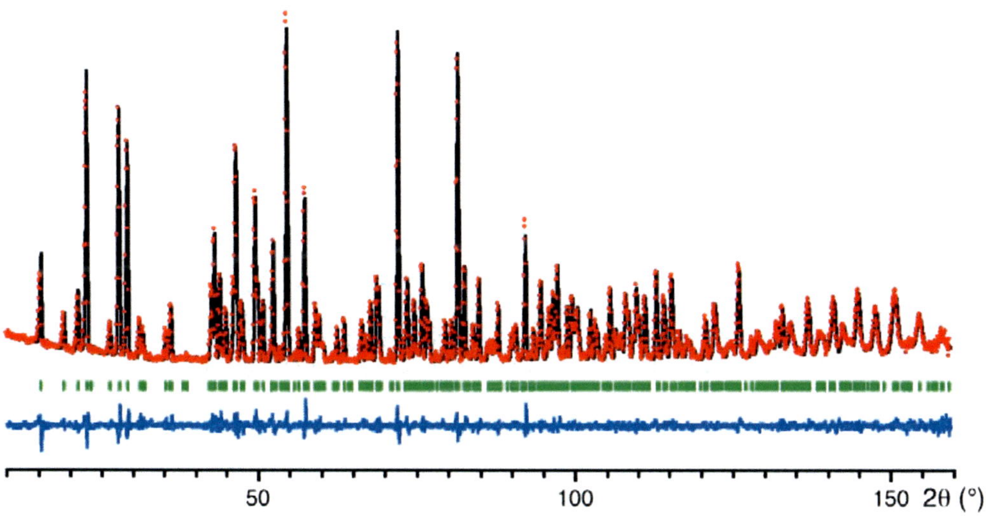

Figure 3. neutron Rietveld plot for β-$Zr_2O(PO_4)_2$ at 800 °C.

A Paradoxical Contraction Effect

The most important effect arises from the sharing of the O-O edges by the close Zr-P pairs and Zr-Zr infinite zig-zag chains (Figure 2), which generate strong Coulombic repulsions. On heating, while the framework softens, these interactions become the driving force of a dramatic expansion of the inter cations distances (respectively 16 and 12 10^{-6} K^{-1} between room temperature and 1000 °C, that is, one order higher than the mean expansion of the cell itself). These movements result in various distortions of the structure.

i) The *a*-directed O-O edge of the phosphate tetrahedron contracts, thus moderating the expansion along this axis: α_a = +3.9 10^{-6} K^{-1} (considering a O-P-O-Zr quadrilateral with constant edges lengths owing to the strong P-O and Zr-O bonds, if one diagonal expands, the other one correlatively shortens): this effect can be assimilated to the principle of bond of expansion with contraction.

ii) The linear Zr-O-Zr unit rotates in the (100) mirror layer around the bridging oxygen atom which acts as a pivot (Figure 4), thus reducing its size along the *b* axis by 0.026 Å between 20 and 800 °C. Simultaneously, the *b* edge of the cell contracts by nearly the same value (α_b = -3.6 10^{-6} K^{-1}).

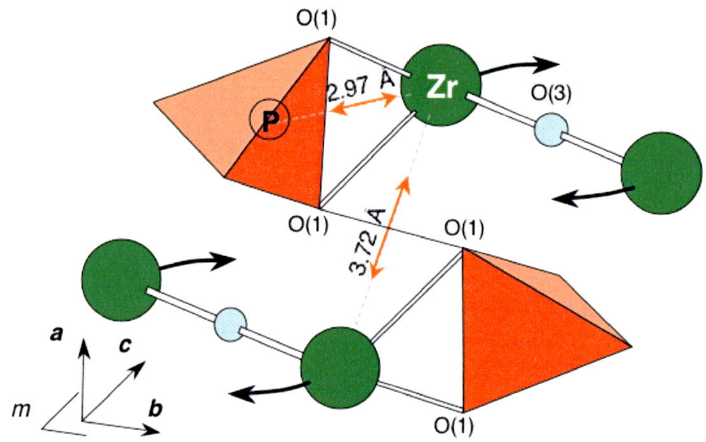

Figure 4. Coulombic Zr-P and Zr-Zr repulsions across the O(1)-O(1) edges in β-$Zr_2O(PO_4)_2$

iii) Normally, the expected drawback of the previous effect would be an increased expansion of the *c* edge, insofar as the Zr-O-Zr unit tends to orientate in this direction. In fact, the phosphate tetrahedron is simultaneously pushed and pulled by the three zirconium neighbours in the (100) plane towards the empty spaces of the crystal cell (Figure 5). This feature of the structure allows to reduce considerably the expansion of the *c* parameter (α_c = +4.3 10^{-6} K^{-1}).

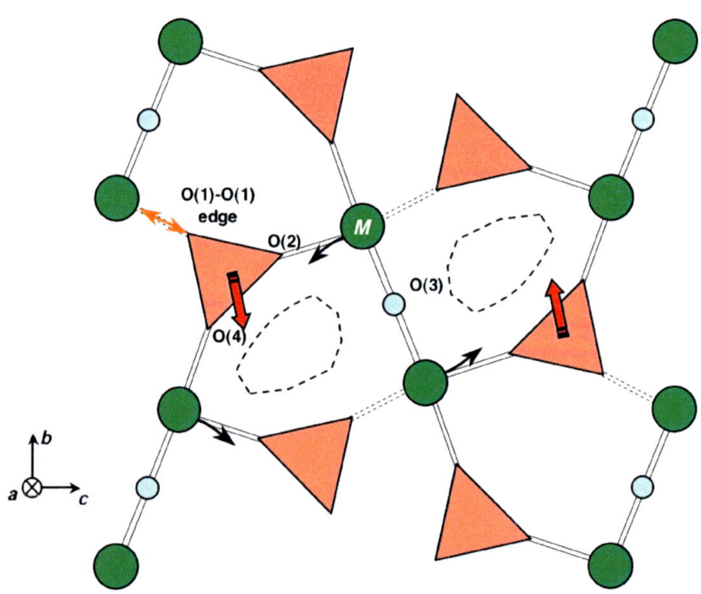

Figure 5. push-pull mechanism in the (100) plane of β-$Zr_2O(PO_4)_2$, note the shift of the phosphate tetrahedra toward the microcavities (dashed line).

The ultra low overall linear expansion: $\alpha_l = (\alpha_a + \alpha_b + \alpha_c) / 3 = +1.5\ 10^{-6}\ K^{-1}$ is the global result of all these effects.

To appreciate the role of the size of the tetravalent cation, it is interesting to observe how the same effect operates in $U_2O(PO_4)_2$ [12], the uranium isotype of β-$Zr_2O(PO_4)_2$. Uranium IV being bigger than Zr^{IV} ($r(Zr^{IV})$ = 0.78 Å; $r(U^{IV})$ = 0.95 Å in seven-fold coordination [13]), the U-U distance is longer (4.04 Å vs. 3.72 Å), so thermal expansion due to Coulombic repulsions is lower ($7\ 10^{-6}\ K^{-1}$ vs. $12\ 10^{-6}\ K^{-1}$). The same tendency can be observed with Thorium IV ($r(Th^{IV})$ = 1.00 Å) in isotypic $Th_2O(PO_4)_2$ [14,15].

Thorium Phosphate Diphosphate: a Related Mechanism in a Different Structure

Another phenomenon of ultra low thermal expansion driven by inter cations repulsions has been observed in β-$ThPO_4(P_2O_7)_{1/4}$ (or TPD for thorium phosphate diphosphate), a material known for its promising performances as a host matrix for the immobilization of actinides nuclear wastes [16]. A low thermal expansion is an asset for this class of materials insofar as it prevents the formation of cracks during the cooling of the ceramic, thus reducing its vulnerability towards leaching and increasing its durability.

This orthorhombic (space group *Pcam*) compound [18] is made up of infinite *c*-directed blocks with a bell-shaped cross-section, connected to each other *via* double Th-O-P bridges, and across structural cavities through weaker Th-O-P linkages (Figure 6). Like the zirconium IV cations in β-$Zr_2O(PO_4)_2$, thorium IV here form infinite zig-zag chains of polyhedra connected by edges, resulting in short inter cations distances.

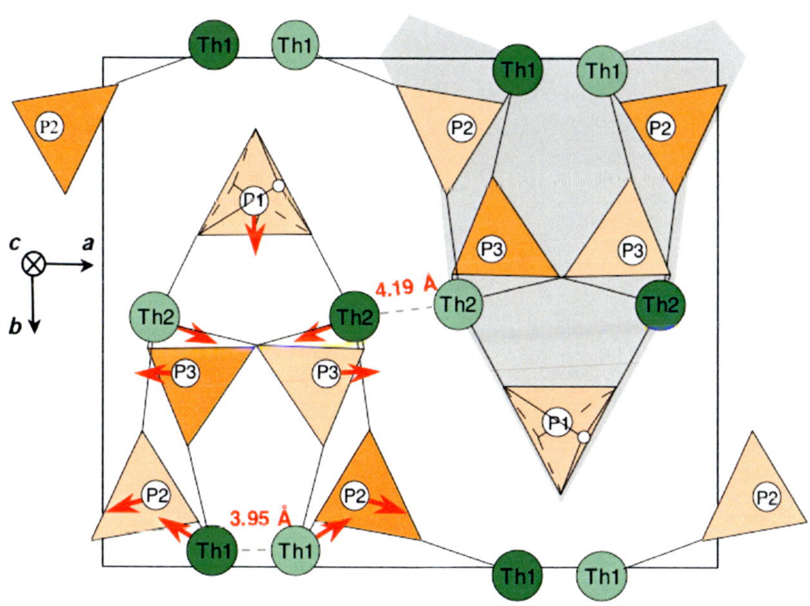

Figure 6. crystal structure of β-$ThPO_4(P_2O_7)_{1/4}$, the shaded area represents a "bell" structural unit, note the thermal shifts of the atoms (arrows).

On heating, under the effect of Coulombic repulsions, both the Th(1)-Th(1) and Th(2)-Th(2) distances expand at 13 10^{-6} K^{-1}, thus displacing the neighbour phosphate tetrahedra nearly the same a rate and ruling the axial expansion coefficients [18].

i) The bells flatten, giving a near zero expansion following *b* (α_b = 0.3 10^{-6} K^{-1}).
ii) The foot of the bells broadens, simultaneously with a shearing movement following *a* and a reduction of the cavities of the structure. Therefore, the thermal expansion following *a* is also very low (α_a = 1.0 10^{-6} K^{-1}).
iii) The moderate expansion of the *c*-axis (α_c = 3.8 10^{-6} K^{-1}) results probably of a rocking of the diphosphate groups.

The mean linear thermal expansion of the cell between 25 and 800 °C is remarkably low (α_l = 1.7 10^{-6} K^{-1} (Figure 7). It offers an excellent agreement with the macroscopic expansion of the sintered ceramic, probably because the faint anisotropy prevents the formation of microcracks.

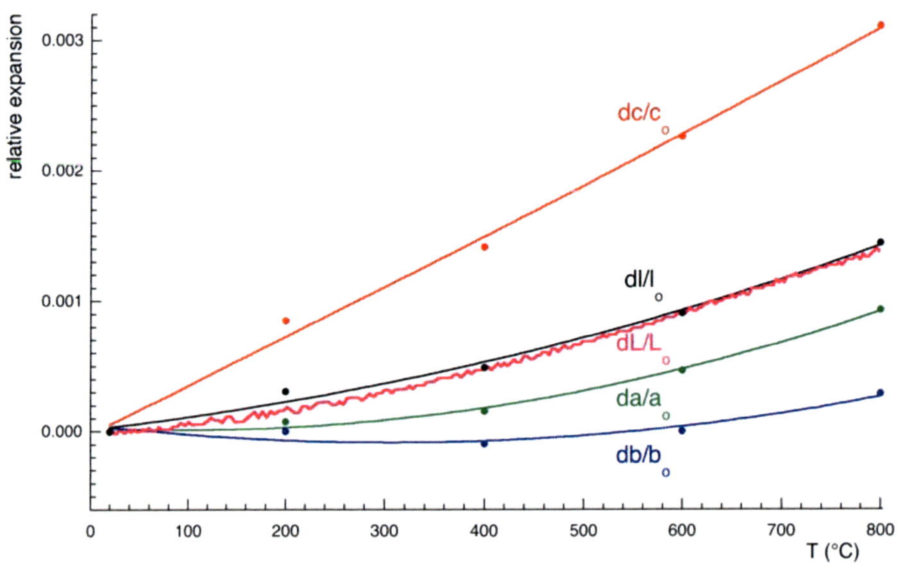

Figure 7. relative thermal evolution of the cell parameters of β-$ThPO_4(P_2O_7)_{1/4}$; the dL/L_o curve represents the macroscopic expansion of the sintered ceramic.

Toward a Concept of "Contraction Through Repulsion" ?

Although built up of geometrically different frameworks, β-$M_2O(PO_4)_2$ (M = Zr, U, Th) and β-$ThPO_4(P_2O_7)_{1/4}$ present mechanisms of ultra-low expansion based upon similar principles: in both compounds, the tetravalent cations are located at short distances, while the structures present structural cavities. More generally, we can say that such an original contraction phenomenon may arise from inter-cations Coulombic repulsions providing that some criteria are satisfied.

i) The oxidation state of the cation is primordial. In addition to a low charge, monovalent and divalent ones are generally big also, therefore their reciprocal Coulombic interactions are too weak to drive such a mechanism; furthermore, their bonds with anions are weak, therefore subject to strong expansion. On the contrary, cations with high oxidation state (V or more) are small, their coordination number is low and their oxygen polyhedra share only vertices, giving also long inter cations distances and weak interactions. Trivalent and tetravalent cations of transition and lanthanide elements seem to be a good compromise, considering both their charge and their size.
ii) Another positive aspect of the tri- and tetravalent cations is their ability to build pairs or infinite chains of oxygen polyhedra by edge sharing. Such an array, very common among the structures reported in the literature, allows the cations to be close enough to yield significant Coulombic repulsions.
iii) At last, the crystal structure must be non compact, with structural cavities likely to act as "contraction reservoirs".

The global mechanism can be summarized as follows: the cations repulse each other and push the neighbour anions or oxyanions so that they shift towards the empty spaces, as shown on Figure 8. One can also expect that the stretching of the polyhedra along the direction of the repulsion will be balanced by a perpendicular contraction, following a classical coupling mechanism.

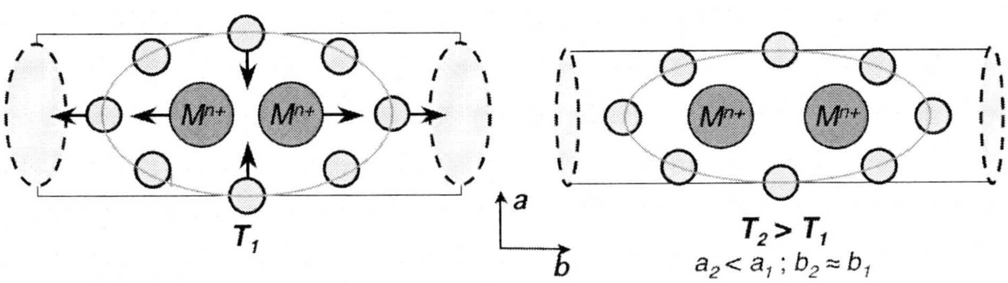

Figure 8. generalized principle of negative expansion driven by Coulombic repulsions between cations; the empty spaces (dashed lines) contract following *b* while the polyhedron contracts following *a*.

Polyhedra Rocking in the β-$M_2O(PO_4)_2$ Family (*M*= Zr, U, Th)

In addition to the previous phenomenon, these isotypic compounds exhibit transverse oscillations of the oxygen atoms, as shown by the shape of their thermal ellipsoids (Figure 9): those in two-fold coordination (linear or nearly linear) are free to oscillate in the two normal directions, whereas that in three-fold coordination moves only perpendicular to the plane of the three cations.

Although this mechanism is clearly related to a rocking of polyhedra, it differs from the classical model by the fact that the polyhedra are connected not only by vertices, but also by edges. Such a framework is generally considered as unsuitable for contraction by oscillation for three reasons: first, because a shared edge compels both polyhedra to deform while rocking, thus restraining considerably the amplitude of the motions; secondly because the bonds of an oxygen anion in three-fold coordination are longer and weaker than in two-fold coordination, therefore more likely to expand; thirdly because such an array favours the repulsions between cations.

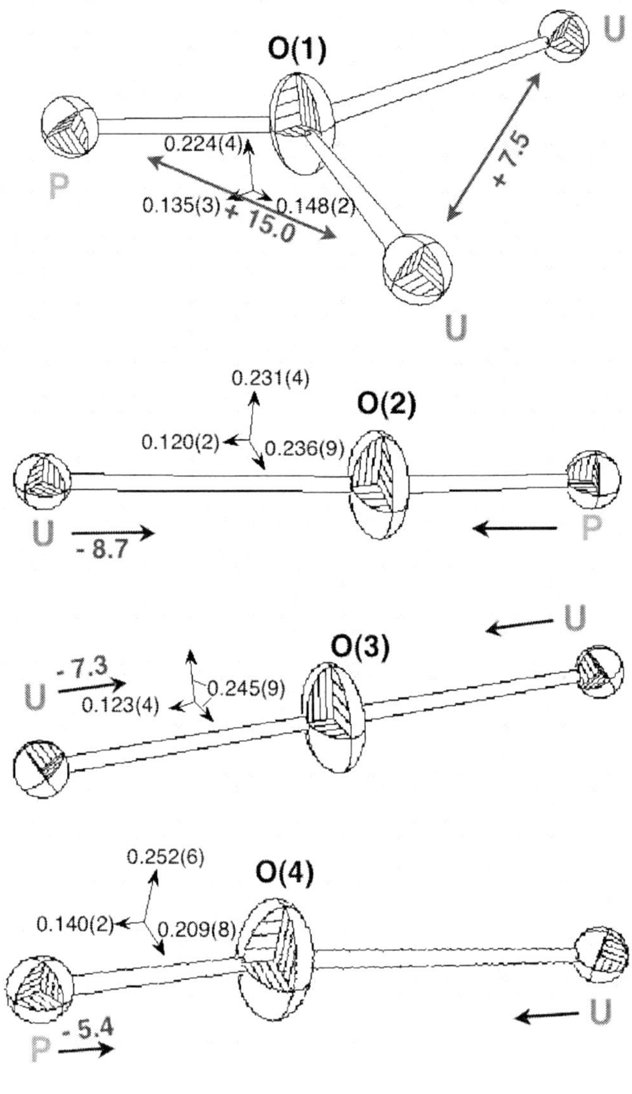

Figure 9. thermal ellipsoids of the bridging oxygen atoms in $U_2O(PO_4)_2$ and thermal expansions of the inter-cations distances (10^{-6} K^{-1}).

Actually, the linkages involving oxygens in three-fold coordination stretch strongly with temperature (mean rates are 13 10^{-6} K^{-1} for $Zr_2O(PO_4)_2$, 9 10^{-6} K^{-1} for $U_2O(PO_4)_2$), this expansion resulting also from the Coulombic interactions as seen before.

On the opposite, all the *M*-O-*M* and *M*-O-P linkages involving oxygens in two-fold coordination contract on heating, (–3 10^{-6} K^{-1} for $Zr_2O(PO_4)_2$, –7 10^{-6} K^{-1} for $U_2O(PO_4)_2$). While this effect appears of minor importance in the zirconium compound, it increases with the size of the tetravalent cation. It is noteworthy that big ones like uranium IV or thorium IV, although poorly efficient regarding the mechanism of contraction through repulsions, allow a stronger rocking of the polyhedra.

The micro Raman spectra of these compounds show that the bands of lattice vibrations shift toward low energies on heating, whereas the bending and stretching wavenumbers of the P-O bonds remain unchanged. In other terms, the rocking of the MO_7 polyhedra increases but the expansion of the phosphate tetrahedra remains negligible.

Taking $U_2O(PO_4)_2$ as example, we can observe positive definite expansions in the (010) plane parallel to the shared edges: $\alpha_a = 0.3\ 10^{-6}\ K^{-1}$; $\alpha_c = 2.8\ 10^{-6}\ K^{-1}$. On the opposite, the thermal evolution of the *b* axis is ruled only by the contraction of the linear linkages, therefore strongly negative: $\alpha_b = -7.4\ 10^{-6}\ K^{-1}$. The overall linear expansion is negative: $\alpha_l = -1.5\ 10^{-6}\ K^{-1}$. Logically, the highest contraction is reached for $M^{IV} = Th^{IV}$ ($\alpha_l = -1.7\ 10^{-6}\ K^{-1}$), the biggest of all tetravalent cations.

Indeed, a bigger cation favours contraction insofar as long bonds are less directed and rigid than small ones, but also because the distances between anions of the same polyhedron increase, thus leaving room for individual oscillations. Note that the first reason applies also to rigid polyhedra, in agreement with the empirical remark often made that for a given structural family (for example, the $M^{IV}P_2O_7$ diphosphates), the thermal expansion coefficient decreases while the cell edge increases [1].

To our knowledge, $U_2O(PO_4)_2$ and $Th_2O(PO_4)_2$ are the first known examples of compounds that exhibit a negative expansion through a rocking of polyhedra that share edges. Compared to famous materials in which the polyhedra are connected only by vertices (ReO_3: 2.3 $10^{-6}\ K^{-1}$, fused SiO_2: 0.5 $10^{-6}\ K^{-1}$, Nb_2O_5: -1.0 $10^{-6}\ K^{-1}$, $Sc_2(WO_4)_3$: -2.2 $10^{-6}\ K^{-1}$), their performances are at least surprising. This remark points out the fact that the sharing of some edges, although not favourable, does not hamper the rocking of polyhedra, nor the related contraction effect, providing that the cation is big enough. In such a case, the mechanism should probably be thought of as independent oscillations of the loosely packed bridging oxygens, rather than as the correlated movements of the vertices of a rigid polyhedron.

Big cations are generally believed to increase the expansion coefficient because of their weak bonds with anions. Although this is true with alkali and earth alkali cations, those with higher oxidation states are less subject to thermal expansion. For example, despite the large size of Th^{IV}, thorianite (ThO_2) shows roughly the same α coefficient as alumina, below 10 $10^{-6}\ K^{-1}$ [19].

The Zirconium Phosphates: Multiple Ways to Low Expansion

The ZrO_2-P_2O_5 phase diagram (Figure 10) [5] has long been considered by ceramists as a fruitful field of researches, because of the high chemical and thermal stability of the zirconium phosphates and their derivates, their low toxicity, moderate cost and easiness of elaboration. Another strong point is their very low thermal expansion, which gives the misleading impression of a structural unity. In fact, the zirconium phosphates adopt different structural types (Table 1) resulting in different expansion modes.

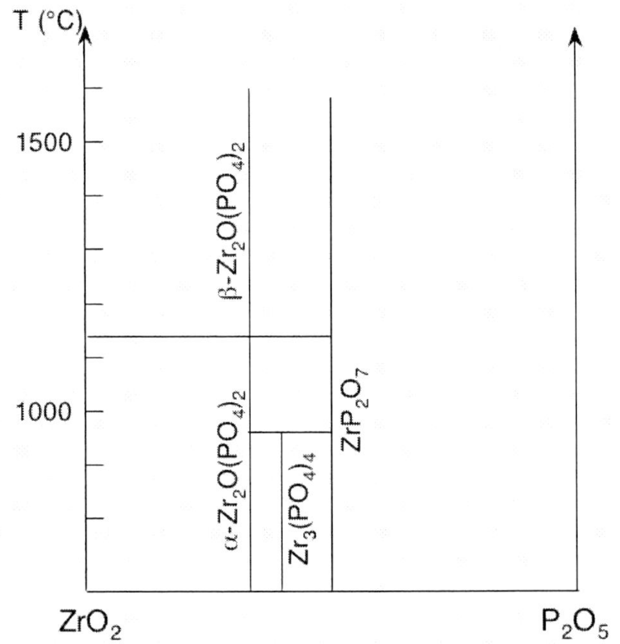

Figure 10. subliquidus ZrO_2-P_2O_5 binary phase diagram, from Alamo and Roy [5].

Table 1. main structural and dilatometric features of the zirconium phosphates

	α-$Zr_2O(PO_4)_2$	**β-$Zr_2O(PO_4)_2$**	**$Zr_{2.25}(PO_4)_3$ and NZP's**	**ZrP_2O_7**
Zr^{IV} coordination	6 and 7	7	6	6
polyhedra connection	vertices, edges	vertices, edges	vertices, faces	vertices
$10^6\ \alpha_l\ (K^{-1})$	2.5	1.5	unknown, variable for substituted forms	12* (LT form) 6.5* (HT form)
Low expansion mechanism	poly. rocking + coupled deformations	reduction of cavities (+ poly. rocking)	coupled deformations	poly. rocking

* calculated from the expansion curve of [26]

α-$Zr_2O(PO_4)_2$ has an original structure (Figure 11) in which the Zr^{IV} cation is half in six-fold, half in seven-fold coordination [7]. The longitudinal expansion of the ribbons is very low (α_a = 1.7 10^{-6} K^{-1}) thanks to the rocking of the roughly linear Zr-O-P linkages [20]. The repulsions between the phosphate groups of neighbour ribbons generate a strong expansion following *c* (α_c = 7.0 10^{-6} K^{-1}). The contraction of the *b* edge (α_b = -1.1 10^{-6} K^{-1}) results from a coupling with the previous effect in the polyhedra cycles, together with a rocking effect of the Zr-O-P bridges between superimposed ribbons.

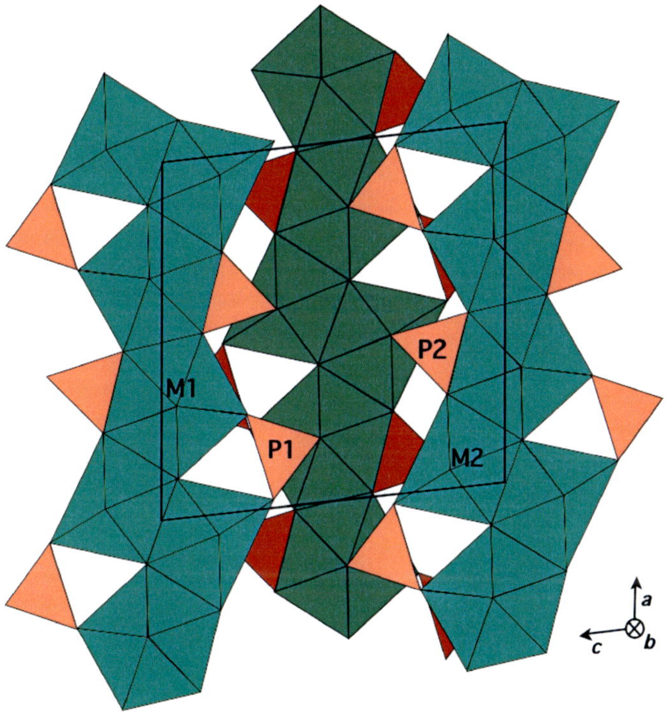

Figure 11. crystal structure of α-$Zr_2O(PO_4)_2$.

This dual mechanism, like in the β form, evidences one more time the non negligible effect of polyhedra rocking in structures involving anions in three-fold coordination. Note that $Hf_2O(PO_4)_2$, the only known isotype, has a slightly lower expansion coefficient (α_l = 2.4 10^{-6} K^{-1}) and, because of the absence of α-β phase transition, a domain of stability extending up to 1500 °C. However, considering the peculiar coordination of the tetravalent cation, the existence of other isotypes of α-$Zr_2O(PO_4)_2$ seems very dubious.

Very little is known about $Zr_{2.25}(PO_4)_3$ itself [5], which expansion mechanism is supposed to be similar to those of the isostructural compounds with formulae $M^IZr_2(PO_4)_3$ (like $NaZr_2(PO_4)_3$ or NZP) or $M^{II}_{0.5}Zr_2(PO_4)_3$, that is, an axial stretching resulting from the sharing of faces between M and Zr anionic polyhedra, coupled with a perpendicular contraction [21-24]. The NZP family has gained a considerable notoriety thanks to the ability of the $[Zr_2(PO_4)_3]^-$ framework to host M cations of various sizes and charges, thus allowing the mean expansion coefficient to take either positive or negative values, with a notable anisotropy.

In both its low and high temperature forms, the ZrP_2O_7 structure appears as a rock salt array of Zr and P_2O_7 units sharing only vertices, therefore the low expansion is purely ruled by the rocking of polyhedra [25,26].

Improving β-$Zr_2O(PO_4)_2$?

A well-known way to achieve zero expansion consists in making composites with phases showing opposite coefficients, like in glass-ceramics. It is also possible to monitor this property in a single-phase material, providing that the same structure can exhibit positive and negative expansion, depending on its composition. This attractive route has been explored with phases that offer wide domains of solid solutions, like the NZP and the $Zr(P_{2-x}V_x)_2O_7$ [27].

Considering that the expansion coefficient of the β-$M^{IV}_2O(PO_4)_2$ compounds varies contrarily with the size of the M^{IV} cation, it was appealing to try substituting big ones for part of the Zr^{IV}. Unfortunately, attempts to improve this compound *via* isovalent substitutions appear nearly hopeless, for two reasons.

i) Among all the stable tetravalent cations, the only one bigger than Zr^{IV} is thorium IV, otherwise undesirable in ceramics for safety reasons.
ii) $Th_2O(PO_4)_2$ is not a stable compound insofar as it only forms by soft dehydration of $ThOHPO_4$ [14] and decomposes as low as 1050 °C (vs. 1600 °C for $Zr_2O(PO_4)_2$). The literature dedicated to other thorium IV compounds always reports environments of at least eight oxygen anions, so this inedite seven-fold coordination seems poorly suitable for steric reasons. Indeed, all our attempts to incorporate thorium (by solid state reaction and from precipitated precursors) only yielded to a mixture of phases including unsubstituted β-$Zr_2O(PO_4)_2$, according to the powder diffraction patterns. Actually, the only tetravalent cation that can be easily substituted for Zr^{IV} is Hf^{IV}, slightly smaller (0.76 Å vs. 0.78 Å), leading to β-type solid solutions with increased thermal expansion.

We also considered replacing Zr^{IV} by a cation of lower oxidation state, bigger than Zr^{IV} but smaller than Th^{IV} to fit in the seven-fold polyhedron (all those with lower oxidation state are smaller). Di- and trivalent cations (Mg^{II}, In^{III} or Ln^{III}) appeared suitable, providing that their lower charge was balanced by substituting corresponding quantities of W^{VI} for P^V in the tetrahedral sites. Failure to synthesize the solid solution, this time, resulted probably from the nature of the oxygen environment itself, not the coordination number. In the β-$M^{IV}_2O(PO_4)_2$ form, the non-phosphate oxygen builds a strong *M*-O-*M* linkage that requires the sharing of one valence unit by each of the two cations. While cations with high oxidation states are likely to form such oxo units, di- and trivalent ones are notoriously unable to do so.

By comparison, in the $NaZr_2(PO_4)_3$ and $Sc_2(WO_4)_3$ structures, Zr^{IV} and Sc^{III} occupy octahedral sites with homogeneous cation-oxygen distances and bond strength distribution. Therefore, the substitution can be made on the only basis of the radius of the cation, not its oxidation state. For this reason, these compounds allow numerous and fruitful substitutions, on the opposite with β-$Zr_2O(PO_4)_2$.

Substitution of bigger arsenic V for P^V (0.335 Å vs. 0.17 Å) was also attempted, but it only yielded a α-type phase. For obvious safety reasons, substitution by vanadium V (0.355 Å) would have been more interesting, but the expected $Zr_2O(VO_4)_2$, never reported in the literature, does not form. ZrV_2O_7, observed in the synthesized powders, seems to be the only stable zirconium vanadate.

Conclusion

Beta zirconium oxide phosphate has revealed an inedite contraction mechanism based mostly on the shrinkage of structural cavities, driven by the Coulombic repulsions between neighbour Zr^{IV} cations. From a practical point of view, there are little chances to succeed in lowering its expansion coefficient by chemical substitutions.

Several other phosphates of tetravalent cations have revealed a low or negative thermal expansion ruled by atomic mechanisms or combinations of mechanisms never reported before. Based on simple and common structural features, these effects are likely to be found in many other compounds, so we propose here some hints inspired by the present work and destined to guide the research of new compounds with controlled expansion or high resistance to thermal shocks.

i) Mechanisms of contraction driven by Coulombic repulsions can appear in various structures, providing that tri- or tetravalent cations are located close enough and that the framework is not compact.
ii) Previous works have often focussed on compounds with a framework built of small or medium cations like Al^{III}, Si^{IV}, Ti^{IV}, Zr^{IV}, P^{V}, Ta^{V} and W^{VI}, known to form strong bonds with negligible thermal expansion and rigid polyhedra. Indeed, bigger cations have revealed unexpected *virtues* in building deformable polyhedra of loosely packed anions, thus allowing independent and efficient rocking motions of the vertices. Providing that these cations have a high oxidation state, their bonds can still be strong enough to expand moderately.
iii) Therefore, the common idea that polyhedra rocking can only be effective if the units are only edge-connected needs to be somewhat revised, even if the sharing of edges does not favour such movements.
iv) It is probable that other contraction mechanisms or combinations of mechanisms are still to be discovered. In particular, attention should be paid to the role of the structural cavities, almost never explored before. Non-compact frameworks are enough common among the oxides to provide a wide field of investigations.

Indeed, the analytical exploration of the crystal structure and its evolutions is a faster and safer way to resolve problems related to ceramics matters than empirical try and error experiments. Sometimes also, it can lead to the conclusion that chances of success are actually little. In both cases, time and efforts are saved, and conclusions, either positive or negative, can prove useful for further researches. Thermal expansion, that results basically from very faint atomic shifts, is a typical exemple of field where high-temperature powder diffraction and recent advances like Rietveld analysis open wide perspectives.

References

[1] Roy, R.; Agrawal, D. K.; McKinstry, H. A. *Annu. Rev. Mater. Sci.* 1989, 19, 59.
[2] Sleight, A. W. *Inorg. Chem.* 1998, 37, 2854.
[3] Mary, T. A.; Evans, J. S. O.; Sleight, A. W.; Vogt, T. *Science* 1996, 272, 90.
[4] Dove, M. T.; Heine, V.; Hammonds, K. D. *Mineral. Magn.* 1995, 59, 629.

[5] Alamo, J.; Roy, R. *J. Am. Ceram. Soc.* 1984, 67, 80.

[6] Yamai, I.; Oota, T. *J. Am. Ceram. Soc.* 1985, 68, 273.

[7] Wallez, G.; Souron, J.-P.; Quarton, M. *Solid State Sciences* 2006, 8, 1061.

[8] Wallez, G.; Launay, S.; Souron, J.-P.; Quarton, M.; Suard, E. *Chem. Mater.* 2003, 15, 3793.

[9] Omori, M.; Hsu, K. C.; Tsunekawa, S.; Hirai, T. *J. Mater. Sc.* 1992, 27, 408.

[10] Bender, B. A.; Jessen, T. L.; Browning, S. *Ceram. Eng. Sci. Prod.* 1995, 16, 613.

[11] Gebert, W.; Tillmanns, E. *Acta Crystallogr.* 1975, *B*31, 1768.

[12] Wallez, G.; Launay, S.; Quarton, M.; Dacheux, N.; Soubeyroux, J.-L. *J. Solid State Chem.* 2004, 177, 3575.

[13] Shannon, R. D. *Acta Crystallogr.* 1976, *A*32, 751.

[14] Clavier, N.; Dacheux, N.; Wallez, G.; Quarton, M. *Solid State Sciences*, under press.

[15] Wallez, G.; Quarton, M. *to be published.*

[16] Dacheux, N.; Clavier, N.; Robisson, A.-C.; Terra, O.; Audubert, F.; Lartigue, J.-E.; Guy, C. *C. R. Chimie* 2004, 7, 1141.

[17] Bénard, P.; Brandel, V.; Dacheux, N.; Jaulmes, S.; Launay, S.; Lindecker, C.; Genet, M.; Louër, D.; Quarton, M. *Chem. Mater.* 1996, 8, 181.

[18] Launay, S.; Wallez, G.; Quarton, M. *Chem. Mater.* 2001, 13, 2833.

[19] Wu, Z.; Farges, F. *Physica B* 1999, 266, 282.

[20] Wallez, G.; Quarton, M. to be published.

[21] Boilot, J. P.; Salanie, J. P.; Desplanches, G.; Potier, D. L. *Mater. Res. Bull.* 1979, 14, 1469.

[22] Alamo, J. *Solid State Ionics* 1993, 63-65, 547.

[23] Pet'kov, V. I.; Orlova, A. I. *Inorg. Mater.* 2003, 39, 1013.

[24] Khosrovani, N.; Sleight, A. W. *J. Solid State Chem.* 1996, 121, 2.

[25] Harrison, D. E.; McKinstry, H. A.; Hummel, F. A. *J. Am. Ceram. Soc.* 1954, 37, 277.

[26] Withers, R. L.; Tabira, Y.; Evans, J. S. O.; King, I. J.; Sleight, A. W. *J. Solid State Chem.* 2001, 157, 186.

[27] Korthuis, V.; Khosrovani, N.; Sleight, A. W.; Roberts, N.; Dupree, R.; Warren, W. W. *Chem. Mater.* 1995, 7, 412.

In: Ceramic Materials Research Trends
Editor: Paul B. Lin, pp. 197-208
ISBN: 1-60021-769-9

Chapter 8

INTRODUCTION OF DIOPSIDE TO IMPROVE THE PERFORMANCE OF PRESSURELESS SINTERED ALUMINA MATRIX CERAMIC MATERIALS

Changxia Liu [a, b]***, Jianhua Zhang*** [a, *]***, Junlong Sun*** [a, b] ***and Xihua Zhang*** [c]

[a]Department of Mechanical Engineering, Shandong University, Jinan 250061, Shandong Province, P.R.China

[b]Communications institute, Ludong University, Yantai 264025, Shandong Province, P.R. China

[c]Department of Materials Science and Engineering, Shandong University, Jinan 250061,Shandong Province, P.R. China

Abstract

In this paper, diopside was introduced into alumina matrix ceramic materials as a sintering aid and fine structural alumina matrix ceramic materials were fabricated by the technology of pressureless sintering, during which new phases were produced. The relative density, hardness, fracture toughness and bending strength of the composites were tested. The experiment results show that the introduction of diopside into alumina matrix can enhance the densification rate due to the generation of liquid phase in the sintering process, which leads to the improvement of the mechanical properties of the composites. The best mechanical properties were achieved when 3vol.% diopside was added into the composite, the hardness, bending strength and fracture toughness of which are 15.57GPa, 417MPa and 5.2MPa• $m^{1/2}$, respectively. The effects of diopside on mechanical properties, fracture mechanism and interspaces of fine structural alumina matrix ceramic materials were analyzed together with the toughening mechanism of the composites.

Key words: Alumina; Diopside; Mechanical properties; Microstructures; Pressureless sintering

1. Introduction

Fine structural ceramics are widely applied in the fields of fine measuring implement, fine sliding way, working platform and high-speed bearing, etc., where high precision durability and long working life are required. Among these fine structural ceramics, alumina matrix ceramic materials are found to have the widest application due to their virtues of high hardness, good chemical inertness, high wear resistance, low coefficient of thermal expansion and friction coefficient. However, the brittleness of pure alumina limits its potential application as a structural material. In order to solve this problem, second phases are always introduced into alumina matrix. Excellent mechanical properties can always be obtained companied with a cost increase on account of the expensive second phases [1~5]. Diopside ($MgCa(SiO_3)_2$) has the virtue of low-cost compared with other additives. Being composed of SiO_2, CaO and MgO phases, it is expected to decrease the temperature of pressureless sintering, reduce interspaces or cavities of alumina matrix ceramic materials, improve their mechanical properties and thus exhibit its high ratio of toughening effect versus cost, especially for those fine structural ceramics with large dimension.

In this paper, fine structural alumina matrix ceramic materials with different content of diopside, acting as a sintering aid during the sintering process, are fabricated by the pressureless sintering technology in a furnace in N_2 protection. The aim of this work is to investigate the effects of diopside on mechanical properties and microstructures of fine structural alumina matrix ceramic materials, and then search after an approach of fabricating large dimension fine structural ceramics with high performances and a decreased manufacturing time and cost.

2. Experimental Procedure

Commercial Al_2O_3 powder of high purity (99.99%) and small grain size (0.5~1μ m) was used as the starting materials. Diopside ($MgCa(SiO_3)_2$), composed of SiO_2 (55wt.%), CaO (24wt.%) and MgO (18wt.%), was used as an additive. The content of diopside ranged from 1vol.% to 18vol.% is listed in table 1. (The suffix in AD_0, AD_1, AD_3, AD_6, AD_{12} and AD_{18} represents the volume content of diopside. For example, AD_0 means the volume content of diopside is zero.)

Table 1. Compositions and relative density of pressureless sintered alumina matrix ceramics

Specimens	Compositions (vol.%)	Relative density(%)
AD_0	100%Al_2O_3	88.6
AD_1	99% Al_2O_3 + 1 % diopside	94.65
AD_3	97% Al_2O_3 + 3 % diopside	97.17
AD_6	94% Al_2O_3 + 6 % diopside	97.5
AD_{12}	88% Al_2O_3 + 12% diopside	97.94
AD_{18}	82% Al_2O_3 + 18% diopside	98.25

Firstly, the raw materials were blended together according to certain proportions and ball milled for 60h in an alcohol medium to obtain a homogeneous mixture. Secondly, the slurry was dried in vacuum and screened. Thirdly, green bodies (50mmx50mmx50mm) were shaped using isostatic cool pressing technology in rubber molds. Lastly, these dense green bodies were sintered at 1520°C (heating rate: 20°C/min) for 140 min in N_2 atmosphere using pressureless sintering method.

The sintered bodies were cut into specimens using an inside diameter slicer and then standard test pieces (3mmx4mmx36mm) for mechanical properties measuring were ground and polished with diamond paste. Three-point-bending mode was used to measure the bending strength on an electronic universal experimental instrument (WD-10) with a span of 20mm at a crosshead speed of 0.5mm/min. Twelve specimens, chosen from one body with the same composition, were used to measure the bending strength in air at room temperature. Vickers hardness was measured on polished surface with a load of 9.8N for 5s with a micro-hardness tester (MH-6). Fracture toughness measurement was performed using indentation method. The indentations on the sample surfaces were generated in a hardness tester (Hv-120), and the formula proposed by Cook and Lawn [6] was used to calculate the final fracture toughness. The microstructures of the specimens were studied on fracture surfaces and polished surfaces by scanning electron microscopy (HITACHI S-570).

3. Results and Discussion

3.1. X-ray Diffraction Phase Analysis

The x-ray diffraction analysis of AD_3 powders before sintering and AD_3 specimen sintered at 1520°C for 140min, are respectively shown in Fig.1 and Fig.2. It is can be seen from Fig.1 that there exist Al_2O_3, CaO and SiO_2 phases in the ball-milled powders. Theoretically, there should exist the peak of MgO in respect that diopside is composed of MgO, CaO and SiO_2 phases, but there is no trace of MgO in Fig.1. MgO is not found in Fig.1 for the reason that the content of MgO is too small to be detected by XRD instrument. Further, there exist in AD_3 specimen the phases of Al_2O_3, mullite ($3Al_2O_3$ $2SiO_2$), anorthite CAS_2 (CaO $3Al_2O_3$ $2SiO_2$) and CA_6 (CaO $6Al_2O_3$). There are no traces of CaO and SiO_2 phases,, which may react with alumina, forming interface reactions and strengthening the grain boundaries.

The sintering temperature is 1520°C, which is higher than the melting point of diopside (1300~1390°C). Diopside will produce liquid phase and the following chemical reactions, yielding mullite, anorthite and CaO $6Al_2O_3$, may occur during the sintering process:

$$3\,Al_2O_3 + 2SiO_2 \rightarrow 3Al_2O_3 \cdot 2SiO_2 \quad (1)$$

$$CaO + Al_2O_3 + 2SiO_2 \rightarrow CaO \cdot Al_2O_3 \cdot 2SiO_2 \quad (2)$$

$$6Al_2O_3 + CaO \rightarrow CaO \cdot 6Al_2O_3 \quad (3)$$

Based on thermodynamic analysis of the reactions, we can get the following ranking [7]:

$$\Delta G^{\theta}{}_{T_i} < 0, \qquad i=1,2,3 \tag{4}$$

where, $\Delta G^{\theta}{}_{T_i}$ represent the Gibbs free energy of the reactions i (i=1,2,3). XRD analysis shows that mullite, anorthite and $CaO{\cdot}6Al_2O_3$ phases actually exist in specimen of AD_3 as previously mentioned. The existing of mullite in fine structural alumina matrix ceramic materials may be helpful for the composite to have a high potential for wear resistance applications [8].

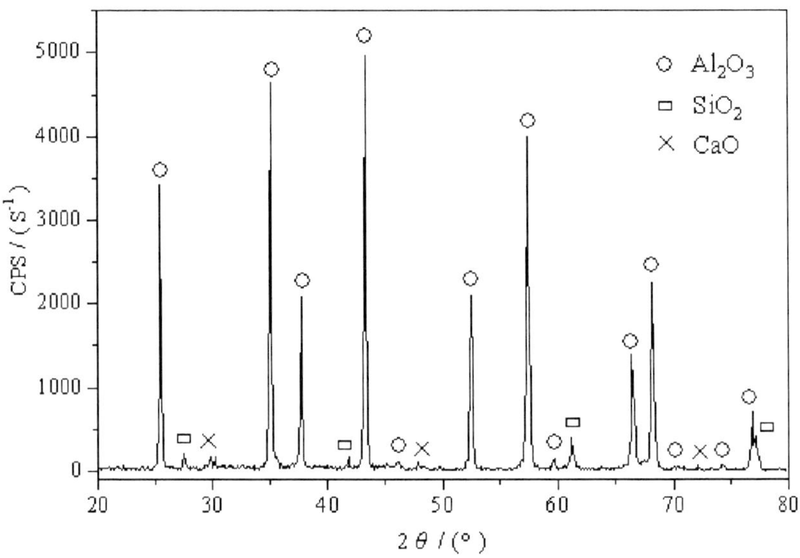

Figure 1. XRD pattern of AD_3 powder before sintering.

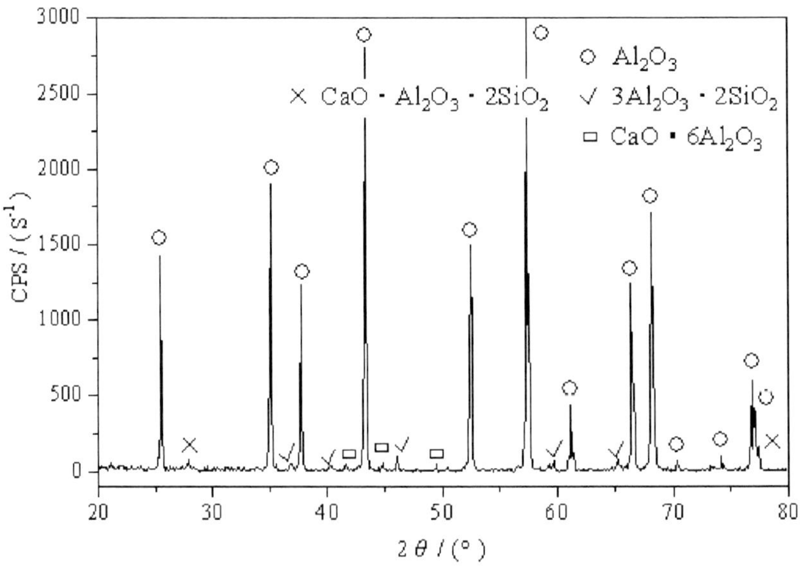

Figure 2. XRD pattern of AD_3 specimen sintered at 1520°C for 140min.

3.2. Mechanical Properties

Mechanical properties of the composites with different volume content of diopside are shown in Fig.3, Fig.4 and Fig.5. As can be seen from Fig.3, the hardness of the composites increases with the raise of the diopside content from 0 to 3vol.%, and then decreases from 3vol.% to 18vol.%. There may be two factors influencing the hardness of composites, one is "hardness effect" due to the addition of diopside, the other is "densification effect" of the composites, which is evaluated by relative density as listed in table1. The hardness of the composites may increase when the densification effect is stronger than the hardness effect, otherwise the hardness of the composites may decrease. In the first stage (0 to 3vol.% addition), the densification effect is dominant and therefore the hardness is enhanced, while the hardness is reduced as the hardness effect is in turn becoming dominant (3vol.% to 18vol.% addition). The effect of diopside content on bending strength of the composites is similar with that on hardness. The bending strength increases sharply with increasing the amount of diopside before 3vol.%, at which point it reaches the maximum values of 417 MPa, and then decreases after 3vol.%. The fracture toughness increases as diopside content is raised from 0 to 6vol.% and reaches its maximum value of 5.3 MPa•m$^{1/2}$, then decreases from 6vol.% to 18vol.%. The trend of bending strength and fracture toughness correlates with the microstructures of the composites, which will be discussed in the following section.

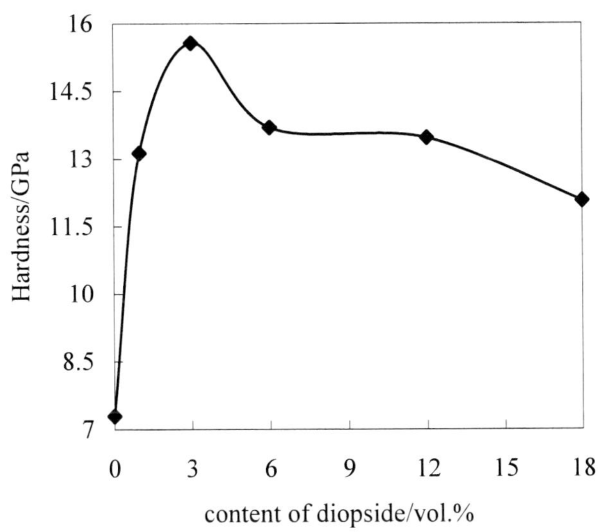

Figure 3. Relation of diopside content and hardness.

It is obvious that the fabricated fine structural alumina matrix ceramic materials, pressureless sintered at 1520°C for 140min in N_2 protection, exhibit significant improvements in mechanical properties than pure Al_2O_3. Composite with the addition of 3vol.% diopside shows better comprehensive mechanical properties, the hardness, bending strength and the fracture toughness of the composite reach 15.57GPa, 417MPa and 5.2MPa•m$^{1/2}$, respectively, which are enhanced by 113.3%, 43.3% and 27.8%, respectively, with respect to pure Al_2O_3 pressureless sintered under the same conditions.

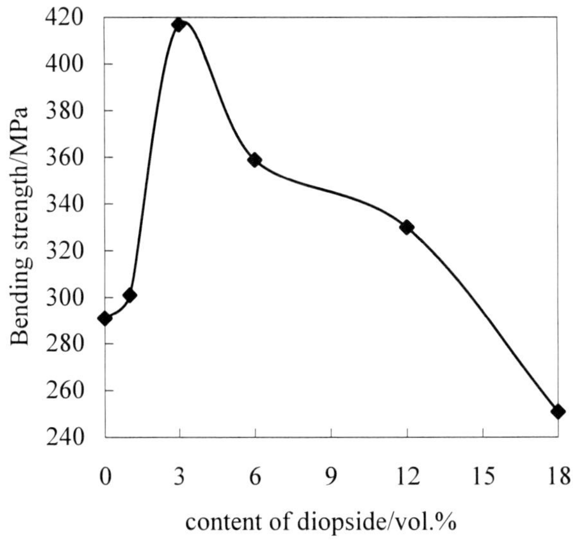

Figure 4. Relation of diopside content and bending strength.

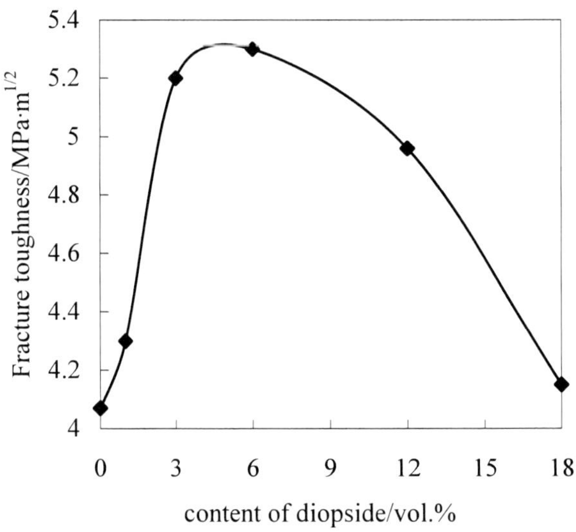

Figure 5. Relation of diopside content and fracture toughness.

3.3. Analysis of Microstructures

SEM photograph on fracture surface of pure alumina pressureless sintered at 1520°C for 140min is shown in Fig.6, and that of AD_3 is shown in Fig.7. As seen from Fig.6, the grain shapes of pure alumina are regular and almost circular. Interspaces appear among the grains owing to incomplete developing of alumina grains. The grain boundaries of pure alumina are observable and the fracture mode is mainly intergranular failure. There are significant microstructural differences among the composites and pure alumina. Evidently, composite with 3vol.% diopside have a fine matrix grain size compared with pure Al_2O_3, which shows the introduction of diopside may restrain the growth of Al_2O_3 grains. This ability of "grain growth restraining" may be in virtue of the existence of magnesia (MgO), which is reported to be good at suppressing abnormal grain growth of alumina during sintering, forming eutectic liquids at the sintering temperature, promoting enhanced densification rate and thus accelerating the sintering rate [9~11], and as a result, making the microstructures of fabricated

Figure 6. SEM photomicrograph on fracture surface of pure Al_2O_3 (5000 times).

Figure 7. SEM photomicrograph on fracture surface of AD_3 specimen (5000 times).

alumina matrix ceramic composites finer and more homogeneous. It can be clearly seen from Fig.7 that the fracture mode becomes combination of intergranular failure and transgranular failure owing to the introduction of diopside. The grain bondings are strengthened when the fracture mode turns to the combination of transgranular failure and intergranular failure. This "strengthened bondings" may be formed by adding diopside, which produces liquid phases and may provide interface reactions with alumina in the sintering process. A high strength cement binder consumes cavities or interspaces, resulting in enhanced strength of the composites.

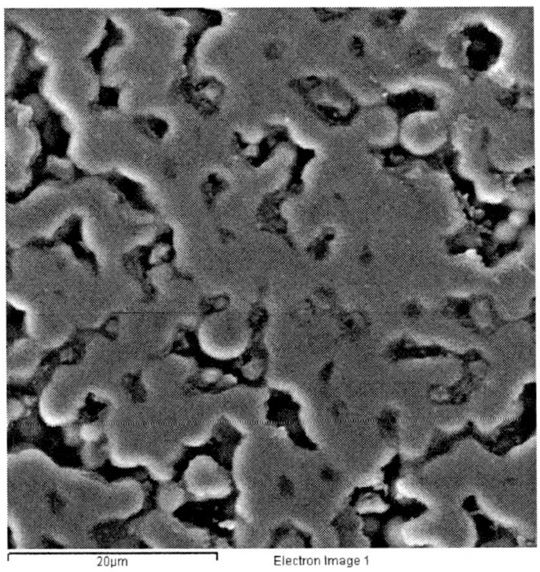

Figure 8. SEM photomicrograph on polished surface of pure Al_2O_3 (2000 times).

Figure 9. SEM photomicrograph on polished surface of AD_3 specimen (3000 times).

Fig.8 and Fig.9 respectively show the SEM photomicrographs on polished surface of pure alumina and AD_3 specimen. It can be clearly seen from Fig.8 that apparent interspaces exist on the polished surface of pure alumina, which is usually difficult to be avoided in pressureless sintering technology. Interspaces, however, on polished surface of AD_3 specimen are almost eliminated (Fig.9), making it clear that the introduction of diopside removes interspaces or air holes in fine structural alumina matrix ceramic materials.

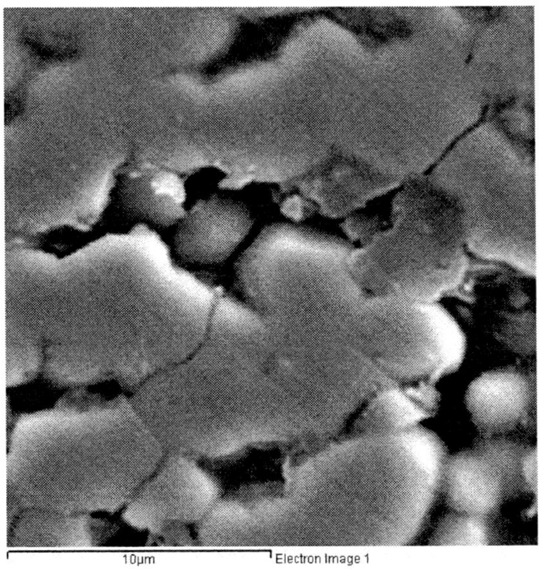

Figure 10. SEM micrographs of the crack on polished surface of monolithic Al_2O_3 (5000 times).

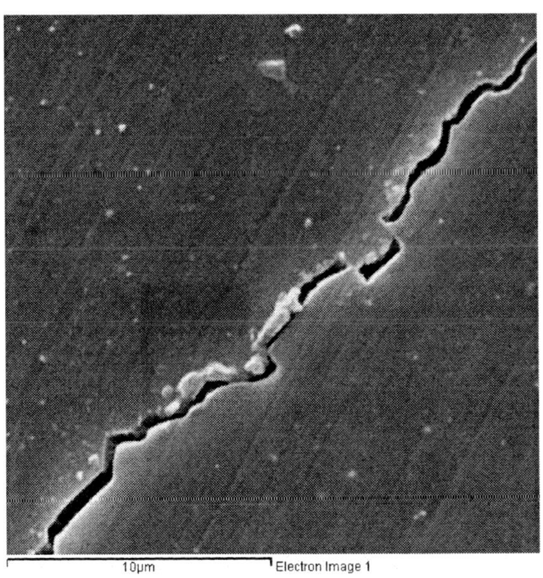

Figure 11. SEM micrographs of the crack on polished surface of AD_3 specimen (5000 times).

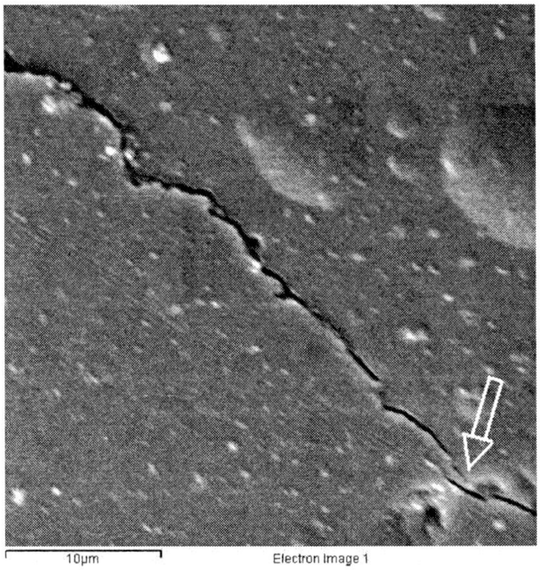

Figure 12. SEM micrographs of the crack on polished surface of AD_6 specimen (3000 times).

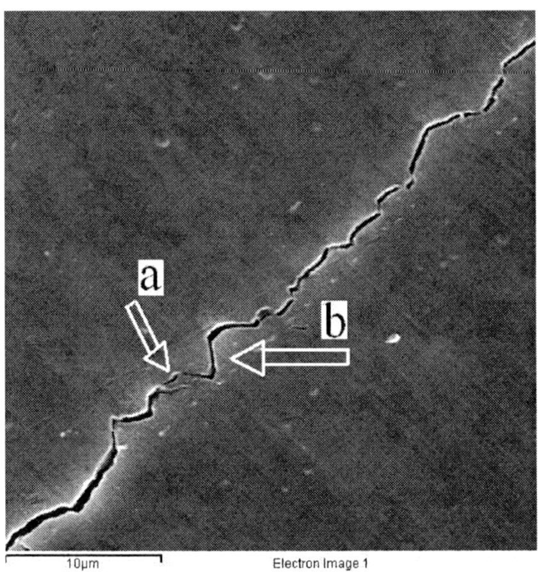

Figure 13. SEM micrographs of the crack on polished surface of AD_{12} specimen (3000 times).

The typical cracks on polished surface of monolithic alumina, AD_3, AD_6 and AD_{12} specimen are shown in Fig.10, Fig.11, Fig.12 and Fig.13, respectively. It can be clear seen form Fig.10 that the propagating route of crack in monolithic alumina is fairly straight, while those of alumina matrix ceramic materials toughened by diopside are tortuous. There exists crack branching at the bottom of crack in Fig.12 (traced by arrow). In Fig.13, there also exists bifurcation of crack (arrowed by a), besides which there appears crack deflection (arrowed by b). The SEM observations clearly indicate that monolithic alumina is characterized essentially by an intergranular fracture mode (Fig.10), while AD_3, AD_6 and AD_{12} specimen are mainly characterized by the combination of transgranular failure and intergranular failure mode (Fig.11, Fig.12and Fig.13), indicating that the addition of diopside causing toughening of the alumina grain boundaries, inducing transgranular fractures, and then, resulting in an increment in fracture toughness of AD_3, AD_6 and AD_{12} specimen. The toughening mechanisms of presureless sintered alumina matrix ceramic materials are crack branching and crack deflection.

4. Conclusion

Fine structural alumina matrix ceramic materials are fabricated by pressureless sintering technology. The introduction of diopside can obviously prompt the sintering of alumina matrix ceramic composites, and consequently improve the mechanical properties of the resulted composites. The best mechanical properties were achieved when 3vol.% diopside was added into the composite, the hardness, bending strength and fracture toughness of which are 15.57GPa, 417MPa and 5.2MPa• $m^{1/2}$, respectively.

The fracture mode of pure Al_2O_3 is mainly intergranular failure. Addition of diopside makes the fracture mode of composites become the combination of transgranular failure and intergranular failure. Simultaneously, introduction of diopside removes interspaces or air holes in fine structural alumina matrix ceramic materials, which may contribute to the improvements in mechanical properties of the composites. The toughening mechanisms of presureless sintered alumina matrix ceramic materials are crack branching and crack deflection.

Acknowledgements

The work described in this paper is supported by the Ministry of Education, P. R. of China (No. 03101) and the Outstanding Young Scientist Rewards of Shandong Province (No. 03BS103).

References

[1] S.K.C. Pillai, B.Baron, M.J. Pomeroy, S.Hampshire, Effect of oxide dopants on densification, microstructure and mechanical properties of alumina-silicon carbide nanocomposite ceramics prepared by pressureless sintering, *Journal of the European Ceramic Society* 24 (2004) 3317-3326.

[2] J Chandradass, M Balasubramanian, Sol-gel based extrusion of alumina–zirconia fibres, *Materials Science and Engineering: A* 408(1-2) (2005) 165-168.

[3] Wataru Nakao, Masato Ono, Sang-Kee Lee, et al, Critical crack-healing condition for SiC whisker reinforced alumina under stress, *Journal of the European Ceramic Society* 25 (16)(2005) 3649-3655.

[4] Eric Laarz, Mats Carlsson, Benoît Vivien, et al, Colloidal processing of Al_2O_3-based composites reinforced with TiN and TiC particulates, whiskers and nanoparticles, *Journal of the European Ceramic Society* 21 (8)(2001) 1027-1035.

[5] Bo Zhang, Freddy Boey, The phases and the toughening mechanisms in (Y) ZrO_2-Al_2O_3- (Ti, W) C ceramics system, *Materials Letters* 43 (4)(2000) 197-202.
[6] R.F.Cook, B.R. Lawn, A modified indentation toughness technique, *J. Am. Ceram. Soc.* 66 (11) (1983) 200-201.
[7] D.L.Ye, J.H.Hu, Handbook of The Thermodynamic Data of Inorganic Substances, Metallurgical Industry Press, *Peking of China*, 2002, pp.57-1061.
[8] Eugene Medvedovski, Alumina–mullite ceramics for structural applications, *Ceramics Internationa* 32 (4)(2006) 369-375.
[9] K. A. Berry and M. P.Harmer, Effect of MgO solute on microstructure development in Al_2O_3, *J. Am. Ceram. Soc*. 69(2)(1986)143-149.
[10] S. Baik, J.H. Moon, Effect of magnesium oxide on grain-boundary segregation of calcium during sintering of alumina, *J. Am. Ceram. Soc*. 74 (4) (1991) 819-822.
[11] S.I. Bae, S. Baik, Critical concentration of MgO for the prevention of abnormal grain growth in alumina, *J. Am. Ceram. Soc*. 77(10) (1994) 2499-2504.

In: Ceramic Materials Research Trends
Editor: Paul B. Lin, pp. 209-225

ISBN: 1-60021-769-9

Chapter 9

PREPARATION AND PERFORMANCE OF NANOPOROUS HYDROPHILIC AND HYDROPHOBIC CERAMIC MEMBRANES FOR USE IN NON-AQUEOUS NANOFILTRATION

B. Van der Bruggen[1*], A. Buekenhoudt[2**], B. Verrecht[2], H. Van den Dungen[1], J. Geens[1] and R. Leysen[2]

[1]K.U.Leuven, Department of Chemical Engineering, W. de Croylaan 46, B-3001 Leuven, Belgium

[2]VITO (Flemish Institute for Technological Research), Environmental and Process Technology, Boeretang 200, B-2400 Mol, Belgium

Abstract

The preparation of nanoporous ceramic membranes and their performance in non-aqueous nanofiltration is discussed. Starting from the typical manufacturing method for hydrophilic nanofiltration membranes, an evaluation is made of the performance of these membranes, relative to other membrane types. Fluxes for polar and non-polar solvents are studied for TiO_2 membranes with changes in the preparation procedure (e.g., the firing temperature). Rejections for some typical organic compounds are given. Suitable membranes in the nanofiltration range, i.e., with a molecular weight cut-off around 500 g/mole, could be obtained.

Possibilities for obtaining a (more) hydrophobic surface are described. One of the most attractive options is the introduction of silane groups into a mesoporous membrane structure. It can be shown that after silane coupling, a hydrophobic surface is obtained suitable for filtration of non-polar solvents. Fluxes and rejections (and, correspondingly, pore sizes) depend on the chain length of the silane coupling agent. Furthermore, the hydrophobic nature and the stability of the membrane depends on the conditions used during the modification procedure. In particular, it was shown that contact with water in some cases partly reverses the modification.

* E-mail address: bart.vanderbruggen@cit.kuleuven.be. Tel: +32 16 322340 / Fax: +32 16 322991
** E-mail address: anita.buekenhoudt@vito.be. Tel:+32 14 335637

The permeabilities of hydrophilic and hydrophobic membranes were analysed in terms of (non-)viscous behaviour. It was found that for hydrophilic membranes with typical nanofiltration separation characteristics, permeabilities deviate from viscous flow behaviour due to polarity effects. Mesoporous hydrophilic membranes are less influenced by polarity effects and are generally closer to viscous behaviour. Hydrophobic membranes also deviate from viscous behaviour, although the conclusions for these membranes are less obvious.

Keywords: ceramic membrane, nanofiltration, solvent filtration, nanoporous materials

Introduction

Separation of relatively small organic compounds (molar mass above 150) from organic solvents by nanofiltration has received considerable attention during the last decade [1-5]. The use of nanofiltration membranes originates from the success of this process in water treatment, where it is increasingly used for simultaneous removal of ions and small organic pollutants [6,7]. For aqueous solutions, polymeric membranes are often stable enough, although membrane fouling remains an unsolved issue for more demanding applications such as in wastewater treatment [8]. Most polymers applied in membranes for water treatment are relatively inexpensive and therefore attractive, even when their performance is not ideal. However, polymeric membranes developed for high-temperature applications or for solvent filtration tend to be much more expensive, while their stability is not always guaranteed and their performance depends on the membrane structure [9-11]. Swelling, cracking or degradation of the membrane material is often observed, leading to a loss of separation power [12]. In these conditions, ceramic nanofiltration membranes are possible alternatives with advantages in terms of membrane stability, fouling resistance and easy cleaning.

Nanoporous ceramic membranes are made of alumina, titania or zirconia and are intrinsicly hydrophilic. Therefore, their relevance is restricted to aqueous applications and polar solvents. Unlike microfiltration membranes, where hydrophobic materials can be used for filtration of polar solvents and vice versa, non-polar solvents can not be transported through a hydrophilic nanofiltration membrane; due to the small size of available pathways through the material. Hydrophobic-hydrophilic repulsion forces dominate the membrane-solvent interactions. The action radius of these forces covers the whole 'pore' area, making transport of a non-polar molecule impossible. In addition, hydration of membrane pores is absent in organic solvents; the reduced solvation capacity of these solvents (in particular for non-polar solvents) results in a larger effectively available pore size and therefore, lower rejections. These aspects have to be taken into account when considering ceramic nanofiltration membranes. In addition to the difficulties in obtaining nanoporous ceramic membrane structures with the desired separation potential, unexpected behaviour in terms of fluxes and rejections is another impediment for successful application of this technology.

Starting from the development of hydrophilic nanofiltration membranes, the aim of this chapter is to describe the capacities of ceramic materials to retain solutes from organic solvents as a function of membrane characteristics and solute properties. Furthermore, new findings on the possibility of developing hydrophobic ceramic nanofiltration membranes by applying a post-treatment procedure within the membrane pores, are presented. For both hydrophilic and hydrophobic membranes solvent transport will be evaluated in terms of viscous behaviour and polarity effects.

Manufacturing of Hydrophilic Nanofiltration Membranes

Successful procedures for manufacturing hydrophilic ceramic nanofiltration membranes have been developed since the late 1990s and resulted in a steadily improved membrane structure [13-15]. Membranes combining a high flux and a high separation factor can only be obtained in an asymmetric multilayer configuration [16]. A succesful multilyer membrane consists of an open porous support, appropriate mesoporous interlayers, and high-quality microporous top layers. A typical picture of the mulitlayer structure of a ceramic nanofiltration membrane is shown in figure 1.

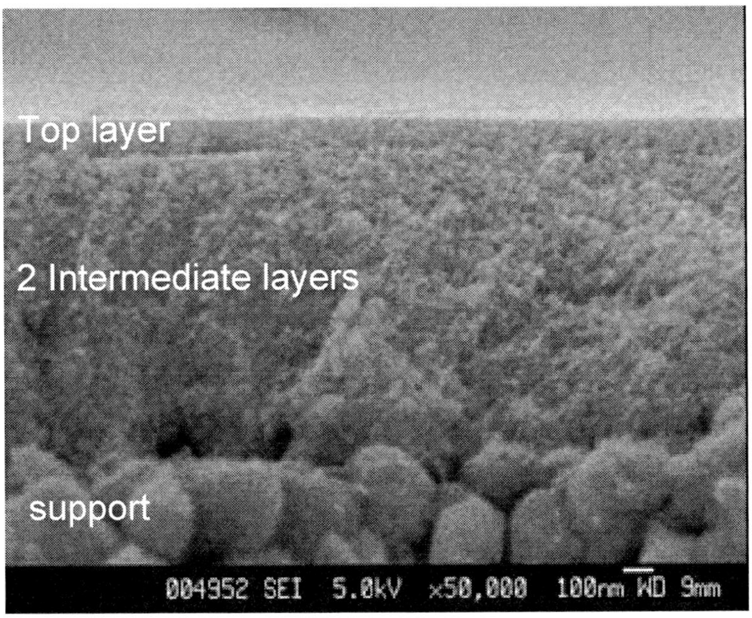

Figure 1. SEM picture of the multilayer structure of a ceramic nanofiltration membrane.

A good support provides the mechanical strength of the final membrane, and has a sufficiently open pore structure in order to minimize the fluid resistance over the support. In order to combine these two contradictory requirements, current supports for commercial fine-porous ceramic membranes are in itself asymmetric structures. They consist of an extruded body containing coarse pores in the range of a few μm, with on top of that a slip-casted layer with pores in the order of 50 to 100 nm. This structure is the same as the one of commercially available open ultrafiltration membranes with cut-off values of 200 kD and more. These commercially available open UF membranes (mostly in tubular form) are also the ones that are used as supports in many labs developing their own fine-porous membranes [13-15]. In some cases of early membrane development, flat laboratory-made supports prepared by slip-casting are used. These supports have a symmetric structure, with the same pores of the order of 50 to 100 nm all over the structure. These supports have a much higher fluid resistance, but show a much lower surface roughness and a smaller amount of defects. Alumina (in the α-alumina phase) is the support material most often used. However, nowadays more and more supports are made of titania, due to its higher chemical stability and its non-toxicity [17].

The intermediate layers are used to gradually decrease the pore size and the surface roughness of the membrane. Multiple interlayers are normally used, since inevitable large surface irregularities and defects in the supports can only be covered properly by a multiple coating procedure. The interlayers are prepared by the colloidal sol-gel procedure [16]. They are mainly produced of alumina (in the γ-alumina phase) or titania, or, less frequently used, of silica or zirconia. Also mixtures of these metal oxides are frequently used. The aqueous sol is prepared from metal organic (alkoxide) precursors as i.a. Al-tri-sec-butoxide, or Ti-tetra-isopropoxide. The sol is hydrolised and peptised in order to form medium-sized ceramic nanoparticles in the order of 20 to 100 nm in the solution. This solution is possibly mixed with typical binder materials as polyvinyl alcohol (PVA) or hydroxypropyl cellulose (HPC). An intermediate layer is formed by dip-coating the support in this sol. It is the porosity of the support that leads to the gelling of the sol. Each gel layer is carefully dried and calcined at an appropriate temperature. The pore size of the formed layers depends on the details of the hydrolising/peptisising process and of the calcination temperature. Good intermediate layers for fine-porous membranes have pore sizes in the order of 3 to 5 nm. In order to get these small pore sizes, calcination temperatures have to be kept away from the phase-transition temperatures of the used metal oxide (i.a. the anatase-rutile transition of titania in the temperature region 500 – 700°C) [13]. γ-alumina interlayers can be made relatively thick (~2 μm) while titania or zirconia interlayers show an average thickness of only 0.5 μm. As a consequence, γ-alumina layers can better mend support irregularities and defects. However, γ-alumina layers decrease the flux of the final membrane due to their higher fluid resistance. Moreover, γ-alumina is chemically unstable in acids and bases (only stable for $3 < pH < 11$) [13].

The final step in the preparation of a multilayer nanofiltration membranes is the coating of the top layer(s). The top layer contains the smallest pores and defines the real selectivity of the porous membrane. The top layers are produced from alumina (in the boehmite phase), titania, zirconia, silica, or mixtures of these metal oxides. Formation of such a layer with very small pores in the order of 1 nm, requires an appropriate sol-gel method called the polymeric sol-gel method [16]. A polymeric sol is produced starting from the common alkoxide precursors, but now partially hydrolysed with a less than stoechiometric amount of water in order to obtain a precipitate-free sol containing small nanometer-sized polymeric structures (fractals). For good nanofiltration layers, average fractal sizes need to be 5 nm are smaller. The sols are used for dip-coating the support+interlayer system, without the addition of any binder. The formed gel layer is again carefully dried, and calcined. For nanofiltration membranes, the calcination temperature needs to be low enough, in order to avoid sintering effects, and the accompanied pore growth. The calcination temperature does not only determine the final pore size of the top layer, but also the phase structure of the top layer material. This is illustrated in table 1 for microporous titania powder made in the same way as the membrane toplayers [13]: for calcination temperatures as low as 200°C titania is in an amorphous state with very small pores, from 300°C onwards titania is in an anatase phase having somewhat bigger pores (remark: the pore size figures for the powder are somewhat larger than expected for the membrane toplayers). This has also a great influence on the chemical stability of the top layer: an amorphous phase is more easily attacked by corrosive liquids, and shows a lower acid and base stability than the crystalline phase [13].

Table 1. pore size and phase structure of microporous titania powder produced by the polymeric sol-gel method, as function of the calcination temperature

Phase	Calcination temperature (°C)	Mean pore size (nm)
Amorphous	200	1.6
Anatase	300	2.4
Anatase	400	3.4

Using the procedures described above, many small-scale nanofiltration membranes have been made, in different labs. In many cases γ-alumina intermediate layers have been used, causing their limited chemical stability. If broad application and commercial use is envisioned, making chemical stability an important issue, a material as titania or zirconia, in a crystalline phase, both for interlayers and toplayers, is a better choice. For the moment, one commercial nanofiltration membrane with titania interlayers and toplayers, is on the market, at large scale. It was developed by Voigt et al. [14] and is commercialised by the company Inocermic in Germany. The membrane has a MWCO of 500. As all these nanofiltration membranes are manufactured from metal oxides, they are intrinsically hydrophilic. This property is greatly determining the performance of these membranes, as will be shown further.

Surface Modification of Hydrophilic Membranes

By surface modification of hydrophilic membranes, it is possible to obtain hydrophobic ceramic nanofiltration membranes. The objectives for surface modifications on hydrophilic membranes are twofold: firstly, to obtain a structure that allows filtering non-polar solvents such as toluene or hexane, and secondly, to allow a precise control of pore sizes so that the separation properties of the membrane can be tuned as a function of the envisaged application. This can be done by binding specific organic compounds to the surface of relatively tight mesoporous membrane (pores in the range of 5 nm). Most frequently used is a silane coupling treatment using various hydrophobic tail lengths. The technique is based on state-of-the-art modification processes used for silica in order to produce appropriate chromatography columns.

This silane coupling method was previously applied for other purposes on macroporous (MF) and mesoporous (UF) membranes using organochlorosilane coupling reagents [18-22]. This resulted in modified membranes with either a decreased pore size or a modified hydrophilic/hydrophobic character. Leger et al. [18] and Javaid et al. [19] modified a mesoporous γ-Al_2O_3 UF membrane (SCT, membralox) with long alkyl chain reagents (R=C18-C28) according to a wet-chemical method in toluene, yielding microporous membranes with gas separation properties. Miller et al. [20] reported comparable results for a mesoporous γ-Al_2O_3 UF membrane (membralox), treated with long alkyl chain reagents (R=C18-C28) according to a vapor phase deposition method. Simon et al. [21] focused on a modification of the hydrophilic character of α-Al_2O_3 MF membranes (0.15 – 0.4 μm), with a short alkyl chain reagent (R=C1). Picard et al. [22] described a modification of the hydrophilic character of ZrO_2 and $KTiOPO_4$ UF membranes (0.23 μm – 6 nm) with C6 and C8 alkyl chain reagents according to a wet-chemical method in chloroform.

Modification of tight mesoporous membranes in order to enhance their performance in apolar solvents was first studied by Tsuru *et al.*, Van Gestel *et al.*, and Voigt *et al.* [15,23,24]. Tsuru et al. modified mixed silica/zirconia membranes with a trimethylchlorosilane in the gas phase [15]. Van Gestel *et al.* [23] modified a γ-Al_2O_3/anatase-TiO_2 multilayer membrane using a short and a long alkyl chain chlorosilane reagent (R=C1,C8) using a wet-chemical route in ethyl acetate. This solvent was chosen because it is a relatively polar solvent that permeates through fine-porous hydrophilic structures. Van Gestel used bi-functional chlorosilanes, with two reactive chloor groups, due to their higher reactivity with the membrane surface. The higher reactivity of the multifunctional chlorosilanes was confirmed by Sah *et al.* [25]. However, these multifunctional chlorosilanes may also result in pore blocking of the micropores due to polymerisation reactions (hydrolysation and condensation between the silane molecules), having a negative influence on the solvent fluxes. That is the reason why Voigt *et al.* [24] worked with silanes with only one reactive group. Their reaction was done with reagents with different chain lengths, in liquid n-heptane. All groups [15, 23, 24, 25] showed that in order to modify the complete specific surface of the membrane, the pore size had to be large enough for the specific silane to enter. In all the studies mentioned, hydrophobisation of the inner membrane surface by the organochlorosilane reaction was clearly shown, e.g. by permeation measurements of polar and apolar solvents.

According to Leger *et al.* [18], the organochlorosilane modification route involves a chemical reaction between the hydroxyl groups on the membrane surface and the silane coupling reagent (Figure 2).

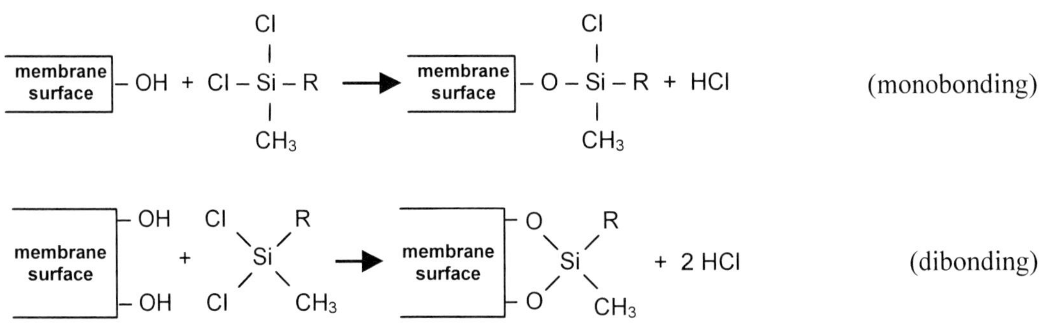

Figure 2. Reactions between membrane surface hydroxyl groups and silane coupling reagent.

However, the precise interaction between the chlorosilanes and the membrane surface is very complex and there is in many cases no evidence that it involves a chemical reaction rather than an adsorption reaction. For instance, Impens *et al.* [26] and Hair and Tripp [27] state that a reaction of alkyltrichlorosilanes with a silica surface is not possible below 300°C. At room temperature the chlorosilane reacts only with water present in the solvent, or adsorbed at the membrane surface. A polymeric chain is then formed that can adsorb or attach to the OH-groups of the membrane surface by hydrogen bonding, without direct reaction with the membrane surface. However, a direct reaction with the membrane surface at room temperature is possible using a two-step procedure, involving an extra chemical agent as catalyst. In the first step, a strong base such as triethylamine (TEA) is bound to hydroxyl groups on the surface. This creates a strongly nucleophilic oxygen atom on the hydroxyl site, which can react with the chlorosilane and then form a chemical bond.

It is also clear from the previous studies that the pretreatment temperature, determining the amount of reactive hydroxyl sites, and a curing process after the silanation can largely influence the final modified surface [26]. The nature of the metal oxide working on, is also important for the reactivity and the success of the modification. It is clear from previous studies [28] and from our own recent work that silica is the easiest material to modify.

Besides chloroalkylsilanes, it is also possible to use phosphonic acids or fluoroalkylsilanes for the modification. Phosphonic acids were up to now only used to change UF membranes to gas separation membranes [28, 29]. Caro *et al.* [28] showed that there is a clear reaction between the γ-alumina surface of the UF membrane and the phosphonic acids. Picard *et al.* [30] used fluoroalkylsilanes in order to make open zirconia UF membranes more hydrophobic. Using ^{1}H NMR, they proved that also for the fluoroalkylsilanes there is a real reaction between fluoroalkylsilanes and the membrane surface. These reagentia were also succesfully used by Voigt *et al.* in order to make hydrophobic nanofiltration membranes [24].

Due to the lack of a real chemical reaction between organochlorosilanes and the membrane surface, the stability of the modification with these reactants is probably limited. Our own recent measurements [31] show that each exposure of the hydrophobic membrane to water systematically decreases the hydrophobicity of the membrane. This is most probably caused by a rehydroxylation of the surface. The stability of the modifications with phosphonic acids and fluoroalkylsilanes is expected to be better, thanks to the existing chemical bond between the membrane surface and the modifying agent.

Performance of Ceramic Nanofiltration Membranes

This section will describe the performance of ceramic nanofiltration membranes, both hydrophilic and hydrophobic.The performance of the hydrophilic membranes in water is sufficiently clear and is mainly governed by the viscosity (for the flux) and the molecular weight of the solutes compared to the MWCO of the membrane (for the rejection) [13,14]. In this respect, ceramic nanofiltration membranes do not differ from polymeric nanofiltration membranes. However, there is no such clear picture for the behaviour of nanofiltration membranes in non-aqueous solvents. The following gives an overview of our own knowledge, complemented by the current knowledge of other groups working on nanofiltration membranes, and especially ceramic nanofiltration membranes in non-aqueous solvents. It will be clear that the behavior is considerably complex, and that a lot of additional research, both experimentally and theoretically, will be necessary in order to clarify this performance. This extra insight will also be needed in order to facilitate the application of these membranes in the industry.

Hydrophilic Membranes: Determination of Transport Mechanisms

Experimental results for polymeric membranes [32] clearly indicate that solvent transport through solvent resistant nanofiltration membranes is influenced both by viscosity and polarity effects. However, conclusions based on experiments with polymeric membranes might not be valid for ceramic nanofiltration membranes. The difficulty in extending the findings towards ceramic membranes is due to the fact that fluxes for polymeric membranes might be strongly affected by swelling effects. Swelling is essentially the result of the instability of the polymer material in (some) organic solvents and should be considered as a shortcoming of the membrane material; although good results were obtained for separations in organic solvents using polymeric membranes as well [33-35], membrane stability is a point of special attention for these membranes [36]. Ceramic membranes are not susceptible to swelling effects and thus possible changes in solvent permeability with changing feed composition can not be attributed to this phenomenon.

Therefore, the performance of ceramic nanofiltration membranes (flux and rejection) should be studied separately, and can bring a better understanding of the performance of all nanofiltration membranes. A performance study was carried out for a ceramic nanofiltration membrane manufactured at Vito (Mol, Belgium), denoted as FSTi-209. This membrane has a molecular weight cut-off (MWCO) of 430, measured in water, and an initial pure water flux of 22 l/m².h.bar. This membrane is representative for the current ceramic nanofiltration membranes. Experiments were carried out on a cross-flow filtration set-up, at a transmembrane pressure (TMP) of 10 bar and a temperature of 20 °C. A feed flow of 2 l/min was used.

The surface tension of the FSTi-209 membrane could not be measured as this is a tubular ceramic membrane, and flat surfaces are required for the determination of the surface tension with contact angle measurements. However, it is known that FSTi-209 consists of a TiO_2 toplayer, which is very hydrophilic. The results obtained with this membrane might therefore be compared with literature results of hydrophilic polymeric membranes.

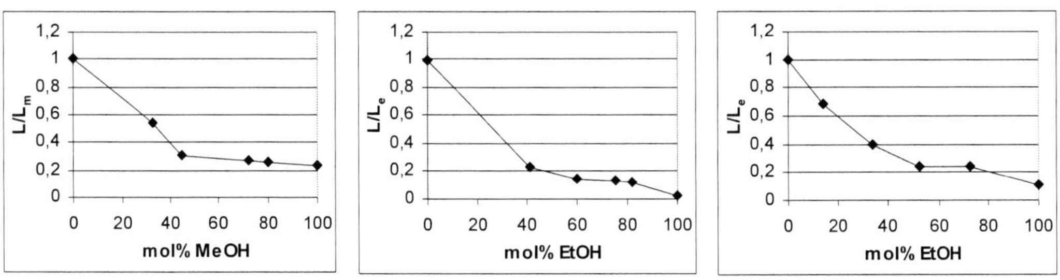

(a) (b) (c)

Figure 3. Normalised permeability of FSTi-209 for (a) water-methanol, (b) water-ethanol and (c) methanol-ethanol (absolute values in l/h.m².bar, for respectively water, methanol and ethanol are: 1.59, 0.37 and 0.04)

Figure 3 shows the normalised permeabilities through the FSTi-209 membrane for (a) water-methanol, (b) water-ethanol and (c) methanol-ethanol mixtures. In all cases, a monotonous decrease of the permeability was observed. Darcy's law suggests an inverse proportionality of the flux with viscosity, so that the flux profile as a function of composition should follow this profile. However, viscosity is not a monotonous increasing/decreasing function of composition (Figure 4), which clearly indicates that for hydrophilic ceramic nanofiltration membranes Darcy's law is not met and that the viscosity can not be the sole transport parameter. Similar deviations from Darcy's law with hydrophilic ceramic nanofiltration membranes (silica-zirconia) were also reported by Tsuru *et al.* [37] for a series of measurements with methanol, ethanol and 1-propanol at different temperatures.

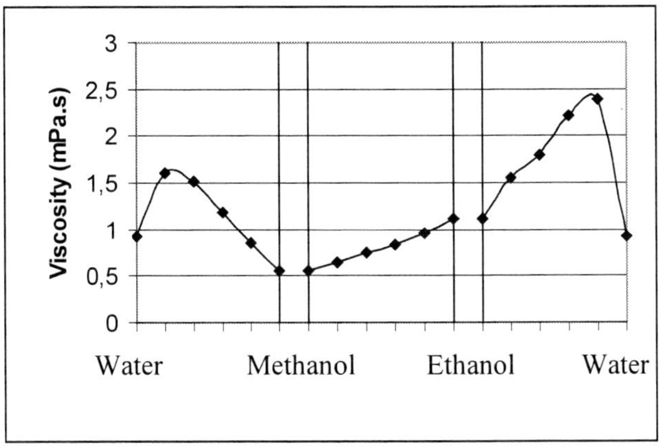

Figure 4. Viscosity of binary mixtures: water-methanol, methanol-ethanol and ethanol-water.

A similar flux study of water-methanol, water-ethanol and methanol-ethanol mixtures was performed on hydrophilic polymeric nanofiltration membranes [32]. The results do show a minimum in flux of the water-methanol and water-ethanol mixtures, corresponding to the maximum in the viscosity in these mixtures [32]. In other words, hydrophilic polymeric nanofiltration membranes show a more viscous-like behavior than the corresponding ceramic membranes. This is unexpected, since ceramic membranes generally have larger pores than polymeric membranes [38], and therefore more convective flow might have been expected for them. The reason is probably the much stronger hydrophilicity of the ceramic membranes.

Non-polar solvent measurements on hydrophilic ceramic membranes are rather scarce. However, Garcia *et al.* [39] found reduced permeability for n-hexane through zirconia-aluminia tight ultrafiltration membranes. Van Gestel *et al.* [23] even found zero n-hexane fluxes for a γ-alumina/titania nanofiltration membrane. This suggests a strong effect of the polarity of the solvent on the transport thorugh hydrophilic ceramic nanoporous membranes. Since the toplayers of untreated ceramic nanofiltration membranes consist of very hydrophilic materials, it is indeed expected that less polar solvents are liable to repulsive interactions at the membrane-solvent interface. For the mixtures we studied, the polarity of the water-methanol, water-ethanol and methanol-ethanol mixtures decreases with increasing fraction of the latter component of each mixture. The reduced polarity would result in lower fluxes. As the permeability curves for FSTi 209 TiO_2 membrane indeed present a monotonous decrease (figure 3), it appears that for ceramic nanofiltration membranes, the polarity effects strongly dominate the other transport parameters, in particular the solvent viscosity.

The partial permeabilities of the different mixtures through the ceramic nanofiltration membranes FSTi 209 TiO_2 are presented in Figure 5. It appears that the partial permeability of the most polar component of a binary mixture (i.e. water for the water-alcohol mixtures and methanol for the methanol-ethanol mixture) is strongly affected by the less polar component, whereas in reverse the most polar component has only a minor effect on the partial permeability of the second component. This can again be explained by the high degree of hydrophilicity of the ceramic membrane, and the high effect of the polarity on the solvent transport.

The partial permeabilities of the same mixtures were also measured for hydrophilic solvent resistant polymeric nanofiltration membranes [40], and the same effects have been seen as for the ceramic membranes. This means that also for the hydrophilic polymeric membranes, the solvent viscosity is not the only parameter influencing the transport through nanofiltration membranes, and other aspects, such as the solvent polarity and the solvent-membrane interactions, play a role in solvent permeation as well. For ceramic membranes, due to their high degree of mechanical and chemical stability, they are not liable to swelling effects and the observed phenomena can therefore not be attributed to swelling. From the similar experimental data for hydrophilic polymeric and ceramic nanofiltration membranes, it can thus be concluded that the results for the polymeric membranes are neither the consequence of membrane instability or swelling effects.

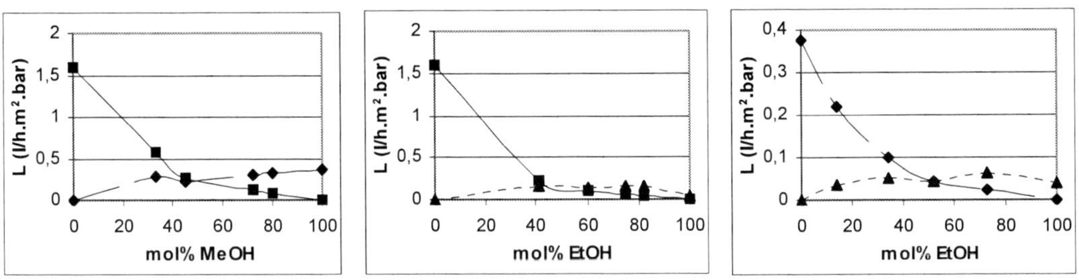

Figure 5. Partial permeabilities for (a) water-methanol, (b) water-ethanol and (c) methanol-ethanol mixtures (FSTi-209). (■: water; ◆: methanol; ▲: ethanol)

It should be noted that these findings are valid for typical nanofiltration membranes such as the FSTi-209 membrane; membranes with larger pores (larger MWCO) are generally closer to viscous behaviour with a smaller effect of polarity. This is shown in Figure 6 for a TiO_2 membrane with a MWCO of 1,400 (measured with a PEG mixture in water): the solvent permeability (flux divided by trans membrane pressure) is proportional to the inverse viscosity, as expected from known equations reflecting viscous transport. Hexane, a highly non-polar solvent, is an exception, which again shows that transport occurs through more than one single mechanism, the dominating mechanism being dependent on the membrane structure.

When the absolute values of the fluxes for non-aqueous solvents through the ceramic membrane are compared with those for the hydrophilic polymeric membranes, it can be seen that the permeability is lower through the ceramic membrane, which can mainly be explained by the low degree of affinity between the strongly hydrophilic TiO_2 toplayer and the less polar solvents. The lower permeability can also partly be attributed to the thickness of the membrane toplayer. The permeability is indeed inversely proportional to the membrane thickness and polymeric TFC-membranes have a very thin toplayer, which is significantly smaller than those of ceramic membranes [38]. The thicker toplayers of ceramic nanofiltration membranes are due to the manufacturing process and are therefore inherent in the technique.

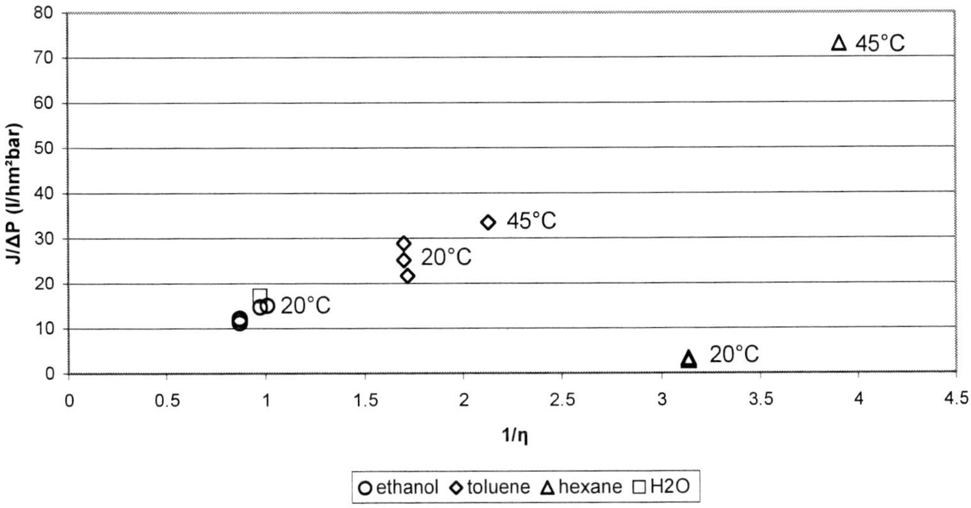

Figure 6. solvent permeability (flux divided by trans membrane pressure) as a function of inverse viscosity for a mesoporous hydrophilc TiO_2 membrane

However, when comparing with polymeric membranes, the differences between hydrophilic polymeric membranes and ceramic membranes (FSTi-209) are smaller than the differences between hydrophilic and hydrophobic polymeric membranes.

Literature data on solute rejection in non-aqueous nanofiltration with ceramic nanofiltration membranes are still scarce. Tsuru *et al.* [41] reported that the rejection of low molecular weight ethylene glycols was lower in ethanol than in methanol, which is in correspondence with the experimental results mentioned above. Figure 7 shows the raffinose rejection for the water-methanol and the water-ethanol mixtures for the FSTi-209 membrane. It was observed that the rejection is almost constant for the water-methanol mixture (93 – 89 %). In contrast, a clear decrease in solute rejection is observed for the water-ethanol mixture (93 – 63 %). A smooth decrease in solute rejection was observed for the methanol-ethanol mixture. These decreases can not be explained by swelling of the ceramic membrane material. The decrease in solute rejection is largely due to solute-solvent and solvent-membrane interactions. With increasing alcohol fraction, raffinose is rather solvated instead of hydrated. Solvation is less effective than hydration, since different intermolecular forces are involved: whereas strong interaction forces between permanent dipoles (hydrogen bonds) or between an induced and a permanent dipole occur for hydration, solvation is based on weaker interaction forces between induced dipoles (and thus based on the polarisability), also known as London dispersion forces [42]. These are typically one order of magnitude smaller than hydration forces [43]. Because of the phenomenon of solvation, the effective solute size, and thus the solute rejection, decreases with increasing alcohol fraction. Moreover, the effect is expected more pronounced in ethanol than in methanol, as ethanol is less polar than methanol; solvation by ethanol is less effective than by methanol. It was calculated that the effective diameter of the solute is smaller for methanol and even more for ethanol. A similar effect can be presumed for the membrane pores, being more effectively hydrated than solvated for a hydrophilic material, resulting in smaller effective pore diameters for more polar solvents. Based on the experiences with ceramic nanofiltration membranes, swelling can be excluded as the most important phenomenon with respect to solute permeation, and the term pore wall solvation may be more appropriate.

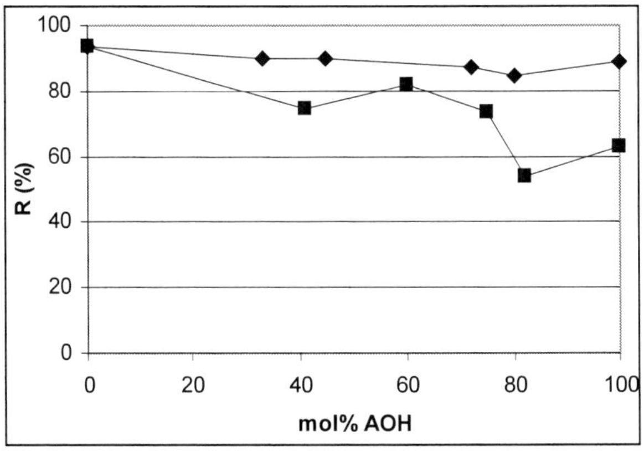

Figure 7. Rejection of raffinose (100 mg/l) in binary mixtures of water-methanol (◆) and water-ethanol (■) with the ceramic FSTi-209 membrane

The extrapolation of these conclusion towards other solvents would obviously lead to rejection values depending significantly on both the solute and the solvent for a given membrane. This is shown in Table 2 for FSTi-128, a membrane similar to FSTi-209 (molecular weight cut-off 420, initial pure water flux 21 l/m².h.bar) for acetone and ethyl acetate.

Table 2. Rejection of six reference components with FSTi-128 in acetone and ethyl acetate (-°: not determined)

FSTI-128	Formula	MW (g/mol)	Acetone	Ethyl acetate
Eusolex	$C_{14}H_{12}O_3$	228.25	15	30
2,2-methylenebis-(6-tert-butyl-4-methyl-phenol)	$C_{23}H_{32}O_2$	340.51	52	73
Victoria Blue	$C_{33}H_{32}ClN_3$	506.09	80	85
DL-α-tocopherol hydrogen succinate	$C_{33}H_{54}O_5$	530.79	-°	22
Bromothymol Blue	$C_{27}H_{28}Br_2O_5S$	624.40	95	42
Erythrosine B	$C_{20}H_6I_4Na_2O_5$	879.92	91	90

It can be concluded that solute transport through hydrophilic ceramic nanofiltration membranes is clearly affected by three parameters: the nominal pore diameter, typically specified by the MWCO, the affinity between the solvent and the solute, resulting in a solvent dependent effective diameter of the solute, and the affinity between the solvent and the membrane, leading to a solvent dependent effective pore diameter.

Experience with the Performance of Hydrophobic Membranes

The performance of hydrophobic ceramic nanofiltration membranes can be variable, depending on the membrane synthesis. Van Gestel *et al.* [23] reported fluxes of non-polar solvents such as *n*-hexane that were highly dependent on the membrane's modification treatment (i.e., silane coupling treatment with various initial pore sizes and various silane chain lengths), with typical zero-fluxes observed for *n*-hexane through untreated ceramic membranes. This confirms the importance of polarity effects acting during the nanofiltration process. The applicability of hydrophobic ceramic NF membranes in organic solvents was also studied by Tsuru *et al.* [37, 41] for SiO_2-ZrO_2 membranes without modification and after modification in the gas phase with trimethylchlorosilane (TMS). Permeability- and rejection measurements with alkanes and alcohols confirmed that the viscous flow mechanism does not apply to unmodified hydrophilic membranes, while the transport mechanism through a 3 nm membranes modified with TMS obeys better the viscous flow mechanism. The latter is probably due to the relatively large pore size of this membrane; membranes with a larger MWCO are generally much closer to viscous behaviour, as explained above.

The combined effect of pore size and polarity, observed for hydrophilic membranes, was also found for hydrophobic membranes. Figure 8 shows fluxes for a 600 Da hydrophobic membrane (cut-off measured in toluene), prepared by HITK, Germany, by a silane coupling treatment of a 1,200 Da ZrO_2 membrane, which effectively reduced its pore size and hydrophilic character. The 600 Da HITK hydrophobic membrane was characterized by dead-end flux measurements with a series of solvents. Results show that fluxes of non-polar organic solvents increase, while water permeability is very low after modification, providing proof that successful modification is possible. The solvent permeabilities have a tendency to be proportional to the inverse viscosity but the deviations from the straight line through the origin (in particular for water) show that viscous transport is influenced by polarity effects.

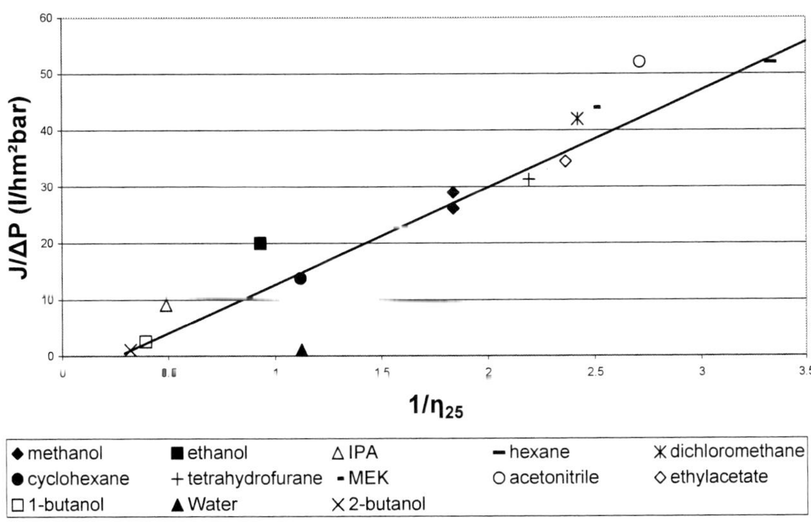

Figure 8. Permeability vs viscosity^{-1} for a 600 Da hydrophobic HITK membrane.

The stability of the current hydrophobic ceramic nanofiltration membranes is still under debate. Our own recent measurements [31] show that each exposure of the hydrophobic membrane to water, systematically increases the water flux, and simultaneously decreases the non-polar solvent fluxes. After sufficient water exposure, the fluxes stabilise on the values found for the unmodified membrane. We tentatively explain this by a rehydroxylation of the membrane surface, leading to a gradual loss of hydrophobicity of the membrane.

Conclusions

Ceramic nanofiltration membranes have become more mature in terms of synthesis methods and performance, in particular for hydrophilic membranes. Manufacturing of hydrophilic membranes with a MWCO of around 500 is possible in a controlled way, leading to a reliable performance for polar solvents. Transport of polar solvents occurs through a combination of viscous and polarity effects; the dominant mechanism depends on the membrane pore structure. The polarity effects are in general stronger than for the corresponding hydrophilic polymeric nanofiltration membranes. The solute transport through hydrophilic ceramic nanofiltration membranes is affected by three parameters: the nominal pore diameter, typically specified by the MWCO, the affinity between the solvent and the solute, resulting in a solvent dependent effective diameter of the solute, and the affinity between the solvent and the membrane, leading to a solvent dependent effective pore diameter

Efficient nanofiltration of non-polar solvents at room temperature is not possible with hydrophilic membranes. By using a silane coupling reaction starting with membranes with a relatively large MWCO (UF type membranes), hydrophobic nanofiltration membranes can be obtained, allowing nanofiltration of non-polar solvents such as n-hexane and toluene with sufficient flux. While these membrane types are still under development (in particular in view of increasing membrane stability and determination of transport mechanisms), it is clear that they can fill the gaps where hydrophilic membranes cannot be used.

References

[1] Cuperus, P. ; Ebert, K. Non-aqueous applications of NF. Chapter 21 in Nanofiltration – Principles and Applications, *Elsevier Advanced Technology*, Oxford, 521-536 (2005).

[2] Ebert, K.; Cuperus, F.P. Solvent resistant NF membranes in edible oil processing. *Membrane Technology* 1999, 107, 5-8.

[3] Guizard, C.; Ayral, A.; Julbe, A. Potentiality of organic solvents filtration with ceramic membranes. A comparison with polymer membranes. *Desalination* 2002, 147, 275-280.

[4] Whu, J.A.; Baltzis, B.C.; Sirkar, K.K. Nanofiltration studies of larger organic microsolutes in methanol solutions. *Journal of Membrane Science* 2000, 170, 159-172.

[5] Yang, X.J.; Livingston, A.G.; Freitas dos Santos, L. Experimental observations of nanofiltration with organic solvents. *Journal of Membrane Science* 2001, 190, 45-55.

[6] Van der Bruggen, B.; Vandecasteele, C. Removal of pollutants from surface water and groundwater by nanofiltration: overview of possible applications in the drinking water industry. *Environ. Pollut.* 2003, 122 (3), 435-445.

[7] Van der Bruggen, B.; Vandecasteele, C.; Van Gestel, T.; Doyen, W.; Leysen, R. Pressure driven membrane processes in process and waste water treatment and in drinking water production. *Environ. Progr.* 2003, 22 (1), 46-56.

[8] Schäfer, A.I.; Andritsos, N.; Karabelas, A.J.; Hoek, E.M.V.; Schneider, R.; Nyström, M. Fouling in nanofiltration membranes. Chapter 8 in Nanofiltration – Principles and Applications, *Elsevier Advanced Technology*, Oxford, 521-536 (2005).

[9] Silva, P.; Han, S.J.; Livingston, A.G. Solvent transport in organic solvent nanofiltration membranes. *Journal of Membrane Science* 2005, 262 (1-2), 49-59.

[10] Robinson, J.P.; Tarleton, E.S.; Millington, C.R.; Nijmeijer, A. Solvent flux through dense polymeric nanofiltration membranes. *Journal of Membrane Science*, 2004, 230, 29-37.

[11] Bhanushali, D.; Kloos, S.; Kurth, C.; Bhattacharyya, D. Performance of solvent-resistant membranes for non-aqueous systems: solvent permeation results and modeling. *Journal of Membrane Science* 2001, 189, 1-21.

[12] Van der Bruggen, B.; Geens, J.; Vandecasteele, C. Influence of organic solvents on the performance of polymeric nanofiltration membranes. *Separ. Sci. Technol.* 2002, 37 (4), 783-797.

[13] Van Gestel, T.; Vandecasteele, C.; Buekenhoudt, A.; Dotremont, C.; Luyten, J.; Leysen, R.; Van der Bruggen, B.; Maes, G. Alumina and titania multilayer membranes for nanofiltration: preparation, characterization and chemical stability. *Journal of Membrane Science* 2002, 207, 73-89.

[14] Voigt, I.; Fischer, G.; Puhlfurss, P.; Schleifenheimer, M.; Stahn, M. TiO2-NF-membranes on capillary supports. *Separation and Purification Technology* 2003, 32 (1-3), 87-91.

[15] Tsuru, T.; Miyawaki, M.; Kondo, H.; Yoshioka, T.; Asaeda, M. Inorganic porous membranes for nanofiltration of nonaqueous solutions. *Separation and Purification Technology* 2003, 32 (1-3), 105-109.

[16] Burggraaf, A.J.; Keizer, K. Inorganic Membranes : Synthesis, Characterization and Applications, Bhave, R. (Ed.), Van Nostrand Reinhoud, New York, 1991

[17] Voigt, I.; Fischer, G.; Puhlfürss, P.; Seifert, D. New filtration ceramics – from support to NF-membrane completely of TiO_2, *Proceedings of the 5th International Conference on Inorganic Membranes,* Nagoya, Japan, June 22-26, 1998, 42-45

[18] C. Leger, H. De L. Lira, R. Paterson, Preparation and properties of surface modified ceramic membranes. Part III: Gas permeation of 5 nm alumina membrane modified by trichloro-octadecylsilane, *J. Membrane Sci*. 120 (1996) 187-195

[19] Javaid, M.P. Hughey, V. Varutbangkul, D.M. Ford, Solubility-based gas separation with oligomer-modified inorganic membranes, *J. Membrane Sci*. 187 (2001) 141-150

[20] J.R. Miller, W.J. Koros, The formation of chemically modified □-alumina microporous membranes, *Separation science and technology* 25 (1990) 1257-1280

[21] Simon, R. Bredesen, C. Denonville, Characterization of surface modified Al2O3 membranes, *Proceedings of the 5th International Conference on Inorganic Membranes ICIM*, Nagoya, Japan, June 22-26 1998, pp. 416-419

[22] Picard, A. Larbot, F. Guida-Pietrasanta, B. Boutevin, A. Ratsimihety, Grafting of ceramic membranes by fluorinated silanes: hydrophobic features, *Separation and Purification Technology* 25 (2001) 65-69

[23] Van Gestel, T.; Van der Bruggen, B.; Buekenhoudt, A.; Dotremont, C.; Luyten, J.; Vandecasteele, C.; Maes., G. Surface modification of γ-Al2O3/TiO2 multilayer membranes for applications in non-polar organic solvent streams. *J. Membr. Sci*. 2003, 224 (1-2), 3-10.

[24] Voigt, I.; Puhlfürss, P.; Holborn, T. Ceramic nanofiltration membranes for applications in organic solvents, Proceedings of the Aachener Membrane Colloquium, Aachen, 2003

[25] Sah, A.; Casricum, H.L.; Bliek, A.; Blank, D.H.A.; Ten Elshof, J.E. Hydrophobic modification of γ-alumina membranes with organosilanes. *J. Membr. Sci.* 2004, 243 (1-2), 125-132.

[26] Impens, N.R.E.N.; Van der Voort, P.; Vansant, E.F. Silylation of micro-, meso- and non-porous oxides: a review. *Microporous and Mesoporous Materials* 1999, 28 (2), 217-232.

[27] Hair, M.L.; Tripp, C.P. Alkylchlorosilane reactions at the silica surface. *Colloids and Surfaces A: Physicochemical and Engineering Aspects* 1995, 105 (1), 95-103.

[28] Caro., J.; Noack, M.; Kölsch, P. Chemically modified ceramic membranes, *Microporous and Mesoporous Materials*, 1998, 22, 321-332

[29] Randon, J.; Paterson, R. Preliminary studies on the potential for gas separation by mesoporous ceramic oxide membranes surface modified by alkyl phosphonic acids, *J. Membr. Sci* 1997, 134, 219-223

[30] Picard, C.; Larbot, A.; Tronel-Peyroz, E.; Berjoan, R. Characterisation of hydrophilic ceramic membranes modified by fluoroalkylsilanes into hydrophobic membranes. *Solid State Sciences* 2004, 6 (6), 605-612.

[31] Verrecht, B.; Leysen, R.; Buekenhoudt, A.; Vandecasteele, C.; Van der Bruggen, B. Solvent flux behaviour of hydrophilic and hydrophobic mesoporous γ-Al_2O_3, TiO_2 and ZrO_2 membranes, *Proceedings of the International Conference on Inorganic Membranes* 9, Lillehammer, Norway, 2006

[32] Geens, J.; Van der Bruggen, B.; Vandecasteele, C. Transport model for solvent permeation through nanofiltration membranes. *Separ. Purif. Technol.* 2006, 48 (3), 255-263.

[33] Han, S.J.; Wong, H.T.; Livingston, A.G. Application of organic solvent nanofiltration to separation of ionic liquids and products from ionic liquid mediated reactions. *Chem. Eng. Res. Design* 2005, 83 (A3), 309-316.

[34] Scarpello, J.T.; Nair, D.; dos Santos, L.M.F.; White, L.S., Livingston, A.G. The separation of homogeneous organometallic catalysts using solvent resistant nanofiltration. J. Membr. Sci. 2002, 203 (1-2), 71-85.

[35] Van der Bruggen, B.; Jansen, J.C.; Figoli, A.; Geens, J.; Boussu, K.; Drioli, E. Characteristics and performance of a 'universal' membrane suitable for gas separation, pervaporation and nanofiltration applications. *J. Phys. Chem. B* 2006, 110 (28), 13799-13808.

[36] Van der Bruggen, B.; Geens, J.; Vandecasteele, C. Influence of organic solvents on the performance of polymeric nanofiltration membranes. *Separ. Sci. Technol.* 2002, 37 (4), 783-797.

[37] Tsuru T., Sudou T., Kawahara S., Yoshioka T., Asaeda M., 2000. Permeation of liquids through inorganic membranes. *J. Coll. Int. Sci.* 228: 292-296.

[38] Mulder M., 1996. Basic principles of membrane technology, Second Edition, Kluwer Academic Publishers, Dordrecht, pp. 564.

[39] Garcia A., Alvarez S., Riera F., Alvarez R., Coca J., 2005. Water and hexane permeate flux through organic and ceramic membranes: Effect of pretreatment on hexane permeate flux. *J. Membr. Sci.* 253: 139-147.

[40] Geens, J. Nanofiltration in organic solvents using polymeric and ceramic membranes. PhD thesis, K.U.Leuven, Belgium, February 2006.

[41] Tsuru T., Sudoh T., Yoshioka T., Asaeda M., 2001. Nanofiltration in non-aqueous solutions by porous silica-zirconia membranes. *J. Membr. Sci.* 185: 253-261.

[42] Silberberg M., Gold L.P., 1996. Intermolecular forces: Liquids, solids and changes of state. Chapter 11 in *Chemistry – The Molecular Nature of Matter and Change*. Mosby-Year Book, Inc., St-Louis: 412-462.

[43] Kotz J.C., Purcell K.F., 1991. Intermolecular forces, liquids, and solids. Chapter 13 in *Chemistry and Chemical Reactivity*, 2nd Ed., Saunders College Publishing, Holt, Rinehart and Winston, Inc., Orlando: 501-557.Voigt, I., P. Puhlfur, and T. Holborn, Ceramic Nanofiltration membranes for applications in organic solvents. 2003.

In: Ceramic Materials Research Trends
Editor: Paul B. Lin, pp. 227-273
ISBN: 1-60021-769-9

Chapter 10

NON-OHMIC CONDUCTION IN OXIDE CERAMICS: TIN DIOXIDE AND ZINC OXIDE VARISTORS

*A.B. Glot**

División de Estudios de Posgrado, Universidad Tecnológica de la Mixteca, Huajuapan de León, Oaxaca 69000 México

Abstract

The phenomenon of non-Ohmic conduction (varistor effect) observed in tin dioxide based and zinc oxide based semiconductor oxide ceramics is discussed. Although ZnO varistors are widely known, last years significant progress in the development of highly nonlinear SnO_2-based varistors is reached since in 1986-1987 the first SnO_2 varistors without zinc oxide addition in a system SnO_2-Bi_2O_3-Co_3O_4-BaO-Nb_2O_5 with the nonlinearity coefficient $\beta = 21$ and the electric field E_1=3500 V/cm (A.B.Glot and A.P.Zlobin, Inorganic Materials, **25** (2),1989,274-276) were obtained. Results on SnO_2 and ZnO varistors obtained using a wide range of techniques are discussed. Special attention to newly developed SnO_2-based ceramic varistors with quite high nonlinearity coefficient about 40-60 and higher is given. Barrier nature of electrical conduction in oxide ceramics is discussed. Experimental evidence of grain-boundary potential barriers in oxide varistor ceramics is outlined. Common features and differences between SnO_2 and ZnO varistors are analyzed. Earlier conclusion that electrical properties of SnO_2 varistors are very similar to the properties of ZnO varistors and that non-Ohmic conduction in SnO_2 and ZnO varistors have a common physical nature related to electron transport across the grain-boundary barriers is additionally confirmed. Critical review of recent results in the field of SnO_2 varistors is presented. The approach to oxide varistor ceramics as a system with different grain boundaries is discussed. From fairly general assumptions about the nature of non-Ohmic conduction in oxide ceramics (decrease of the barrier height on the electric field) the relationship $j = \sigma_0 E \exp(\alpha E)$ between the current density j and the electric field E is suggested. This expression can be used for the approximation of current-voltage characteristics of SnO_2 and ZnO varistors. The nonlinearity factor α or the normalized nonlinearity coefficient β_E can be used instead of the traditional

* E-mail address: alexglot@mixteco.utm.mx. Fax: + 52 (953) 5320214 Ext 106; (On leave from Dniepropetrovsk National University, Dniepropetrovsk, Ukraine)

but empirical nonlinearity coefficient β for an estimation of the nonlinearity of current-voltage characteristic of a varistor and for a comparison of the nonlinearity of varistors having different values of the electric field at fixed current density. The relation between them is given by the expression $\alpha \approx \beta_E = (\beta / E_1)$. The problems in the area of tin dioxide varistors and desirable future developments are outlined.

Keywords: electrical conduction, grain boundary, grain-boundary potential barrier, nonlinear current-voltage characteristic, non-Ohmic conduction, SnO_2 ceramics, SnO_2 varistor, ZnO ceramics, ZnO varistor

1. Introduction

Last 30-40 years a great progress in the field of oxide varistors has been reached. Zinc oxide varistors are now used as powerful voltage-surge suppressors in high-voltage and low-voltage equipment in electrical engineering and electronics [1]. Zinc oxide varistor is ceramic semiconductor device with highly nonlinear current-voltage characteristic. Increase of voltage causes the decrease of resistance, therefore, current is increased strongly. This characteristic is symmetrical - it is the same for both voltage polarities. Such semiconductor ceramics with sharp dependence of the electrical conductivity on the electric field strength are fabricated by a sintering process. For example, ZnO is mixed with several oxide additives like Bi_2O_3, CoO, MnO, Cr_2O_3, Sb_2O_3, then obtained mixture is pressed and sintered in air at about 1520 K [1]. The resultant ceramic material is comprised of relatively highly conducting n-type ZnO grains and a small amount of other phases.

The results of research and development are published in numerous original papers. Several reviews are devoted to different aspects of preparation, structure, electrical properties and possible mechanisms of electrical conduction and non-Ohmic phenomenon in oxide ceramics [1-12]. This tremendous amount of information is explained not only by the fundamental interest of non-Ohmic conduction in oxide ceramics but the important applications of varistors.

For practical application, current-voltage characteristic of a varistor is frequently approximated by the empirical power-law relation:

$$I = BU^{\beta}, \tag{1}$$

where I is the current through a varistor, U is the voltage applied to a varistor, B and β are the constants of the approximation. Eq.(1) can be used as well in a form $j = AE^{\beta}$, where j is the current density, E is the electric field, A and β are the constants.

The degree of the nonlinearity is frequently estimated by the empirical nonlinearity coefficient:

$$\beta = \frac{R_s}{R_d} = \frac{U}{I}\frac{dI}{dU} = \frac{Edj}{jdE} = \frac{d(\ln j)}{d(\ln E)}, \tag{2}$$

giving the ratio of the static resistance $R_s = U/I$ of a varistor to the differential resistance $R_d = dU/dI$. The integration of Eq. (2) at a condition $\beta = const$ gives the expression (1) (or, respectively, $j = AE^{\beta}$). The nonlinearity coefficient β and the electric field strength E_1 calculated at the fixed current density (for example, at 1 mA cm^{-2}) are the main parameters of a varistor. In the literature different notations exist as well, namely, $j = AE^{\alpha}$ or $E = Cj^{\alpha}$ but in this chapter only Eqs.(1) and (2) will be used.

There are some reasons for writing this chapter.

Last years the nonlinearity of current-voltage characteristic of SnO_2 based varistors has been improved significantly due to a finding of new additives. Now the nonlinearity of SnO_2 based varistors is comparable with the nonlinearity of ZnO based varistors. However, there are still too many unknown important details in the behaviour of SnO_2 based varistors. But varistors on the basis of ZnO and SnO_2 have definitely many common properties and, therefore, non-Ohmic conduction and related phenomena in these materials can be considered on a common basis.

The importance of the research in the varistor area for industry and science was the reason why such developments took place and now are continued in many countries, sometimes quite independently. For that reason, for example, some results on ZnO varistors were obtained and published in Russian even earlier than similar results were published in English scientific and technical literature. Sometimes authors of papers and reviews published in English, even if they declare that they wish to give whole independent review of this field (see [12]), do not give more than several references to initial publications on varistors in Russian performed in the 1950s-1960s. Naturally, not only language problems can be responsible for such a situation. Therefore, it would be interesting to give more balanced image of research performed in varistor area.

Quite similar situation is observed in the literature on tin dioxide varistors. In spite of SnO_2-based varistors (for example, $\beta = 3-10$, E_1=350-1500 V/cm [13-18] and $\beta = 21$, E_1=3500 V/cm [20]) were studied in 1963-1991 and results were published in Russian and English [13-22], years before than other researchers started to study varistor effect in SnO_2-based ceramics, these results are still poorly known. There was not any attempt to analyze critically the state of research in the field of SnO_2-based varistors. Some authors wrote reviews on SnO_2 varistors (see, for example, [23-25]) but these works are rather some compilations of their own results. Therefore, it would be useful to analyze the state of research in the area of tin dioxide ceramic varistors with some more attention to publication of different authors.

Electronic processes in ZnO varistors are quite complex [1-12]. The same situation is expected in SnO_2 varistors. It explains why there were no attempts to get an analytical expression for current-voltage characteristic of a single grain boundary and a whole ceramic varistor. Only numerical calculations were performed [26-33]. However, it should be mentioned that Eq.(1) was obtained analytically in [34] under some particular assumptions about the energy distribution of the grain-boundary states. These assumptions can be not valid if take into account that varistor effect is reproducible for a wide range of impurities and processing parameters in ZnO and SnO_2 varistors. In these conditions one should expect significant variation of the energy distribution of the grain-boundary states.

However, in many cases an analytical expression for current-voltage characteristic is desirable. That is why traditional empirical expression (Eq.(1)) is still used in practice frequently. It is good for a presentation of experimental data in a wide current range (in double logarithmic scale) but it is not good for an analysis. The Ohm's law cannot be obtained from Eq.(1) in the limit of low voltages. The nonlinearity coefficient β is not related to any mechanism of non-Ohmic conduction and, therefore, current-voltage characteristics in a wide range of current can be approximated by Eq.(1) with several empirical β values.

Additionally, a difference in static resistances of low-voltage and high-voltage ZnO varistors at highly nonlinear region is up to one decade. As a result, usually the nonlinearity coefficient of a high-voltage varistor is much higher than the nonlinearity coefficient of a low-voltage varistor, in spite of the nonlinearity of current-voltage characteristics of a single grain boundaries in both devices can be comparable. Therefore, the nonlinearity coefficient β cannot be used to compare the nonlinearity of current-voltage characteristics of those varistors which have quite distinguishing E_1-parameters.

Therefore, it would be interesting to obtain analytical expression for current-voltage characteristic of a varistor taking into account the main aspects of the electrical conduction in ZnO and SnO_2 non-Ohmic ceramics.

Recently the phenomenological model of non-Ohmic conduction in ZnO varistors and SnO_2 varistors was suggested [35-37]. Using such phenomenological approach to varistor effect an analytical expression for current-voltage characteristic is obtained and the nonlinearity factor α instead of the nonlinearity coefficient β is introduced. This approach allows to understand why the nonlinearity coefficient in a high-voltage varistor is usually higher than the nonlinearity coefficient in a low-voltage varistor and how to compare the nonlinearity of current-voltage characteristics of varistors having different values of the electric field at fixed current density.

In this chapter the analysis of non-Ohmic conduction (varistor effect) in SnO_2 based and ZnO based semiconductor ceramics using literature data, recently suggested phenomenological approach [35-37] and some other author's results is presented.

2. Historical Aspects of Non-Ohmic Oxide Ceramics

The nonlinearity of current-voltage characteristic of oxide semiconductors firstly was observed in ZnO-TiO_2 by Valeev and Mashkovich [38] and slightly later in ZnO-Bi_2O_3 ceramics by Kosman and Pettsold [39]. Then ZnO-SiO_2 and ZnO-SnO_2 ceramics were studied [40]. All these ZnO-based varistor materials had very low nonlinearity coefficient $\beta \leq 7$.

Among the first varistor materials, the ceramics on a basis of tin dioxide with addition of 20 mol.% of zinc oxide and a small amount (about 2 wt.%) of bismuth oxide was developed in 1963-1966 by Dr.B.K.Chernyi in Dniepropetrovsk State University (Ukraine) [13-15]. The varistor of such a system SnO_2-ZnO-Bi_2O_3 had a high enough value of the nonlinearity coefficient ($\beta \cong 10$) and it was used for the improvement of some parameters of the electroluminescent matrix screen [41].

Later in SnO_2-ZnO-Bi_2O_3 ceramics we have observed the strong effect of additional heat treatment on electrical conductivity and nonlinear properties and the concept on the role of amorphous intergranular phase in electrical properties was developed [16]. On the basis of this concept a group of new varistor materials on the basis of SnO-ZnO with addition of glass-forming oxides with the nonlinearity coefficient $\beta \cong 10$ was obtained [17]. Later titanium dioxide [42] based varistors were reported.

In 1969-1970 M. Matsuoka, T. Masuyama and Y. Iida have suggested a new ZnO based varistor material ZnO-Bi_2O_3-Sb_2O_3-CoO-MnO-Cr_2O_3 with the nonlinearity coefficient $\beta \cong 50$ [43,44]. The composition in addition to zinc oxide as a main ingredient contained small additives of other mentioned oxides. This achievement was applied in practice very soon and many researchers have contributed to the understanding of the nature of this effect and further development of zinc oxide based varistors (see reviews [1-12]).

Zinc oxide with additives of rare earth oxides and transition metal oxides is alternative varistor composition [45]. However, compositions with bismuth oxide are still very important due to their application in varistor industry. The nonlinearity coefficient β of ZnO varistors may achieve 40-60 or higher values and the electric field E_1 can be varied in a wide range about 200-5000 V/cm.

Since the earlier 1970-s zinc oxide based varistors have been investigated in great detail and are widely used. It was important to study electrical properties and phase composition of ZnO-Bi_2O_3-Sb_2O_3-CoO-MnO-Cr_2O_3 and similar ceramics, to understand the role of additives, effect of the variation of technological parameters and etc. Soon it was understood that electrical conduction in zinc oxide varistors is controlled by charge transfer processes across the grain-boundary potential barriers formed during sintering in oxidizing atmosphere and grain-boundary barriers can be satisfactorily modeled by double Schottky barriers with depletion regions spreading into ZnO grains [26-28,34,46-56].

In spite of tremendous progress in ZnO varistors, the question arises whether zinc oxide is a unique material for varistors or if any other oxide may also be used as a basis for varistor ceramics with a high value of the nonlinearity coefficient in the range of at least 40-50. However, the nonlinearity coefficient of tin dioxide [13-19] and titanium dioxide [42] varistors remained low for a long time up to 1980-s and the answer to the aforesaid question was open.

Additionally, SnO_2-based varistors known in the earlier 1980-s (SnO-ZnO-Bi_2O_3 [13-16], SnO_2-ZnO-GeO_2 (SiO_2, B_2O_3, Sb_2O_3, As_2O_3, P_2O_5) [17], SnO_2-ZnO-CoO-Bi_2O_3 (SiO_2, GeO_2) [19] contained quite high amount (20 mol.%) of zinc oxide, therefore, at such a high concentration of zinc oxide in this material, the possibility existed that electrical conduction was controlled by the charge transport across the potential barriers formed at the grain boundaries of ZnO grains but not at the SnO_2 grain boundaries. Therefore, the sintering of highly nonlinear ceramics on a basis of tin dioxide without zinc oxide addition might prove the possibility of a formation of sufficiently high barriers at SnO_2 grain boundaries and would help to exclude this indetermination.

The first SnO_2–based varistors (without zinc oxide addition) in a system SnO_2-Bi_2O_3-Co_3O_4-BaO-Nb_2O_5 with quite high nonlinearity coefficient $\beta \cong 21$ and the electric field

E_1=3500 V/cm (at current density 10^{-3} A cm^{-2}) were developed quite long ago in 1987-1989 [20]. The additions of now frequently used cobalt and niobium oxides for SnO_2 ceramics with high nonlinearity coefficient $\beta > 20$ were first suggested in [20]. A range of SnO_2 based varistor ceramics was studied during these years [20-22,57].

It was revealed that electrical properties of SnO_2 varistors are very similar to the properties of ZnO varistors (but they are not identical) and that non-Ohmic conduction in ZnO and SnO_2 varistors have a common physical nature related to the transport of electrons across the grain-boundary barriers [20,21]. This conclusion is supported by experimental results: electrical properties of different ZnO and SnO_2 varistors [16-22,54,58,59], the existence of Bi-O based thin films at the surface of SnO_2 grains detected by X-ray photoelectron spectroscopy (XPS) [21], the effect of "negative capacitance" accompanying nonlinear current-voltage characteristic of SnO_2 varistor [21,60], non-Arhenius dependences of conductivity versus temperature [18,59,60] and others.

Several years later in a SnO_2-CoO-Nb_2O_5-Bi_2O_3 system some lower nonlinearity coefficient β=10 was obtained [61]. However, then SnO_2 varistors with higher nonlinearity β = 72 and extremely high values of the electric field strength E_1=18200 V/cm in a SnO_2-CoO-Al_2O_3-Nb_2O_5 system were reported [62].

Later in a SnO_2-CoO-Nb_2O_5-Cr_2O_3 system the values of β=41 and E_1=3990 V/cm [63] and β=33 and E_1=3920 V/cm [64] were reached. Some higher values of nonlinearity coefficient even β = 142 and very high electric field strength E_1=11525 V/cm in a SnO_2-CoO-Nb_2O_5-Cr_2O_3-La_2O_3 system [65] were reported. Although observations of quite high nonlinearity coefficients at extremely high values of the E_1=18200 V/cm [62] or E_1=11525 V/cm [65] allow us to assume that such materials had very high static resistance R_s due to a small average grain size but the differential resistance R_d was not low enough. Meanwhile, just the reduction of a differential resistance is a general way to obtain varistor with a highly nonlinear current-voltage characteristic.

The role of Cr_2O_3 addition in the achievement of high value of the nonlinearity coefficient β is not well-grounded in the paper [63] because the increase of Cr_2O_3 addition from 0 to 0.05 mol.% causes the increase of β but samples with higher Cr_2O_3 concentrations have a different shape of current-voltage characteristic with considerably lower values of the nonlinearity coefficient. The explanation of this effect in [63] is not enough and obvious differences in the shape of current-voltage characteristics presented in [63] and [64] can be explained as poor reproducibility of electrical properties of SnO_2-Co_3O_4-Nb_2O_5-Cr_2O_3 ceramics. At such low concentrations of Cr_2O_3 addition its distribution throughout a sample could be very heterogeneous causing the mentioned effect. In this case liquid phase sintering would be useful.

Therefore, creation of tin dioxide based ceramics not only with the high nonlinearity coefficient $\beta > 40$ but with $\beta > 40$ and a comparatively low electric field strength $E_1 < 10^4 V \cdot cm^{-1}$, fairly reproducible, and satisfactorily controlled by the concentration of used admixtures, could be quite interesting. One of the ways for solving this problem might be the use of bismuth oxide Bi_2O_3 additive, which, in zinc oxide based systems promotes liquid-phase character of sintering and provides an achievement of high nonlinearity coefficient [10]. Thus, it is worthwhile to use the SnO_2-Bi_2O_3-Co_3O_4-Nb_2O_5 system studied earlier [20] and add to it, for example, chromium oxide which could improve the nonlinearity.

Thinking in this way we have prepared new tin dioxide based non-Ohmic ceramics [66,67] in a SnO_2-Bi_2O_3-Co_3O_4-Nb_2O_5-Cr_2O_3 system with the nonlinearity coefficient in the range 50-60 and not high (in comparison to known highly nonlinear SnO_2 based systems [62,65]) electric field about $3500 V \cdot cm^{-1}$. Further addition of B_2O_3 to a SnO_2-Bi_2O_3-Co_3O_4-Nb_2O_5-Cr_2O_3 system [67] leads to the increase of both the nonlinearity coefficient and the electric field at fixed current density.

It was revealed as well that non-Ohmic conduction in SnO_2-based ceramics is accompanied by a sensitivity of electrical conductivity to a variation of relative humidity and, therefore, tin dioxide ceramic materials with nonlinear current-voltage characteristics can be used as humidity sensors too [19,66,68,69]. On the basis of SnO_2 the devices with combined properties of a varistor and a sensor of relative humidity in the systems SnO_2-ZnO-CoO-Bi_2O_3 (SiO_2, GeO_2) [19] and SnO_2-Bi_2O_3-Co_3O_4-Nb_2O_5-Cr_2O_3 [66,69] were developed. Another devices with combined properties of a varistor and a gas sensor, varistor-type gas sensors, on the basis of ZnO and SnO_2 ceramics were suggested [70-72].

In any case, our idea about the common physical and chemical nature of ZnO and SnO_2 based varistor ceramics [20,21,60] was supported independently in some recent publications [64,65,73]. Earlier obtained results in physics of non-Ohmic conduction in ZnO varistors can be useful for the understanding of a strong varistor effect in SnO_2 based ceramics which is studied now very intensively [35,37,62-67,69,73]. It can be assumed that the mechanism of non-Ohmic conduction in SnO_2 varistors, especially in the cases of high nonlinearity coefficients ($\beta > 30$), should be quite similar to the mechanism of nonlinearity in ZnO varistors [21,35,37,64,66,69,73].

3. Experimental Details of Varistor Effect in SnO_2 Ceramics

The details of the preparation procedures and measurement techniques can be found in respective original papers. In this section some remarks on the experimental study of varistor effect in SnO_2 ceramics will be addressed.

In Fig.1 current-voltage characteristics of SnO_2-Bi_2O_3-Co_3O_4-Nb_2O_5-Cr_2O_3 varistor sample at different temperatures are presented. This material is the same as reported earlier (with nonlinearity coefficient about 50-55 and electric field about 3400 V/cm) [37,66]. Data were obtained avoiding overheating owing to Joule effect.

In SnO_2 based high-voltage varistor ceramics some authors reported very high values of the nonlinearity coefficient β = 142 and very high electric field strength E_1=11525 V/cm [65]. However, it should be noted that even if authors give double logarithmic presentation of dc current-voltage characteristic it is not obvious that Joule heating was avoided at high dc current. In this case in a sample with typical sizes (thickness about 1 mm and cross-section about 1 cm^2) power about 10 W was dissipated. Similar situation can be seen in other publication [74]. The last point at about $j = 2.5\ mA \cdot cm^{-2}$ and $E \approx 1.2 \cdot 10^4\ V \cdot cm^{-1}$ shows current increment at nearly the constant electric field. It is quite different in comparison to the previous points. It can be if some Joule heating took place. In this case power dissipated in the sample (about 1 cm diameter and 0.1 cm thick) was $3\ W$. It is quite huge power for dc measurements. These conditions are very heavy if take into account that samples are not homogeneous and this power is dissipated in a vicinity of grain boundaries. As a result of parasitic Joule heating in semiconducting sample desirable current could be observed at some lower voltage and, therefore, the nonlinearity coefficient could be overestimated.

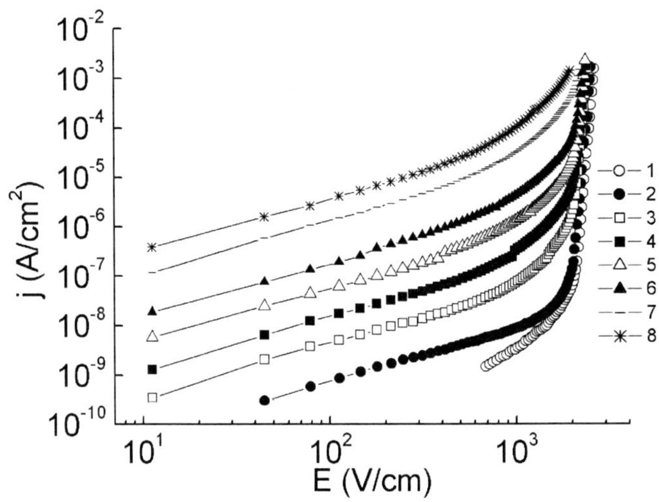

Figure 1. Current density versus electric field for SnO_2-Bi_2O_3-Co_3O_4-Nb_2O_5-Cr_2O_3 varistor measured at different temperatures, K: 1 – 292, 2 – 332, 3 – 378, 4 – 400, 5 – 424, 6 – 444, 7 – 476, 8 – 494. Samples were obtained and measurements were performed by I.A.Skuratovsky.

In many other cases some authors even do not present double logarithmic plots and give the values of the nonlinearity coefficient obtained probably by two points measurements. In these cases it is rather difficult to avoid some Joule heating at high dc current densities like several $mA \cdot cm^{-2}$. In SnO_2 ceramics Joule heating at dc current densities about 10^{-2} $A \cdot cm^{-2}$ can be more probable than in ZnO ceramics due to usually higher voltages applied. Additionally, conditions of current flowing related to microstructure can be different in SnO_2 and ZnO ceramics. Therefore, very high values of the nonlinearity coefficient sometimes reported in SnO_2 based varistors especially if they are accompanied by very high electric field (like, for example β =142 and E_1=11525 V/cm [65]), can be overestimated due to the influence of Joule heating.

In this relation, it would be interesting to stress that ac (60 Hz) current-voltage characteristics of SnO_2 varistors are quite different from ZnO varistors (Fig.2): for SnO_2 varistors the traces of $I(U)$ dependence for increase and decrease of voltage in the highly nonlinear region are not coincided though for ZnO varistors they are coincided. It can be if in SnO_2 varistors some Joule heating takes place. Then increase of voltage causes some additional micro heating and current is increased more strongly than it should be if there is no over heating. If voltage is decreased, then current is decreased by other way at the screen (current is slightly upper than it was when voltage was increased) due to grain-boundary regions are still overheated. In ZnO varistors heat exchange between grain-boundary regions and grains is probably better and, therefore, this effect usually is not seen. Even if liquid-phase sintering of SnO_2 varistor takes place as in the case of SnO_2-Bi_2O_3-Co_3O_4-Nb_2O_5-Cr_2O_3 varistor [66], probably, the moistening of SnO_2 grains by Bi_2O_3-based liquid phase is less than the moistening of ZnO grains by Bi_2O_3-based liquid phase. It can affect the area of grain-boundary regions and current density in the case of SnO_2 varistor can be higher.

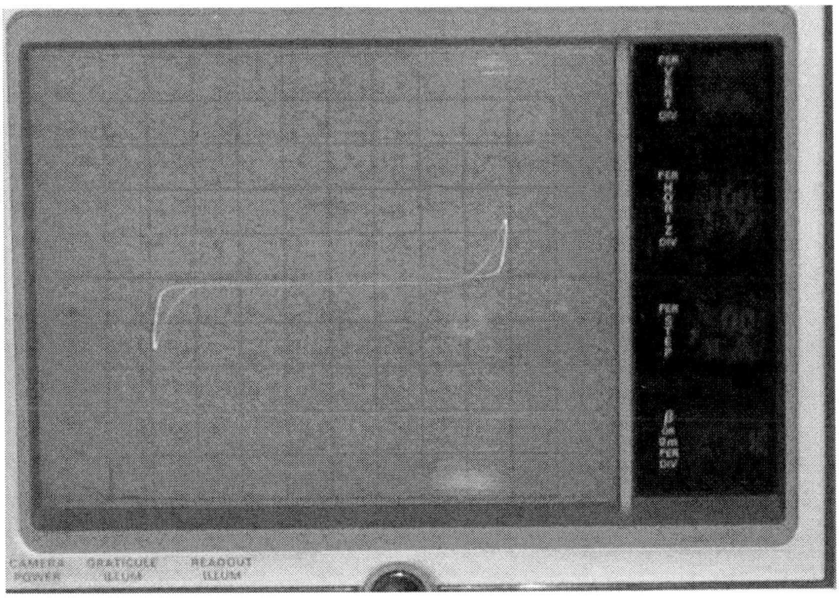

Figure 2. Ac current-voltage characteristic of SnO_2-Bi_2O_3-Co_3O_4-Nb_2O_5-Cr_2O_3 varistor. Horizontal: 100 V/div; vertical: 1mA/div. Sample was obtained by I.A.Skuratovsky.

In highly nonlinear region of SnO_2 varistors, according to our preliminary results, capacitance is decreased and even "negative capacitance" is observed. However, due to possible influence of Joule heating, observed decrease of capacitance can be related to increasing contribution of inductive-like current behavior directly related to Joule heating. This investigation is in progress. Earlier "negative capacitance" effect was observed in SnO_2 varistors with low nonlinearity [21,60].

If Joule heating is negligible, then current-voltage characteristics obtained in dc and pulse regime (using short pulses in a μs range) should be the same. Unfortunately, such measurements are not published and only some preliminary results are reported [37]. In this study grain resistivity $\rho_g \cong 4\ Ohm \cdot cm$ of SnO_2 in highly nonlinear ceramic material was estimated [37].

The absence of published data on pulse current-voltage characteristics of SnO_2 ceramic varistors in spite of a great amount of published papers (see, for example, [25]) can mean that SnO_2-based varistors developed by this group exhibit poor pulse properties. Experiments on pulse measurements performed in our laboratory revealed that at least our samples SnO_2-Bi_2O_3-Co_3O_4-Nb_2O_5-Cr_2O_3 (prepared by I.A.Skuratovsky, see [66]) and samples SnO_2-

Co_3O_4-Nb_2O_5-Cr_2O_3 with composition identical to studied in [63] (prepared by A.V.Gaponov) cannot withstand high-current pulses and frequently breakdown is occurred.

At low voltages in SnO_2-based varistor ceramics current decay can take place due to capture of electrons at the grain-boundary states. Therefore, non steady-state measurements of current-voltage characteristics can give some errors: conductivity can be decreased with increase of voltage instead to be constant. Such artefact, for example, can be seen from original data presented in [74].

4. Grain Boundaries in Varistor Ceramics

Electrical conductance in ZnO or SnO_2 varistor ceramics is determined by highly-resistive grain-boundary regions. Increase of voltage leads to decrease of their resistance and, therefore, current in a sample is increased nonlinearly with voltage leading to varistor effect. This concept is supported by comparison of the grain resistivity estimated from pulse measurements [32,75,124-127] and low-field dc resistivity of varistor. The grain-boundary resistance (resistance of a sample) is much higher (about ten decades) than the grain resistance. Typical values of the grain resistivity in ZnO varistors is less than $1\ Ohm \cdot cm$ [32,75,124-127].

For SnO_2-based varistors SnO_2-ZnO-Bi_2O_3 with $\beta \cong 10$ (see [15-18]) quite high grain resistivity about $40\ Ohm \cdot cm$ was found [126]. In the case of highly nonlinear tin dioxide based SnO_2–Bi_2O_3–Co_3O_4–Nb_2O_5–Cr_2O_3 varistor ($\beta \cong 50$) the study of the high-current region of current-voltage characteristics gives the grain resistivity about 4 $Ohm^{-1}cm^{-1}$ [37]. This value is at least nine decades less than the low-field resistivity of varistor. This fact confirms the assumption about the existence of highly-resistive grain-boundaries in tin dioxide based varistor ceramics.

4.1. Initial Ideas on the Nature of Grain Boundary in Oxide Varistors

Now the concept of Schottky-type grain-boundary barriers in ZnO varistor ceramics and SnO_2 varistor ceramics is accepted by all researchers but in the early 1970s the nature of grain-boundary area was not known.

Some authors thought that intergranular phase is responsible for grain-boundary barrier [16-18,44,75,76]. Highly nonlinear current-voltage characteristics observed in ZnO varistors could be explained by the mechanism of space charge limited current in a regime of trap filled limit [44]: electrons injected from ZnO grain to intergranular dielectric layer with traps could be captured at these traps and complete filling of traps would give strong current rise due to mentioned traps will not be able to capture electrons more. Trap filled limit voltage depends on trap density.

However, microcontact measurements of single grain boundary current-voltage characteristics showed that high nonlinearity was observed at about 3.4 V for many grain boundaries in varistors with different additives and processing parameters [27,46,49,52,53]. It was impossible to assume that in all cases the trap density and thickness of dielectric layer could be nearly the same. Therefore, the mechanism of space charge limited current was ruled out.

The mechanism of Fowler-Nordheim tunnelling across very thin intergranular dielectric layer could explain some properties of varistors [75] but it was as well in contradiction with results of microcontact measurements revealed quite narrow distribution of threshold varistor voltage (about 3.4 V) [27,46,49,52,53]. Thickness of intergranular dielectric layer in different materials could not be so constant. This mechanism was ruled out as well.

The attention was focussed on zinc oxide because only base component was unchanged for different ZnO varistors and the model of ZnO homojunction was suggested with band-to-band Zener tunnelling mechanism (at voltage slightly more than band gap E_g in ZnO $U > E_g / q = 3.2\ V$ electrons from valence band at negative side could directly tunnel to conduction band at positive side of junction [49,53]. Additionally, it was observed that many grain boundaries in a sample of ZnO varistor do not contain some intergranular phase [50].

4.2. Switching Effect in Bismuth Oxide and Non-ohmic Oxide Ceramics

Independently, in the early 1970s understanding the importance of bismuth oxide for tin dioxide based [14,16] and zinc oxide based [44] varistors, we decided to study non-Ohmic effects in high electric fields in Bi_2O_3-based structures [77-79].

Considering an important role of bismuth oxide (Bi_2O_3) in a formation of non-Ohmic properties of ceramics many efforts were made to study electrical properties of bismuth oxide containing materials (polycrystalline layers, thin films, glasses) at high electric fields [77-79]. In various Bi_2O_3-based materials switching effect was observed. At high electric field E_b electrical breakdown occurred causing sharp decrease of voltage drop and current increase if it was not limited by an external circuit. A sample was turned into an on-state (low-resistive state). If current was properly limited by an external circuit (resistor with some high resistance was connected in series), then subsequent current decrease allowed a sample to return to an off-state (high-resistive state). Due to this effect was observed in Bi_2O_3-based materials of different preparation (polycrystalline layers, thin films, glasses), it was suggested that something similar could take place in non-Ohmic oxide ceramics.

Though in various Bi_2O_3-based materials with metallic electrodes switching effect was observed [77], however, in ZnO-Bi_2O_3-ZnO structures (where Bi_2O_3-based materials were placed between previously sintered ZnO pellets and this structure was heat treated again to melt Bi_2O_3-layer) only nonlinear current-voltage characteristics without switching [78,79] were observed. From these studies it was revealed that depletion surface layers of zinc oxide grains in varistors can contribute to grain boundary resistance [51,55]. This independent conclusion was useful to realize that grain-boundary barrier in ZnO varistors consists not only of the intergranular phase layer but mostly from depletion surface layers of zinc oxide [51,55].

For the approximation of current-voltage characteristic of SnO_2-ZnO-Bi_2O_3 varistor [16], some other ceramic varistors on the basis of SnO_2 [17], ZnO-Bi_2O_3 varistor [18] and ZnO based varistors with Bi_2O_3, CoO and SnO_2 addition (with the nonlinearity coefficient β=40-50) [55] the semi-empirical equation

$$j(E) = \sigma_0 E \exp(\alpha E) \tag{3}$$

(σ_0 and α are parameters) was used. Initially the idea to use this equation for the approximation of current-voltage characteristic of oxide varistor ceramics was suggested in [14] taking into account possible tunneling through grain-boundary contacts in SiC varistors [80].

Mentioned Eq.(3) was obtained under the assumption of reversible switching at high electric fields of thin intergranular layers in varistors [81]. This assumption was based on the experimentally observed reversible switching in Bi_2O_3-rich layers [55,77]. An appearance of an "additional voltage" applied to a sample due to a decrease of voltage drop at switching of intergranular layer to on-state was suggested as the reason of an appearance of current increment leading to nonlinearity of current-voltage characteristic. For current-voltage characteristic of a varistor the equation

$$j(E) = \sigma_0 E \exp(\alpha E^n) \tag{4}$$

(j is the current density, E is the electric field, σ_0, α and n are constants) was obtained. Here α is related to the breakdown field of selected intergranular layer and distribution function of intergranular layers in a sample on breakdown voltage and in the frames of assumed breakdown mechanism [82] it is temperature dependent: $\alpha = A\exp(\Delta V / 2kT)$, where ΔV is the bandwidth of localized states in intergranular layer. The parameter $n > 0$ is used in the approximation of the distribution function $f(E) \propto E^n$, $E < E_{max}$ and $f(E) \propto E^{-n}$, $E > E_{max}$. At $E > E_{max}$ the number of switching intergranular layers is decreased with electric field and nonlinearity of current-voltage characteristic is gradually decreased. In this model electrical conductivity of ceramic sample is obtained in general form:

$$\sigma(T, E) = \sigma_0 \exp\left(-\frac{E_\sigma(T, E)}{kT}\right), \quad E < E_{max}, \tag{5}$$

with the activation energy of electrical conduction E_σ:

$$E_\sigma(T, E) = E_{\sigma 0} - 2AkTE^n \exp\left(\frac{\Delta V}{2kT}\right), \tag{6}$$

which is decreased with increase of electric field and decrease of temperature according to experiment. Eqs. (5) and (6) explain experimental results.

However, taking into account the absence of the switching effect in sandwich structures with ZnO-"electrodes" [78], observed electroluminescence in ZnO varistors [55,83] and the whole range of electrical properties of ZnO-ceramics, the contribution of the depletion ZnO layers into the grain-boundary resistance has been recognized essential [51,55]. This conclusion was supported independently by the publications appeared at that time [27,28,47,53].

Therefore, the hypothesis on reversible switching at the high electric fields in thin intergranular layers in varistors was rejected [55]. Meanwhile, the fact of the approximation of current-voltage characteristics of SnO_2 and ZnO varistors by the Eq.(3) looks curious.

5. Grain-Boundary Potential Barriers in SnO_2 and ZnO Varistor Ceramics

An idea that a double Schottky barrier model with infinitely thin interganular phase can be applicable not only to SiC but as well to ZnO-based varistors was presented firstly in [34]. The first experimental confirmation that a double Schottky barrier model with quite thin (about 2 nm) bismuth oxide segregation layer can give correct explanation of basic properties of ZnO varistors was shown in [47].

It was observed that an additional heat treatment caused the increase of electrical conductivity of ceramic varistor material [16,17] and it was accompanied by increase of the dielectric permittivity and by the decrease of the activation energy of the electrical conduction. The same situation was observed in highly nonlinear ZnO varistors [51,55]. These facts was interpreted as if grain-boundary barrier region in ZnO varistor consists of two surface depleted layers and decrease of the barrier height as a result of a heat treatment leads to the increase of the electrical conductivity and grain-boundary capacitance (due to the decrease of the width of the depletion layers) [51,55].

At that time it was understood that the electrical conduction in zinc oxide varistors is controlled by charge transfer processes across the grain-boundary potential barriers formed during sintering in oxidizing atmosphere and grain-boundary barriers can be satisfactorily modeled by barriers with depletion regions spreading into ZnO grains [27,28,47,53].

Quite interesting confirmation of the existence of highly resistive ZnO surface layers in ZnO varistors was obtained in 1979 when we decided to explore luminescent properties of varistor ceramics: it was known that zinc oxide is luminescent material. In such a way the red electroluminescence of ZnO varistors was observed [55,83]. About 3-5 years later the red and ultraviolet electroluminescence in ZnO varistors was observed by other authors [84,85]. The observation of ultraviolet electroluminescence in ZnO varistors is considered as confirmation of holes participation in non-Ohmic conduction [8,27,84,85,131].

Now it is reliably shown that electrical conduction in zinc oxide varistors is controlled by the charge transfer across the double Schottky grain-boundary potential barriers formed at the surface of zinc oxide grains during sintering in oxidizing atmosphere [27-28,47,56].

Formation of the grain-boundary potential barriers in ZnO varistors is related to oxygen in atmosphere during sintering [56]. The development of structure during sintering of ZnO-based samples are studied in numerous papers, for example, [86-91]. Quite important is the cooling stage. The same takes place in SnO_2 varistors [73]. Additionally, it was shown that heat treatment in vacuum [92] or even electron beam irradiation in vacuum [93] or even heat treatment in air [94] causes the degradation of electrical characteristics of ZnO varistors.

Independently and nearly at the same time it was demonstrated by X-ray photoelectron spectroscopy (XPS) that excessive amount of oxygen at the grain boundary exists [95,96]. Between the excessive amount of oxygen at the grain boundary and the barrier height certain correlation takes place [95]. Earlier the segregation of bismuth (Bi) (but not oxygen) within a 20 Å layer at the grain boundary was detected by Auger electron spectroscopy [47].

XPS study of SnO_2 based varistor was performed in [21]. It was shown that relative concentrations of Bi/Sn, O/Sn and Bi/O decrease at a distance of about 20 Å and then become constant. So it is possible to assume that Bi-O-phase can cover grains by thin layer as well as in ZnO varistor ceramics (see [47,95,96]). The observed tendency to bifurcation of $Bi4f_{5/2}$ and $Bi4f_{7/2}$ lines at ion etching shows the partial reduction of bismuth. Therefore, bismuth at the grain boundary is in oxide form. Kinetic energies of photoelectrons related to Sn3d and O1s lines are approximately equal. Therefore, conditions of electron emission are rather the same. Due to this reason the decrease of $O1s/Sn3d_{5/2}$ ratio on etching time may indicate some excess of oxygen at the grain boundary in comparison to SnO_2 grains bulk [21]. Adsorption nature of the grain-boundary barriers in SnO_2 based varistor is assumed [21].

Potential barriers at zinc oxide ceramic surface can be formed even at room temperature using special Ag-paste possibly providing necessary access of oxygen [97]. It additionally shows the role of oxygen in the formation of barrier at the grain boundary in ZnO varistors.

Electrical conduction in SnO_2 based and ZnO based varistors at low and high electric fields is controlled by grain-boundary potential barriers and these grain-boundary barriers can be satisfactorily modeled by double Schottky barrier (two back-to-back Schottky barriers) with thin (about 10-20 Å) intergranular layer between them or without such a layer. This statement is confirmed by a totality of the next experimental evidence:

(i) thermally-activated electrical conduction with quite high low-field activation energy of electrical conduction $E_\sigma \cong 0.6 - 1\ eV$ [75] and experimentally observed decrease of the activation energy (the barrier height) on electric field [27,36,37]:

(ii) appearance of non-Ohmic conduction in oxide ceramics at relatively low electric fields; for example, in ZnO-SnO_2-CoO-Bi_2O_3 ceramic varistor with β=40 and $E_1 = 2080\ V \cdot cm^{-1}$ deviations from Ohm′s law are appeared at electric field about 250 $V \cdot cm^{-1}$ [36];

(iii) observation of quite high values of relative dielectric permittivity due to the depletion width is significantly lower than the grain size [75,98];

(iv) decrease of capacitance with dc voltage [47,99];

(v) admittance spectroscopy (dielectric spectroscopy) – frequency dependence of real (ε') and imaginary (ε'') parts of complex dielectric permittivity $\dot{\varepsilon} = \varepsilon' - j\varepsilon''$ [54,84,100,101];

(vi) coordinated changes of $\sigma(T)$, E_σ, capacitance and ac conductance at additional heat treatments (decrease of E_σ causes not only an increase of dc and ac conductance but as well an increase of capacitance) [51,55];

(vii) step-like distribution of the brightness (electrostatic potential) of ZnO grains in the scanning electron microscope (SEM) (detection of the voltage step at the grain boundary) if dc voltage is applied (voltage contrast technique in the SEM) [52,93,102];

(viii) observation of current maximum if electron beam passes grain boundary with dc voltage applied [93] and using this current to modulate the intensity of the picture - electron beam induced current (EBIC) technique [52,103];

(ix) observed in SnO_2 based varistors reversible increase of capacitance with increase of relative humidity [68,69].

For the consideration of alternative conduction mechanisms in ZnO varistors not only the barrier height but the barrier width is important (for an estimation of tunnelling probability). Therefore, it was noted that deep donor levels can give some contribution to positive charge of the depletion region and, therefore, the barrier width can be significantly less than in the case of barrier with only shallow donor levels [104]. Later the concept of deep donor levels was used in detailed model of the grain boundary [8,84].

For the estimation of the barrier height φ_0 in SnO_2 varistor ceramics temperature dependence of electrical conductivity $\sigma(T)$ was studied and the activation energy E_σ was found [66]. It is expected that the barrier height φ_0 is slightly less than the activation energy E_σ, $\varphi_0 \cong E_\sigma$. It was found that at temperatures below approximately 373 K, the conductivity is decreased with a temperature rise, and is increased at higher temperatures. This effect is reversible: cooling of a sample causes the minimum of the conductivity situated below 373 K. Such an anomaly of the electrical conduction is related to the desorption of water molecules at heating and subsequent adsorption of water molecules at cooling because the above-mentioned anomaly was not found if the measurements were repeated at low air pressure. The adsorption of water molecules at the surface of SnO_2 grains causes the lowering of the barrier height and subsequent increase of the conductivity [69]. Some difference between $\sigma(T)$ curves at heating and cooling can be attributed to the partial desorption of oxygen from a sample that causes the decrease of the barrier height.

The barrier height φ_0 is equal to the energy difference between the conduction band edge at the boundary and far from it inside the grain. However, the activation energy E_σ is equal to the energy difference between the conduction band edge and the Fermi level at the boundary. It is assumed that the Fermi level in studied SnO_2 materials (far from the boundary) is situated about several tenths of eV below the conduction band edge [37,105,106]. Then it is expected that the barrier height φ_0 is only slightly less than the activation energy E_σ, $\varphi_0 \cong E_\sigma$. The variation of the chemical composition can cause the alteration of the Fermi level position in respect to the conduction band edge but it is assumed that mentioned relation $\varphi_0 \cong E_\sigma$ between φ_0 and E_σ is not changed significantly for all studied SnO_2 ceramics.

In the case of the electrical conduction controlled by the grain-boundary barriers one should expect some correlation between the barrier height φ_0 (or the activation energy E_σ) and the electrical conductivity for different samples if the grain size is not changed strongly. The highest activation energy is observed for the varistor with Cr_2O_3 addition having the lowest electrical conductivity and the highest nonlinearity coefficient [37]. Therefore, some correlation between the activation energy (as a measure of the barrier height), the electrical conductivity and the nonlinearity coefficient exists. This correlation is an additional fact in favour of the key role of the grain-boundary barriers in the electrical conduction of tin dioxide based ceramics.

Therefore, the barrier height in SnO_2-Bi_2O_3-Co_3O_4-Nb_2O_5-Cr_2O_3 ceramics can be about several tenths of eV less than obtained activation energy E_σ=1.2 eV, i.e. about φ_0=0.9-1.0 eV. This estimation is comparable to the values in the range 0.7-1.1 eV found in [64]. Additionally, the same authors studying earlier current-voltage characteristics at different temperatures have found that the barrier height for SnO_2-CoO-Nb_2O_5-Bi_2O_3 ceramic varistor is 0.63 eV [61].

From low-voltage part of capacitance-voltage characteristics very high values 2.48-2.57 eV for the barrier height were found for SnCoNbCrLa$_2$O$_3$ ceramics in [65]. This values 2.48-2.57 eV have been even included by the same authors in the review of their own results [25]. However, such high values of the barrier height, in our opinion, do not look correct because, in first, they are more than a half of the band gap of SnO_2 (E_g=3.6 eV). In second, it is easy to find from the experimental data shown at the same paper [65] that the activation energy of the ac conductance is about 0.86 eV. This is significantly less than the barrier height 2.48-2.57 eV mentioned by the authors of [65]. However, they do not see such contradiction in their data.

It is rather quite difficult to expect a formation such a high barriers at SnO_2 surface. Typical values of Schottky barrier at SnO_2 surface and barrier between SnO_2 grains in sintered tin dioxide are usually about 1 eV [107-109]. The value 0.86 eV (our estimation from [65]) or 0.63 eV [61] are close to our estimation of the barrier height in [37,66] about φ_0=0.9-1.0 eV. Therefore, such value of the barrier height looks more likely.

Considering quite wide scattering of the barrier height values obtained from low-voltage part of capacitance-voltage characteristics for different SnO_2 varistor ceramics [25,64,65] and taking into account that in this method it is necessary to assume that barriers at all grain boundaries are equal (what is not obvious), it can be concluded that this method can be used rather for qualitative estimations. Additionally, according to our measurements, capacitance of SnO_2 varistor ceramics at low voltages is only slightly decreased. It was found as well that capacitance of SnO_2 varistor ceramics is decreased on time at dc bias and it is increased after an elimination of dc bias similarly to the behavior of capacitance in ZnO varistor ceramics [110]. Mentioned capacitance relaxation allows to estimate the depth of the grain-boundary interface capture states $E_t \cong 0.4\ eV$ (our result). Therefore, mentioned factors should be taken into account in the interpretation of rather strange results on quite high barriers found in [25,65]. Some critics in respect of the estimation of grain boundary parameters from C-V measurements were expressed as well in [111].

6. Current-Voltage Characteristic of a Single Grain Boundary

Now it is well known that typical current-voltage characteristic of a single grain boundary in ZnO varistors is “varistor-like” (nonlinear and symmetrical) and current increase is observed at about 3.5 V [27,46,52,53]. However, this voltage can vary even in the same sample

[52,53]. In [52] the average voltage 3.6 V (at 1 μA) with standard deviation 0.2 V was found. At least one contribution to observed variation of voltage can be originated from the fact that voltages were measured at constant current but a large spread exists in the contact areas between two grains, therefore, current density could differ widely [52].

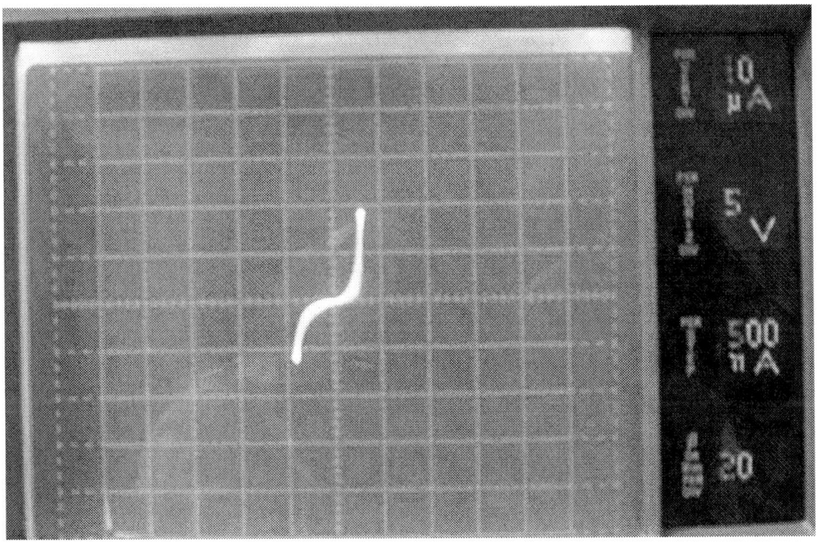

Figure 3. Ac current-voltage characteristic of assumed single grain boundary in commercial ZnO varistor (CH2-1). Horizontal: 5 V/div; vertical: 10 μA /div.

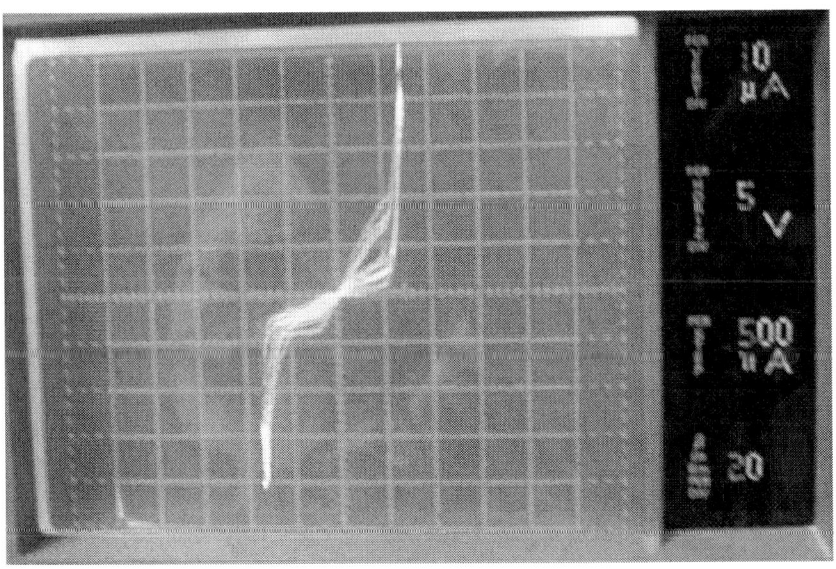

Figure 4. Ac current-voltage characteristic of assumed single grain boundary in SnO_2-Bi_2O_3-Co_3O_4-Nb_2O_5-Cr_2O_3 varistor. Horizontal: 5 V/div; vertical: 10 μA /div. Samples were obtained by I.A.Skuratovsky.

It would be quite interesting to compare current-voltage characteristic of a single grain boundary in ZnO varistors and in SnO_2 varistors. Here we present some preliminary results of such a comparison. In Fig.3 current-voltage characteristic of assumed single grain boundary in ZnO varistor and in Fig.4 current-voltage characteristic of assumed single grain boundary in SnO_2 varistor are shown. Both current-voltage characteristics are obtained by the same method (curve tracer Tektronix 576) using metal probe (one contact) positioned near the Ag electrode and Ag electrode as another contact.

It is clear that nonlinearity takes place at voltages below 5V in ZnO varistor (Fig.3). This observation is in accordance to published data [27,46,52,53]. However, in the case of SnO_2 varistor highly nonlinearity for assumed single grain boundary is observed at voltages slightly above 5V (Fig.4). Taking into account performed additional calibration of the curve tracer, SnO_2 single grain boundary voltage at current 50 μA (at nonlinear region) can be 5.2-5.5 V. Now we do not know any other result on current-voltage characteristic of a single grain boundary in SnO_2 varistor. Unfortunately, now we cannot reliably exclude the possibility that in the preliminary experiment shown in Fig.4 exist, for example, two SnO_2 grain boundaries between electrodes (though mentioned voltage was minimal for SnO_2 sample), alternate conduction path or studied grain boundary is special with higher voltage than typical one. This investigation is continued.

If all grain boundaries in the samples studied in [37,66] are characterized by the same single grain boundary voltage, then this value can be estimated as $U_S = E_1 \cdot l_G$, where electric field E_1 calculated at 1 mA cm^{-2} is 3400 V·cm^{-1} and average grain size is about 5 μm [37,66]. Thus, $U_S \cong 1.7\ V$ what is significantly less than 5.2-5.5 V estimated from experiment. It means that grain boundaries in studied SnO_2-Bi_2O_3-Co_3O_4-Nb_2O_5-Cr_2O_3 ceramics are not identical. Similar situation takes place in ZnO varistors [59] and it is possible expect that in SnO_2-based ceramics obtained by other researchers grain boundaries are not identical too.

From results presented in Fig.3 and Fig.4 it can be assumed that higher single barrier voltage for SnO_2 varistor can be related to higher energy gap in SnO_2 ($E_g = 3.7\ eV$ at 300 K) than in ZnO ($E_g = 3.2\ eV$ at 300 K). If single barrier voltage for SnO_2 varistor is more or less constant for many grain boundaries as it takes place in ZnO varistors, then nearly the same mechanism of non-Ohmic conduction in SnO_2 based and ZnO based varistors could be assumed.

7. Varistor Like a Totality of Non-identical Grain Boundaries

The inhomogeneous distribution of red electroluminescence at the surface of ZnO varistors can visualize the current paths [59,112,113]. This our result was soon confirmed by other researchers [8]. This fact is important for the understanding that current-voltage characteristics of ZnO varistors can be understood taking into account two factors: non-Ohmic conduction in single grain boundary and the variability of electrical parameters of grain boundaries and the grain size in a sample [12,33,52,59,112,113].

Figure 5. Surface distribution of red (700 nm) electroluminescence at the surface of ZnO-SnO_2-CoO-Bi_2O_3 varistors obtained at two dc voltages: 390V (a) and 410V (b). Distance between electrodes situated vertically to the left and to the right of light emitting area is 0.27 cm.

In Fig.5 the example of the image of the light emitting varistor ZnO-SnO_2-CoO-Bi_2O_3 is shown [59,112,113]. Intensity distribution is quite inhomogeneous and grain boundary areas emit the light. From electrode to electrode nearly parallel current paths are seen. These paths are consisted of electrically-active grain boundaries connected in series. Increase of applied voltage causes an enlargement of luminescent area due to more current paths become visible.

Such a concept of parallel current paths is confirmed by inhomogeneous distribution of current and nonlinearity coefficient over cross-section of ZnO varistor studied using multi-contact electrode [114]. By this method it was observed as well that after electrical or thermal degradation outer areas of ZnO varistor becomes more conductive [114]. Similar effect (outer areas of a varistor becomes more conductive) takes place in SnO_2 varistor in humid air [69]. Therefore, in oxide varistors at certain conditions macroscopic redistribution of current can be observed. It should be noted that after the sintering surface areas of a sample due to be enriched of oxygen exhibit fairly different properties from that ones in the bulk [115].

It was estimated by electrostatic force microscopy technique that amount of electro-active grain boundaries in SnO_2-varistor is around 85% (much higher than 15-30% found in ZnO-varistors) [116]. It looks rather questionable that all of these barriers can participate in electrical conduction because the grain-boundary barriers in the sample can be non-identical. It takes place in ZnO varistors [12,33,52,59,112,113] and can take place in SnO_2 varistors too.

Taking into account that not only non-Ohmic conduction in single grain boundary is important factor of varistor effect but as well large-scale electric inhomogeneity (non-identity of grain boundaries in a sample) [59,113], some attempts were made to obtain suitable expression for current-voltage characteristic of varistor [32,113,117]. It was revealed that conduction process in oxide varistors is non-trivial. For example, the effect of current branching should be considered [117]. The application of percolation approach to conduction in oxide varistors could be useful [29,118-120].

The problem of non-Ohmic conduction in oxide varistors is fundamental and practical at the same time. Therefore, new publications still appear. The attention to varistor as to a nonlinear system with non-identical grain boundaries last years is become rather stronger. Some approaches to modeling of current-voltage characteristics of varistors with different grain boundaries were developed [33,121,122].

Taking into account our idea that explanation of electrical properties of varistor ceramics must be based on the joint consideration of two factors - non-Ohmic conduction of single grain boundary and non-identity of grain boundaries in a sample [59], some technological solutions for improvement of electrical parameters of ZnO varistors were earlier suggested, for example, the application of surface-active substances to improve a homogeneity of ZnO non-Ohmic ceramics [113]. Reached level of technology in ZnO varistors allows to control microstructure and prepare ceramics with controllable imperfections [123].

8. Non-ohmic Conduction in SnO_2 and ZnO Ceramics

Earlier it was revealed that electrical properties of SnO_2 varistors are very similar to the properties of ZnO varistors and that non-Ohmic conduction in ZnO and SnO_2 varistors have a common physical nature related to the transport of electrons across the grain-boundary barriers appeared as a result of sintering at oxidizing atmosphere [20,21,60]. Mentioned similarity in the properties of ZnO and SnO_2 varistors was confirmed later [73]. Therefore, it can be assumed that mechanism of non-Ohmic conduction in ZnO and SnO_2 varistors could be very similar.

The non-Ohmic conduction in ZnO varistor is attributed to sufficiently complex processes which are less temperature-dependent than processes of Ohmic conduction in high-temperature range [75]. The similar situation is found in SnO_2 varistors [37]. To illustrate this point, current-voltage characteristics of SnO_2–Bi_2O_3–Co_3O_4–Nb_2O_5–Cr_2O_3 varistor were obtained at different temperatures and shown in Fig.1. Such behaviour of $j(E)$ characteristics with respect to temperature in ZnO varistors can be caused by the contribution of thermionic [36,51,55,59,75,129] or thermionic-field [36,51,55,59,128,130] emission across the grain-boundary barriers at low fields and tunneling from valence to conduction band [49,53,129], tunneling across the depletion region in ZnO [27,48,51,54,55,58,130] or impact ionization in the depletion region of grains [30,131-133] at high fields. The strong decrease of the barrier height at high fields due to tunneling from grain-boundary states [48] or holes appearance in the grain-boundary region [27,30,58,131-133] can lead to quite high nonlinearity of current-voltage characteristic.

For the understanding of non-Ohmic conduction in ZnO varistors it is important to take into account that the nonlinearity of current-voltage dependence is accompanied by relatively strong red [55,83-85] and quite weak ultra-violet [84,85,138] electroluminescence and by the phenomenon of "negative capacitance" [59,84,131,136,137]. The phenomenon of "negative capacitance" initially observed in [136,137] was confirmed in subsequent investigations [59,84,131]. Both mentioned phenomena (electroluminescence and "negative capacitance") were discussed in [8] in the frames of the mechanism of impact ionization in the depletion region.

In the case of impact ionization mechanism the requirements to the width of the grain-boundary barriers are not so strict as for tunneling-related mechanisms. However, impact ionisation with optical phonon scattering can lead to positive temperature coefficient of electric field at fixed current density $\nu = (1/E_1)(dE_1/dT)$ but this coefficient in varistors is negative. Therefore, it is rather difficult to exclude completely the contribution of alternative tunnel mechanisms. The existence of the grain boundaries with the barrier voltage lower than 3.2 V [52] proves that in ZnO varistors several mechanisms of high nonlinearity can take place.

The barrier height in ZnO varistors is weakly decreased on voltage below about 3-3.5 V per grain boundary (the energy gap of ZnO at room temperature is $E_g \cong 3.2\ eV$) because due to electrons capture some increase of the negative interface charge by absolute value prevents sharp barrier lowering and at higher voltages the barrier height is decreased stronger due to the accumulation of holes or ionization of the interface or bulk states [8,27,48,131]. The importance of the filling of the interface states at application of voltage to grain boundary with double Schottky barrier (semiconductor bicrystal) was demonstrated firstly in [150]. Later this approach was confirmed by others [151]. Participation of tunnel processes in the voltage-dependent filling of the interface states in semiconductor bicrystal was considered in [152].

At decrease in temperature (below room temperature) the continuous decrease of the activation energy of Ohmic conduction from about 0.8 eV to about 0.02 eV in ZnO varistors [18,36,51,55,59,118,119,128] and in SnO_2 varistors [18,60,119] is observed. It is due to the domination (in ZnO varistors) of thermally-activated tunneling in the range about 150-300 K and tunneling or hopping conduction below about 150 K.

If impact ionization mechanism is assumed at high fields in ZnO varistors [131], then mentioned change of thermionic emission across the grain-boundary barriers at low fields and at high temperature to field emission at low temperature [18,36,51,55,59,118,119,128] can need some increase of electric field because at low temperature electrons cross the barrier at lower energy. Therefore, it can give some negative contribution in temperature coefficient of electric field ν.

It was observed that an application of voltage in the range of highly nonlinearity to ZnO varistors at low temperature (about 100 K) caused an increase of low-field conductivity [55,134,135]. This changed state can be kept at low temperature for a long time but heating to room temperature restored initial current-voltage characteristic and thermo-stimulated current was observed. For the explanation of this effect it can be assumed that an application of voltage in the range of high nonlinearity caused the decrease of the barrier height due to decrease of negative interface charge (on absolute value) and at low temperature such a change could not be restored quickly. At high temperature such a restoration of the interface negative charge takes very short time. This observation indirectly confirms the model of double Schottky barrier and the idea that the interface negative charge is changed at high fields (the barrier height is decreased at high fields).

The whole range of electrical properties of ZnO varistors can be realized if additionally to the grain-boundary phenomena to take into account that the grain boundaries in the sample are not identical causing (together with grain size distribution) the large-scale electric inhomogeneity [59,113,114,117,139] which is proved particularly by the inhomogeneous space distribution of red electroluminescence [8,59].

9. Role of Additives

The effect of each additive in ZnO varistor ceramics on their electrical properties was studied in numerous papers. For example, in [135] dc and pulse current-voltage characteristics and thermo stimulated currents were studied in ZnO-Bi_2O_3-Co_3O_4-Sb_2O_3-MnO_2 ceramics and ceramics obtained without one of four additive oxides. The highest nonlinearity coefficient

$\beta = 50$ and lowest low-field conductivity were found for ZnO-Bi_2O_3-Co_3O_4-Sb_2O_3-MnO_2 sample. The absence of any additive leads to the decrease of the nonlinearity coefficient. The most significant decrease of the nonlinearity coefficient was observed in composition without bismuth oxide ($\beta < 6$). Therefore, addition of Bi_2O_3 is very important for the formation of nonlinear current-voltage characteristics. Elimination of Sb_2O_3 gives the shift of current-voltage characteristic to lower fields due to observed additional grain growth. The absence of MnO_2 practically does not change electrical properties, however, Co_3O_4 addition is important in reaching high nonlinearity coefficient. Therefore, an addition of bismuth oxide in ZnO varistors promotes the formation of electrically-active grain boundaries. However, without transitional metal oxide (like Co_3O_4) the nonlinearity of current-voltage characteristic is not high.

Cobalt oxide can dissolve in the ZnO lattice (giving to material dark-green colour) and form deep centers in ZnO which probably can be ionized at high electric field or tunneling from such centers at high field can take part in narrowing of grain-boundary barrier. The possibility of impact ionization of deep Co-centers in ZnO varistors was suggested in [140]. The increase of electron-hole generation rate with cobalt oxide concentration was suggested [133].

Consider the role of oxide additives in the formation of electrical properties of SnO_2-Bi_2O_3-Co_3O_4-Nb_2O_5-Cr_2O_3 varistor. The addition of bismuth oxide significantly decreases the electric field E_1 due to some increase of the average grain size. The grain size (about 3-7 μm) in SnO_2-Bi_2O_3-Co_3O_4-Nb_2O_5-Cr_2O_3 ceramics sintered at 1470 K [66] is slightly higher than the grain size (2.4 μm) of SnO_2-CoO-Nb_2O_5-Cr_2O_3 ceramics with 0.5 mol.% Cr_2O_3 sintered at 1570 K in [63] possibly due to Bi_2O_3 addition. This observation becomes even stronger if to take into account the higher sintering temperature (1570 K) in the latter case that gives higher grain size. Therefore, bismuth oxide addition can promote SnO_2 grain growth and to some extent it can also promote the formation of the grain-boundary barriers due to the additional adsorption of oxygen.

Cobalt oxide can segregate in the grain-boundary region [20] and, bearing in mind different oxidation states of Co-ions, it is assumed that cobalt oxide can participate in the formation of the grain-boundary barriers. In addition, cobalt oxide promotes the densification of SnO_2 ceramics [63] which is also in accordance to other observations [66].

The addition of Nb_2O_5 leads to the increase of the electrical conductivity of SnO_2 ceramics [141]. In the crystalline lattice of SnO_2 (band gap of SnO_2 is about 3.6 eV) singly and doubly ionized oxygen vacancies and such impurities as Nb^{5+}, Sb^{5+}, As^{5+} and others are known as donors with different values of the ionization energy in the range of 0.024-0.72 eV [105,106,142]. Therefore, Nb can be the donor impurity in the crystalline SnO_2 grains causing the decrease of the grain resistivity and promoting additional concentration of electric field in the vicinity of the grain boundary. Some confirmation of this assumption can be found from the recent publication [143]. Thus, in SnO_2-Bi_2O_3-Co_3O_4-Nb_2O_5-Cr_2O_3 ceramics the impurity of Nb and the oxygen vacancies can perform the role of donors with fairly low ionization energy (possibly no more than about 0.2-0.3 eV). Taking into account that at temperature around 300 K shallow donors can be partially or completely ionized it is assumed that the Fermi level in the bulk of tin dioxide grains is situated about several tenths of eV below the conduction band edge. Therefore, the barrier height φ_0 in SnO_2 varistors is only slightly less than the activation energy E_σ, $\varphi_0 \cong E_\sigma$ (see section 5).

The addition of Cr_2O_3 leads to an increase of the nonlinearity coefficient β in SnO_2-Co_3O_4-Nb_2O_5-Cr_2O_3 [63] and SnO_2-Bi_2O_3-Co_3O_4-Nb_2O_5-Cr_2O_3 [66] ceramics. The mechanism of the influence of Cr_2O_3 addition on the nonlinearity is still not clear in details. Tin vacancy occupied by chromium having negative charge can participate in charge exchange with oxygen species and help to fix oxygen at the grain boundary [25]. The stronger electrically-active role of chromium impurity in the SnO_2 varistor in comparison with Mo, W and Ti impurities is not related to the difference in the ionic radiuses because those for Cr^{3+}, Mo^{6+}, W^{6+} and Ti^{4+} are comparable [37].

Up to 28 mol.% of chromium oxide can be dissolved in the crystalline lattice of SnO_2 during heat treatment at temperature 1570 K forming a solid solution by substituting Sn^{4+} ions for Cr^{3+} (due to ions Sn^{4+} and Cr^{3+} have comparable ionic radiuses) [144]. As a result of this process, some deep levels within the band gap of SnO_2 can appear. The nature of these levels (donor or acceptor) is not known exactly.

Taking into account the existence of different oxidation states of Cr-ion it can be assumed that Cr-ion at the surface of SnO_2 can have a higher oxidation state than in the bulk and, thus, it can promote the additional chemisorption of oxygen during the cooling of a sample after sintering. As a result of this process, the barrier height in the varistor with Cr_2O_3 addition can be higher and consequently the conductivity can be lower. This assumption is in accordance with obtained experimental data [37].

Assume that deep energy levels of Cr at the boundary of SnO_2 grains can be filled with electrons. With an increase of electric field the voltage drop at certain grain boundaries can reach some critical value at which the ionization of chromium states at the grain boundary is initiated. This process can be due by the field emission of electrons (in the sense of mechanism suggested in [48]) or can be due by the recombination of electrons filling the chromium states with holes appeared as a result of the impact ionization of valence or Cr_2O_3-related deep bulk states in the reverse biased depletion region and subsequently drifted to the interface (in the sense of the mechanism suggested in [131,140]). Both mentioned options can give a strong decrease of the barrier height in the high electric field and lead to non-Ohmic behavior of this material. Therefore, in SnO_2 ceramics with Cr_2O_3 addition the electric field dependence of the barrier height is enhanced and consequently the nonlinearity of current-voltage characteristic is increased. The role of Cr_2O_3 addition in SnO_2 varistors can possibly be similar to the role of CoO and MnO in ZnO varistors.

10. Temperature Shift of Highly-Nonlinear Current-Voltage Characteristic

Increase of temperature causes relatively strong shift of the low-field part of $j(E)$ characteristic to higher current densities and relatively small shift of high-field (highly-nonlinear) part of $j(E)$ characteristic to lower fields with the temperature coefficient of electric field (at fixed current density) [27,75]:

$$\nu = \frac{1}{E_1}\frac{dE_1}{dT} = \frac{d(\ln E_1)}{dT} \cong \frac{\Delta(\ln E_1)}{\Delta T} \cong -5\cdot 10^{-4}\,K^{-1}. \qquad (7)$$

It is explained as thermionic emission over the barrier at low fields and as an influence of temperature on tunnel currents at high fields [27].

However, a small temperature shift of the highly-nonlinear part of $j(E)$ characteristic to lower fields can be understood simply as increase of thermionic current with temperature if the barrier height becomes fairly low at high fields [35-37,131].

To verify this assumption estimate at first the increment of current density in a logarithmic scale $\Delta(\ln j)$ for the increment of temperature ΔT at constant electric field in a high-field region. From Eq.(2) and Eq.(7) we have:

$$\Delta(\ln j) \cong -\nu\beta\Delta T\,. \qquad (8)$$

From the other side, the value $\Delta(\ln j)$ in a high-field region can be estimated independently from the approximate expression for the current density in the case of thermionic emission across the grain-boundary barrier:

$$j = j_0 \exp(-\frac{\eta + \varphi(U)}{kT}), \qquad (9)$$

where j_0 is a constant, η is the distance between the conduction band edge and Fermi level inside the grain, k is the Boltzmann′s constant, T is the absolute temperature. Differentiating on temperature and assuming $\eta << \varphi(U)$, one can obtain:

$$\Delta(\ln j) \cong \frac{\varphi(U)}{kT^2}\Delta T\,. \qquad (10)$$

The comparison of Eqs.(8) and (10) gives the expression for the barrier height at high fields:

$$\varphi(U) \cong -\nu\beta kT^2\,. \qquad (11)$$

Around $T = 300\ K$ typical values are: $\beta = 50$ and $\nu \cong -5\cdot 10^{-4}\,K^{-1}$ [27]. Then from Eq.(11): $\varphi(U) \cong 0.195\ eV$. This value is close to the barrier height at high fields reached in the experiment in [27,36].

This estimation supports the initial assumption that temperature shift of current-voltage characteristic at high fields can be considered as an increase of thermionic current (at relatively high temperatures) over the barrier if the barrier height becomes quite low at high fields. However, tunnel currents (and/or impact ionization) can be responsible for a decrease of the barrier height at high bias.

11. Simple Approach to Non-ohmic Conduction in Oxide Ceramics

Estimation performed in the previous section actually means that thermionic current (at relatively high temperatures) or tunnel current (at relatively low temperatures) can be modulated by significantly lower current originated from the processes controlling the decrease of the barrier height (like tunnel and/or impact ionization). In some sense, it is assumed that it is possible to separate the processes responsible for the control of the barrier height from higher nonlinear current flowing across the grain boundary.

The expression (1), used often for the approximation of current-voltage characteristic of the varistor, is an empirical one and also the nonlinearity coefficient (2) is an empirical parameter. Therefore, it would be interesting to obtain the expression for the approximation of current-voltage characteristic more closely related to non-Ohmic conduction process and to find some parameter to characterize the quality of a varistor. With this aim, let us consider the phenomenological model of non-Ohmic conduction of ceramic varistor [35-37]. The basic idea is: the barrier height in oxide varistor material is decreased on electric field allowing respective nonlinear current density increase.

Generally speaking, the super linear dependence of current on voltage is due to the existence of the nonlinear part of the increment of the current density Δj_n in addition to the linear one Δj_l in the response to the increment of the average electric field ΔE. It is assumed that decrease of the barrier height with electric field is the reason of the appearance of Δj_n value and non-Ohmic conduction in a varistor. Then, the increase of the current density Δj at the increase of the average electric field ΔE can be written as

$$\Delta j = \Delta j_l + \Delta j_n , \tag{12}$$

where the linear (Ohmic) part of the current density increment Δj_l can be presented as:

$$\Delta j_l = \sigma(E)\Delta E = \frac{j}{E}\Delta E . \tag{13}$$

Here, $\sigma(E)$ is the conductivity of material at electric field E.

The value of Δj_n is assumed to be proportional to ΔE and to the current density j, and can be written in quite general form as:

$$\Delta j_n = \alpha j \Delta E , \tag{14}$$

where α is a constant coefficient, the nonlinearity factor. This assumption will be discussed below. Then, from Eqs.(12)–(14) it is possible to write the equation for the increment of the current density in differential form:

$$dj = \frac{j}{E}dE + \alpha j dE . \tag{15}$$

After an integration of this equation the relationship between the current density and the electric field in varistor can be obtained in a form of Eq.(3), where the constant of an integration σ_0 is the conductivity of material at low electric field. The dependence of the conductivity $\sigma(E)$ on the electric field can be written as:

$$\sigma(E) = \sigma_0 \exp(\alpha E) . \tag{16}$$

The nonlinearity factor

$$\alpha = \frac{d}{dE}\ln\frac{j(E)}{E} = \frac{d(\ln\sigma(E))}{dE} \tag{17}$$

determines quantitatively the nonlinearity of current-voltage characteristic. If $\alpha = 0$, current-voltage characteristic becomes linear, Ohm's law $j = \sigma_0 E$ takes place and vice versa.

12. Percolation Theory Based Approach

In a development presented above Eq.(14) ($\Delta j_n = \alpha j \Delta E$) can be not very obvious. Therefore, let us to obtain this equation by slightly other way than in [35-37] using percolation theory ideas [145,146].

Varistor sample contains many grain boundaries and they are rather different [12,21,32,33,52,59,117,121,139]. Assume that each grain boundary in the sample is characterized by certain barrier height φ and φ-values in the sample are distributed homogeneously from $\varphi_{\min}$ to $\varphi_{\max}$. If the barrier height scattering $\Delta\varphi = \varphi_{\max} - \varphi_{\min}$ in the sample is high, $\Delta\varphi \cong \varphi_{\max}$, the approach based on the percolation theory [145,146] can be applied. Substitute the grain boundaries and grains in the sample by the resistor network. Replace each grain boundary with the barrier height φ by some resistor with resistance $R = R_0 \exp(\varphi / kT)$. Grains are assumed to be quite conductive, therefore, they are nodes in the resistor network.

Assume that initially all resistors are removed out of the sample. Then some resistors with $\varphi < \varphi_1$ (where $\varphi_1 \cong \varphi_{\min}$) are imported into the sample in random places. They cannot allow conduction between electrodes. There is no percolation in the system. At certain

$\varphi_1 < \varphi_k$ imported resistors can belong to some finite cluster (group of resistors connected together but not connected to both electrodes at the same time). If increase φ_1 gradually and discretely importing into the sample at each step some resistors with increasing φ and allowing them to occupy randomly vacant places in the resistor network, then at $\varphi_1 = \varphi_k$ percolation in the system will appear: the last portion of imported resistors with $\varphi \cong \varphi_k$ will make electrical connection between the electrodes. At this point conductivity of the sample will jump from zero to $\sigma \propto \exp(-\varphi_k / kT)$, because resistors with $\varphi \cong \varphi_k$ give the main contribution to the conductance of the sample. Resistors with lower barrier height φ have exponentially lower resistance. If this process is continued, then resistors with exponentially higher resistances will be connected in parallel to the critical subnet of infinite cluster (group of resistors which have ensured percolation in the system at $\varphi \cong \varphi_k$) but the value of conductivity of the sample $\sigma \propto \exp(-\varphi_k / kT)$ will not be changed significantly.

Grain-boundary barriers with $\varphi \cong \varphi_k$ are called as key barriers. Therefore, the conduction of a sample σ is controlled by key barriers with barrier height φ_k ($\varphi_{\min} < \varphi_k < \varphi_{\max}$): $\sigma \propto \exp(-\varphi_k / kT)$. Therefore, in the sample key barriers are connected in series to barriers with lower barrier heights. Barriers with higher barrier heights are connected in parallel to the critical subnet of infinite cluster.

This situation can take place in varistors at low electric fields but at high electric fields grain boundaries in ZnO varistors can be more identical [21,32]. Therefore, the distance between key barriers L in general case can be greater or equal to the grain size l_g, $L \geq l_g$. In other words, if all grain boundaries are equal, then the average distance between them is equal to the grain size, $L = l_g$. If the variation of the barrier height $\Delta\varphi$ is high ($\Delta\varphi \cong \varphi_{\max}$), then $L > l_g$.

Percolation theory can be applied for consideration of systems with non-Ohmic conduction [145,147-149]. It was shown that polycrystalline ZnO layers obtained at low-temperature heat treatment exhibit grain-boundary-controlled conduction with quite wide distribution of grain-boundary barrier heights $\Delta\varphi \cong \varphi_{\max}$ [147-149]. However, the nonlinearity in polycrystalline ZnO layers is very low ($\beta \cong 5$) in comparison with $\beta \cong 50$ observed in SnO_2 based and ZnO based varistors. This fact is related to the difference in the mechanisms of non-Ohmic conduction in the grain boundary in different systems.

For the purposes of the verification of Eq.(14) ($\Delta j_n = \alpha j \Delta E$) it would be useful to estimate the relation between the increment of electric field ΔE and subsequent increment of the barrier height $\Delta\varphi$. For that it would be necessary in the non-Ohmic regime two times repeat the procedure similar to key barriers finding in the Ohmic regime (see above). It is necessary to stress that repeating this procedure at each electric field can lead, in principle, to different percolation paths at each electric field. Assume that condition of an applicability of percolation theory (wide distribution of grain-boundary barrier heights in the sample) is performed and respective values, $\varphi(E)$ and $\varphi(E+\Delta E)$ are found.

It means that the current density through a sample of ceramic varistor (but not only the current density through a single grain boundary) can be presented in a form:

$$j = j_0 \exp(-\frac{\varphi(E)}{kT}). \quad (18)$$

Here $\varphi(E)$ is the barrier height of the sample at electric field E. In spite the sample contains barriers with quite different barrier heights (ranging from φ_{min} to φ_{max}) exist some value $\varphi(E)$ which determine conductivity (see Eq.(18)).

If ΔE is infinitely small, then one can expect that $\varphi(E)$ and $\varphi(E+\Delta E)$ are very close. Therefore, $\Delta\varphi = \varphi(E) - \varphi(E+\Delta E)$ is infinitely small. Then

$$\Delta\varphi \cong \left(-\frac{d\varphi}{dE}\right)\Delta E = A(E)\Delta E. \quad (19)$$

For simplicity, the function $A(E)$ can be treated in a first approximation as a constant.

Then "nonlinear" part of current density increment related to the decrease of the barrier height (Eq.(14)) can be obtained from Eqs.(18) and (19), where the nonlinearity factor is:

$$\alpha = \frac{1}{kT}\left(-\frac{d\varphi}{dE}\right). \quad (20)$$

It is necessary to stress that in Eq.(20) only parameters related to a whole sample are presented: here φ is key barrier height, the barrier height of those barriers which are responsible for a percolation in a sample at electric field E.

It would be interesting to compare obtained $j(E)$ characteristic (Eq.(3)) with calculations of $j(E)$ dependence performed by other authors using percolation theory [145,149].

Earlier the expression $\sigma = \sigma_0 \exp\left[(CV_0/kT)(qEa/\delta V)^{1/\nu}\right]$ for conductivity in disordered polycrystalline semiconductor in the case of a filling of grain-boundary states under applied voltage was obtained [149]. Here V_0 is the amplitude of random variations of the barrier height for different grain boundaries in a sample, δV is the increase of the barrier height due to the capture of electrons, C is a constant and $\nu \cong 0.9$ is the critical index of the correlation length in percolation theory. Field dependence of conductivity in this formula is similar to one obtained here (see Eq.(16)) and it is fairly different from equation $\sigma = \sigma_0 \exp(D \cdot E^{1/(1+\nu)})$ [145] obtained in the assumption that grain-boundary states are already filled. In ZnO varistors grain-boundary states are indeed filled at low electric fields [8,27]. Therefore, simple approach presented here is in agreement with detailed theory [149].

13. Parameters of Nonlinearity

In this section the relation between the nonlinearity factor α and the nonlinearity coefficient β will be discussed and then the normalized nonlinearity coefficient will be introduced.

13.1. Nonlinearity Factor and Normalized Nonlinearity Coefficient

The nonlinearity factor (see Eqs.(3) and (16))

$$\alpha = \frac{d}{dE} \ln \frac{j(E)}{E} = \frac{d(\ln \sigma(E))}{dE} \cong \frac{\Delta(\ln \sigma(E))}{\Delta E} \tag{21}$$

determines quantitatively the nonlinearity of $j(E)$ characteristic. The value of α is numerically equal to the increment of the logarithm of conductivity $\ln \sigma(E)$ at the increment of the electric field $\Delta E = 1\ V\ cm^{-1}$. If $\alpha = 0$, then conductivity does not depend on electric field, $j(E)$ characteristic becomes linear, Ohm's law $j = \sigma_0 E$ takes place, and vice versa.

The relation of the nonlinearity factor α with the nonlinearity coefficient β can be found from Eq.(3) using Eq.(2):

$$\beta = 1 + \alpha E\,. \tag{22}$$

In the case of linear $j(E)$ characteristic, $\alpha = 0$ and $\beta = 1$. If $\alpha = const \neq 0$, then from Eq.(22), the nonlinearity coefficient β is increased with the electric field. This fact is frequently observed experimentally. If the nonlinearity coefficient $\beta >> 1$, then from Eq.(22) we have:

$$\alpha \approx \frac{\beta}{E_1} = \beta_E\,, \tag{23}$$

where β_E is the normalized nonlinearity coefficient. In Eq.(23), the fixed value of the average electric field E_1 is used instead of the variable value E to avoid field dependence of the normalized nonlinearity coefficient β_E. This is reasonable because the variation of the electric field ΔE in a highly nonlinear region ($\beta >> 1$) is much less than the electric field,

$\Delta E << E$. Therefore, the value of the normalized nonlinearity coefficient $\beta_E = \beta / E_1$ can be used for the characterization of varistors.

Experimental $j(E)$ dependences in ZnO and SnO_2 varistors can be linearized in the scale $\lg\sigma - E$ [35-37]. As an example, the $j(E)$ characteristics of SnO_2 varistor studied at different temperatures (see Fig.1) are presented in the scale $\lg\sigma - E$ in Fig.6. It is seen that obtained experimental data at different temperatures can be approximated by Eq.(16) with two values of the nonlinearity factor, α_1 and $\alpha_2 > \alpha_1$. In ZnO varistors experimental points are distributed quite well along the straight line (with the slope α_2) for about seven decades of conductivity [36]. The empirical power-law relation (1) formally also gives satisfactory approximation of experimental current-voltage characteristics but parameters of the approximation (B and β) are not related directly to the physical properties of non-Ohmic ceramics.

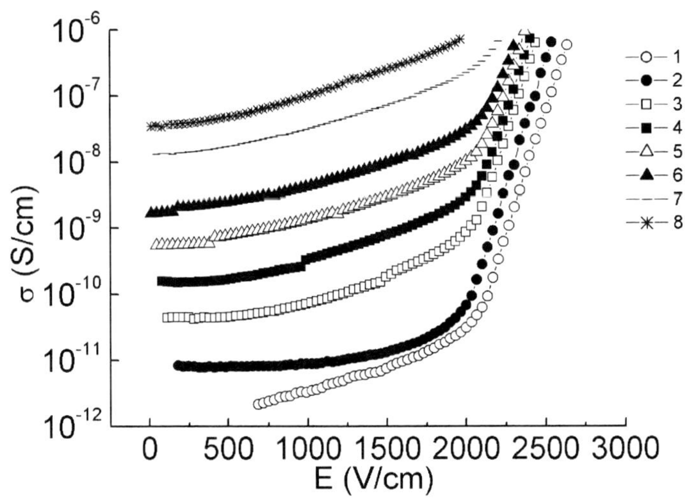

Figure 6. Conductivity versus electric field in SnO_2-Bi_2O_3-Co_3O_4-Nb_2O_5-Cr_2O_3 varistor measured at different temperatures, K (re-plotted from Figure 1): 1 – 292, 2 – 332, 3 – 378, 4 – 400, 5 – 424, 6 – 444, 7 – 476, 8 – 494. Samples were obtained and measurements were performed by I.A.Skuratovsky.

Table 1. Electrical parameters of varistor ceramics

Ceramic system	Nonlinearity coefficient β	Electric field E_1 at current density 10^{-3} A cm^{-2}, V·cm^{-1}	Nonlinearity factor α_1, cm·V^{-1}	Nonlinearity factor α_2, cm·V^{-1}	Normalized nonlinearity coefficient $\beta_E = \frac{\beta}{E_1}$	Refe rence
ZB	6.7	160	-	3.6 10^{-2}	4.19 10^{-2}	[18,36]
ZBCMT	7	130	-	6.32 10^{-2}	5.38 10^{-2}	[153]
ZBCMT	18	89	2.6 10^{-2}	1.16 10^{-1}	2.0 10^{-1}	[154]

Table 1. Continued

Ceramic system	Nonlinearity coefficient β	Electric field E_1 at current density 10^{-3}A cm^{-2}, V·cm^{-1}	Nonlinearity factor α_1, cm ·V^{-1}	Nonlinearity factor α_2, cm ·V^{-1}	Normalized nonlinearity coefficient $\beta_E = \frac{\beta}{E_1}$	Refe rence
ZBCMT	21	275	2.2 10^{-2}	6.9 10^{-2}	7.5 10^{-2}	[155]
ZSCB	40	2080	2.3 10^{-3}	1.7 10^{-2}	1.92 10^{-2}	[55]
ZSCB2	34	1740	1.6 10^{-3}	1.8 10^{-2}	1.95 10^{-2}	[55]
ZSBCC	34	1208	-	2.77 10^{-2}	2.8 10^{-2}	[75]
ZBCMS	31	1202	2.9 10^{-3}	2.63 10^{-2}	2.6 10^{-2}	[135]
ZBCMS	50	1585		3.16 10^{-2}	3.15 10^{-2}	[35]
ZX	70	2512	5.1 10^{-4}	3.0 10^{-2}	2.8 10^{-2}	[123]
SBiCBN	21	3500			6 10^{-3}	[20]
SBCoN	15	3800		$3.59 \cdot 10^{-3}$	$3.86 \cdot 10^{-3}$	[37]
SBCoM	16	4400		$3.44 \cdot 10^{-3}$	$3.76 \cdot 10^{-3}$	[37]
SBCoW	17	4000		$4.11 \cdot 10^{-3}$	$4.28 \cdot 10^{-3}$	[37]
SBCoNT	26	5900		$4.35 \cdot 10^{-3}$	$4.38 \cdot 10^{-3}$	[37]
SBCoNC	53	3400		$1.59 \cdot 10^{-2}$	$1.56 \cdot 10^{-2}$	[37]
SCAN	72	18200			$3.96 \cdot 10^{-3}$	[62]
SCNC	41	3990			$1.03 \cdot 10^{-2}$	[63]
SCNCL	142	11525			$1.23 \cdot 10^{-2}$	[65]
SBCNCB	80	5000			$1.60 \cdot 10^{-2}$	[67]

For ZB and ZBCMT the single value of the nonlinearity factor is shown as α_2.

ZB - $ZnO-Bi_2O_3$
ZBCMNT - $ZnO-Bi_2O_3-Co_3O_4-MnO_2-NiO-TiO_2$
ZBCMT - $ZnO-Bi_2O_3-Co_3O_4-MnO_2-TiO_2$
ZSCB1 - $ZnO-SnO_2-CoO$-1 mol.% Bi_2O_3
ZSCB2 - $ZnO-SnO_2-CoO$-1.5 mol.% Bi_2O_3
ZSBCMC - $ZnO-Sb_2O_3-Bi_2O_3-Co_3O_4-MnO_2-Cr_2O_3$
ZBCMS - $ZnO-Bi_2O_3-Co_3O_4-MnO_2-Sb_2O_3$
ZX – $ZnO-Bi_2O_3-Sb_2O_3-CoO-NiO-Mn_2O_3-Cr_2O_3$
SBiCBN - $SnO_2-Bi_2O_3-Co_3O_4-BaO-Nb_2O_5$
SBCoN - $SnO_2–Bi_2O_3–Co_3O_4–Nb_2O_5$
SBCoNM - $SnO_2–Bi_2O_3–Co_3O_4–Nb_2O_5-MoO_3$
SBCoNW - $SnO_2–Bi_2O_3–Co_3O_4– Nb_2O_5-WO_3$
SBCoNT - $SnO_2–Bi_2O_3–Co_3O_4– Nb_2O_5-TiO_2$
SBCoNC - $SnO_2-Bi_2O_3-Co_3O_4-Nb_2O_5-Cr_2O_3$
SCAN - $SnO_2-CoO-Al_2O_3-Nb_2O_5$
SCNC - $SnO_2-Co_3O_4-Nb_2O_5-Cr_2O_3$
SCNCL - $SnO_2-Co_3O_4-Nb_2O_5-Cr_2O_3-La_2O_3$
SBCNCB - $SnO_2-Bi_2O_3-Co_3O_4-Nb_2O_5-Cr_2O_3-B_2O_3$

In Fig.7 the dependence $\lg \sigma(E)$ is presented schematically (curve 1). In a general case the dependence of $\lg \sigma$ on electric field consists of three parts: (i) an Ohmic part with constant conductivity, (ii) a low-field part with the nonlinearity factor α_1 and a high-field part with the nonlinearity factor $\alpha_2 > \alpha_1$.

In Table 1 electrical parameters of ZnO and SnO_2 varistor ceramics are presented [35-37]. Sufficiently clear correlation between the values α_2 and β_E exists.

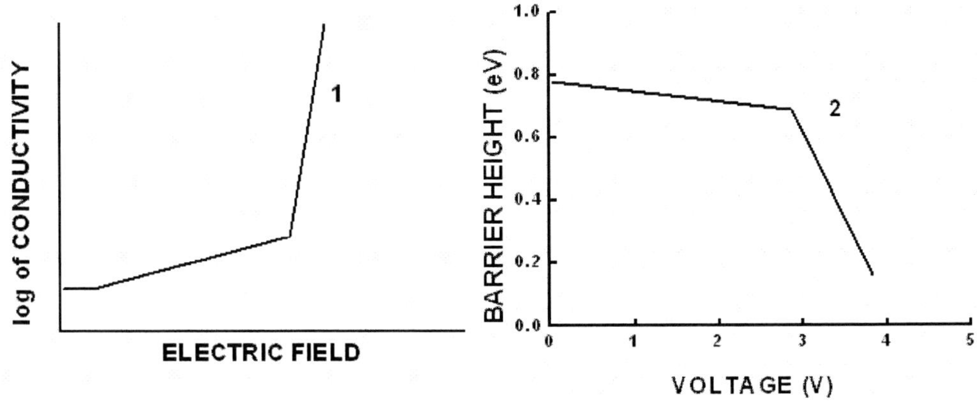

(a) (b)

Figure 7. Schematical presentation of the logarithm of conductivity in a varistor vs electric field (curve 1) and the barrier height vs voltage per grain boundary (curve 2).

It is necessary to stress that experimental values of α_1 and α_2 for low-voltage varistors [154,155] are slightly higher than the experimental values of α_1 and α_2 for high-voltage varistors [75,123,135] in spite of the experimental values of the nonlinearity coefficients, on the contrary, are higher in the case of the high-voltage varistors (Table 1). Eq.(22) explains why it takes place: the nonlinearity coefficient is, approximately, the product of the nonlinearity factor α and the average electric field E. The electric field E_1 (at fixed current density) is much higher in the case of the high-voltage varistors and it dominates over the variation of the α-values. Therefore, the value of the nonlinearity coefficient β is rather some joint parameter combining (i) the nonlinearity of $j(E)$ characteristic and (ii) the position of nonlinear $j(E)$ characteristic at the axis of electric field.

From Eq.(22) it is seen that if the nonlinearity factors for two varistors are equal, $\alpha_A = \alpha_B$, but they exhibit highly nonlinear $j(E)$ characteristics in different ranges of electric field, then their nonlinearity coefficients should be different. This explains why the values of the nonlinearity coefficients of low-voltage varistors are systematically lower than the values of the nonlinearity coefficients of high-voltage varistors.

13.2. Temperature Dependence of the Nonlinearity Factor

The nonlinearity factor α is some phenomenological constant (see Eqs.(3) and (16)). The expression for α in Eq.(20) was obtained for thermionic emission across the grain-boundary barrier (probable mechanism of the barrier-controlled transport at relatively high temperatures). However, at temperatures near 300 K the change of electron transport mechanism in ZnO varistors takes place: above about 300 K thermionic emission across the grain-boundary barrier is observed and at temperatures below about 300 K thermally-activated tunneling across the barrier is likely [36,55,59,128]. Above about 300 K the nonlinearity coefficient is inversely proportional to temperature [36,76] and at lower temperatures it weakly depends on temperature [36]. Therefore, it is necessary to analyse the nonlinearity factor in both regions separately.

13.3. Nonlinearity Factor in High-temperature Range

At temperatures around and above 300 K thermionic emission is the main mechanism of electron transport across the grain-boundary barrier in ZnO varistors [8,26-28,59]. In this case the current density j at voltages $U >> kT / q$ (where q is the elementary charge) is proportional to the probability that electron can overcome the potential barrier with the barrier height $\varphi(U)$ (see Eq.(18)). Then Eq.(20) gives expression for the nonlinearity factor.

Assume that the relation between the average voltage U applied to one key barrier and the electric field E in a sample is:

$$U \approx LE . \tag{24}$$

Then the expression for the nonlinearity factor can be presented in the form:

$$\alpha = \frac{L}{kT}\left(-\frac{d\varphi}{dU}\right), \tag{25}$$

where $d\varphi / dU < 0$, because the barrier height is decreased on voltage and, therefore, $\alpha > 0$. It is clear from Eq.(25) that the nonlinearity factor α is proportional to the rate of change of the barrier height on voltage (the absolute value of the derivative $d\varphi / dU$) and it is inversely proportional to temperature.

The dependence of the barrier height $\varphi(U)$ in ZnO varistor [27] can be approximated by two straight lines with different slopes. In SnO_2 varistor the situation can be similar. Then experimentally observed two-slope curve in the scale $\lg\sigma - E$ in Fig.6 and schematically presented in Fig.7 (curve 1) can be understood as a consequence of the two-slope curve $\varphi(U)$ in ZnO varistor. It is shown schematically in Fig.7 (curve 2). The horizontal axis for each curve has different sense: the transition to more sharp field dependence of $\lg\sigma(E)$ takes place if the barrier height of key grain boundaries become to drop with voltage more strongly.

Possibly, Cr_2O_3 addition in SnO_2 varistors (see section 9) makes the $\varphi(U)$ dependence at high voltages stronger leading (according to Eq.(25)) to higher nonlinearity.

For numerical estimation of α value at T=300 K it is necessary to use the functional dependence of the barrier height on voltage. Let us use the simple model of the grain-boundary barrier with the fixed interface charge (in respect to the voltage change). In this case the barrier height at a given voltage is [156]:

$$\varphi(U) = \varphi_0 \left(1 - \frac{qU}{4\varphi_0}\right)^2, \tag{26}$$

where φ_0 is the barrier height at zero voltage $U = 0$. The function $\varphi(U)$ in ZnO varistor is more complex [8,27]. A similar situation is expected in SnO_2 varistor, though Eq.(26) can be used as a first-order approximation. It is possible to write from Eq.(25) and Eq.(26):

$$\alpha = \frac{qL}{2kT}\left(1 - \frac{qU}{4\varphi_0}\right). \tag{27}$$

Eqs.(26) and Eq.(27) are applicable in the range $qU < 4\varphi_0$. The nonlinearity factor α in Eq.(27) depends on voltage though it is the consequence of used approximation (Eq.(26)) for $\varphi(U)$. Actually, it is enough to have in Eq.(25) $d\varphi / dU = const$. Let us assume, for example $qU \approx \varphi_0$ and use (in a first-order approximation) the average grain size for the average distance L between key barriers in SnO_2 varistor, $L \cong l_g \approx 5\mu m$. It means that identical barriers are situated at all grain boundaries. With these assumptions, one can obtain the calculated nonlinearity factor $\alpha_{cal} \approx 7 \cdot 10^{-3} cm \cdot V^{-1}$ [37]. Experimental values of the nonlinearity factor α_2 (Table 1) are higher and it can be due to the higher distance between key barriers.

In Eq.(25) the nonlinearity factor α is inversely proportional to temperature if the $d\varphi / dU$ value is nearly constant. It was shown recently that the barrier height in ZnO varistors is almost constant from room temperature up to similar to 670 K [157]. Therefore, at temperatures above 300 K the nonlinearity factor can be decreased with temperature as $\alpha \propto 1/T$. It is confirmed by the experimental data [36,76]. According to Eqs.(22) and Eq.(23) temperature dependences of α and β are the same due to negligible temperature dependence of E_1.

13.4. Nonlinearity Factor in Low-Temperature Range

Near temperature 300 K the crossover between two mechanisms of electron transport across the grain boundary in ZnO varistors can take place [51,55,59]. Therefore, it is desirable to obtain the expression for the nonlinearity factor α at $T \leq 300\ K$ (in a low-temperature range).

Below about 300 K the dependence of Ohmic conductivity in ZnO varistors on absolute temperature plotted in the scale $\lg\sigma - T^{-1}$ is not a straight line: the slope of this curve (the activation energy of electrical conduction E_a) is decreased gradually at the lowering of temperature from about $E_a = 0.8\ eV$ at 340 K up to about $E_a = 0.03\ eV$ at 100 K [36,51,55,59,118,128]. This fact was explained in the assumption of thermally-activated tunneling across the grain-boundary barriers in ZnO varistors in temperature range below about 300 K and nontraditional expression for conductivity versus temperature $\sigma(T) \propto \exp(T/T_0)$ was obtained [36,51,55,59,128].

For ZnO varistor experimental points $\sigma(T)$ forms the straight line in the range of about 150-340 K if they are presented in the scale $\lg\sigma - T$ [36,51,55,59]. Taking into account that in the case of thermally-activated tunneling across double Schottky barrier in GaAs bicrystall [158] the dependence $\sigma \propto \exp(T/T_0)$ as well takes place [128], it means that in the range of about 150-300 K thermally-activated tunneling can give the main contribution to electron transport in ZnO varistors [36,51,55,59,128]. At lower temperatures (below about 150 K) and at low voltages tunneling across the barrier (field emission at the Fermi level) or hopping conduction in the intergranular phase can be the main mechanisms of conduction in ZnO varistors.

The same behaviour of $\sigma(T)$ is occurred for varistor structure with a single grain boundary: the linearization of experimental data is better in the scale $\lg\sigma - T$ than in the scale $\lg\sigma - T^{-1}$ [36]. Therefore, it could be that in spite of percolation processes in oxide ceramics, conduction at different temperatures is controlled practically by the same key barriers.

In the case of thermally-activated tunneling across the double Schottky barrier the current density j across the single grain boundary at low voltages at $T \leq 300\ K$ can be obtained in a form [128]:

$$j = j_0 \exp\left(\frac{T}{T_0}\right)\exp\left(-\gamma\sqrt{\frac{8\varepsilon\varepsilon_0\varphi}{q^2 N_D}}\right), \tag{28}$$

where

$$T_0 = \frac{2q^2 N_D}{3\gamma^2 \varepsilon\varepsilon_0 k}, \tag{29}$$

N_D is the donor density in the grain, ε is the relative dielectric permittivity in ZnO, ε_0 is the dielectric permittivity of vacuum, φ is the barrier height, γ is the parameter used in the approximate expression for the probability of the tunneling across the Schottky barrier ($p \propto \exp(-\gamma l_T)$, l_T is the tunneling length). The γ values estimated for GaAs bicrystal and ZnO bicrystal from experimental data and from the approximate expression $\gamma \cong 2\pi(2m^*\varphi_0)^{1/2} h^{-1}$ (m^* is the effective mass of electron, $m^* = 0.2m_0$ is assumed, m_0 is the free electron mass, h is the Planck´s constant) are in the range

$\gamma \cong 8.2\cdot10^6 - 2.3\cdot10^7\,cm^{-1}$ [128]. Therefore, here for the estimations $\gamma = 1\cdot10^7\,cm^{-1}$ is used.

If in Eq.(28) the barrier height φ is assumed to be voltage-dependent, $\varphi = \varphi(U)$, the current density increment dj can be found differentiating Eq.(28). Substituting $dU = LdE$ (see Eq.(24)), it can be presented in the form $dj = \alpha j dE$, where

$$\alpha = \gamma L \sqrt{\frac{2\varepsilon\varepsilon_0}{q^2 N_D \varphi(U)}} \left(-\frac{d\varphi}{dU} \right). \tag{30}$$

From comparison Eq. (25) with Eq. (30) it is seen that in the case of thermionic emission and in the case of thermally-activated tunneling across the grain-boundary barrier in a ZnO varistor the nonlinearity factor α is proportional to the rate of change of the barrier height on voltage (the absolute value of the derivative $d\varphi / dU$), $\alpha \propto (-d\varphi / dU)$. The decrease of the barrier height on voltage is confirmed directly by the experimental data [27, 36].

Eq. (25) and Eq. (30) support the phenomenology of suggested model – the complex processes of the decreasing of the barrier height with voltage are presented by the value $(-d\varphi / dU)$. Therefore, for numerical estimation of α value in the phenomenological model it would be useful to use the value $(-d\varphi / dU)$ found from the experiment.

The barrier height is decreased with voltage slightly at low voltages and sharply at high voltages [27]. This dependence can be approximated by two straight lines with the slopes $(d\varphi / dU)_1 = -0.027\ eV/V$ and $(d\varphi / dU)_2 = -0.48\ eV/V$, respectively. Due to that $j(E)$ characteristic in the scale $\lg(j/E) - E$ according to Eq.(25) should have two regions with different slopes (different values of the nonlinearity factor α_1 and $\alpha_2 > \alpha_1$). Approximation of the $j(E)$ curve by two straight lines gives not very different values $(d\varphi / dU)_1 = -0.083\ eV/V$ and $(d\varphi / dU)_2 = -0.42\ eV/V$ [36].

From Eq. (25) and Eq. (30) it is seen that the nonlinearity factor is proportional to the average distance between key barriers L, $\alpha \propto L$. Therefore, α for low-voltage varistors should be higher than for high-voltage varistors. Just this takes place in the experiment for ZnO varistors (see Table 1).

The nonlinearity factors α_1 and α_2 were calculated in [36] at high-temperature (Eq.(25)) and low-temperature (Eq.(30)) regions using estimated quantities $(d\varphi / dU)_1 = -0.027\ eV/V$ and $(d\varphi / dU)_2 = -0.48\ eV/V$, respectively. It is assumed that $L = l_g$ because it is difficult to estimate the exact value of L. The next numerical values were used: $\gamma = 1\cdot10^7\,cm^{-1}$; $l_g = 20\ \mu m$; $\varepsilon = 8.5$; $N_D = 3\cdot10^{17}\,cm^{-3}$; $\varphi = 1\ eV$; $T = 300\ K$. For both temperature ranges α_1 and α_2 have the same order of value (the difference is less than the factor 1.5). That means the consistency of obtained estimations.

It is necessary to make some remarks in respect of Eq.(25). In Eq.(25) the average distance L between key barriers actually is not a constant and depends on electric field: $L = L(E)$. In the frames of a percolation theory $L(E)$ is the correlation radius of infinite cluster. Increase of electric field causes the decrease of $L(E)$. It means that with increase of electric field more and more grain-boundary barriers begin to contribute in a conduction process – at high electric fields the most of grain boundaries participate in non-Ohmic conduction. In other words, with increase of electric field $L(E)$ decreases from $L(E) >> l_g$ to l_g (l_g is the average grain size). Actually, in Eq.(25) the nonlinearity factor α depends on electric field through two values: the average distance $L(E)$ between key barriers and the absolute value of the rate of a change of the barrier height on voltage $(-d\varphi / dU)$: $\alpha \propto L(E)(-d\varphi / dU)$.

Consider the dependence $\alpha \propto L(E)$. If electric field is low and $L(E) >> l_g$, then a few key barriers exist in the sample and an increment of external voltage is applied to key barriers causing quite strong current increment. With increase of electric field the network of barriers becomes denser and an increment of external voltage is applied to higher number of barriers – the nonlinearity should become lower. The dependence $\alpha \propto (-d\varphi / dU)$ is quite obvious: the higher $(-d\varphi / dU)$ value the lower every barrier at the same increment of external voltage and the stronger a nonlinearity of current-voltage characteristic. Finally, the dependence $L(E)$ is relatively weak and $(-d\varphi / dU)$ plays the main role.

14. Comparison with Experiment

Eq.(3) as a semi-empirical one was used earlier for the approximation of current-voltage characteristic of SnO_2-ZnO-Bi_2O_3 varistors [14-16], some other ceramic varistors on the basis of SnO_2-ZnO [17], ZnO-Bi_2O_3 varistors [18] as well as ZnO based varistors with addition of Bi_2O_3, CoO and SnO_2 (with the nonlinearity coefficient β=40-50) [36,51,55].

Consider additional experimental data for verification of conclusions from suggested approach.

14.1. ZnO Varistors

According to Eq.(3), $j(E)$ characteristic should be the straight line if plotted as $\lg(j / E)$ versus E. It was found that current-voltage characteristics of varistors in ZnO-SnO_2-CoO-Bi_2O_3 system indeed can be approximated by Eq.(3) and three regions can be seen: Ohmic region, where $\alpha = 0$, the region with relatively low α value and the region with high α value which corresponds to a highly nonlinear part of current-voltage characteristic [51,55]. For ZnO-Bi_2O_3 (0.5 mol.% of Bi_2O_3) ceramics ($\beta \cong 6$) alone straight line is obtained at low-field region [18].

To verify additionally the conformity of $j(E)$ dependence (see Eq.(3)) to the experimental results, in [36] some published data [75,123,135,154,155] are presented in the scale $\lg\sigma - E$. It was found that the $\sigma(E)$ dependence of varistor ceramics consists of two parts: a low-field part with the nonlinearity factor α_1 and a high-field part with the nonlinearity factor $\alpha_2 > \alpha_1$. In these data authors have not reached the Ohm's law at low fields. Electrical parameters of these varistors are shown in Table 1. It is seen that the experimental $\sigma(E)$ dependences at least are not in a contradiction to Eq.(3) and Eq.(16). The experimental points are distributed quite well along the straight line for about seven decades of conductivity [36].

In [36] it was shown that estimated values of α_1 and α_2 are fairly close to α_1 and α_2 found for high-voltage varistors [75,123,135] (Table 1). However, the experimental values of α_1 and α_2 for low-voltage varistors [154,155] (Table 1) are even slightly higher than estimated values [36]. This can be explained by the larger distance between key barriers in low-voltage varistors than used for the estimations.

In Eq.(30) the nonlinearity factor α is not dependent on temperature directly. In a wide range of temperature the nonlinearity factor α (and the nonlinearity coefficient β as well) should be only slightly temperature dependent at low temperatures (see Eq.(30)) and should decrease on temperature above about 300 K (see Eq.(25)). This behaviour of $\beta(T)$ was found in the experiment [36].

It is possible to perform additional verification of Eq.(25). Combining Eq.(23) and Eq.(25) and using the slope of high-temperature part of curve $\beta = \beta(T^{-1})$ (19 K), $L = 20\ \mu m$, $E_1 = 2080\ V \cdot cm^{-1}$ for $(-d\varphi / dU)_2$ one can have 0.39 eV/V what is fairly close to the quantity 0.48 eV/V found by other authors for different varistor [27] and to the quantity 0.42 eV/V obtained from $\varphi(U)$ dependence for ZnO varistor structure with a single grain boundary [36].

It is worth to note that the varistor developed in [123] is characterized by the lowest value of $\alpha_1 = 5.1 \cdot 10^{-4}$ (Table 1). It means that the barrier height is only slightly decreased at low fields and, therefore, current in the pre-breakdown region is kept at quite low level. From the practical point of view, varistors should have the value of the nonlinearity factor α_1 as low as possible and the value of α_2 - as high as possible.

It is interesting to note that low-voltage varistors ([154,155]) have higher values of the nonlinearity factor α_2 than high-voltage varistors ([75,123,135]) though the nonlinearity coefficients of low-voltage varistors are lower (Table 1). The nonlinearity factor for ZnO-Bi_2O_3 is even higher than the α_2-values for ZnO varistors with high β-value but the nonlinearity coefficient for ZnO-Bi_2O_3 is only 6.7 (Table 1). Therefore, the nonlinearity coefficient reflects not only the nonlinearity of current-voltage characteristic but also the range of electric field, where the nonlinearity is observed.

Eqs. (3) or Eq.(16) were obtained in the assumption that the barrier height is decreased under applied electric field. To verify additionally this fundamental assumption, in [36] current-voltage characteristic of ZnO varistor structure with single grain boundary ($\beta = 25$; $U = 4\ V$ at $I = 10^{-3}\ A \cdot cm^{-2}$) was recorded at different fixed temperatures in the range 293-410 K. This varistor structure was obtained by fracturing of preliminary thinned sample of large-grain ZnO ceramic varistor with two electrodes. Then these data were plotted in a scale $\log I - 1/T$ to get the activation energy of electrical conduction E_a (conductivity σ in the range 350-410 K can be approximated by the expression $\sigma = \sigma_0 \exp(-E_a / kT)$ with constant value E_a). Then the decrease of the activation energy on voltage was obtained. An over-barrier activation took place in the range 350-410 K and the barrier height φ was only slightly less than the activation energy E_a, $\varphi \cong E_a$. Therefore, the barrier height in ZnO varistor is indeed decreased on voltage.

14.2. SnO_2 Varistors

To verify the conformity of $\sigma(E)$ dependence in Eq.(16) to the experimental data, current-voltage characteristic of SnO_2 varistors were presented in double logarithmic scale $\lg j - \lg E$ and in a scale $\lg \sigma - E$ [37]. It was seen that $\sigma(E)$ dependences at least are not in a contradiction to Eq.(16). As an example, in Fig.1 and in Fig.6 current-voltage characteristics of SnO_2–Bi_2O_3–Co_3O_4–Nb_2O_5–Cr_2O_3 ceramics obtained at different temperature are shown in both mentioned scales, respectively. Measurements were taken in air, therefore, at temperatures below 373 K behaviour of low-field conductivity is rather complicated (curve 2 in both figures) due to possible influence of relative humidity (see [66,69]).

In a scale $\lg \sigma - E$, dependence $\sigma(E)$ at room temperature consists of two parts (Fig.6): a low-field part with the nonlinearity factor $\alpha \approx 5\ 10^{-4}$ cm V^{-1} and a high-field part with $\alpha \approx 1.5\ 10^{-2}$ cm V^{-1} . The experimental points (at highly nonlinear high-field region) are distributed quite well along the straight line for more than about four decades of conductivity (Fig.6).

The existence of two mentioned parts of $\sigma(E)$ dependence is related to two slopes of the straight lines which can approximate $\varphi(U)$ dependence known for ZnO varistor [27]. For SnO_2 varistor such data are available in [37]. By comparing the high-field non-linearity factor obtained from the experiment with its possible theoretical estimation ($\alpha_{cal} \approx 7 \cdot 10^{-3}\ cm \cdot V^{-1}$), one can see that calculations give at least the same order of its magnitude. Eq.(16) gives reasonable approximation of $\sigma(E)$ dependence for SnO_2 varistors.

In Table 1, the parameters β , α and β_E for the characterization of the nonlinearity of current-voltage characteristic are shown. There is reasonable conformity between α and β_E. It means that the normalized nonlinearity coefficient β_E is the physically grounded parameter too.

On the basis of obtained results it is interesting to compare non-Ohmic properties of SnO_2 varistors having different values of the electric field E_1 at fixed current density. In Table 1, the parameters β , E_1 and β_E are presented for some SnO_2 varistors reported in the literature. It is obvious that SnO_2-Bi_2O_3-Co_3O_4-Nb_2O_5-Cr_2O_3 varistor studied in [37,66,69] has practically the highest value of the normalized nonlinearity coefficient β_E among those reported in the literature. Some slight increase of the normalized nonlinearity coefficient β_E has been reached in [67] using addition of B_2O_3 to the composition SnO_2-Bi_2O_3-Co_3O_4-Nb_2O_5-Cr_2O_3.

It is necessary to stress that even varistors with higher nonlinearity coefficients β=72 [62] and 142 [65] but with very high values of the electric field E_1 have some lower values of β_E (Table 1). In the frames of presented model, SnO_2-Bi_2O_3-Co_3O_4-Nb_2O_5-Cr_2O_3 varistor [37,66,69] exhibits higher nonlinearity of current-voltage characteristic than materials obtained in [62,65].

Consider temperature shift of current-voltage characteristic (Fig.6). It is seen that obtained experimental data at the high field region can be satisfactorily approximated by the straight lines and therefore $j(E)$ dependence at different temperatures can be described by Eq.(3). The slope of these lines is gradually decreased with temperature. It means the lowering of the nonlinearity factor α with temperature. Indeed, according to Eq.(25), the nonlinearity factor α must decrease with temperature as $\alpha \propto 1/T$. Obtained temperature dependence of the nonlinearity factor α shows fairly good linear relationship between α and $1/T$ [37]. Therefore, $\alpha(T)$ dependence predicted in Eq.(25) is correct.

From Eq.(27) it is clear that increase of the zero-voltage barrier height φ_0 causes the growth of the nonlinearity factor α . Indeed ceramics with the highest activation energy E_σ of the low-field electrical conduction (with Cr_2O_3 addition) have the highest nonlinearity coefficient [37]. The addition of Cr not only leads to the increase of the barrier height φ_0 but causes higher sensitivity $d\varphi/dU$ of the barrier height to voltage and, therefore, gives according to Eq.(25) higher nonlinearity factor $\alpha \propto d\varphi/dU$.

Data on current-voltage characteristic of different SnO_2 varistors obtained by other researchers as well can be linearized in the scale $\ln\sigma - E$.

Conclusions

Varistor effect (decrease of resistance on voltage independently on voltage polarity) observed in tin dioxide based and zinc oxide based ceramics is related to the decrease of the barrier height of grain-boundary barriers with increase of the electric field. The nature of varistor effect is the same in these materials. In spite of the reported values of the nonlinearity coefficient in SnO_2 varistors are comparable to the nonlinearity coefficients in ZnO varistors, modern tin dioxide varistors exhibit some lacks in comparison with zinc oxide varistors, for example:

i) very high values of the nonlinearity coefficient (more than about 60) reported for SnO_2 varistors in some publications are obtained only in dc regime and, therefore, they needed to be reliably confirmed to exclude possible influence of Joule heating;
ii) practical absence of published pulse current-voltage characteristics of SnO_2 non-Ohmic ceramics can certify about poor usability of these materials as a varistors for protection of equipment from voltage surges;
iii) strong humidity sensitivity of low-field electrical conductivity in SnO_2 varistors should be considered as non-desirable feature of varistor material and respective encapsulation would be necessary;
iv) not high reproducibility of current-voltage characteristics and not high producibility of SnO_2 varistor ceramics due to necessity of low-level additions;
v) impossibility at present time to obtain substantial SnO_2 grain growth for a variation of electric field (at fixed current density) to lower values.

From fairly general assumptions about the nature of the non-Ohmic conduction in SnO_2 and ZnO varistor ceramics (decrease of the barrier height with increase of the electric field) the relationship $j = \sigma_0 E \exp(\alpha E)$ between the current density j and the electric field E is obtained. This expression can be used for the approximation of current-voltage characteristic in ZnO and SnO_2 varistors. Estimated numerical value of the nonlinearity factor α is in agreement with experimental data.

It is shown that the nonlinearity factor α is proportional to the rate of a change of the barrier height on voltage (the absolute value of the derivative $d\varphi / dU$). This result gives simple and clear explanation of varistor effect in oxide ceramics.

The normalized nonlinearity coefficient $\beta_E = (\beta / E_1) \cong \alpha$ is introduced and grounded. The nonlinearity factor α or the normalized nonlinearity coefficient β_E can be used instead of traditional but empirical nonlinearity coefficient β for the estimation of the nonlinearity of current-voltage characteristic of a varistor and for a comparison of the nonlinearity of varistors having different values of the electric field E_1 (at fixed current density).

In spite of the simplicity, the suggested model gives useful quantitative results confirmed by experimental data: the expression for the approximation of $j(E)$ characteristic of a varistor, the expression for the nonlinearity factor α showing the dependence of α on temperature and on the rate of a change of the barrier height on voltage, the approximate expression for the normalized nonlinearity coefficient β_E allowing easily compare the nonlinearity of current-voltage characteristic of varistors with different values of the electric field at fixed current density.

Acknowledgments

It is my great pleasure to acknowledge my co-authors and colleagues with whom we discussed different physical and technological aspects of non-Ohmic oxide ceramics, in particular, Dr. A.M. Chack, who many years ago has introduced me in this research area, Dr. B.K. Chernyi, Dr. A.Ya. Yakunin, Dr. A.I. Ivon, Prof. I.M. Chernenko, Prof. A.S.

Tonkoshkur, Prof. C.A. Hogarth, Dr. R. Bulpett, Prof. E. Traversa, Dr. E. Di Bartolomeo, Dr. G. Behr, S.V. Mazurik, Dr. A.N. Bondarchuk, I.A. Skuratovsky, A.V. Gaponov.

The support from the National Science and Technology Council (CONACYT), Mexico under the project SEP-2003-C02-42821 is gratefully acknowledged.

References

[1] Matsuoka, M. In *Advances in Varistor Technology.* Levinson, L.M.; Ed,; *Ceramic Transactions*, Am Ceram Soc: Westerville, OH, 1989; Vol.3, pp 3-9.

[2] Philipp, H. R.; Levinson, L. M. In *Additives and interfaces in electronic ceramics. Advances in ceramics,* 1983; Vol.7, pp 1-21.

[3] Valeev, Kh.S.; Kvaskov, V.B. Nonlinear metal-oxide semiconductors. Energiya, Moscow, 1983, pp 1-160.

[4] Einzinger, R. In *Polycrystalline semiconductors. Physical properties and applications.* Harbeke, G.; Ed.; Springer-Verlag: Berlin, Heidelberg, New York, Tokyo, 1985; pp 228-244.

[5] Levinson, L. M.; Philipp, H. R. *Am Ceram Soc Bull.* 1986, 65, 639-646.

[6] Einzinger, R. In *Amorphous and polycrystalline semiconductors.* Heywang, W.; Ed.; Mir: Moscow, 1987; pp 123-147.

[7] Kvaskov, V.B. Semiconductor devices with bipolar conduction. Energoatomizdat, Moscow, 1988, pp 1-128.

[8] Greuter, F.; Blatter, G.; Rosinelly, M.: Stucky, F. In *Advances in Varistor Technology.* Levinson, L.M.; Ed,; *Ceramic Transactions*, Am Ceram Soc: Westerville, OH, 1989; Vol.3, pp 31-53.

[9] Glot, A.B. Non-Ohmic oxide ceramics, Dniepropetrovsk State University: Dniepropetrovsk. 1989, pp 1-60.

[10] Gupta, T. K. *J Am Ceram Soc.* 1990, 73, 1817-1840.

[11] Glot, A.B.; Ivon, A.I.; Tonkoshkur, A.S.; Chernenko I.M. Oxide ceramic varistors. Dniepropetrovsk State University: Dniepropetrovsk. 1998, pp 1-140.

[12] Clarke, D. R. *J Am Ceram Soc.* 1999, 82, 485-502.

[13] Chernyi, B.K. In *Physical and technological problems of cybernetics*. Kiev, Ukraine, 1966; Vol.2, pp 66-91.

[14] Chernyi, B.K. In *Materials and Devices of Radioelectronics*. Dniepropetrovsk University: Dniepropetrovsk, Ukraine, 1968; pp 66-75.

[15] Chernyi, B.K. PhD thesis. Dniepropetrovsk State University: Dniepropetrovsk, Ukraine, 1968.

[16] Glot, A. B.; Chakk, A. M.; Chernyi, B. K.; Yakunin, A. Ya. *Inorganic Mater.* 1974, 10, 1866-1868.

[17] Yakunin, A. Ya.; Chernyi, B. K.; Chakk, A. M.; Glot, A. B. *Inorganic Mater.* 1976, 12, 803-805.

[18] Glot, A. B.; Chernyi, B. K.; Yakunin, A. Ya. *Inorganic Mater.* 1977, 13, 1316-1318.

[19] Glot, A. B. *Inorganic Mater.* 1984, 20, 1522-1523.

[20] Glot, A. B.; Zlobin, A. P. *Inorganic Mater.* 1989, 25, 274-276.

[21] Glot, A.B.; Proshkin, Yu.N.; Nadzhafzade, A.M. In *Ceramics Today –Tomorrow's Ceramics, Materials science monographs,* 66C, Vincenzini, P.; Ed.; Elsevier: 1991; pp 2171-2180.
[22] Nadzhafzade, A. M.; Glot, A. B. *Inorganic Mater.* 1994, 30, 1382-1384.
[23] Kale, A. S.; Seal, S.; Date, S. K.; Santhosh, P. N.; Barve, R. N. *J Vac Sci Techn, A: Vac, Surf, Films*. 1999, 17, 1196-1200.
[24] Bueno, P. R.; Oliveira, M. M.; Cassia-Santos, M. R.; Longo, E.; Tebcherani, S. M.; Varela, J. A. Ceramica (Sao Paulo), 2000, 46, 124-130.
[25] Cássia-Santos, M. R; Sousa, V. C; Oliveira, M. M.; Sensato, F. R.; Bacelar, W. K.; Gomes, J. W.; Longo, E.; Leite, E. R.; Varela, J. A. *Mater Chem Phys.* 2005, 90, 1-9.
[26] Emtage, P. R. *J Appl Phys.* 1977, 48, 4372-4384.
[27] Mahan, G. D.; Levinson, L. M.; Philipp H. R. *J Appl Phys*. 1979, 50, 2799-2812.
[28] Hower, P. L.; Gupta, T.K. *J Appl Phys*. 1979, 50, 4847-4855.
[29] Tonkoshkur, A. S. *Solid State Phys.* 1990, 32, 2260-2263.
[30] Manuel, P. *Rev Phys Appl*. 1987, 22, 971-983.
[31] Tonkoshkur, A. S. *Electronnaya Tekhnika. Ser.5. Radiodetali i Radiocomponenty.* 1991, No.2(83), 15-19.
[32] Ivon, A.I.; Glot, A. B.; Chernenko, I.M. In *Fourth Euro-Ceramics. Electroceramics*, Gusmano, G.; Traversa, E.; Ed.; Gruppo Editoriale Faenza Editrice: Faenza, Italy, 1995; Vol.5. pp 475-482.
[33] Bartkowiak, M.; Mahan, G. D.; Modine, F. A.; Alim, M. A.; Lauf, R.; McMillan, A. *J Appl Phys*. 1996, 80, 6516-6522.
[34] Levine, J. D. *CRC Crit Revs Solid State Sci.* 1975, 5, 597-608.
[35] Glot, A. B. *J Mater Sci*. 2006, 41, 5709-5711.
[36] Glot, A. B. *J Mater Sci. Mater in Electronics*. 2006, 17, 755-765.
[37] Glot, A. B.; Skuratovsky, I. A. *Mater Chem Phys.* 2006, 99, 487-493.
[38] Valeev, Kh. S.; Mashkovich, M. D. *Zhurnal Tekhnich Fiz.* 1957, 27, 1649-1651.
[39] Kosman, M. S., Pettsold, E.G. *Uchenye Zapiski Leningr Gos Pedagogich Inst im A.I. Gertsena.* 1961, 207, 191-197.
[40] Valeev, Kh. S.; Knyazev, V. A.; Drozdov, N. G. Elektrichestvo. 1964, 4, 72-76.
[41] Chakk, A.M. PhD thesis. Dniepropetrovsk State University: Dniepropetrovsk, Ukraine, 1972.
[42] Yan, J. F., Rodes, W. W. *Appl Phys Lett.* 1982, 40, 536-537.
[43] Matsuoka, M.; Masuyama, T.; Iida, Y. *Suppl J Jap Soc Appl Phys.* 1970, 39, 94-101.
[44] Matsuoka, M. *Jap J Appl Phys*. 1971, 10, 736-746.
[45] Mukae, K.; Tsuda, K.; Nagasawa, I. *Jap J Appl Phys*. 1977, 16, 1361-1368.
[46] Einzinger, R. *Ber Dt Keram Ges*. 1975, 52, 244-246.
[47] Morris, W.G. *J Vac Sci Techn.* 1976, 13, 926-931.
[48] Bernasconi, J.; Strassler, S.; Knecht, B.; Klein, H. P.; Menth, A. *Solid State Communs.* 1977, 21, 867-870.
[49] Einzinger, R. *Appl Surf Sci.* 1978, 1, 329-340.
[50] Clarke, D.R. *J Appl Phys*. 1978, 49, 2407-2411.
[51] Glot, A.B. In *Proc Nat Conf Phys dielectr and new areas of their applications. Electrophys process in multicomponent dielectrics;* Tonkonogov, M.P.; Ed.; Karaganda Politechnical Institute: Karaganda, 1978; pp 88-89.
[52] van Kemenade, J. T. C.; Eijnthoven, R. K. *J Appl Phys*. 1979, 50, 938-941.

[53] Einzinger, R. *Appl Surf Sci.* 1979, 3, 390-408.
[54] Glot, A.B.; Tonkoshkur, A.S.; Chernyi, B.K.; Yakunin, A.Ya. *Electronnaya Tekhnika. Ser.5. Radiodetali i Radiocomponenty.* 1979, No.4(35), 63-67.
[55] Glot, A.B. PhD thesis. Dniepropetrovsk State University: Dniepropetrovsk, Ukraine, 1979.
[56] Senos, A. M. R.; Baptista, J.L. *J Mat Sci Lett.* 1984, 3, 213-216.
[57] Glot, A.B.; Nadzhafzade, A.M., Ivon, A.I. In *Proc. 4th Int. Conf. on Electronic Ceramics and Applications.* Augustinus Buch handlung: Aachen, Germany, 1994; Vol.2, pp 797-800.
[58] Glot, A.B. *Electronnaya Tekhnika. Ser.5. Radiodetali i Radiocomponenty.* 1982, No.3(48), 16-18.
[59] Glot, A.B. In *Advances in Varistor Technology.* Levinson, L.M.; Ed,; *Ceramic Transactions*, Am Ceram Soc: Westerville, OH, 1989; Vol.3, pp 194-203.
[60] Glot, A.B. *Dielectr Semicond.* 1988, No.33, 69-73.
[61] Pianaro, S. A.; Bueno, P. R.; Olivi, P.; Longo, E.; Varela, J.A. *J Mat Sci Lett.* 1997, 16, 634-638.
[62] Santosh, P. N.; Potdar, H. S.; Date, S. K. *J Mat Res.* 1997, 12, 326-328.
[63] Pianaro, S. A.; Bueno, P. R.; Longo, E.; Varela, J.A. *Ceram Intern.* 1999, 25, 1-6.
[64] Bueno, P. R.; de Cassia-Santos, M. R.; Leite, E.R.; Longo, E.; Bisquert, J.; Garcia-Belmonte, G.; Fabregat-Santiago, F. *J Appl Phys.* 2000, 88, 6545-6548.
[65] Bueno, P. R.; Oliveira, M. M.; Bacelar-Junior, W. K.; Leite, E. R.; Longo, E.; Garcia-Belmonte, G Bisquert, J. *J Appl Phys.* 2002, 91, 6007-6014.
[66] Skuratovsky, I.; Glot, A.; Di Bartolomeo, E.; Traversa, E.; Polini, R. *J Eur Ceram Soc.* 2004, 24, 2597-2604.
[67] Skuratovsky, I. A.; Glot, A. B. *Func Mater*, 2003, 10, 314-316.
[68] Glot, A.B.; Bulpett, R.; Nadzhafzade, A. M.; Skuratovsky, I. A. *Photoelectronics.* 2001,No.10, 47-49.
[69] Skuratovsky, I.; Glot, A.; Traversa, E. *Mater Sci Eng B*, 2006, 128, 130-137.
[70] Egashira, M.; Shimizu, Y.; Fukuyama, Y.; Takao, Y. *Sens. Actuators B.* 1996, 33, 89-95.
[71] Egashira, M.; Shimizu, Y.; Takao, Y.; Sako, S. *Sens. Actuators B.* 1996, 35-36, 62-67.
[72] Wada, K.; Egashira, M. *Sens. Actuators B.* 1996, 62, 211-219.
[73] Bueno, P. R.; Leite, E. R.; Oliveira, M. M.; Orlandi, M.O.; Longo, E. *Appl Phys Lett.* 2001, 79, 48-50.
[74] Oliveira, M. M.; Bueno, P. R.; Longo, E.; Varela, J. A. *Mat Chem Phys.* 2002, 74, 150-153.
[75] Levinson, L.M.; Philipp, H. R. *J Appl Phys.* 46 (1975) 1332-1341.
[76] Fix, M.; Soln, J. *Appl Phys Lett.* 1975, 26, 519-521.
[77] Yakunin, A.Ya.; Glot, A. B. *Russ Phys J.* 1976, 19, 646-647.
[78] Glot, A. B. *Russ Phys J.* 1977, 20, 105-106.
[79] Glot, A. B.; Odnorozhenko, V. B.; Chernyi, B. K.; Yakunin, A. Ya. *Russ Phys J.* 1977, 20, 621 – 623.
[80] Kurchatov, I. W.; Kostina, T. Z.; Rusinov, L. J. *Phys Zs Sowjetunion.* 1935, 7, 192-197.
[81] Glot, A.B. In *Interaction of elementary particles.* Dniepropetrovsk University: Dniepropetrovsk, Ukraine, 1976; pp 106-108.
[82] Frohlich, H. *Proc Roy Soc.* 1947, A188,521-529.

[83] Glot, A.B.; Firsin, S.V.; Yakunin, A.Ya. Izvestiya Vysshikh Uchebnykh Zavedenii Fizika, 1981, No.5, 101-102.

[84] Rossinelli, M.; Blatter, G.; Greuter, F., In *Electrical Ceramics. British Ceramic Proc.* No.36, Steel. B.C.H.; Ed.; Inst Ceramics: Shelton, Stoke-on-Trent, Staffs., U.K., 1985; pp 1-17.

[85] Pike, G. E.; Kurtz, S. R.; Gourley, P. L.; Philipp, H. R.; Levinson, L. M. *J Appl Phys.* 1985, 57, 5512-5518.

[86] Wong, J. *J Appl Phys*. 1975, 46, 1653-1659.

[87] Wong, J.; Rao, P.; Koch, E. F. *J Appl Phys*. 1975, 46, 1827-1830.

[88] Inada, M. *Jap J Appl Phys.* 1978, 17, 1-10.

[89] Inada, M. *Jap J Appl Phys.* 1980, 19, 409-419.

[90] Avdeenko, B. K.; Katkov, V. F.; Kushnereva, A. K.; Makarov, V. O.; Chernenko, I. M. *Inorganic Mater.* 1985, 21, 1064-1065.

[91] Katkov, V. F.; Ivon, A. I.; Makarov, V. O.; Chernenko, I. M. *Inorganic Mater.* 1988, 24, 1358-1363.

[92] Makarov, V. O.; Tonkoshkur, A. S.; Chernenko, I. M. *Inorganic Mater.* 1987, 23, 2016-2020.

[93] Glot, A. B.; Hogarth, C. A.; Bulpett, R. *Int J Electronics*. 1988, 65, 797-804.

[94] Avdeenko, B. K.; Glot, A. B.; Chernenko, I. M.; Schelokov, A. I. *Inorganic Mater.* 1981, 17, 378-379.

[95] Stucki, F.; Brüesch, P.; Greuter, F. *Surf Sci.* 1987, 189/190, 294-299.

[96] Glot, A. B. *Inorganic Mater.* 1987, 23, 155-157.

[97] Glot, A. B.; Hogarth, C. A. *Int J Electronics*. 1985, 58, 117-121.

[98] Avdeenko, B. K.; Ivon, A. I.; Schelokov, A.I. *Inorganic Mater*. 1978, 14, 1162-1163.

[99] Mukae, K.; Tsuda, K.; Nagasawa, I. *J Appl Phys*. 1979, 50, 4475-4476.

[100] Tonkoshkur, A. S. *Ukr Phys J.* 1978, 23, 2030-2038.

[101] Avdeenko, B. K.; Ivon, A. I.; Chernenko, I. M.; Schelokov, A. I. *Inorganic Mater.* 1981, 17, 1098-1101.

[102] Chopra, K. L.; Bhushan, B.; Kashyap, S. C. *J Appl Phys*. 1983, 54, 1610-1612.

[103] Bernds, A.; Löhnert, K.; Kubalek, F. *J de Phys*. 1984, 45, C2-861-C2-864.

[104] Glot, A. B. *Electronnaya Tekhnika. Ser.5. Radiodetali i Radiocomponenty.* 1983, No.3(52), 9-11.

[105] Jarzebski, Z. M.; Marton, J. P. *J Electrochem Soc.* 1976, 123, 299C-310C.

[106] Jarzebski, Z. M.; Marton, J. P. *J Electrochem Soc.* 1976, 123, 333C-346C.

[107] Erickson, J. W.; Fryberger, T. B.; Semancik, S. *J Vac Sci Technol A*, 1988, 6, 1593-1598.

[108] Varma, S.; Rao, K.V.; Kar, S. *J Appl Phys*. 1984, 56, 2812-2822.

[109] Rantala, T. S.; Lantto, V.; Rantala, T. T. *J Appl Phys*. 1996, 79, 9206-9212.

[110] Glot, A.B.; Perepelitsa, Yu.A. *Electronnaya Tekhnika. Ser.5. Radiodetali i Radiocomponenty.* 1988, No.2(71), 35-37.

[111] Fernandez-Hevia, D.; de Frutos, J.; Caballero, A. C.; Fernandez, J. F. *J Appl Phys.* 2002, 92, 2890-2898.

[112] Glot, A.B.; Dudko, A.A. *Deposited in VINITI (Moscow) 11.08.88. No.6490-B88.* Abstract: Izvestiya Vysshikh Uchebnykh Zavedenii Fizika, 1989, No.3, 126.

[113] Glot A.B. *Dielectr Semicond.* 1989, No.35, 57-62.

[114] Ivon, A. I.; Glot, A.B.; Mozharovsky, L.A.; Chernenko, I. M. *Inorganic Mater*. 1998, 34, 1285-1290.

[115] Katkov, V.F.; Glot, A.B.; Ivon, A. I. *Inorganic Mater*. 1989, 25, 464-466.

[116] Vasconcelos, J. S.; Vasconcelos, N. S. L. S.; Orlandi, M. O.; Bueno, P. R.; Varela, J. A.; Longo, E.; Barrado, C. M.; Leite, E. R. *Appl Phys Lett*. 2006, 89, 152102-1-152102-3.

[117] Glot, A.B.; Bulpett, R.; Ivon, A. I.; Hogarth, C.A. In *Fourth Euro-Ceramics. Electroceramics*, Gusmano, G.; Traversa, E.; Ed.; Gruppo Editoriale Faenza Editrice: Faenza, Italy, 1995; Vol.5. pp 483-490.

[118] Avdeenko, B.K.; Makarov, V.O.; Chack, A.M.; Chernenko, I.M. *Russ Phys J*. 1981, 24, 109-111.

[119] Glot, A.B. In *Electroceramics 5. Proc. Int.Conf.on Electronic Ceramics and Applications Aveiro, Portugal, 2-4 Sept. 1996*. Baptista, J.L.; Labrincha, J.A.; Vilarinho, P.M.; Ed.; Aveiro, 1996; Vol.2, pp 553-556.

[120] Ivon, A.I.; Florov, A.K; Glot, A.B. *Key Eng Mater*. 1997, 132-136, 1297-1300.

[121] Wen, Q.; Clarke, D. R. In *Grain Boundaries and Interfacial Phenomena in Electronic Ceramics*. Levinson, L.M.; Shin-ichi Hirano, Ed.; *Ceramic Transactions*. Am Ceram Soc: Westerville, OH, 1994; Vol.41, pp 217-230.

[122] Abelard, P.; Smith, D.S.; Baumard, J.F. In *Proc 14th Int. Conf., Canterbury, Sept.7-9,1987. Sci. Ceram. 14:* Stoke-on-Trent,1988; pp 939-944.

[123] Greuter, F.; Christen, T.; Glatz-Reichenbach, J. In *Mat Res Soc Symp Proc*. Material Research Society: 1998; Vol.500 pp 235-246.

[124] Levinson, L.M.; Philipp, H.R. *J Appl Phys*. 1976, 47, 3116-3121.

[125] Ivon, A. I.; Glot, A. B.; Chernenko, I.M. In *Electroceramics 5. Proc. Int.Conf.on Electronic Ceramics and Applications Aveiro, Portugal, 2-4 Sept. 1996*. Baptista, J.L.; Labrincha, J.A.; Vilarinho, P.M.; Ed.; Aveiro, 1996; Vol.2, pp 573-575.

[126] Ivon, A. I.; Glot, A. B.; Gaponov, A. V.; Mazurik, S. V. *Key Eng Mater*. 1997, 132-136, 1289-1292.

[127] Ivon, A. I. *Inorganic Mater*. 2000, 36, 1074-1077.

[128] Glot, A. B. *Sov Phys Solid State*. 1983,25, 121-123.

[129] Vanadamme, L.K.J.; Brugman, J.C. *J Appl Phys*. 1980, 51, 4240-4244.

[130] Eda, K. *J Appl Phys*. 1978, 49, 2964-2972.

[131] Pike, G.E. In *Grain Boundaries in Semiconductors. Proc.Mater. Res. Soc.Ann.Meet.*, ed. Pike, G.E.; Seager, C.H.; Leamy, H.J.; Ed.; Elsevier: 1982, pp 369-379.

[132] Blatter, G.; Greuter, F.; *Phys Rev B,* 1986, 34, 8555-8572.

[133] Kim, M.; Oh, H.; Kim, C. *Jap J Appl Phys*. 1991, 30, L1917-L1920.

[134] Avdeenko, B. K.; Ivon, A. I.; Glot, A. B.; Chernenko, I. M.; Shchelokov, A. I. *Russ Phys J,* 1978, 21, 828-840 (Abstracts of articles deposited at VINITI).

[135] Avdeenko, B. K.; Glot, A. B.; Ivon, A. I.; Chernenko, I.M.; Schelokov, A.I. *Inorganic Mater*. 1980, 16, 1310-1311.

[136] Karachentsev, A.Ya.; Potashev, Yu.N. *Electronnaya Tekhnika. Ser.5. Radiodetali i Radiocomponenty*. 1973, No.2(31), 95-96.

[137] Karachentsev, A.Ya.; Potashev, Yu.N. *Electronnaya Tekhnika. Ser.5. Radiodetali i Radiocomponenty*. 1973, No.4(33), 37-43.

[138] Miralles, A.; Cornet, A.; Morante, J.R. Semicond Sci Techn. 1986, 1, 230-233.

[139] Brückner, W.; Bäther, K.H.; Moldenhauer, W.; Wolf, M.; Lange, F. *Phys Stat Sol (a).* 1980, 59, K1-K4.

[140] Pike, G.E.; Seager, C.H.; Dosch. R.G. In *Mat Res Soc Symp Proc.* Mat Res Soc: 1985, Vol.46, pp 409-414.

[141] Fliegel, W.; Behr, G.; Werner, J.; Krabbes, G. *Sensors and Actuators B,* 1994, 18-19, 474-477.

[142] Vainshtein, V.M.; Fistul, V.I. In *Sums of Science and Technology, Electronics and its Application,* VINITI: Moscow, 1973; Vol.4, pp 108-152.

[143] S.R. Dhage, V. Ravi, *Appl Phys Lett.* 2003, 83, 4539-4541.

[144] Escribano, P.; Guillem, M.C.; Guillem, C.; *Trans J Br Ceram Soc.* 1983, 82, 208-210.

[145] Shklovskii, B.I. *Sov Phys Semicond* 1979, 13, 93-97.

[146] Shklovskii, B.I.; Efros, A.L. Electronic properties of doped semiconductors, Solid State Sci. 45, Springer, 1984.

[147] Vinnikov, A.Ya.; Meshkov, A.M.; Savushkin. V.N. *JETF Pisjma,* 1980, 6, 726-729.

[148] Vinnikov, A.Ya.; Meshkov, A.M.; Savushkin. V.N. *Solid State Phys.* 1980, 22, 2989-2995.

[149] Vinnikov, A.Ya.; Meshkov, A.M.; Savushkin. V.N. *Solid State Phys.* 1982, 24, 1352-1359.

[150] Goldman, E.I.; Zhdan, A.G. *Sov Phys Semicond.* 1976, 10, 1839-1845.

[151] Pike, G.E.; Seager, C.H. *Appl Phys Lett.* 1979, 50, 3414-3422.

[152] Goldman, E.I. *Sov Phys Semicond.* 1987, 21, 95-100.

[153] Glot, A.B.; Mazurik, S.V. *Inorganic Mater.* 2000, 36, 636-639.

[154] Trontelj, M.; Kolar, D.; Krasevec, V. In *Additives and interfaces in electronic ceramics. Advances in ceramics.* 1983; Vol.7, pp 107-116.

[155] Bowen, L.J.; F.J. Avella. *J Appl Phys.* 1983, 54, 2764-2766.

[156] Taylor, W.E.; Odell, N.H.; Fan, H.Y. *Phys Rev.* 1952, 88, 867-875.

[157] Baptista, J.L.; Mantas, P.Q. *J Electroceram.* 2000, 4, 215-221.

[158] Seager, C.H.; Pike, G.E. *Appl Phys Lett.* 1982, 40, 471-474.

In: Ceramic Materials Research Trends
Editor: Paul B. Lin, pp. 275-323
ISBN: 1-60021-769-9

Chapter 11

IMMOBILIZATION OF CHROMIUM ION FROM ASHES GENERATED BY INCINERATION OF TANNED LEATHER SHAVINGS BY MEANS OF VITRIFIED CERAMIC BODIES

T. Basegio**, A.P.B. Leão, D.L.Villanova, J.M. Hohemberger, A.M. Bernardes, L.G. Ionescu and C.P. Bergmann

PPGEM- Programa de Pós-Graduação em Engenharia de Minas, Metalúrgica e de Materiais, Escola de Engenharia, Universidade Federal do Rio Grande do Sul, LACER - Laboratório de Materiais Cerâmicos, Av. Osvaldo Aranha, 99/705, 90035-190 Porto Alegre, RS, Brazil

Abstract

The immobilization of chromium ion in ceramic bodies produced from ashes generated by the incineration of chromium tanned leather shavings (ACS) and using waste of soda-lime glass as raw material is discussed.

The raw materials, leather shavings and soda-lime glass, were mixed in different proportions. In order to investigate the phenomena acting in the chromium ion immobilization process, pure oxides - Na_2O, TiO_2, MgO and CaO – were mixed with soda-lime glass or added to ceramic mixtures with ASC and soda-lime glass. These samples were dry-pressed and fired in an electrical furnace, at temperatures of 750, 800, 950 and 1000°C. The ceramic bodies were evaluated with respect to physical properties and mineralogical composition by x-ray diffraction and EDS mapping. The chromium immobilization was characterized by leaching tests according Brazilian Regulations (NBR 10005).

The higher the firing temperature, the more effective the chromium immobilization in the ceramic bodies investigated. Raising the firing temperature diminishes the open porosity, by glass phase formation, leading to a lower content of chromium in the leaching water. However, when adding only glass to the ACS, the content of chromium in the leaching water increased. The lixiviation decreased when higher temperatures were used. This occurred because there are more leaching phases, such as, sodium chromates, originated from soda-

* E-mail address: basegio@ufrgs.br. Tel.: +55 51 33163637; fax: +55 51 33163405.(Corresponding author)

lime glass and ACS reactions. All ceramic bodies containing sodium chromate investigated were inefficient in immobilizing the chromium ion.

The results confirmed the immobilization of the chromium ion within the allowed limits of Brazilian regulations, NBR 10004 (5 mg/L), by means of incorporation in glass and glazing/densification agents such as titanium and magnesium oxides.

Keywords: Chromiun tanned shavings; hazardous wastes; immobilization of chromium; recycling; vitreous ceramics.

1. Introduction

The industrial development of the last two centuries also led to serious environmental problems. The preoccupation and concern with the environment resulted in the adoption of systematic treatment processes for solid, liquid and gaseous waste during industrial production, in spite of the fact that the preservation of the environment and ecology are sometimes still considered as additional production costs and expenditure.

Undoubtedly, at the present time there is a general awareness related to the environment and inspection, and law enforcement by environmental protection agencies throughout the world has become wide spread. The exportation of pollution is no longer permitted by means of environmentally unsafe technology or by the exportation of ecologically aggressive products. Motivated by such thoughts and vision, the industry seeks new routes to control and reduce environmentally unsafe procedures by developing new appropriate technology aimed to eliminate/or treat industrial residues in order to minimize risks to man and the environment.

The leather industry is inserted into this context. The major leather production centers in the world are found in Mexico, Brazil, Japan, South Korea, China, India and Pakistan. Korea, Japan and Italy import hides from countries which have a large meat production industry i.e. USA, Australia, and the European countries, whereas the South American countries, for example Argentina and Brazil, process their own hides. The leather industry is prevalent in Brazil, especially in the South, in the State Rio Grande do Sul, where more than 50% of Brazilian leather is produced [1].

According to European Commission (EC) the quantities of solid waste produced by tanneries depend on the type of leather processed, the source of hides and skins and the techniques applied. On the average at the end of the process about 20% of the weight of the raw hides is (grain side) leather [2]. On the other hand, in Rio Grande do Sul, approximately 40% of the initial raw material is transformed into solid and liquid wastes [3].

In the tanning industry raw skin is transformed into leather by means of a series of chemical and mechanical operations [4, 5]. Chromium salts (in particular chromium sulfate) are the most widely used tanning substances today. Hides that have been tanned with chromium salts have a good mechanical resistance, an extraordinary dyeing suitability and a better hydrothermal resistance in comparison with hides treated with plant substances. Chromium salts also have a high rate of penetration into the interfibrillar spaces of the skins. This results in savings in terms of production time and better control of the process [6].

In Brazil approximately 90% leather industry uses chromium in hide processing, resulting hazardous chromium wastes. The American [7] and the Brazilian [8] regulations consider that chromium and chromium compounds are hazardous constituents in waste materials. The

German [9] regulations consider hazardous the wastes resulting from the leather and hide processing (group 14). On the other hand, the European Commission (EC) recently rejected a call by some member states of the EU to include tannery wastes containing chromium in the European Hazardous Waste List on the basis that the wastes did not possess the characteristics necessary for classification as a hazardous waste, except degreasing waste containing organic solvents without a liquid phase [2]. These wastes listing in the European Waste Catalogue [10] under group 04 01, Wastes from the leather and fur industry.

Disposal routes for trimmings, fleshings, splits, and shavings differ from tannery to tannery, and depend on the available recycling facilities and the quantity of waste generated. Furthermore, economic feasibility of recycling depends on the demand for the by-products and the cost/revenue from recycling compared to disposal routes such as landfill. For example, in the UK, a large proportion of tannery waste is landfilled because tanneries are dispersed across the country and, in comparison to landfilling, communal waste treatment plants would not be economic. The landfill cost of waste in the UK is significantly lower than in some other countries in Europe due to the wide availability of landfills. Landfilling of raw residues is not allowed in Germany. In Italy it is economically feasible to establish communal waste treatment plants because tanneries are situated together in large numbers (pools) [2]. In Brazil, such residues are disposed of in controlled and semi-controlled industrial residues landfills, maintained by the tanneries themselves.

Among these residues, the most problematic one is the chromium tanned leather sawdust, especially from the shaving operation, which is carried out in order to make the leather thickness uniform [3].

Since tanned hides are too thick for most purposes, they are split using a machine similar to a horizontal band saw. After splitting, the thickness of the hides must be uniform all over the hide. This is achieved with a shaving machine used in the process line. The helical shaped cutting blades level the overall thickness to meet exact specifications and open the fibber structure, in order to improve the response to subsequent chemical processing. This operation is called shaving and results in the removal of small pieces of leather with a consistency similar to very coarse sawdust. In the Federal State Rio Grande do Sul, it is estimated that each hide generates an average of 4 kg of shavings [10]. Since each tannery industry processes 1500 hides every day, the daily generation of shavings is about 6 tons [11] and these shavings must be disposed of.

In this context, considering the nature of this residue and the big amount that is generated, the need for alternatives to its disposal and/or application becomes very important. Recycling is highlighted as an excellent alternative. Besides avoiding soil and water contamination, it has the following advantages: there are no costs associated to controlled disposal of residues; helps in conservation of natural resources; and makes possible the production of lower cost products, since it can be used as a raw material.

Studies to minimize the amount of tannery waste, their treatment and possible exploitation of waste products are very important. According to Rivela et al. [12] the tanned wastes like shavings can be utilized with or without the presence of chromium. The direct application includes the manufacture of bonded leather, leather boards, fibrous sheets grafted with acrylates, insulators and building materials.

Considering the great volume of shavings generated, the direct use is not always possible. Therefore, to avoid the high cost of dumping, shavings can be processed to recover chromium, or thermally treated to reduce the volume to be disposed of. The issue associated with each one of these processes is the problems that they generate. One problem observed by tanners during the recycling of chromium by dissolution from wastes is the poor quality of the chromium due to the presence of organic lipolytic components, metals and other impurities [6]. The thermal treatment of wastes involves incineration, gasification and pyrolysis as a means of disposal, while also recovering energy from waste [8]. Fluidized bed combustion, when optimized and controlled, can be an alternative for treatment and recycling of this kind of residue, because it can reduce in 90-95% the residue amount, and the chromium-rich ash generated in this process can be used as raw material by different industrial segments.

Solidification is another process that has been considered as an alternative solution to the disposal of wastes containing heavy metals. In this process, waste materials are mixed with various binding media to obtain a new product with improved physical properties [13]. Different papers have reported the use of solidification, stabilization and vitrification processes to recycle different wastes such as low-level debris [14], radioactive wastes [15], red mud from aluminum extraction [16] and zinc hydrometallurgy [17], incineration ashes [18–19], steel plant fly ash [20,21,22], coal ash [23] and post-cupric-slag [24]. From these studies it can be seen that technologies involving the transformation of wastes into cement agglomerates, ceramics, glasses or glass–ceramics have acquired economical importance. Immobilization within a cement-matrix does not require expensive thermal treatments to transform hazardous waste into inert and stable residues [17].

Several authors [25–26] have studied the solidification of tannery wastes with cement. Filibeli et al. [27] have studied the solidification of tannery sludges with high chromium content. The sludge was mixed with cement and some additives as fly ash or sand. After 28 days there was a great reduction of chromium and sulphate concentrations, showing their immobilization in the cement phase. The final product is resistant to aggressive environmental agents and can be safely stored in landfill sites. On the other hand, a study of Wang and Vipulanandan [26] about chromium immobilization in cements showed that mixing K_2CrO_4 with cement in concentrations of 0.5, 2 and 5 wt% resulted in a process efficiency of 80%. The immobilization capacity of cement depends on the oxidation state of chromium. Cr3+ is more effectively incorporated, while Cr6+ is not immobilized in the hydrated phases of cement [28]. For this reason, this process is experiencing increasing opposition in many European countries, because the durability of cement mixtures is not proven and the cement-mix containing hazardous wastes is not reusable, especially in the building industry.

The adsorption of copper and lead ions onto regenerated sludge from a water treatment plant was studied by Chung-Hsin Wu and associates [29]. Water-treatment sludges were sintered under various experimental conditions to generate Al-containing adsorbentsThe adsorption of Cu^{+2} and Pb^{+2} onto sintered oxide was examined and the sorption results were modeled using a surface complex model. The triple layer model (TLM) simulation revealed that Cu^{+2} and Pb^{+2} sorption on sintered adsorbent involve inner-sphere interactions and that the dominant surface metal species are monodentate.

More recent studies invove the stabilization of primary chemical sludge from tannery using starved air combustion [30]. The sludge was subjected to flow-through column test to evaluate the concentration of leachable organucs (tannin, COD and TOC) and heavy metal ion Cr^{+3}, Fe^{+2} present in it. The dry sludge was incinerated at 800°C in an incinerator under starved oxygen supply (starved air combustion) to prevent conversion of Cr^{+3} to Cr^{+6}. The efficiency of then starved air combustion was studied under different loading rates of sludge.

The calcined sludge was solidified/stabilized using dry ash and Portland cement/gypsum.\the solidified bricks were tested for unconfined compressive strength and heavy metal leaching. The stabilization of chromium (III) in the cement gel matrix was confirmed with scanning electron microscopy (SEM), and X-ray energy dispersive spectroscopy (EDX). Leachability studies on the solidified bricks were also carried out to determine metal fixate.

We have studied environmental and technical aspects of the utilization of tannery sludge as a raw material for clay products [31]. The raw materials, tannery sludge and clay were dried mixed , ground and ceramic bodies were molded into bars using a hydraulic press., followed by drying and firing at 1000, 1100 and 1180°C. The mechanical properties were evaluated and were similar to those specified for ceramic bricks. Leaching tests showed that chromium was immobilized in the finished ceramic product, while zinc and chlorine were shown to be present in the gas emissions.

The production of materials with alumina and ashes from incineration of chromium tanned leather shavings was also studied. The raw materials, tannery ashes and alumina were mixed together in different proportions. The ceramic bodies were molded using a hydraulic press and fired with a heating rate of 100 K/h until 1400°C for 4 h in a muffle furnace. The ceramic specimens were characterized regarding physical, mechanical and thermal properties. Leaching tests, according to Brazilian, German and Dutch regulations, were performed on ceramic bodies containing different additions of ash. Results show that the ceramic materials produced are acceptable for refractory applications [32].

Electrostatic painting residues were investigated as an alternative raw material for the red clay industry. Different ceramic mass formulations were used and the ceramic bodies prepared were characterized in terms of physical and mechanical properties. Environmental compatibility was tested by the evaluation of gaseous emission and leaching and solubilization tests according to Brazilian standard regulations NBR 10.005 and NBR 10.006. The results showed that the ceramic materials containing the electrostatic painting residues were satisfactory in terms of the required standard regulations [33].

M. Campos et al used recovered waste as raw material for manufacturing sintered structural tiles. A mineralogical characterization by chemical analysis and x-ray diffraction and thermal analysis provided data for the sintering process. The properties of the sintered samples were checked by density, dimensional change, volume loss, mass decrement, three-point bending strength, water absorption and permeability and wear behavior. The experimental results indicated that the materials obtained from slate waste were suitable for the production of ceramic tiles [34].

Ceramic and glass technologies are considered to be the most versatile for rendering hazardous waste inert because they destroy organic matter, immobilize regulated heavy metals in a stable matrix and are able to convert complex chemical compositions into useful materials with potential for market exploitation [17].

According to Bernstein [35], vitrification technology is increasingly becoming the solution of choice to the problem of hazardous waste disposal containing heavy metals and/or organic pollutants. This can be done directly if the waste composition is suitable for vitrification; otherwise the batch composition can be adjusted using cheap raw materials, such as glass cullet as composition correctors. Vitrification technology permits the recovery or reuse of vitrified material, because the toxic substances are inertized due to the destruction of the organic portion during the process and to the immobilization of the inorganic portion that may be chemically bonded to the highly durable gas network.

Among the positive aspects of this inertization technology using vitrification, are process flexibility, (various types of wastes such as sediment and fly ashes can sometimes be treated without preliminary preparation); volume reduction (caused by mass fusion/vitrification), and the process cost that can be lower than their landfill disposal. In addition, vitrification is attractive from ecological point of view, elimination landfills and at the same time producing glass that can be reused, giving aggregated value to the waste.

The generic process to obtain ceramic materials aiming to immobilization of residues in vitrified structures (or structures with high amount of vitreous phase) is based on the incorporation of the residues in masses rich in silica and/or fluxes. Glasses and crystalline silicates lixiviate very slowly under the action of water; so, the obtained materials are generally considered to be safe regarding final disposal and do not require secondary contention.

L. Barbieri [36] and collaborators have studied the use of alkaline and alkaline-earth silicate glasses and glass ceramics for the immobilization of municipal and industrial wastes containig heavy metals. The properties that a glass should exhibit, such as good melting behavior, homogeneity, durability, among others, are closely related not only to chemical composition, but also to the kind and amount of raw materials used during the melting process. In this perspective, they suggest that it is important that wastes of different origin should be simultaneously submitted to vitrification, so that each of them may contribute to supply the appropriate quantities of vitrifying (SiO_2, Al_2O_3), melting (Na_2O, K_2O) and stabilizing (CaO, MgO, ZnO, PbO) agents in the final glass and so suitable and desirable physical-chemical features can be obtained. The vitrification process simulates what is happening in nature: natural glasses produced by volcanic eruptions, such as obsidians (Na2O-K2O-Al_2O_3-SiO_2) and basaltic (CaO-MgO-Al_2O_3-SiO_2) rocks are basically inert materials, even when they contain toxic elements, because they are embedded in an extremely stable glassy matrix.

The behavior of basalt rocks serves as a good example in order to get an idea of the corrosive effects of atmospheric agents. The composition of these rocks can be simulated by mixing different kinds of wastes and their leaching rate never exceeds in 100 000 years. Waste glass, also, will interact with the environment to a negligible extent, not releasing significant amounts of the toxic elements. By this vitrification technique it is possible to obtain monolithic glasses which by controlled thermal treatments can be transformed into glass-ceramics, particularly wollastonite (CaO.SiO_2), diopside (CaO.MgO.2SiO_2), anorthite (CaO.Al_2O_3.2SiO_2) and iron oxides (Fe_2O_3, Fe_3O_4) in order to obtain coating materials with high chemical and mechanical for building or thermomechanical applications [37].

Romero, Rawlings and Rincon developed a new glass-ceramic by means of controlled vitrification of inorganic wastes from urban incineration. The major components of the raw material, which was from a Spanish domestic incinerator, were CaO, SiO_2 and Al_2O_3, but nucleating agents such as TiO_2, P_2O_5 and Fe_2O_3 were also present in reasonable amounts. They found that a relativelky stable glass with a suitable viscosity could be obtained by mixing 65 wt% of raw material with 35 wt% glass cullet. The heat treatment resulted in a glass ceramic containing a number of crystalline phases, the most stable being clinoenstatite. The microstructure contained both fiber like and dendritic crystals. The mechanical properties were acceptable for applications such as tiles for the building industry [38].

Pelino, Karamanov and associates have studied the vitrification and chemical durability of glasses containing industrial wastes such as electric arc furnace dusts, zinc-hydrometallurgy residues, drainage mud and granite mud. Batch compositions were prepared by mixing the wastes with glass, cullet and sand to force the final glass forming region of the SiO_2-Fe_2O_3 (CaO, MgO) system heavy metals such as Cr, Pb, Zn, Cu and Cd were immobilized and leaching tests confirmed the chemical stability [39].

Mônica Ferraris and her collaborators investigated the production glass-ceramic bodies from the bottom ashes of municipal solid waste incinerators that was subsequently used for tile production. Two different mixtures were used for tile production: a) glass from bottom ashes plus corundum-based waste from aluminum foundry and b) glass from bottom ashes plus kaolin based waste from the kaolin ore extraction process. The sintered materials were found to be appropriate for building applications tiles, bricks, etc. and they exhibited bending strengths up to about 60 MPa and Young modulus of up to about 53 GPa [40].

The vitrification of hazardous residues is consolidating as an environment-compatible stabilization process because of the high chemical resistance of the vitreous products in environmental conditions [41].

The immobilization of ACS, rich in chromium, by the incorporation in glasses or ceramics has a special attraction. Both glass and ceramics technologies are consederd potentially interesting in the immobilization of hazardous residues, since they can be able to retain heavy metals in a stable matrix with a significant reduction of volume and to convert complex chemical compositions into usable materials in the market. Besides that, the amount of generated ACS would be compatible to the amount of raw material processed in the production of ceramic materials to be used in the civil construction, for instance.

However, the main concern in the use of a residue as secondary raw material in the formulation of a ceramic mass is its immobilization inside the ceramic body after transformations occurred during ceramic process. The transference of the contaminants present inside the ceramic body to the environment depends on their ability to be lixiviated by water.

An ideal immobilization process turns the hazardous constituents into chemically non-reactive or stable compounds, in such a way that the residue can be safely reused or disposed of. To be totally effective, the treatment process must produce a final product with good dimensional stability, weather and mechanical resistance, and environment compatibility [42].

The efficiency of a treatment is usually evaluated by means of lixiviation tests. These tests measure the ability that a stabilized residue have to release immobilized contaminants during the solidification-stabilization process, and they are based in the comparison between a previously established standard and the lixivia obtained from the exposition of the residue to a lixiviate agent.

Hence, the purpose of this study was to investigate the immobilization of the chromium ion present in ash from incineration of chromium tanned leather shavings (ACS) by means of vitrified ceramic bodies and thus contribute toward a solution for the hazardous solid wastes generated by the leather industry and minimizing the environmental impact caused by these industrial residues. In particular, the study involved the formulation and preparation of ceramic bodies using ACS (ash from incineration of chromium tanned leather shavings) and soda-lime glass.

2. Experimental Procedure

2.1. Raw Materials

The Raw materials used in this work were: ash from the incineration of chromium tanned leather shavings (ACS); soda lime-glass and analytical reagent grade oxides (MgO, TiO_2, Na_2O, CaO, Cr_2O_3).

2.1.1. Ash from Incineration of Chromium Tanned Leather Shavings (ACS)

Table 2.1 shows the results of chemical analysis obtained for ACS, using ICP, by *Activation Laboratories Ltd.*, Canada. As can be seen, the total amount of chromium oxide was 29,4% by weight. Analysis done by GREENLAB, Porto Alegre, RS, Brazil showed that only 2,5% of the chromium was hexavalent.

Table 2.1. Chemical analysis of ash obtaided by the incineration of chromium tanned leather shavings (ACS).

Oxides	(%) Weight
Cr	29,4
Na	8,0
Al	1,5
Si	1,8
Fe	1,9
Ca	2,4
Ti	0,04
K	0,33
P	0,30
Mn	0,07
Mg	19
Loss due to fire	17,0

The ash was obtained experimentally by the incineration of chromium tanned leather shavings (ACS) using fluid bed combustion at 800°C [43]. The raw material was supplied by a leather factory from Rio Grande do Sul.

Figure 2.1 illustrates the mineral composition obtained by X- ray diffration with a PHILIPS Model X-PERT MPD.

As can be seen, the minerals present were: eskolaite (Cr_2O_3), quartz (SiO_2), gypsum ($CaSO_4{:}2H_2O$), anhydrite ($CaSO_4$), $Na_3Fe_2(PO_4)_3$, and $Na_2Ca_3Si_6O_{16}$ $NaCaAlSi_2O_7$.

Figure 2.2, shows differential thermal analysis and gravimetric analysis (DTA/TG) results obtained for ACS with a Harrop ST-736 thermobalance using a heating rate of, 80 K/h and a maximum temperature of 1400°C. The ACS sample exhibited a mass loss of about 20%. The exothermic peak around 500°C is probably due to the formation of calcium sulfate, anhydrite [44].

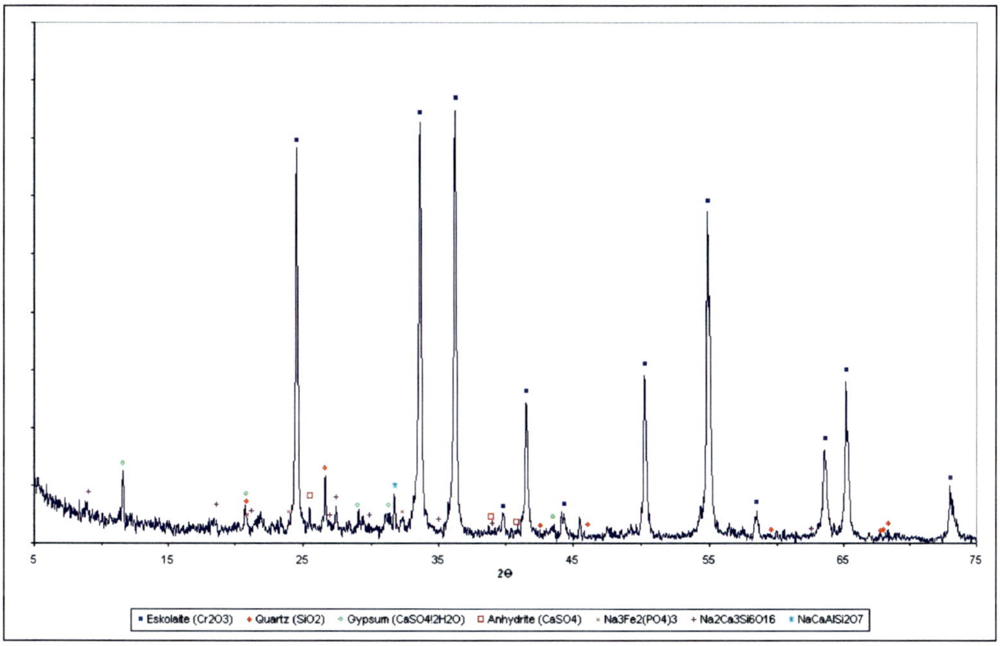

Figure 2.1. Mineral composition of ash from the incineration of crhomium tanned leather shavings (ACS).

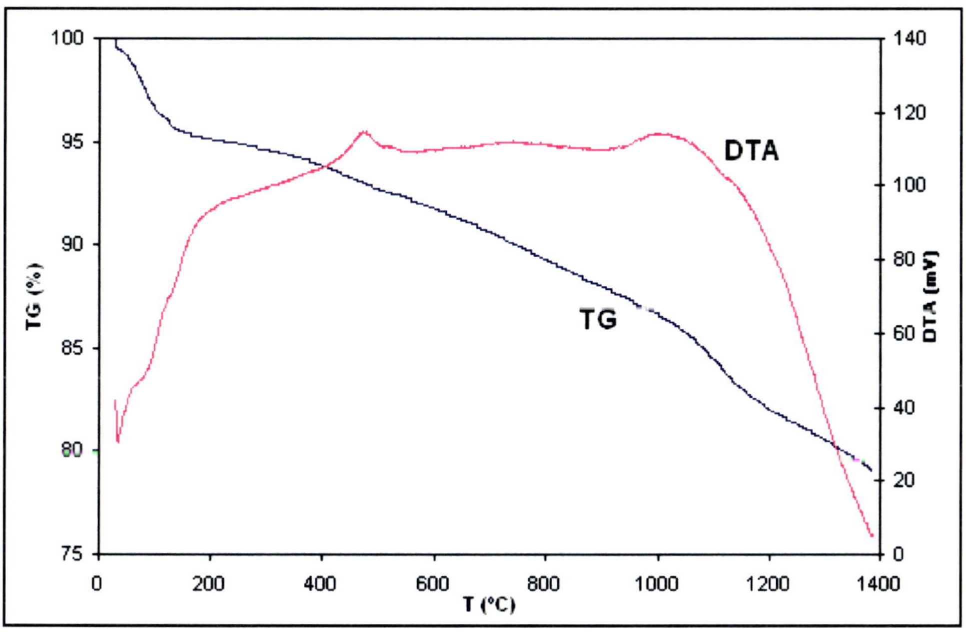

Figure 2.2. Thermal analysis and gravimetrical analysis (DTA/TG) of ash from the incineration of chromium tanned leather shavings (ACS).

Granulometric analysis and obtained using a Cillas – Model-1180 equipment indicated an average size of 7,94 μm for the particles. Values of D_{10} (0,69μm), D_{50} (5,47μm) and D_{90} (19,23μm) were obtained from the graph.

Figure 2.3 illustrates the results obtained for elementary analysis using an EDS microprobe coupled with SEM.

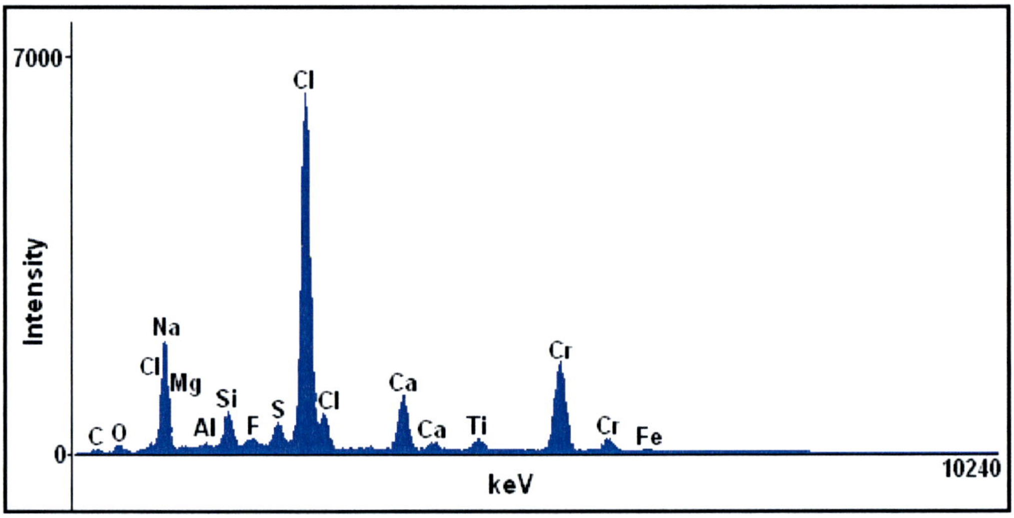

Figure 2.3. Results obtained using an EDS microprobe (same region samples in Table 2.2).

2.1.2. Soda-Lime Glass

The soda-lime glass used in this work for the formation of the vitreous phase during the combustion of the ceramic mass was obtained by dry grounding in a ball mill for 24 hours. It was prepared from commercial glass used for bottels. Fire loss for the soda-lime glass was 0,035%. The Table 2.3 illustrade the results obtained for the chemical analysis of the glass using X-ray fluorescence. Granulometric distribution obtained by laser difraction shows an average diameter of 28,50 μm with the respective D values: D_{10}=2,38, D_{50}=20,24 μm and D_{90}=68,46μm.

The Table 2.2 shows the chemical composition of the soda-lime glass.

Table 2.2. Chemical composition of the soda-lime glass used.

Oxide	(%) by Weight
Fe_2O_3	0,17
MnO	0,038
TiO_2	0,062
CaO	12,38
K_2O	0,92
P_2O_5	0,063
SiO_2	72,00
Al_2O_3	2,03
MgO	0,38
Na_2O	11,97

2.1.3. Pure Oxides

The oxides used as additives were of pure analytical reagent grade. The choice was based due to their presence in the ash (ACS) Na_2O, CaO and MgO. On the other hand, TiO_2, although present in small quantities in the ash was chosen because of its characteristic fusion properties [45]. Pure chromium oxide (Cr_2O_3) was also used in the formulation of the soda-lime glass mainly for comparison purposes. Table 2.3 shows the results obtained for granulometric analysis of the oxide employed using laser diffraction.

Table 2.3. Granulometric analysis of the oxides used.

Oxide	D_{10} (μm)	D_{50} (μm)	D_{90} (μm)	D_{avg} (μm)
Cr_2O_3	0,16	0,99	2,76	1,36
MgO	1,51	4,32	7,68	4,5
TiO_2	0,13	0,33	0,64	0,37
CaO	4,33	14,21	34,68	17,16
Na_2O	3,86	12,20	27,95	14,31

2.2. Preparation of the Test Samples

The glass and the ashes where sieved (n° 80 sieve – 180μm), homogenized under dry conditions for 15 minutes in a high energy mill, using a proportion of 2:1 for the mass of the sample and the homogenizing balls.

The soda-lime glass was added to the ACS (ash from the incineration of chromium tanned leather shavings) in the following perceptual by weight, as shown in Table 2.4

Table 2.4. Formulations of the soda-lime glass with ash from the incineration of chromium tanned leather shavings (ACS), in weight percent.

Formulation	ACS (%)	Glass (%)
100A	100	-
40G	60	40
70G	30	70
80G	20	80
90G	10	90

Table 2.5 illustrates the formulation of the soda-lime glass with pure oxides and Cr_2O_3 as additive. On the basis of the results obtained with the formulations shown about, TiO_2 and MgO were selected to prepare the test specimens containing the glass and the ashes.

Table 2.5. Formulations of soda-lime glass with various pure oxides and Cr_2O_3 as additives (percent by weigth)

Formulation	Glass (%)	Oxide (5%)	Cr_2O_3 additive (%)
5Cr	95	-	5
5Cr5Na	90	Na_2O	5
5Cr5Ca	90	CaO	5
5Cr5Mg	90	MgO	5
5Cr5Ti	90	TiO_2	5
10Cr5Na	85	Na_2O	10
10Cr5Ca	85	CaO	10
10Cr5Mg	85	MgO	10
10Cr5Ti	85	TiO_2	10
15Cr5Na	80	Na_2O	15
15Cr5Ca	80	CaO	15
15Cr5Mg	80	MgO	15
15Cr5Ti	80	TiO_2	15

The formulations used for the preparations of the ceramic vitreous bodies are summarized in Table 2.6

Table 2.6. Formulations for the vitreous ceramic bodies containing the ACS, glass, titanium oxide and magnesium oxide (Weight Percent).

Formulation	Glass Powder (%)	ACS (%)	Oxide used (%)
5Ti35V	35	60	5 TiO_2
5Ti65V	65	30	5 TiO_2
5Ti75V	75	20	5 TiO_2
5Ti85V	85	10	5 TiO_2
5Mg85V	85	10	5 MgO
10Mg80V	80	10	10 MgO

The ceramic masses were humidified with with an aqueous solution containing 10 wt% polyvinylic alcohol (PVA) as a plasticizer agent. The ceramic bodies were molded into 20x60mm2 bars using a hydraulic press (pressure 20 MPa).After forming, the specimens were were dried at room temperature for 24 h and subsequently in an oven at 110°C for an additional 24 h in order to remove any remaining humidity.The samples were then fired at 750°C, 800°C, 950°C and 1000°C with a heating rate of 80K/h and held at this temperature for 2h. The firing conditions used were determined on the basis of pre-tests aiming to optimize firing time, densification and comparison of the different formulations with respect to chromium immobilization/mobilization in the ceramic bodies.

2.. Characterization of the Materials Obtained

2.3.1. Immobilization/Mobilization of Chromium

The degree of immobilization of the chromium in the vitreous ceramic bodied was evaluated by leaching tests; In addition, gas emission was determiined during the firing step. The idea of this determination was to see if chromium was present in the gas along with calcium, sodium, and magnesium and if this interfered in the evaluation of the immobilization of the chromium ion in the ceramic body.The leaching test was performed according to th Brazilian technical standard test NBR 10.005 [46] that is similar to the American Extraction Procedure Toxicity Test [47].This procedure uses deionized water as an extractor fluid. The amount of water added was 16 times the amount of sample. In this study vitreous ceramic body samples, pressed while still green and weighing 100g were used. They were crushed to a particle size of less than 9.5 mm. Subsequently, they were dried in an oven at 110°C for 24 hours and fired under the conditions described above.The pH of the water was adjusted to 5.o using acetic acid.

The leaching test had duration of 24 hours under stirrng. After this period the solid and the liquid was separated by filtration. The lixiviate was chemically analyzed and the results were compared with the maximum values cited in the Brazilian residues specification NBR10.004 [48]. A rediue is classified as hazardous or not, according to this norm and procedure.

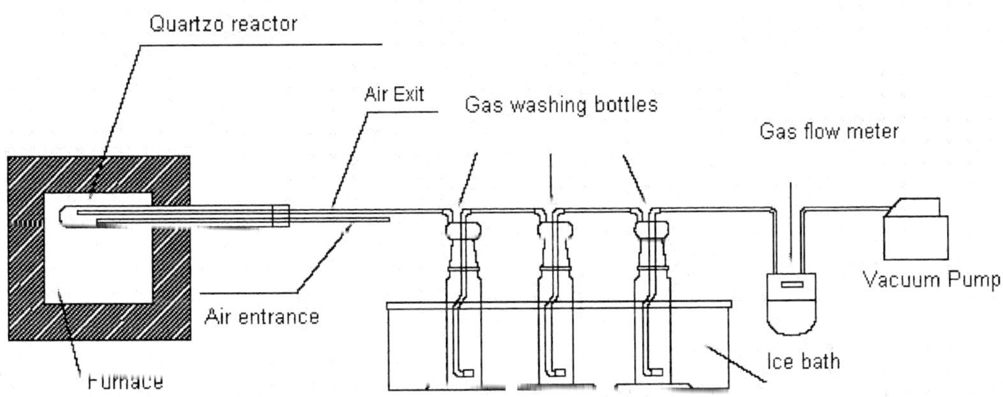

Figure 2.4. Apparatus used to evaluate gaseous emissions generated during firing process.

Gaseous emissions generated during material firing were also evaluated. The apparatus shown schematically in Figure 3.4 [31] was used to perform this evaluation. Specimens subjected to the firing cycle were placed in a quartz reactor inside an electrical oven. This reactor has an air intake (air from the atmosphere is blown over the specimens) and in an outlet (air from the burning atmosphere is collected under vacuum, with the aid of a Teflon diaphragm pump) Collected air was washed in a sequence of three gas washing bottles (500ml each), that were immersed in an ice bath, in order to increase the solubility of gases in water. The he first two bottles were filled with deionized water. The third bottle (also called dry bottle) was empty in order to help process visualization, to help retain any condensation, and to avoid humidity reaching in the vacuum pump. Air collection occurred during the whole process.Water washings were collected and the concentration of the elements present in the samples was determined by means of atomic absorption.

Table 2.7 summarizes the various formulations investigated other respect chemical composition, chromium quantity and firing temperature. As can be seen gaseous chromium emission was always below the experimental limits.

Table 2.7. Determination of the gaseous emission of the ceramic bodies 100A and 90G using firing temperatures of 750°C and 1000°C

Sample	Ca (mg/L)	Na (mg/L)	Mg (mg/L)	Cr_{total} (mg/L)
$ACS_{(pure)}$ - 750°C	0,20	0,12	0,7	< 0,02
$ACS_{(pure)}$-1000°C	0,38	0,33	0,05	< 0,02
90G - 750°C	0,23	0,03	0,7	< 0,02
90G - 1000°C	0,23	0,7	0,07	< 0,02

2.3.2. Physical and Microstructure Characterization

The ceramic specimens were characterized with respect to water absorption e linear shrinkage.

Water absorption was measured by the Archimedes method, in accordance with ASTM C 373/94[49] and the linear shrinkage was determined by the difference in the length of the test bar before and after firing, according to the ASTM-C326-03 [50].

The ceramic bodies were also submitted to mineralogical and microstructural analyses. The mineralogical analysis of CSC and ceramic bodies was carried out by X-rays diffraction in a PHILIPS equipment model X-PERT MPD. This diffractometer uses a X-ray tube with Cu target and was operated at 40kV and 40mA in all measurements made in this work. A graphite monochromator was attached to the secondary optical system.

The samples microstructural analysis was made by the evaluation of present phases. Both optical (OM) and scanning electron microscopy (SEM) were used. The optical microscope was an OLYMPUS BX51M, with a digital camera attached to it, while the used SEM was a JEOL JSM-5800 from the UFRGS Microscopy Center. An EDS microprobe, attached to the SEM, was used to punctual chemical analysis and mapping of chemical elements o interest in the ceramic body.

Samples for both optical and electron microscopy were prepared by conventional grinding and polishing steps. Polishing was made with a suspension of alumina in water or diamond paste. A thin layer of carbon was sputtered in the surface of samples, in order to obtain good electrical conductivity for SEM analysis

3. Results

Figure 3.1 shows the variation in the concentration of total lixiviated chromium, according to NBR 10.005[46] standard, for ceramic bodies obtained with different additions of glass to the ACS, as a function of their firing temperature. As can be observed, for firing temperatures of

750°C and 800°C there was an increase of lixiviated chromium with increasing content of glass up to 80% and 70%, respectively. However, with glass addition of 90%, a decrease in the chromium lixiviation was observed, reaching values similar to those obtained for ACS with no glass additions.

At 950°C, a decrease in the chromium lixiviation occurred, comparing to pure ACS, for glass additions of 80% and 90%. At 1000°C, a drastic reduction in chromium lixiviation was observed, reaching a minimum value of 57 mg/L for 70% of glass content. This value is similar to that attained by the composition containing 90% of glass, at firing temperature of 950°C.

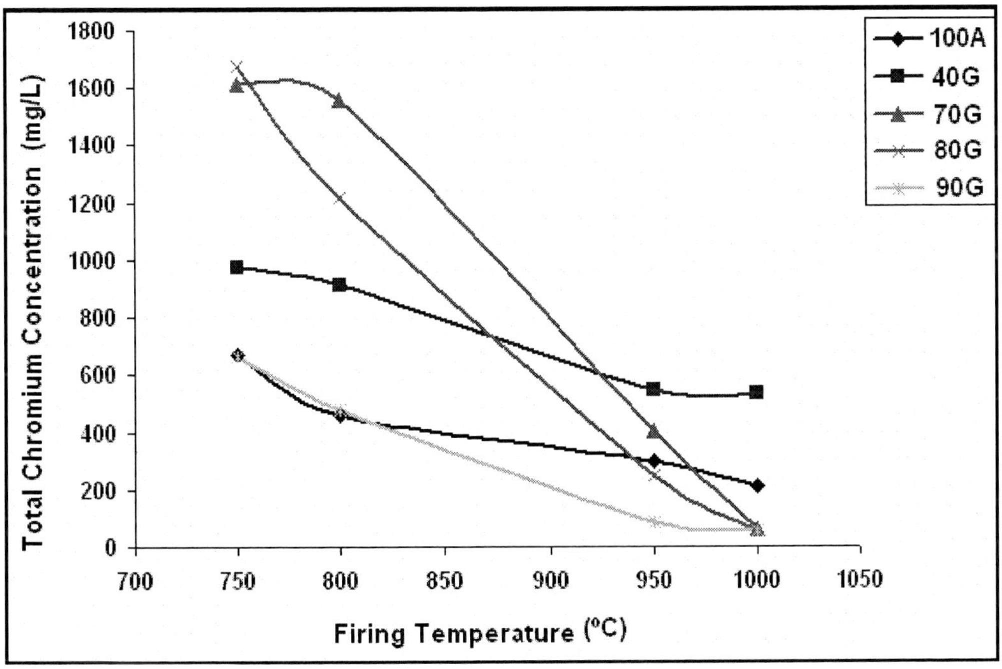

Figure 3.1. Total chromium concentration in the lixivia as a function of firing temperature and glass content in formulations with ACS.

Nevertheless, one must consider that increasing glass additions into the ceramic mass proportionally reduce the ACS amount, and, hence, the total chromium content available for lixiviation in the ceramic body. Thus, the corrected values of lixiviated chromium concentration as a function of the chromium content that is available for lixiviation is shown in Figure 3.2.

With the exception of the ceramic body with 90V, fired at 750°C, it was observed that the glass addition to the ACS led to a clear increase in total chromium lixiviation. The formulation 90V always had the biggest total chromium lixiviation for firing temperatures higher than 750°C. Hence, the bigger the glass addition to the ceramic bodies, the higher the total chromium lixiviation. This lixiviation is always lower for higher firing temperatures.

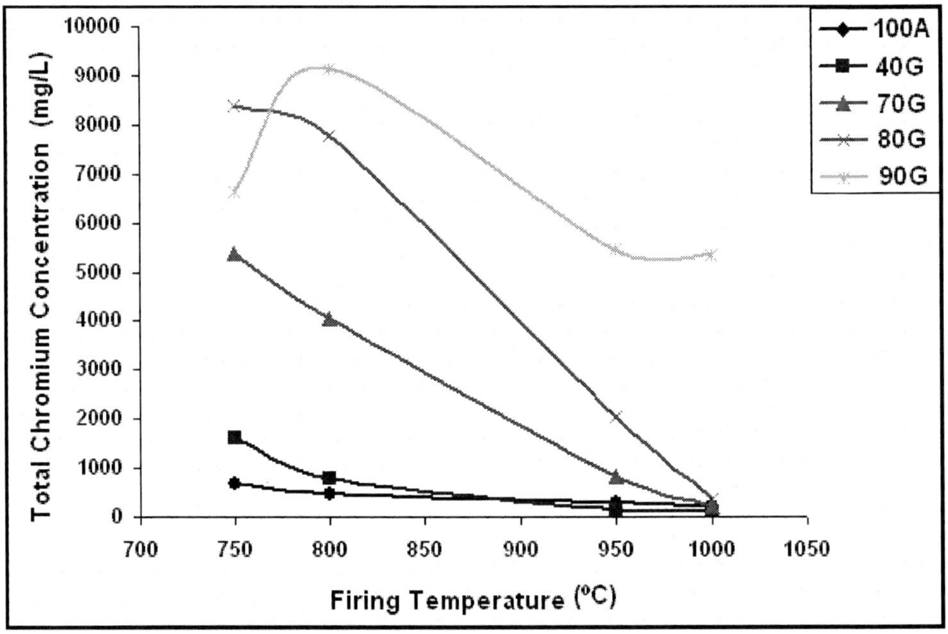

Figure 3.2. Corrected chromium concentration in the lixivia as a function of firing temperature and glass content in formulations with ACS.

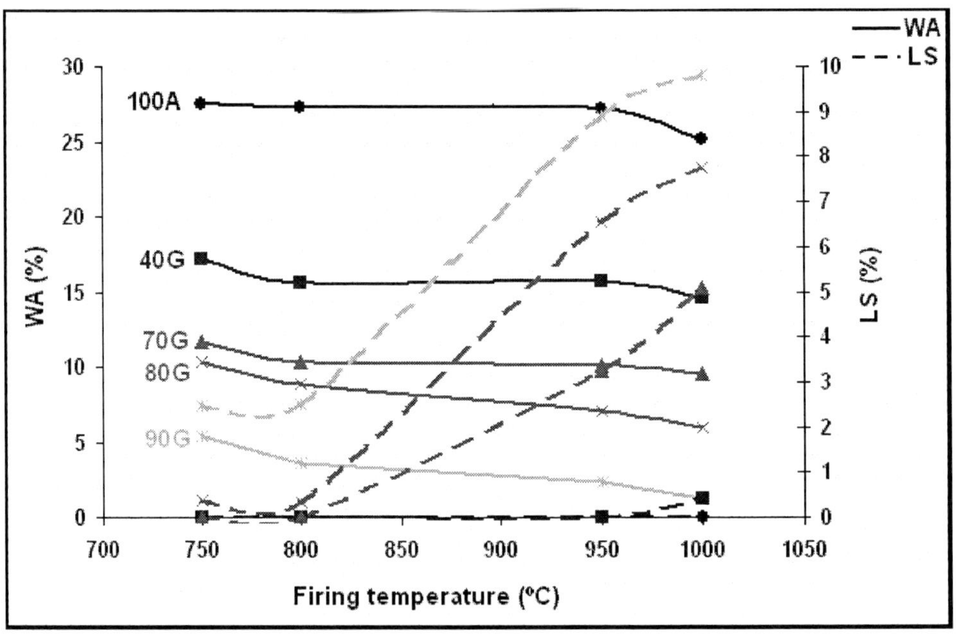

Figure 3.3. Gresification curve (water absorption – WA and linear shrinkage – LS as a function of firing temperature) for formulations 100A; 40G; 70G; 80G e 90G.

Such facts indicate that the higher total chromium lixiviation for ceramic bodies containing glass could be associated to their lower densification during firing caused by the glass additions. So, for the considered firing temperatures the glass did not act as a flux. Here, the glass granulometry, which is about four times higher than ACS granulometry (glass: D_{mean}=28.5 μm, and ACS: D_{mean}=7.94 μm) should be also taken into account.

The graph in Figure 3.3 shows the variation in water absorption of the same ceramic bodies shown in Figure 3.2 as a function of firing temperature and glass addition to the ACS. When this graph is analyzed, it can be seen that a reduction in water absorption and an increase in linear shrinkage occurred for ceramic bodies contain in glass in their formulation. Therefore, the ceramic body densification is intensified, leading to a smaller surface area and lower reactivity, which play a significant role in chromium lixiviation. It means that the chromium lixiviation was higher despite the lower porosity of the ceramic bodies.

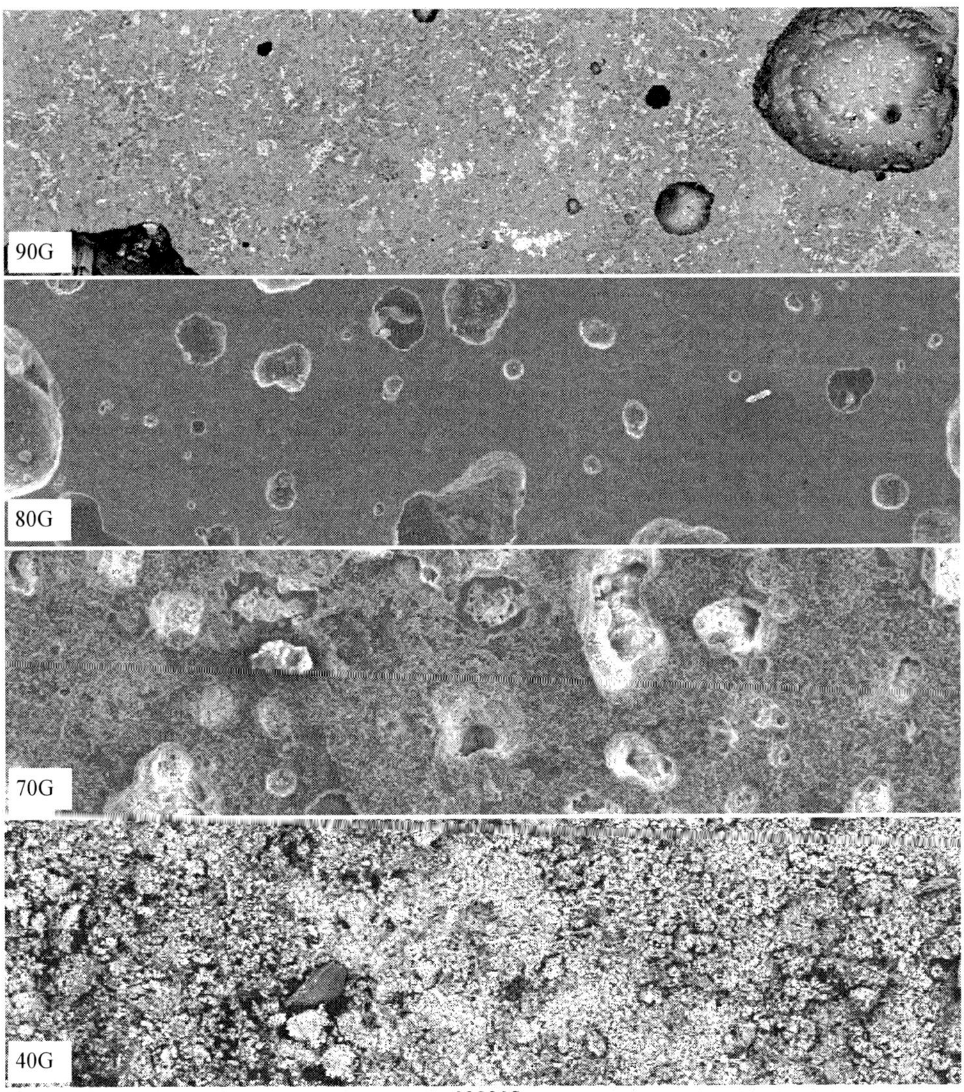

Figure 3.4. Micrographs of ceramic bodies of formulations 40G, 70G, 80G, and 90G fired at 1000°C.

In order to show the densification of ceramic bodies in function of glass amount in the formulation, Figure 3.4 shows the micrographs of ceramic bodies 40G, 70G, 80G, and 90G, all fired at 1000°C. For higher glass content, higher densification and structure vitrification of ceramic bodies can be verified. For 80G and 90G ceramic bodies the pores are rounded, indicating an advanced stage of densification.

On one hand, it is possible to conclude that the open porosity (water absorption being the "opposite" of densification) is not the only determining factor in chromium lixiviation. On the other hand, however, a correlation between the drastic reduction in chromium lixiviation at 950°C and the increase in densification for all formulations was evidenced.

The higher total chromium lixiviation for temperatures up to 950°C in spite of the increase in densification of the ceramic bodies as a function of glass addition (the glass acts as a flux, since higher linear shrinkage and lower water absorption were observed for higher glass amounts) must be explained. One possibility could be related to the phases formed during firing, and the ability of these phases to retain chromium in the ceramic body during the lixiviation process.

Table 3.1 summarizes the present phases in the ceramic bodies for different formulations and different firing temperatures. Phase's constitution was determined by X-rays diffraction. Phases denoted in green color are those where chromium is in principle trivalent (Cr^{+3}), while phases denoted in orange color are those where chromium is in principle hexavalent (Cr^{+6}).

From the data shown in Table 3.1, it can be observed that the phases present in the ceramic body produced from ACS with no glass addition (100C) are quartz (SiO_2), eskolaite (Cr_2O_3), and magnesiochromite ($MgCr_3O_4$) for all tested firing temperatures. These phases were detected in ACS prior to firing, except magnesiochromite.

Despite magnesium was detected in the chemical analysis by ICP, the presence of MgO was not observed in none of the mineralogical phases of *in natura* ACS. This was probably due to the small amount (about 3% according to the chemical analysis), which is bellow the detection limits of X-ray diffraction. Only when the magnesium oxide reacts with the chromium oxide a detectable phase appears (magnesiochromite), which was observed in the mineralogical analysis of the ceramic body 100A fired at 750°C.

From the temperature of 800°C on, the hexavalent chromium in the form of magnesium chromate ($MgCrO_4$), which was present in the original ACS, can be found. On reason for that is that proposed for Bray (1985) [51] and Weyl [45]. According to them, the Cr^{+3} is oxidized to Cr^{+6} in the presence of alkalis at high temperatures.

The formation of $MgCrO_4$ at temperatures above 800°C might justify the higher lixiviation of chromium for samples made from formulation 100A (Figure 3.2) and fired from this temperature, since that compound is very soluble is=n water. However, it can be supposed that this phase, despite not being detected by X-ray diffraction, is present in ceramic bodies fired at 750°C, because of the chromium concentration in the lixivia obtained for bodies of formulation 90V fired at this temperature. This phase was not detected probably because of its small concentration (lower than 3%, which is the detection limit of the equipment).

According to Weyl [45], Cr_2O_3 and SiO_2 are practically immiscible when molten. The solubility of Cr_2O_3 in glasses is typically low, except when the glass has a high content of alkalis. In fact, according to Table 3.1, Cr_2O_3 was never found together with SiO_2, despite these oxides are present in every investigated ceramic body.

The phases present in sample 40G were always the same for all firing temperatures, with the exception of sodium chromate, which is present only up to the temperature of 800°C. From this point on, only sodium dichromate was found, as shown in Table 3.1. The chromium lixiviation of sample 40G (Figure 3.2) decreases with the increase in temperature, in the same way as for ceramic bodies mad from formulations with higher glass content.

According to Nietz *apud* Bernardes [52], the chromate formation starts at 200°C, following the Equation 3.1.

$$2\ Na_2O + Cr_2O_3 + 3/2O_2 \rightarrow 2\ Na_2CrO_4 \qquad \text{Equation (3.1)}$$

At 750°C, the chromate phases can be found, as shown in Table 3.1. The higher chromium lixiviation for temperatures of 750°C and 850°C is probably related to the presence of both sodium chromate and sodium dichromate at these temperatures. As described before, when the firing temperature is increased to 950°C and 1000°C, sodium dichromate is not detected anymore, and only sodium chromate is present. Thus, the decrease in chromium lixiviation observed for ceramic bodies fired at those temperatures must be also associated to a smaller amount of phases containing hexavalent chromium.

Table 3.1. Phases present in ceramic bodies made from different formulations as a function of firing temperature. Phase composition was determined by X-ray diffraction

	Firing temperature (°C)			
	750	**800**	**950**	**1000**
ACS Original phases in ACS: Quartz Anhydrite $Na_2Ca_3Si_6O_{16}$ $CaSO_4.2H_2O$ $Na_3Fe_2(PO_4)_3$ $NaCaAlSi_2O_7$ Eskolaite 100A	Quartz $Na_2Ca_3Si_6O_{16}$ $Na_3Fe_2(PO_4)_3$ Akermanite Anhydrite Eskolaite Magnesiochromite	Quartz $Na_2Ca_3Si_6O_{16}$ $Na_3Fe_2(PO_4)_3$ Akermanite Anhydrite Eskolaite Magnesiochromite Magnesium Chromate	Quartz $Na_2Ca_3Si_6O_{16}$ $Na_3Fe_2(PO_4)_3$ Akermanite Anhydrite Eskolaite Magnesiochromite Magnesium Chromate	Quartz $Na_2Ca_3Si_6O_{16}$ Akermanite Anhydrite Eskolaite Magnesiochromi te Magnesium Chromate
40G	Quartz Clinoenstatite Cristobalite Barium oxide Diopside Wollastonite-1A Potassium sulfate Eskolaite $Na_2Cr_2O_7$ $Na_2CrO_4.H_2O$	Quartz Clinoenstatite Cristobalite Barium oxide Diopside Wollastonite-1A Potassium sulfate Eskolaite $Na_2Cr_2O_7$ $Na_2CrO_4.H_2O$	Quartz Clinoenstatite Cristobalite Barium oxide Diopside Wollastonite-1A Potassium sulfate Eskolaite $Na_2Cr_2O_7$	Quartz Clinoenstatite Cristobalite Barium oxide Diopside Wollastonite-1A Potassium sulfate Eskolaite $Na_2Cr_2O_7$

Table 3.1. Continued

	Firing temperature (°C)			
	750	800	950	1000
70G	Quartz Cristobalite Eskolaite $Na_2CrO_4.H_2O$	Quartz Cristobalite Eskolaite $Na_2CrO_4.H_2O$	$Na_2Ca_3Si_6O_{16}$ Wollastonite-2M Cristobalite Clinoenstatite Eskolaite $Na_2CrO_4.H_2O$	Quartz Wollastonite-2M Cristobalite Clinoenstatite Eskolaite
80G	Cristobalite Aluminum phosphate Quartz $Na_2Ca_3Si_6O_{16}$ Eskolaite $Na_2CrO_4.H_2O$	Cristobalite Aluminum phosphate Quartz $Na_2Ca_3Si_6O_{16}$ Eskolaite $Na_2CrO_4.H_2O$	Cristobalite Wollastonite-2M Quartz $Na_2Ca_3Si_6O_{16}$ Clinoenstatite Eskolaite $Na_2CrO_4.H_2O$	Cristobalite Wollastonite-2M Quartz $Na_2Ca_3Si_6O_{16}$ Clinoenstatite Eskolaite
90G	Cristobalite $Na_2Ca_3Si_6O_{16}$ Quartz Sodium sulfate() Eskolaite $Na_2CrO_4.H_2O$	Cristobalite $Na_2Ca_3Si_6O_{16}$ Quartz Sodium sulfate() Eskolaite $Na_2CrO_4.H_2O$	Cristobalite $Na_2Ca_3Si_6O_{16}$ Quartz Sodium sulfate $Na_3Fe_2(PO_4)_3$ Eskolaite $Na_2Cr_2O_7$	Cristobalite $Na_2Ca_3Si_6O_{16}$ Tridymite Quartz Sodium sulfate Alumina Eskolaite $Na_2CrO_4.H_2O$

Clinoenstatite= $MgSiO_3$ Sodium dichromate= $Na_2CrO_4.4H_2O$ Akermanite= Ca_2MgSiO_7 Aluminum phosphate= $AlPO_4$ Anhydrite= $CaSO_4$ Wollastonite-2M= $CaSiO_3$ Magnesiochromite= $MgCr_2O_4$ Sodium sulfate= $Na_2S_2O_3$ Diopside= (Ca(Mg,Al) (Si, Al)$_2$O$_6$ Chromium/sodium oxide= $Na_2Cr_2O_3$ Magnesium chromate= $MgCrO_4$

The stability of ionic species as a function of pH to the system involved in the lixiviation test for the investigated ceramic bodies is shown in Figure 3.5. This graph aids to understand the lixiviation process in a slightly acid environment – containing acetic acid – which the ceramic bodies with sodium chromate are submitted to.

According to the diagram of Figure 3.5, it can be observed that the species CrO_4^{-2} and Na^+ are soluble in the pH measured during the lixiviation test (5.0 to 6.1). The solubility of both ions is also evident by the analysis of the lixiviation tests for several ceramic bodies containing the same basic constituents of ACS and glass. It will be discussed latter, when the results for ceramic bodies with additions of Na_2O are stated.

Table 3.2 shows the image mapping for elements Na, Cr, and O, detected in ceramic bodies 40G for all investigated firing temperatures. This mapping was obtained by scanning electron microscopy, with 1000x magnification. It can be verified that the distribution of these three elements for temperatures of 800°C and 1000°C are quite similar. This confirms the presence of sodium dichromate, as indicated by X-ray diffraction. When comparing the mapping of ceramic bodies fired at 800°C and 1000°C, a higher homogeneity on the later one can be observed, possibly because of the higher amount of vitreous phase caused by the higher firing temperature.

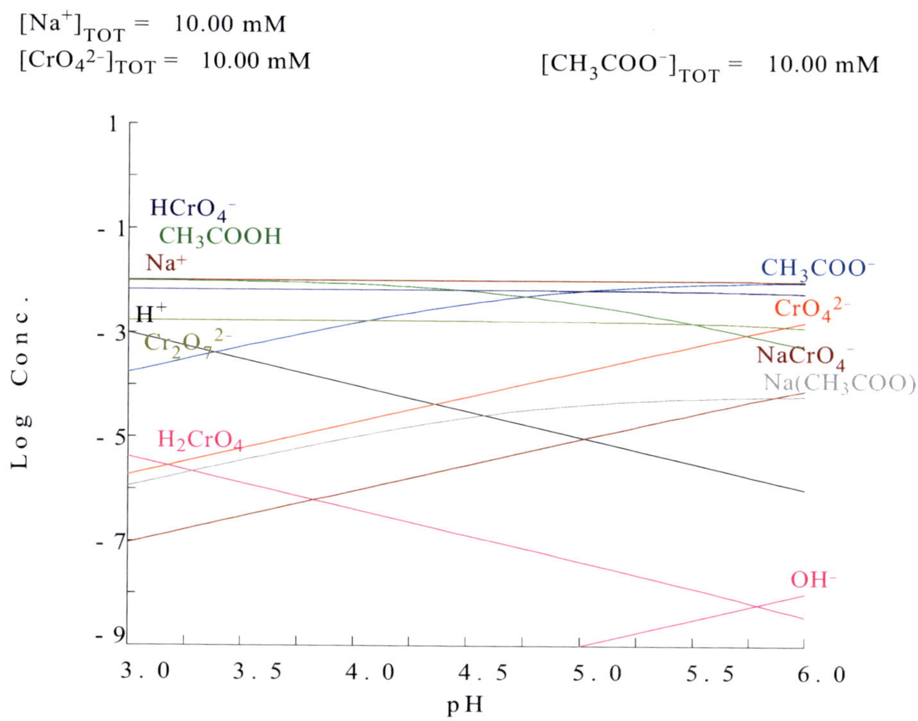

Figure 3.5. Stability diagram for the system involved in the lixiviation of ceramic bodies made from formulation 40G.

Despite the presence of chromates in both ceramic bodies 40G (fired at 800°C and 1000°C), their microstructure, which is closer and with higher amount of vitreous phase, can also explain the lower chromium lixiviation.

Thus, the increasing of phases that can be lixiviate, such as sodium chromates resulting from the reactions of sodium, glass containing sodium and calcium, and chromium of ACS, could explain the increase in lixiviation of the ceramic bodies with the increase of glass content in their formulation. On the other hand, the increase of firing temperature would decrease the open porosity (Figure 3.3) because of the formation of vitreous phase, explaining the decrease in lixiviation with the increase of firing temperature. These processes would not be concurrent to each other, but they might have opposite effects regarding lixiviation.

The interpretation of the results obtained for ceramic bodies made from formulation with 70 wt% of glass (70G) follows the same ideas previously described. For firing temperatures of 750°C and 800°C, the presence of phases eskolaite and sodium chromate is observed. The formation of new phases starts at 950°C, when the phases wollastonite ($CaSiO_3$), clinoenstatite ($MgSiO_3$) and $Na_2Ca_3Si_6O_{16}$ were detected.

The phase sodium chromate is present in ceramic bodies fired up to 950°C, what could explain the high lixiviation showed by samples fired at this temperature (Figure 3.1). The chromium lixiviation of the ceramic body 70G fired at 1000°C (74mg/L) was significantly lower if compared to the sample 70G fired at 750°C, which was about 1700mg/L.

Table 3.2. Image mapping of the ceramic body 40G, fired at 800°C and 1000°C. Elements: Na, Cr and O.

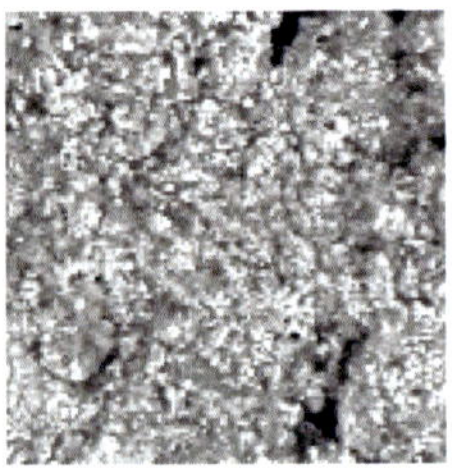	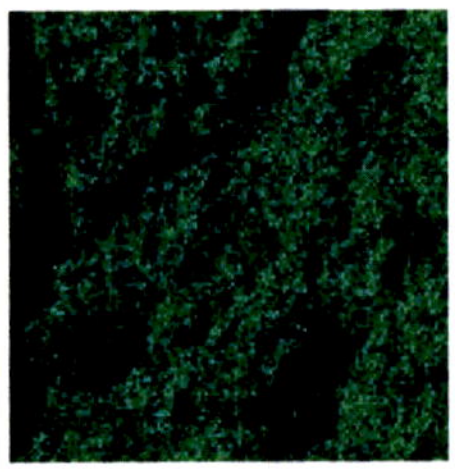		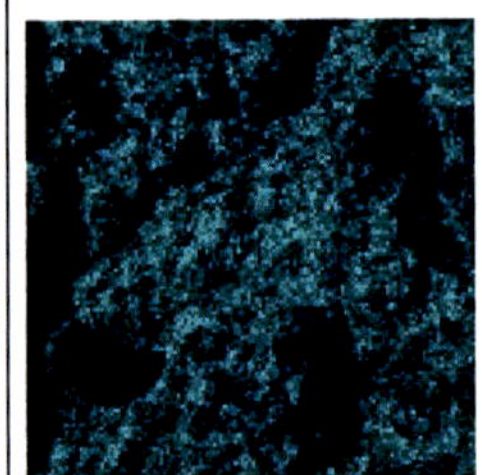
40G - 800°C	O	Cr	Na
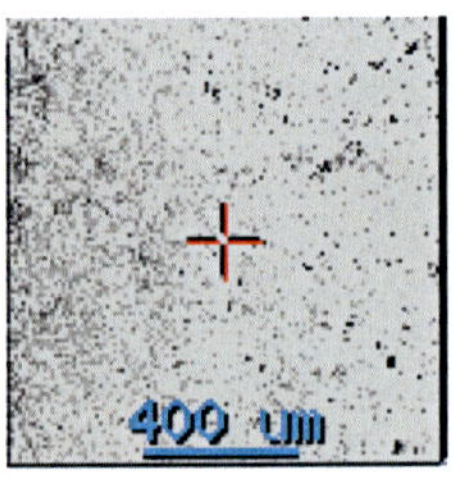	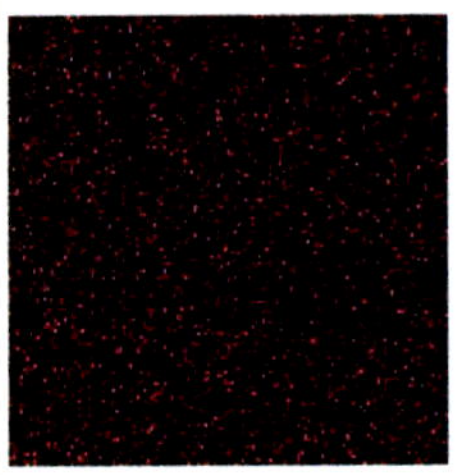		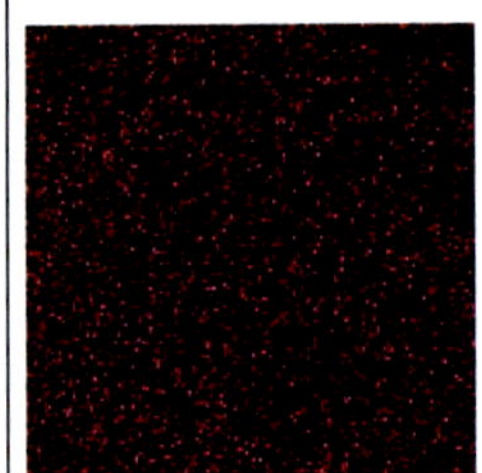
40G - 1000°C	O	Cr	Na

The analysis of the image mapping of elements O, Cr, and Si of the ceramic body 70G fired at 1000°C (Table 3.4) shows that part of the chromium matches silicon. Since no crystalline phase containing chromium and silicon in those ceramic bodies was detected, it is supposed that the chromium that is the same position of silicon is incorporated into the vitreous phase, making it less susceptible to lixiviation.

Table 3.4. Image mapping of ceramic body 70G fired at 1000°C.

Elements: O, Cr, and Si.

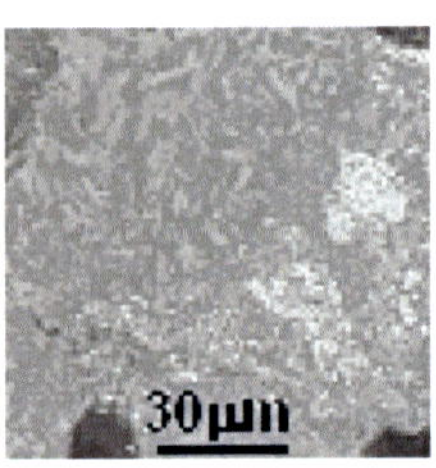		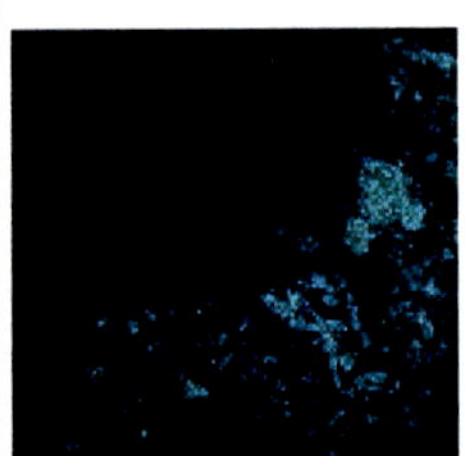	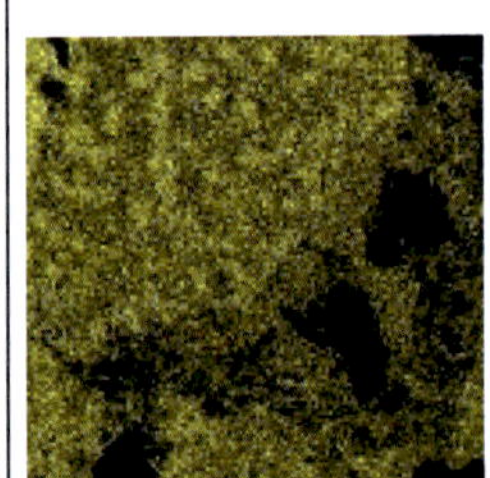
70G - 1000°C	O	Cr	Si

The same observations can be made for ceramic bodies 80G when analyzing Table 3.1. These bodies have the same phase constitution up to 950°C. The absence of sodium chromate (Na_2CrO_4) at 1000°C must be pointed out. Hence, it can be inferred that at this temperature the sodium chromate is incorporated into the vitreous phase, decreasing chromium lixiviation, as can be seen in Figure 3.2.

Table 3.5 shows the image mapping of elements O, Cr, Na, Ca, and Si for the ceramic body 80G fired at 1000°C. This mapping evidences the presence of eskolaite, which is the only chromium-containing phase at this firing temperature. This phase appears in the micrograph of ceramic body 80G of Table 3.5. It is the clearest one, and it constituted solely by Cr and O (Cr_2O_3).

However, there are other regions where the simultaneous detection of chromium and O, Si, Ca, and Na could be observed. Since no other phase containing chromium for samples fired at 1000°C were detected by X-ray diffraction, two hypotheses can be stated: one, that the formed crystalline phase is quantitatively irrelevant to be detected by the X-ray diffraction; the other one, which is more probable, that chromium would be embedded into the vitreous phase formed mainly by Si, O, Na, and Ca. These elements come from the sodium-calcium-containing glass that was used in the formulation of the investigated ceramic bodies. This could explain the lower chromium lixiviation for ceramic bodies 80G at this firing temperature (Figure 3.2), in the same way that was observed for ceramic bodies 40G and 70G.

Table 3.5. Image mapping of ceramic body 80G fired at 1000°C. Elements: O, Cr, Na, Ca, and Si.

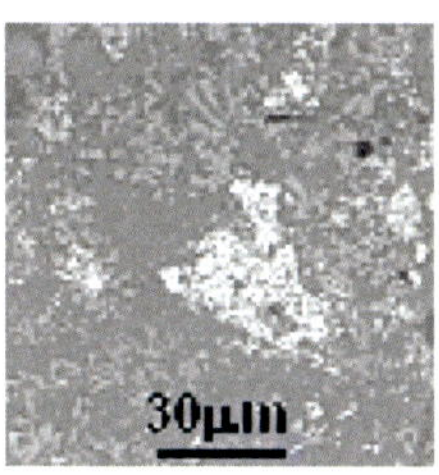		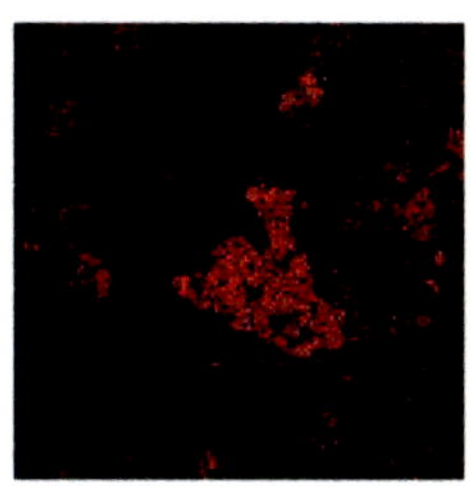	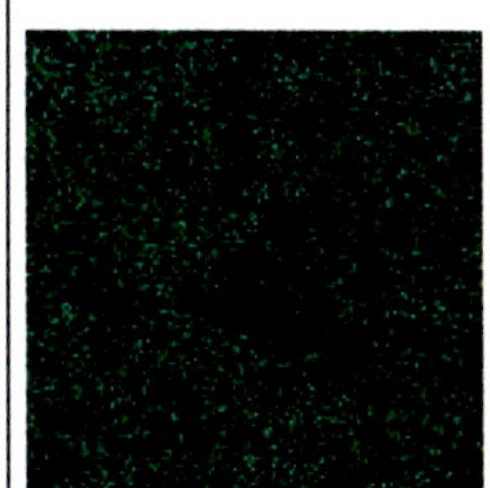
80G - 1000°C	O	Cr	Na
	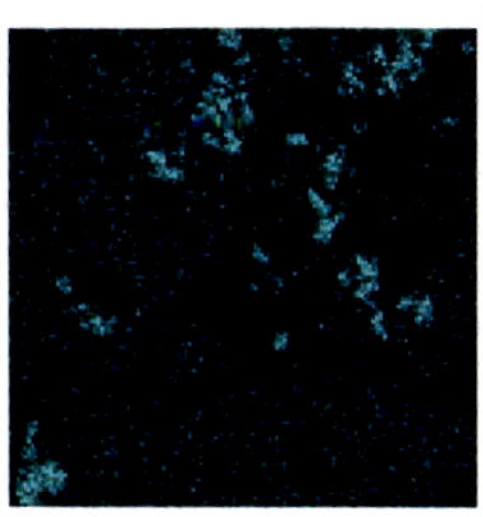		

The results obtained for ceramic bodies 90G are also found in Table 3.1. For all firing temperatures the presence of sodium chromate was verified. Since this compound is soluble, it explains the high lixiviation for this formulation for all investigated firing temperatures.

From the image mapping of elements Cr, Na, and Si of ceramic bodies 90G for all firing temperatures, which is shown in Table 3.6, it is possible to see that for higher firing temperatures (950°C and 1000°C), Na and Cr are more spaced out from each other than for lower temperatures (750°C and 850°C). However, the mineralogical analysis made by X-ray diffraction (Table 3.1) shows that the phase sodium chromate is present for all firing temperatures. The higher spacing between Cr and Na observed for ceramic bodies 90G fired at higher temperatures indicates that the presence of chromate is less intense, leading to a lower chromium lixiviation for these temperatures, which is in agreement with the data shown in Figure 3.2.

Eskolaite is the phase which is present for all investigated firing temperatures, and its main role is to immobilize chromium. The image mapping of Na and Si indicates that these elements are probably constituents of the vitreous phase of the ceramic bodies. This vitreous phase is an important factor in chromium immobilization within ceramic bodies. The Na from sodium chromate would react with the vitreous phase and chromium would be reduced to Cr^{+3}. This fact can be supported by the SEM micrographs of 90G ceramic bodies fired at increasing temperatures shown in Figure 3.6.

From Figure 3.6, it can be observed that the phase eskolaite distributed in the microstructure of the ceramic body 90G. Besides that, the higher surface area for lower firing temperatures can be also observed; this can increase the tendency for lixiviation. Hence, phases able to lixiviate such as sodium chromate would have more reaction area with the lixiviate media.

Table 3.6. Image mapping of the ceramic body 90G fired at 750°C, 850°C, 950°C, and 1000°C. Elements: Cr, Na, and Si.

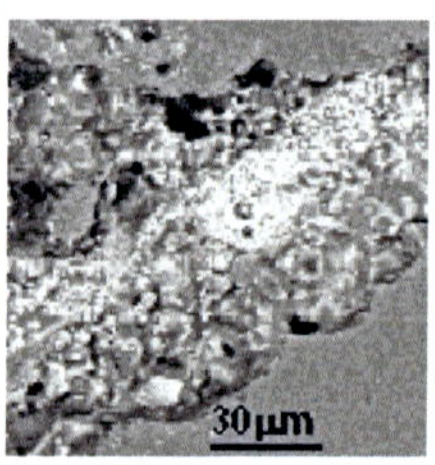	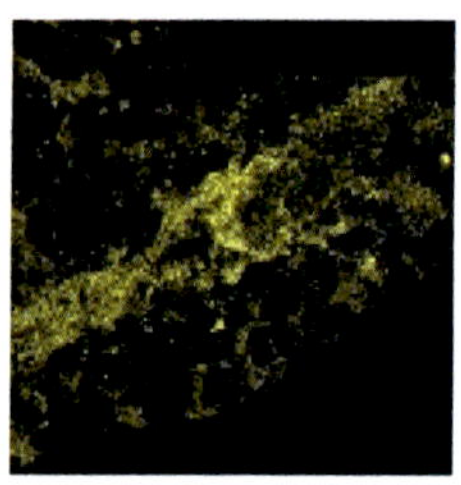	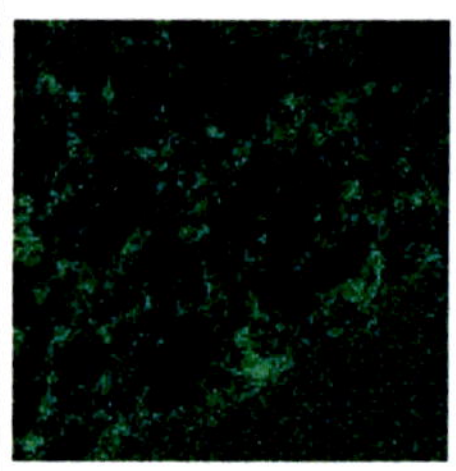	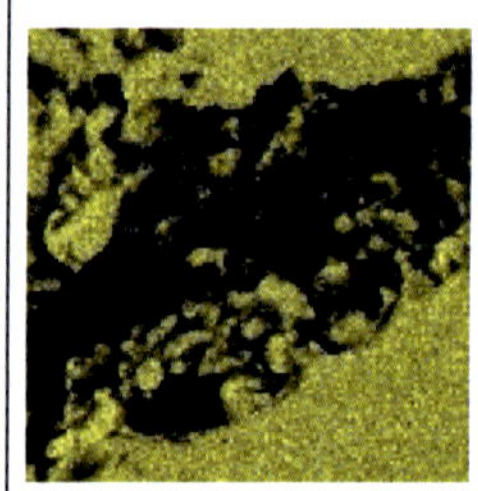
90G - 750°C	Cr	Na	Si
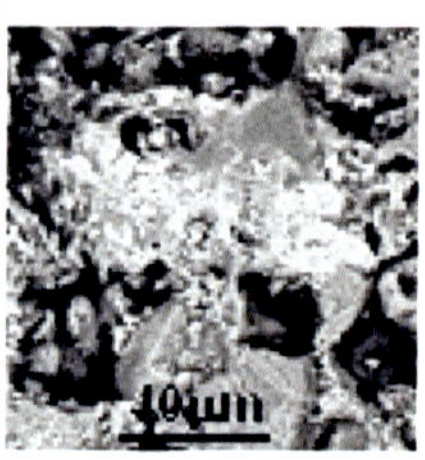	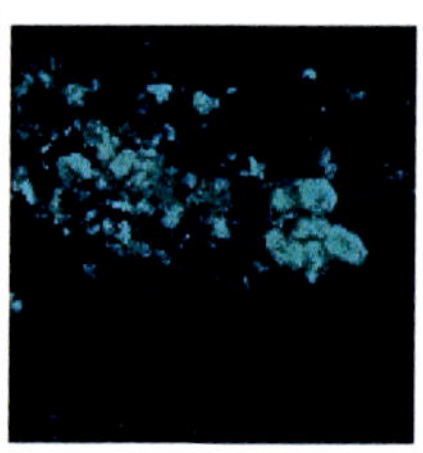	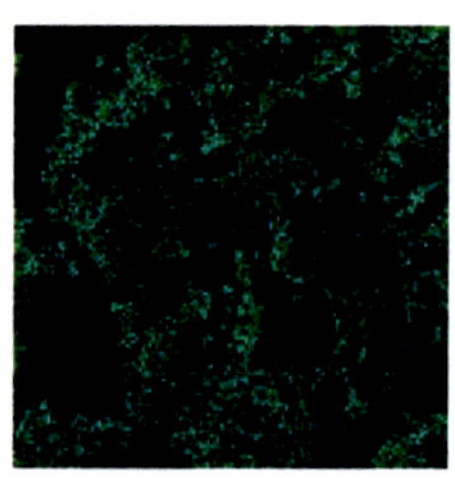	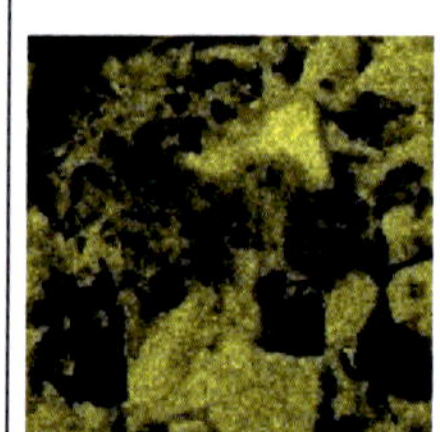
90G – 800°C	Cr	Na	Si

Table 3.6. Continued.

	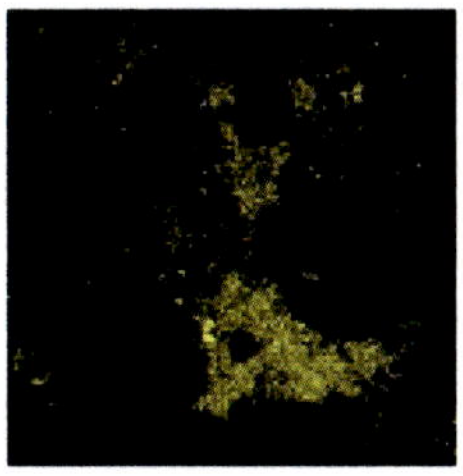	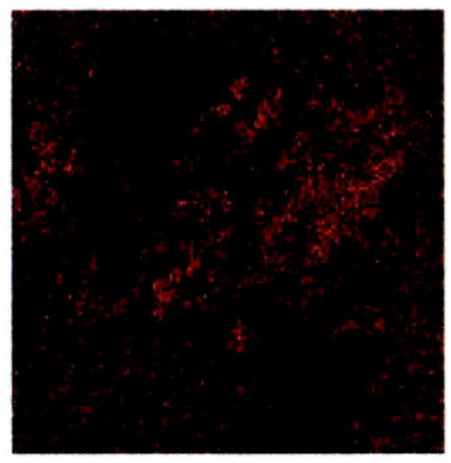	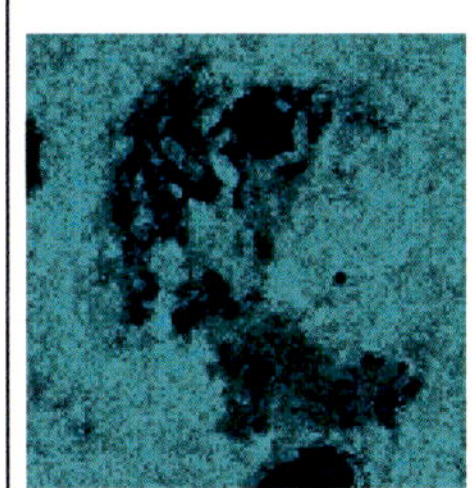
90G - 950ºC	Cr	Na	Si
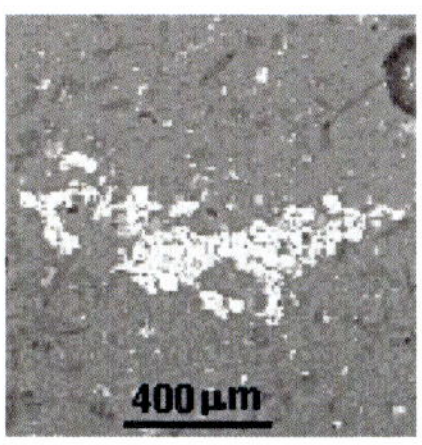	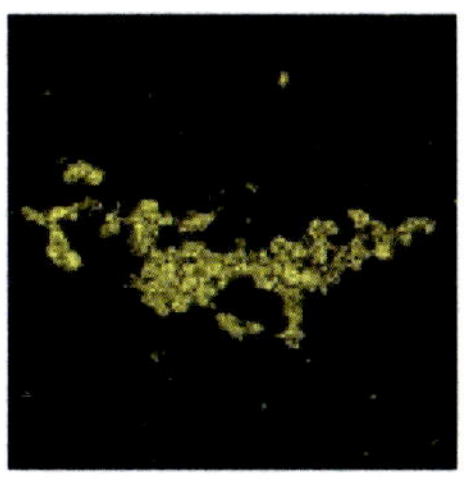	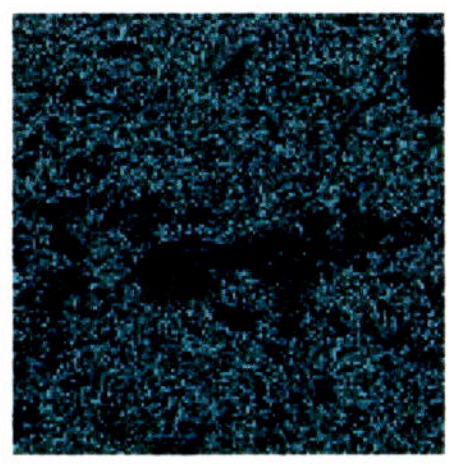	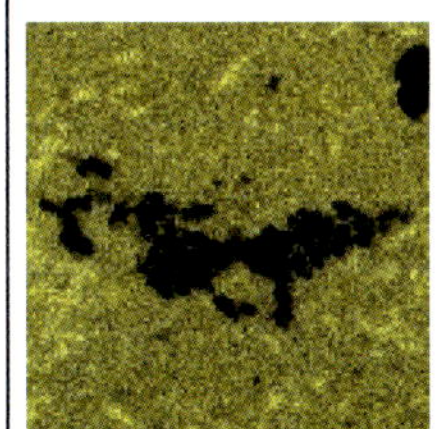
90G - 1000ºC	Cr	Na	Si

According to Anak and Güldal (1980), the Cr^{+3}-to-Cr^{+6} ratio in a sodium-calcium-containing glass is sensitive to other reduction and oxidant agents in the glass. The effectiveness of these agents varies from one to other. Among them, the Fe^{+2} is the most effective reduction agent. It can reduce Cr^{+6} to Cr^{+3} even in the presence of strong oxidant agents such as sodium sulfate. ACS has around 3.8% of iron in its composition, and formulation 90G would have circa 0.38% of iron. This amount would be insufficient to reduce the hexavalent chromium present in glass in the firing conditions. However, from the image mapping of Fe ion in the ceramic body 90G fired at 800ºC, shown in Table 3.7, it is possible to see iron and chromium are very close to each other in the ceramic body, which is necessary for iron to act as a reduction agent of hexavalent chromium.

However, if the effectiveness of Fe as a reduction agent for Cr^{+6} in the vitreous phase of the investigated ceramic bodies depends on the amount of Fe^{+2}, it can be considered irrelevant in the process of immobilization/mobilization of chromium in the ceramic body formulated with ACS and glass. Both Fe^{+2} and vitreous phase amounts are function of the glass amount in the formulation, and the vitreous phase is also dependant on the firing temperature.

K. Fuwa *apud* Weyl [45] replaced CaO for increasing amounts of Cr_2O_3 in different kinds of glasses and found that lithium-containing glasses are better solvents of chromium oxide than sodium-calcium-containing glasses, and that this solubility decreases when the temperature is decreased. A segregation of chromium dioxide from the vitreous phase, forming emerald green thin plates, was also observed.

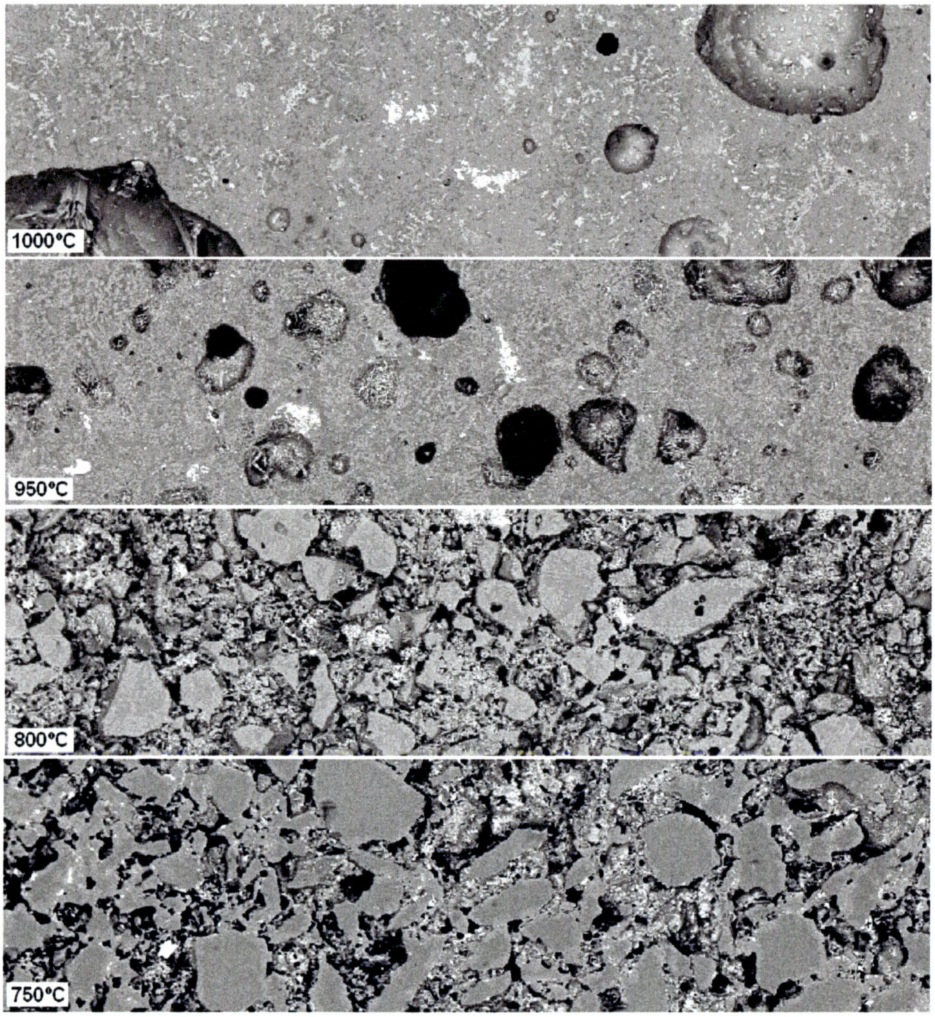

Figure 3.6. Micrographs of ceramic bodies made from formulation 90V with increasing firing temperatures.

Table 3.7. Image mapping of the ceramic body 90G fired at 800°C. Elements: Cr, Fe, and O.

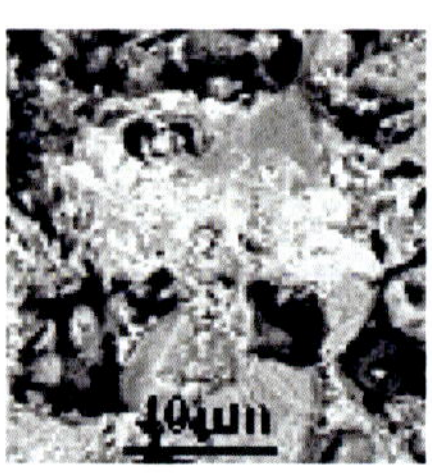	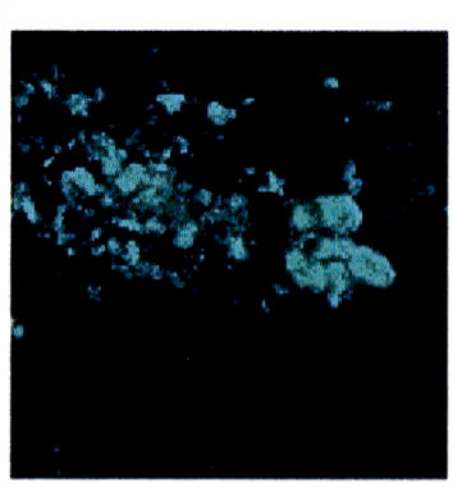		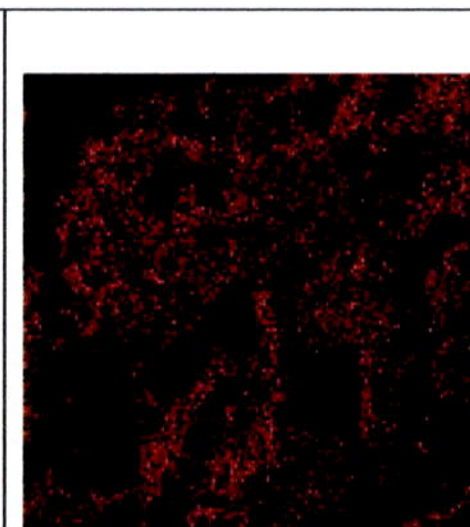
90V-800	Cr	Fe	O

In order to get deeper into the question concerning the main immobilization agents of chromium in ceramic bodies with glass additions, the ACS was replaced by pure chromium oxide (formulations glass-chromium), and other pure oxides were also added. The criterion to define the use of these oxides was the major presence in the chemical composition of ACS, except TiO_2, which was used because it is an excellent flux. Thus, a higher densification of ceramic bodies could be achieved, allowing determining the role of densification in chromium lixiviation.

Figure 3.7 shows the concentration of total lixiviate chromium from the ceramic body formulated with glass and 5% of Cr_2O_3 (5Cr) as a function of firing temperature.

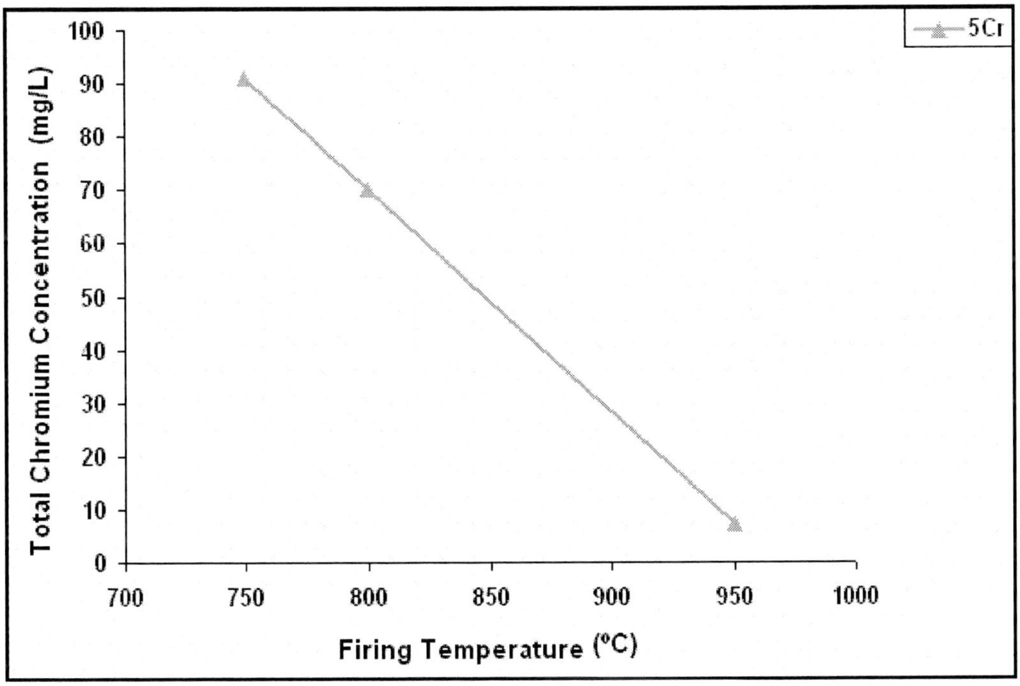

Figure 3.7. Concentration of total lixiviate chromium from ceramic bodies made from formulation 5Cr (5% Cr_2O_3 + 95% glass) as a function of firing temperature.

It can be observed that chromium lixiviation decreases about 12 times with the increase in firing temperature, between 750°C (91 mg/L) and 950°C (7,3 mg/L). In this case, this lower lixiviation could be associated with higher densification of the ceramic body, which is higher for increasing firing temperatures. In fact, when observing Figure 3.8, which shows the variation in water absorption of the same ceramic bodies as a function of firing temperature, the same behavior of the lixiviation can be seen (Figure 3.7). Obtained results indicates that the increase in firing temperature caused reduction in water absorption and increased linear shrinkage, leading to higher densification of the ceramic body and decreasing the contact area available for lixiviation as well.

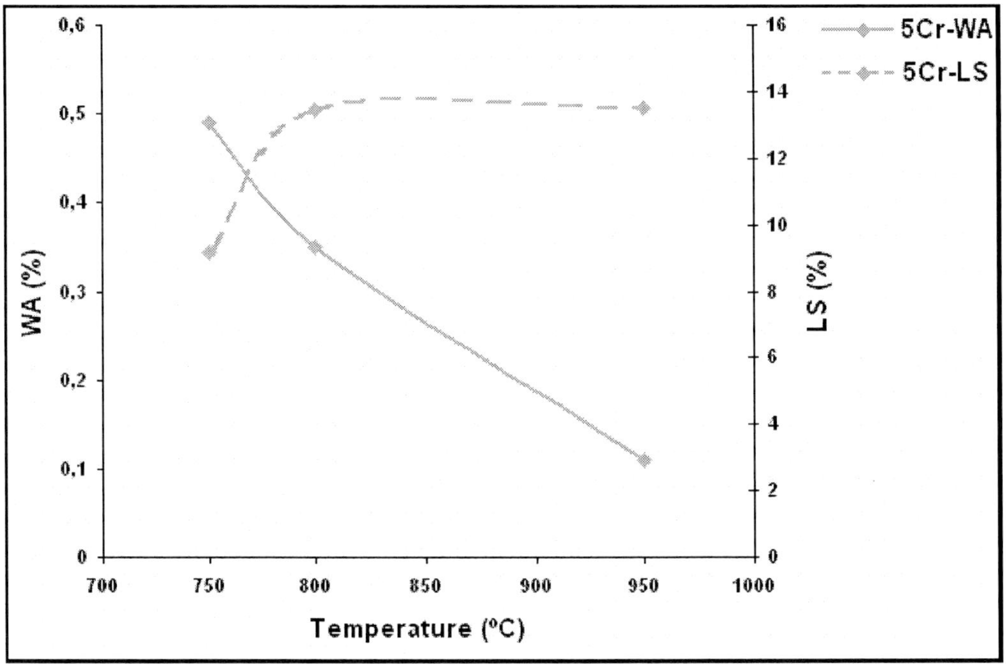

Figure 3.8. Gresification curve (water absorption – WA and linear shrinkage – LS as a function of firing temperature) of ceramic bodies obtained from formulation 5Cr (5% Cr_2O_3 + 95% glass).

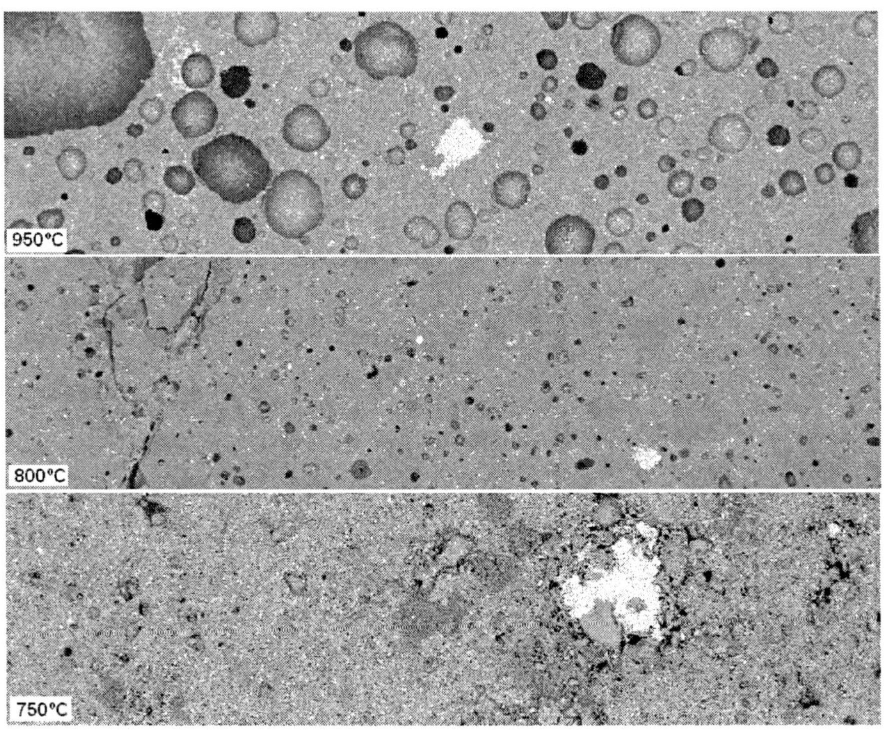

Figure 3.9. Micrographs of ceramic bodies made from formulation 5Cr for increasing firing temperatures.

However, the densification of ceramic bodies would not exclude the role of present phases and their transformation in function of temperature, as seen before. Figure 3.9 shows the micrographs of the ceramic bodies made from formulation 5Cr for increasing firing temperatures. When comparing these micrographs with those of ceramic bodies 90G (Figure 3.6), higher densification and vitrification of the microstructure can be observed, for the same firing temperatures.

In this case, the role of other constituents on the chromium lixiviation can be very significant. Table 3.8 summarizes the present phases, which were determined by X-ray diffraction of the ceramic bodies made from formulations glass-chromium and glass-chromium-oxide at different firing temperatures.

From data shown in Table 3.8, it can be observed that the phases present in the ceramic body 5Cr fired at 750°C are quartz, cristobalite, calcium and sodium silicates, and eskolaite. However, the phase sodium chromate is also present, what could explain the high chromium lixiviation for this firing temperature. The image mapping of sample 5Cr fired at 750°C shown in Table 3.9 supports this interpretation.

Table 3.8. Phases present in the ceramic bodies obtained with formulations glass-chormium-oxide

	Firing temperature (°C)			
	750	**800**	**950**	**1000**
5Cr	Quartz Cirstobalite Ca/Na silicate Eskolaite Sodium chromate	Quartz Cirstobalite Ca/Na silicate Eskolaite	Quartz Cirstobalite Ca/Na silicate Wollastonite Eskolaite	–
5Cr5Na	Quartz Cirstobalite Corundum Ca/Na silicate Eskolaite Sodium chromate	Quartz Cirstobalite Corundum Ca/Na silicate Eskolaite Sodium chromate	Quartz Cirstobalite Corundum Ca/Na silicate Eskolaite Sodium chromate	Quartz Cirstobalite Corundum Ca/Na silicate Eskolaite Sodium chromate
5Cr5Ca	Quartz Cirstobalite Ca/Na silicate Wollastonite Corundum Eskolaite Cosmoclor Sodium chromate	Quartz Cirstobalite Ca/Na silicate Wollastonite Corundum Eskolaite Cosmoclor Sodium chromate	Quartz Cirstobalite Ca/Na silicate Wollastonite Corundum Mullite Eskolaite Cosmoclor	Quartz Cirstobalite Ca/Na silicate Wollastonite Corundum Mullite Eskolaite Cosmoclor

Table 3.8. Continued

	Firing temperature (ºC)			
	750	**800**	**950**	**1000**
5Cr5Ti	Quartz Cirstobalite Ca/Na silicate Corundum Rutile Eskolaite	Quartz Cirstobalite Ca/Na silicate Corundum Rutile Eskolaite	Quartz Cirstobalite Ca/Na silicate Corundum Rutile Eskolaite	–
5Cr5Mg /	Quartz Cirstobalite Diopside Mullite Eskolaite Sodium chromate	Quartz Cirstobalite Diopside Mullite Eskolaite Sodium chromate	Quartz Diopside Mullite Corundum Al/Mg silicate Eskolaite	Quartz Diopside Mullite Corundum Al/Mg silicate Eskolaite

Cosmoclor=$NaCrSi_2O_6$ Diopside=Ca(Mg,Al)(Si,Al)$2O_6$ Wollastonite=$CaSiO_3$ Sodium chromate=$Na_2CrO_4.4H_2O$

Table 3.9. Image mapping of sample 5Cr fired at 750ºC and 950ºC.

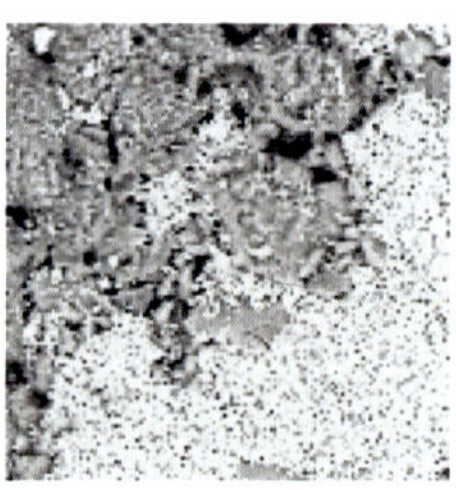	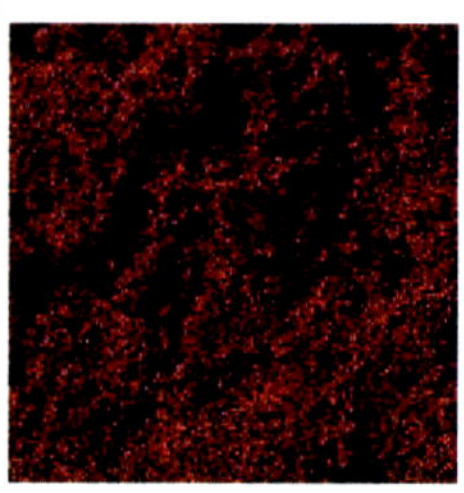	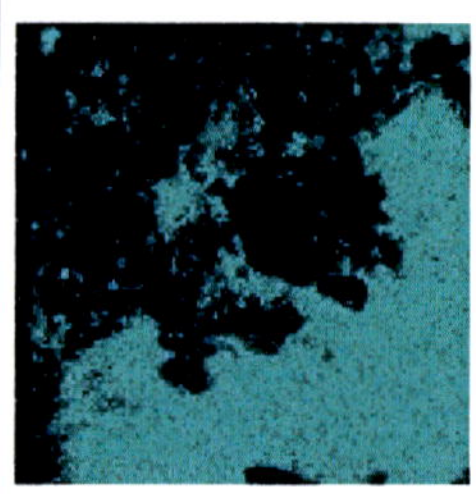	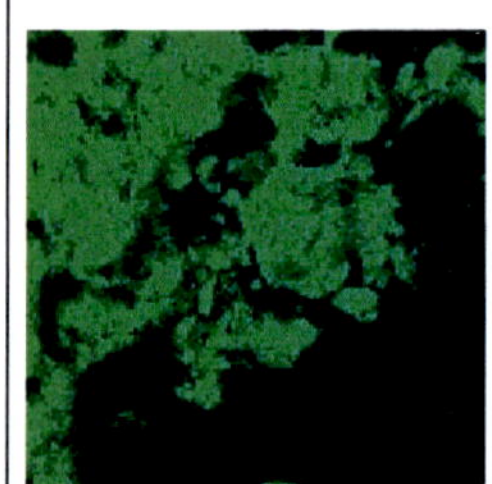
5Cr -750ºC	O	Cr	Si
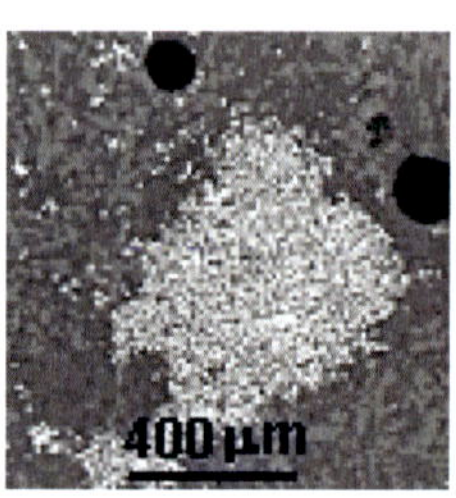	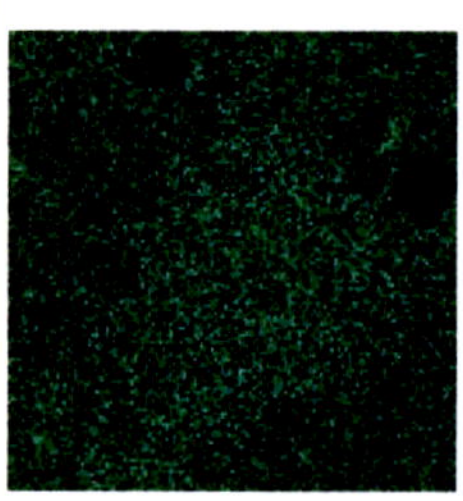	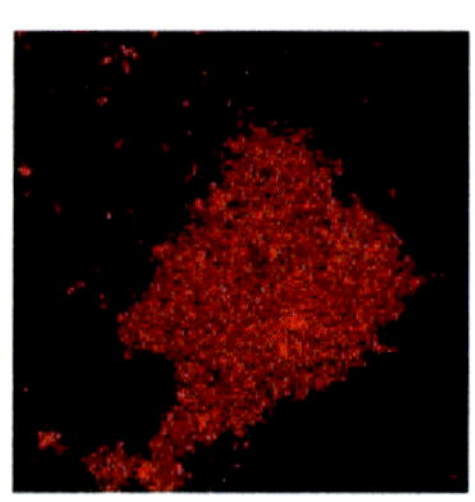	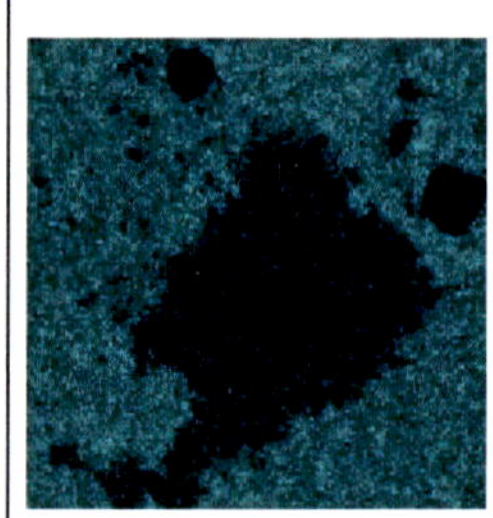
5Cr -950ºC	O	Cr	Si

The image mapping of Si and Cr shows that there is no chromium in the regions where silicon is present for 750°C, when the highest chromium lixiviation occurred. At 950°C – the lowest chromium lixiviation - the presence of Si and Cr is not so clear. This suggests, as expected, an increase in the vitreous phase, which causes a reduction in porosity and minimizes chromium lixiviation.

Formulations glass-chromium-oxide, which 5% of oxides (Na_2O, CaO, TiO_2, MgO) were separately added to, were prepared to estimate the role of these oxides in the chromium immobilization in the investigated ceramic bodies.

Figure 3.10 shows the concentration of total chromium in the lixiviation of ceramic bodies obtained from formulations with addition of 5% of Na_2O in masses with varying percentages of Cr_2O_3 and glass and submitted to different firing temperatures. These formulations are designed as follows: 5Cr5Na (5%Cr_2O_3+5%Na_2O+90G), 10Cr5Na (10% Cr_2O_3+5% Na_2O+85G), 15Cr5Na (15%Cr_2O_3+5%Na_2O+80G).

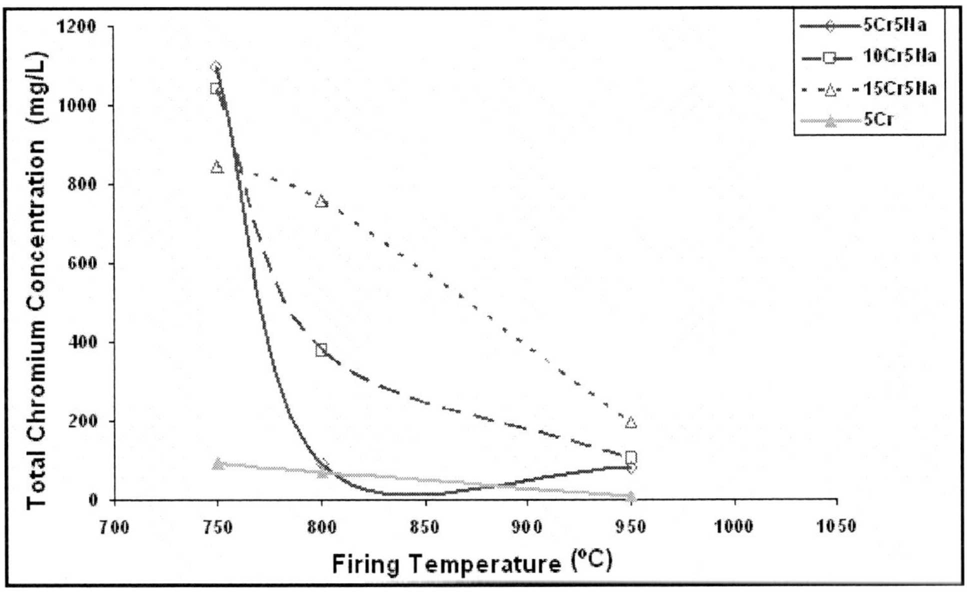

Figure 3.10. Concentration of total lixiviate chromium in ceramic bodies made from formulations 5Cr5Na, 10Cr5Na, 15Cr5Na, and 5Cr as a function of firing temperature.

From the data of the graph shown in Figure 3.10, it can be observed that the concentration of lixiviated chromium of the ceramic bodies made from all formulations decreased for increasing firing temperatures. Besides that, all formulations with addition of sodium oxide lixiviate more chromium ion than that the formulation containing only 5% of chromium oxide (5Cr). Hence, the presence of sodium is an instability factor of chromium in the ceramic body. From the formulations containing sodium oxide, the formulation 5Cr5Na was the one that showed the lowest concentration of lixiviated chromium, for all firing temperatures.

It is interesting to observe from the data of water retention and linear shrinkage (Figure 3.11) of the same ceramic bodies shown in Figure 3.10 that even tough samples with higher chromium content had higher water absorption and lower linear shrinkage, i.e., less densification (which causes higher chromium lixiviation), the same does not occur when the formulation contains sodium oxide. For the ceramic body 5Cr the chromium lixiviation was significantly lower than ceramic bodies 5Cr5Na, even tough they had the same densification. Thus, the strong relationship between the presence of sodium and the lixiviation of chromium ion in glass-containing ceramic bodies arises again.

This relationship becomes more evidenced in Table 3.10 and Figure 3.12, which show the concentration values of Cr^{+6} ion and compare them with the concentration of Na^{+} ion and other ions in the lixivia of ceramic bodies of the Cr-glass-Na series.

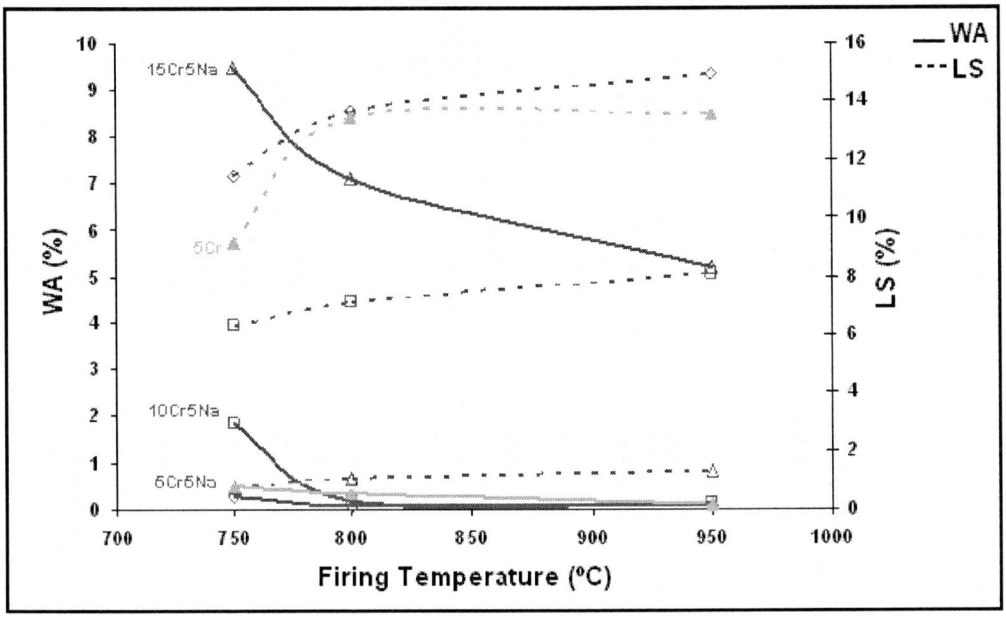

Figure 3.11. Gresification curve (water absorption – WA and linear shrinkage – LS) of formulations 5Cr5Na, 10Cr5Na, 15Cr5Na, and 5Cr.

Table 3.10. Concentration of present ions in the lixivia of ceramic bodies obtained from formulations 5Cr5Na, 10Cr5Na, and 15Cr5Na as a function of firing temperature.

Temperature	Formulations	Cr^{+6}	Na	Ca	Mg	Ti	K	Al	Si
750°C	5Cr5Na	1098	1112	368	1,5	<1	30	<0,2	63
	10Cr5Na	2079	1586	403	0,96	<1	57	<0,2	54
	15Cr5Na	2540	1982	387	0,95	<1	76	<0,2	51
800°C	5Cr5Na	87	103	19	0,25	<1	4,6	<0,2	1,2
	10Cr5Na	756	768	51	1,1	<1	31	<0,2	3,8
	15Cr5Na	2276	2419	292	0,82	<1	118	0,4	18
950°C	5Cr5Na	79	91	23	0,92	<1	3,6	1,9	7,3
	10Cr5Na	209	177	29	0,47	<1	9,3	0,8	5,4
	15Cr5Na	580	537	42	0,76	<1	20	1,2	8,1

From the data shown in Table 3.10 and Figure 3.12, regarding ions Na^{+1} and Cr^{+6}, it is possible to verify a consistent relation between these two ions concerning the lixiviate amounts. The concentration of other analyzed ions did not show similar relation with the concentration of ions Cr^{+6}. This reinforces the hypothesis that the presence of Na_2O causes a higher chromium lixiviation because the formation of sodium chromate, which is highly soluble and remains present in the ceramic bodies even for high firing temperatures. This can be observed in Table 3.11 for formulation 5Cr5Na fired at 950°C.

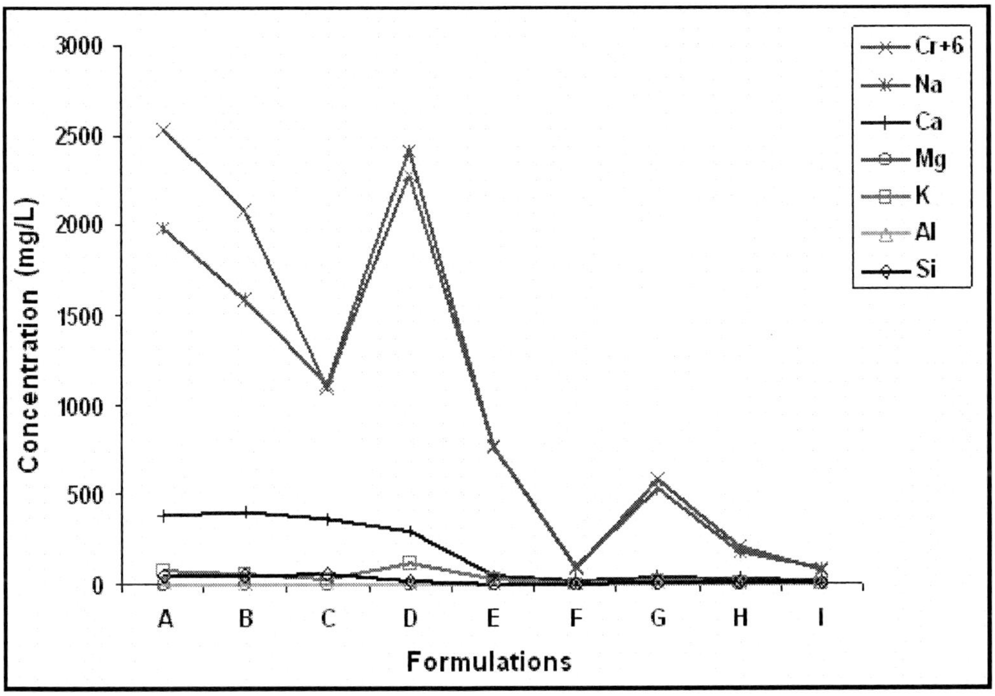

Figure 3.12. Concentration of present ions in the lixivia of ceramic bodies as a function of firing temperature for the following formulations: 750°C (A=15Cr5Na; B=10Cr5Na;C= 5Cr5Na); 800°C (D=15Cr5Na; E=10Cr5Na;F= 5Cr5Na); 950°C(G=15Cr5Na; H=10Cr5Na; I= 5Cr5Na).

In 1640 Van Helmont *apud* John [53] already observed that glass with excess alkalis have high solubility in water. In a study concerning chromium-containing lode incorporation in a vitreous matrix in temperatures between 800°C and 1000°C with holding time of two hours, Kirk *et al* [54] also attributed the chromium lixiviation to the formation of soluble sodium and/or calcium chromates.

Data shown in Table 3.10 indicate that the lixiviation of Ca^{+6} ion varies with the firing temperature of the ceramic bodies and follows the diminution in lixiviation of ions Cr^{+6} and Na^{+}. This could be in agreement with the observations of Kirk *et al.* [54]. However, when comparing the data from Table 3.10 with the lixiviation values for glass only (Table 3.11), it is verified that the proportion between the concentration of ions Na^{+1} and Ca^{+2} in the lixivia does not remain the same as for sodium-calcium-glass (approximately 1 to 1), even if the incorporation of sodium oxide in the formulation of the ceramic body is considered.

Table 3.11. Lixiviation of pure sodium-calcium glass.

	Cr^{+6}	Na	Ca	Mg	Fe	Si	Mn	Al	P	K	Ti
Glass	-	147	70	2,5	<0,03	62	0,04	<0,2	<0,01	8,2	<1

Hence, the assumption that the sodium chromate in the ceramic body plays a role in the solubilisation of chromium ion is here emphasized. From the difference between concentrations of Cr^{+6} and Na^{+1}, verified for ceramic bodies 15Cr5Na (A, Figure 3.12) and 10Cr5Na (B, Figure 3.12) fired at 750°C, one can assume that, because of the lower firing temperature, there is a lower vitrification in the ceramic bodies fired at higher temperatures or with lower content of chromium oxide and sodium oxide.

The addition of CaO to the formulations that contains chromium oxide and glass provides more data to analyze the role of Ca^{+2} ion in the immobilization of Cr^{+6} ion in the ceramic body during firing. Figure 3.13 shows the concentrations of total chromium in the lixiviation of ceramic bodies produced from the formulations with 5% CaO in masses with varying percentages of Cr_2O_3 and glass and submitted to different firing temperatures. These ceramic bodies were named as follows: 5Cr5Ca (5%Cr_2O_3+5%CaO+90G), 10Cr5Ca (10%Cr_2O_3+5%CaO+85G), 15Cr5Ca (15%Cr_2O_3+5%CaO+80G).

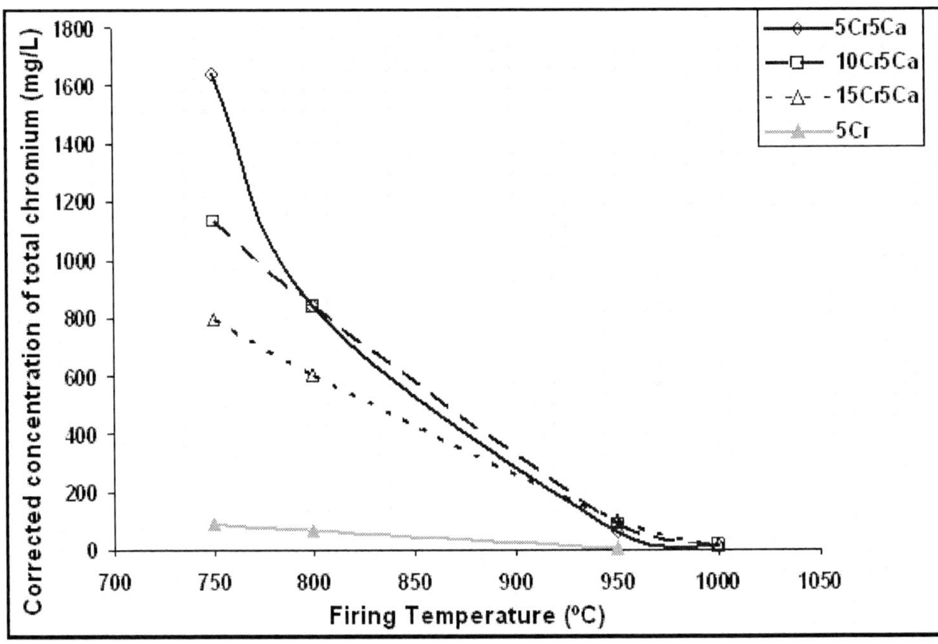

Figure 3.13. Corrected concentration of total chromium in the lixiviation of ceramic bodies produced from formulations 5Cr5Ca, 10Cr5Ca, 15Cr5Ca, and 5Cr as a function of firing temperature.

Likewise the addition of Na_2O in the formulations, the addition of CaO caused a higher chromium lixiviation. However, differently from the sodium oxide, this higher lixiviation can be associated to a lower densification of the ceramic bodies, as can be observed in the graph of the Figure 3.14, where water absorption and linear shrinkage data for all ceramic bodies of Figure 3.13 are shown.

The analysis of the present phases in the formulation 5Cr5Ca shows that there is no formation of calcium chromate despite the addition of CaO. However, the presence of sodium chromate was detected for firing temperature of 750°C, in agreement with the previous results that show that the Na^{+1} ion is the responsible for the chromium lixiviation by the formation of sodium chromate. This does not agree with the study made by Kirk *et al* [54], which states that the chromium lixiviation may be associated to the formation of calcium chromate. However, the role of calcium chromate in the chromium solubilisation does not exclude the role of sodium chromate in such process.

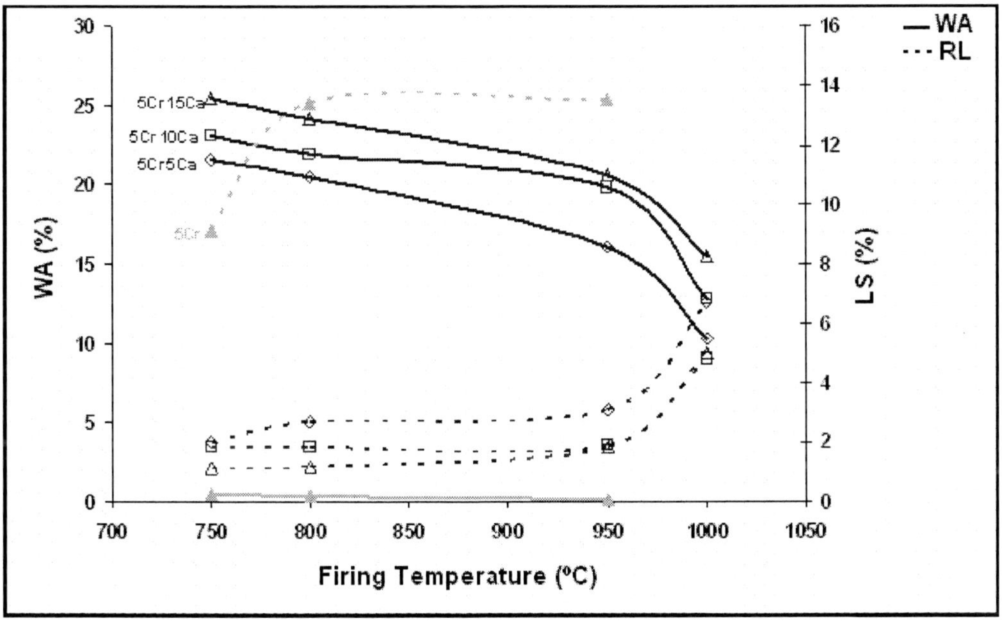

Figure 3.14. Gresification curve (water absorption – WA and linear shrinkage – LS) as a function of firing temperature of formulations: 5Cr5Ca, 5Cr10Ca, 5Cr15Ca, and 5Cr.

One explanation for the non-formation of calcium chromate in those formulations may be based on a study by Iler [55] and Kalousek [56]. According to them, additions of calcium compounds to sodium silicates cause the precipitation of insoluble silicates. This way it is assumed that the calcium present in the formulations is all, or almost all, in the form of wollastonite ($CaSiO_3$) and calcium-sodium silicate ($Na_2Ca_3SiO_8$). However, the detection limits of the X-ray diffraction technique (3%) must be considered.

Thus, the obtained data indicate that the higher chromium lixiviation with the addition of calcium oxide (even higher than the addition of sodium oxide) is related to the lower densification of the ceramic bodies. Hence, in order to increase densification and to evaluate its effect on the immobilization of chromium ion in the vitreous matrix, the role of titanium oxide in formulations containing chromium and glass was investigated.

Figure 3.15 shows the concentrations of total chromium in the lixiviation of ceramic bodies produced form formulations with additions of 5% of TiO_2 in masses with varying percentages of Cr_2O_3 and glass submitted to different firing temperatures. The formulations were named as follows: 5Cr5Ti (5%Cr_2O_3+5%TiO_2+90G), 10Cr5Ti (10%Cr_2O_3+5%TiO_2+85G), 15Cr5Ti (15% Cr_2O_3 + 5%TiO_2+80G). Graphs were normalized in function of chromium content in the samples, which was 5%.

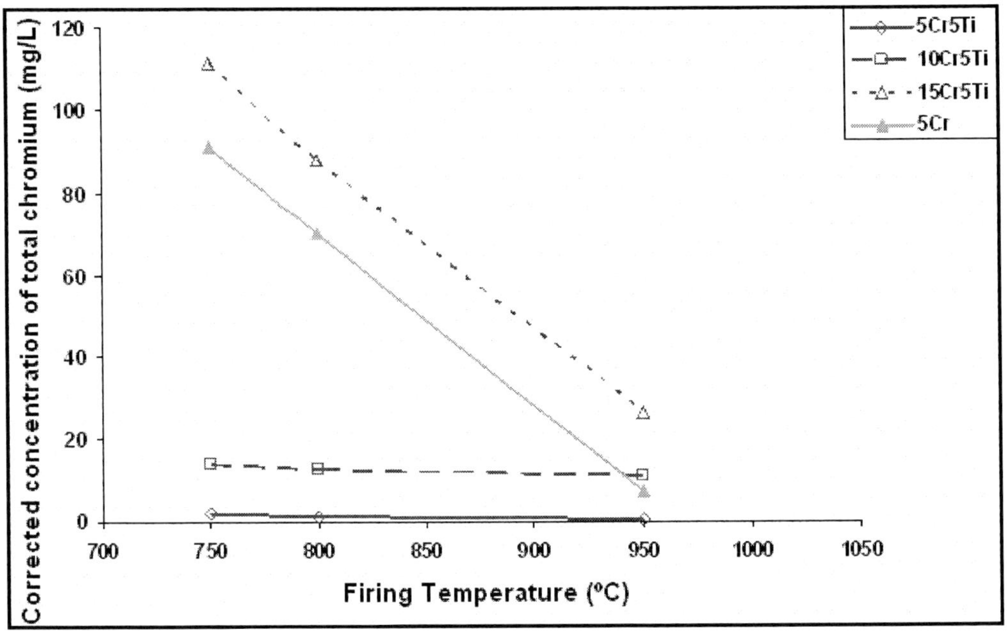

Figure 3.15. Correct concentrations of total chromium in the lixiviation of ceramic bodies produced from formulations 5Cr5Ti, 10Cr5Ti, 15Cr5Ti, and 5Cr as a function of firing temperature.

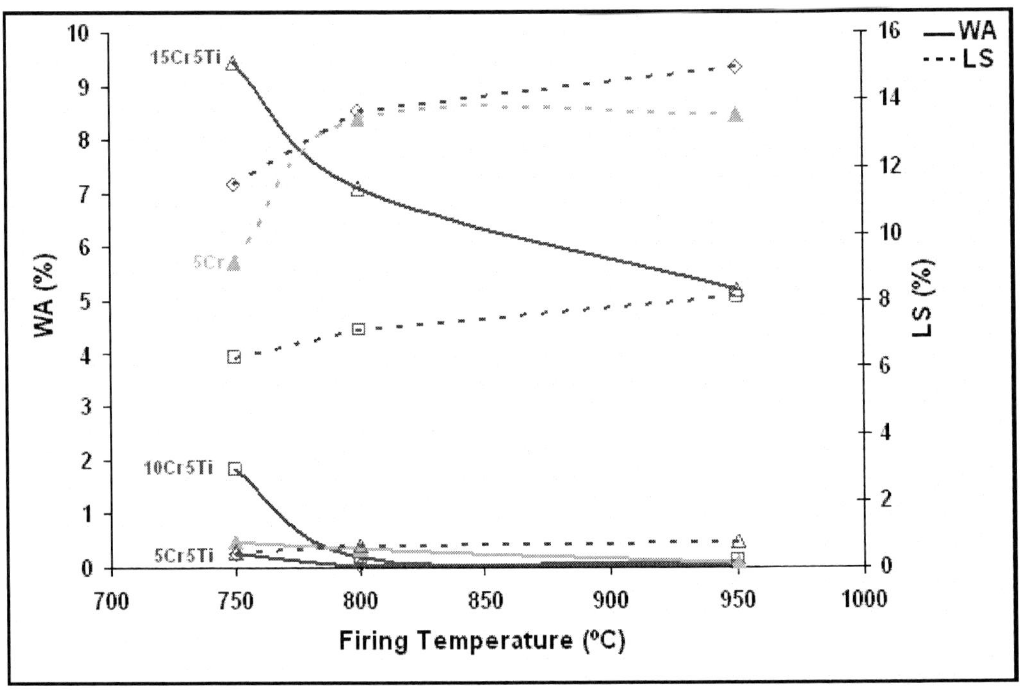

Figure 3.16. Gresification curves (water absorption – WA and linear shrinkage LS) as a function of firing temperature of formulations 5Cr5Ti, 10Cr5Ti, 15Cr5Ti, and 5Cr.

From the data of Figure 3.16, it can be verified that, differently from previous cases, the addition of TiO_2 was effective in the immobilization of chromium ion. For firing temperatures up to 950°C (when fired at 1000°C, these ceramic bodies were deformed), the chromium lixiviation of ceramic bodies 5Cr5Ti and 10Cr5Ti was lower than ceramic body 5Cr. It is interesting to observe that the chromium lixiviation of ceramic bodies 10Cr5Ti and 5Cr5Ti fired at 750°C and 800°C was significantly lower than the ceramic body 5Cr, even though the water absorption all of these bodies is quite similar and very low (<2%).

The data from Table 3.8 show that the crystalline phases present in formulation 5Cr5Ti are the same for any firing temperature. They are: eskolaite (Cr_2O_3), quartz (SiO_2), sodium calcium silicate ($Na_2Ca_3Si_2O_8$), cristobalite (SiO_2), corundum (Al_2O_3), and rutile (TiO_2). The formation of chromates was not observed, what could be expected in presence of Cr_2O_3 and Na_2O (which comes from the calcium-sodium glass). This is, probably, the explanation for the low chromium lixiviation of this formulation. However, the detection limits of the X-ray diffraction technique must be again considered. Phases with contents lower than 3% are hard to detect.

Figure 3.17 shows the concentrations of total chromium in the lixiviation of ceramic bodies produced from formulations with additions of 5% of MgO in masses with varying percentages of Cr_2O_3 and glass submitted to different firing temperatures. They were named as follows: 5Cr5Mg (5%Cr_2O_3+5%MgO+90G), 10Cr5Mg (10%Cr_2O_3+5%MgO+85G), 15Cr5Mg (15%Cr_2O_3+5%MgO+80G). Graphs were normalized in function of chromium content in the samples, which was 5%.

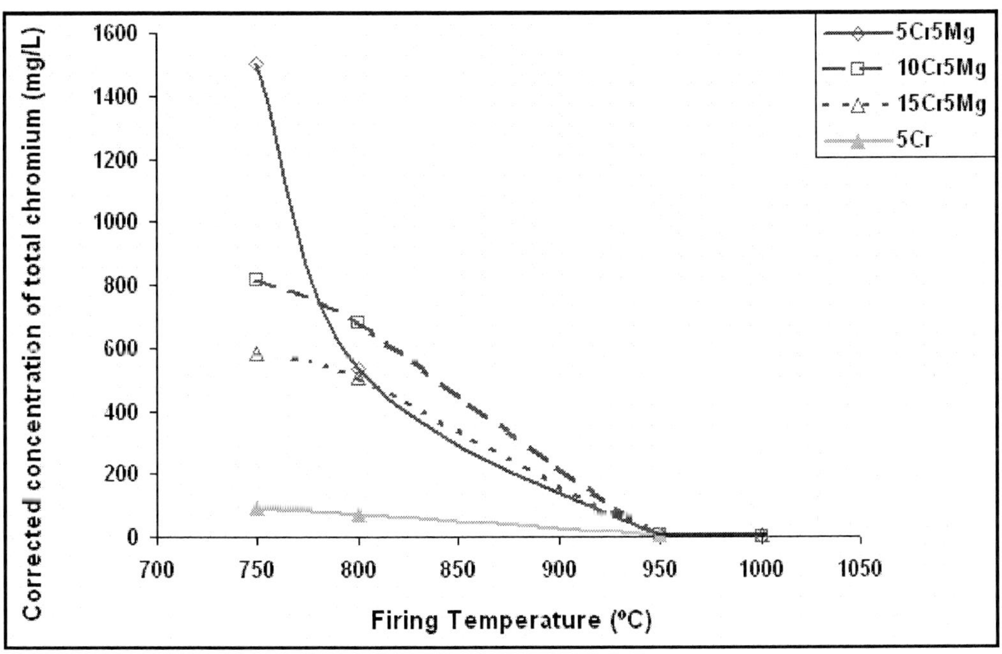

Figure 3.17. Total chromium concentration in the lixiviation of ceramic bodies produced from formulations 5Cr5Mg, 10Cr5Mg, 15Cr5Mg, and 5Cr as a function of firing temperature.

The effect of firing temperature of ceramic bodies is evidenced by the chromium ion lixiviation for firing temperature of 950°C, which reached a minimum value of 4.2 mg/L in the lixivia of ceramic body 5Cr5Mg. This can be seen in Figure 3.17. For 1000°C, the concentration of chromium in the lixivia of formulations 5Cr5Mg and 10Cr5Mg were 2.2.

and 2.7 mg/L, respectively (values below 5 mg/L, which is the limit established by NBR-10004 standard).

The graph of Figure 3.18 shows data concerning water absorption and linear shrinkage of the same ceramic bodies of Figure 3.17. From this graph, the intense densification that occurs for formulations with MgO fired in temperature of 800°C and above can be observed. However, values similar to those obtained to ceramic body 5Cr, regarding water absorption, occurred only for firing temperature of 1000°C.

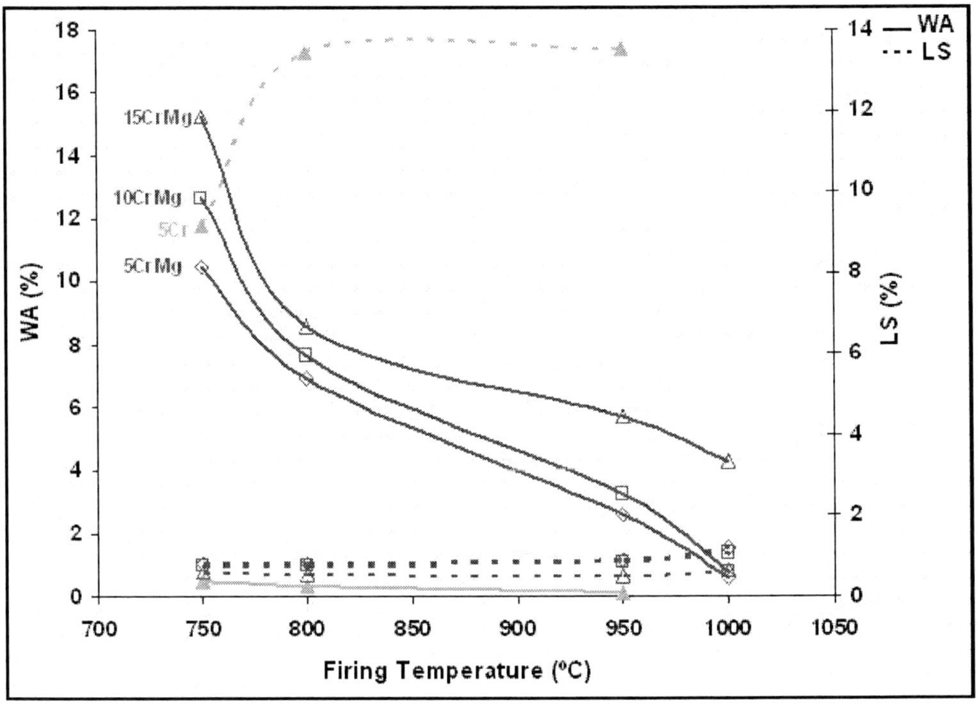

Figure 3.18. Gresification curve (water absorption – WA and linear shrinkage LS) as a function of firing temperature for formulations 5Cr5Ti, 10Cr5Ti, and 15Cr5Ti.

From the analysis of Table 3.8, it can be observed that the presence of sodium chromate in ceramic body 5Cr5Mg was only observed for firing temperatures of 750°C and 800°C, indicating that ability for chromium lixiviation, what was confirmed by the obtained results. From 950°C on, the crystalline phases quartz (SiO_2), mullite ($3Al_2O_3.2SiO_2$), corundum (Al_2O_3), aluminum magnesium silicate ($Al_2MgSi_2O_8$), and diopside ($Ca(Mg,Al)(Si,Al)2O_6$) remained in the structure. The phases cristobalite and sodium chromate were not detected. This is, probably, the explanation for the low chromium lixiviation for ceramic bodies fired in temperatures higher than 850°C, besides the significant decrease in water absorption mentioned before.

Figure 3.19 summarizes the influence of additions of oxides Na_2O, CaO, TiO_2, and MgO in chromium lixiviation. As can be seen, for temperatures of 750°C and 800°C all added oxides, with the exception of TiO_2, caused a higher chromium lixiviation. The chromium lixiviation for the formulation with addition of TiO_2 was lower for all firing temperatures, reaching a minimum of 0.56 mg/L for 950°C. The second lower concentration was obtained for the formulation containing MgO, with concentration of 2.2 mg/L when fired at 1000°C.

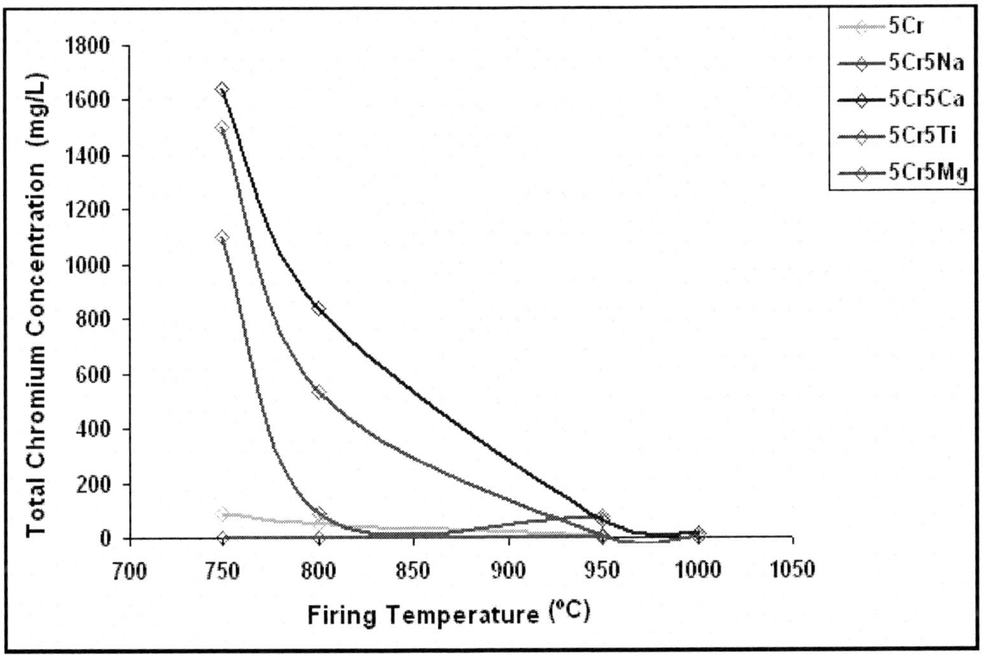

Figure 3.19. Total chromium concentration in the lixiviation of ceramic bodies produced from formulations 5Cr, 5Cr5Na, 5Cr5Ca, 5Cr5Ti, and 5Cr5Mg as a function of firing temperature.

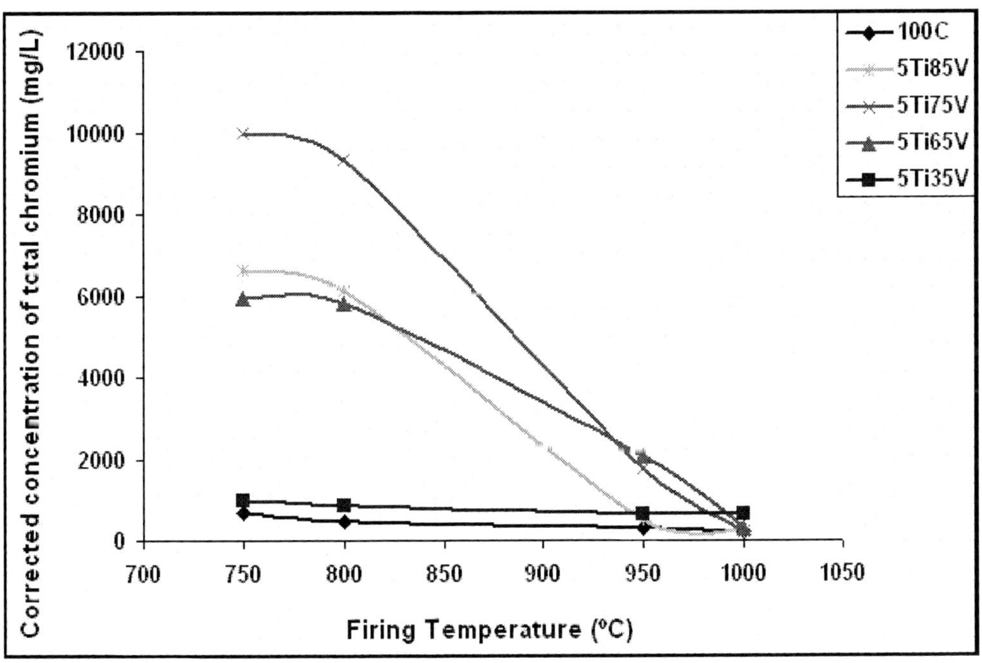

Figure 3.20. Corrected concentration of total chromium in the lixiviation of ceramic bodies produced from formulations 100C, 5Ti35V, 5Ti65V, 5Ti75V, and 5Ti85V as a function of firing temperature.

One possibility to transpose the obtained results for formulations containing pure oxides to ceramic bodies with ACS and sodium calcium glass in the composition is offered by the series whose formulations vary as follows: 10, 20, 30, and 60% of ACS and 5% of TiO_2 (sodium calcium glass: balance). The graph of Figure 3.20 shows the chromium concentration in the lixiviation of those ceramic bodies as a function of firing temperature. In the same graph, data for formulation 100C are also plotted.

As can be seen from figure 3.20, the chromium lixiviation of ceramic bodies produced with the addition of 5% of titanium and glass follow the same trend of ceramic bodies produced only with additions of glass to ACS (Figure3.1).

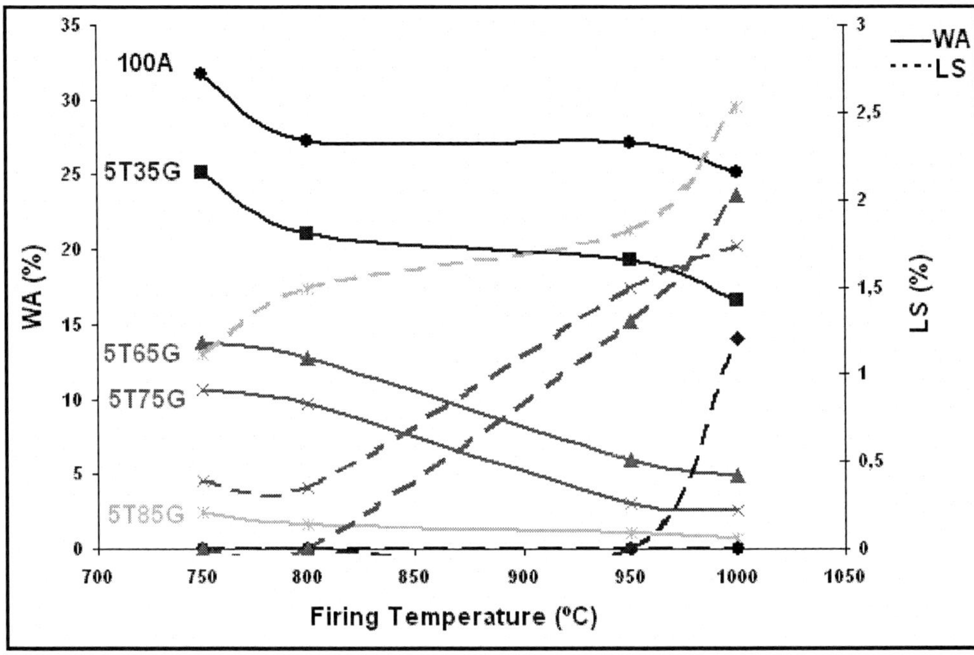

Figure 3.21. Gresification curve (water absorption – WA and linear shrinkage – LS) as a function of firing temperature of formulations 100C, 5Ti35V, 5Ti65V, 5Ti75V, and 5Ti85V.

The behavior of those ceramic bodies, concerning gresification curves, was similar to the ceramic bodies produced with glass and ACS, i.e., the higher the glass addition, the lower the water absorption and the higher the linear shrinkage. Gresification curve is shown in Figure 3.21. Hence, the densification of the ceramic bodies is intensified, leading to a lower reactivity, which favors the lixiviation phenomenon.

Figure 3.22 shows the micrographs of the ceramic bodies of formulation 5Ti85G with increasing firing temperatures. As can be seen, the number of phases in the microstructure of this ceramic body is higher than in ceramic bodies with ACS and glass, and more open pores are observed when compared to ceramic bodies 5Cr (Figure 4.9).

However, a significant increase in chromium immobilization when compared to formulations with ACS and glass was expected, because of the more intense formation of vitreous phase caused by TiO_2 addition. This expectation was based in the obtained results for formulation containing glass and pure Cr_2O_3 and TiO_2 addition, but this is not observed in the data shown in Figure 3.20. Formulations with glass and ACS (Figure 3.1) had a minimum concentration of chromium in the lixivia of 57 mg/L, for the formulation 90G (10% ACS +90% glass) fired at 1000°C. For this temperature, the chromium concentration for the formulation 5Ti85G (5% TiO_2 + 85% glass + 10% ACS) was 23 mg/L. Oppositely, the chromium concentration for the formulation 5Cr5Ti (5% Cr_2O3 + 5% TiO_2 + 90% glass) fired at 950°C was only 0.56 mg/L.

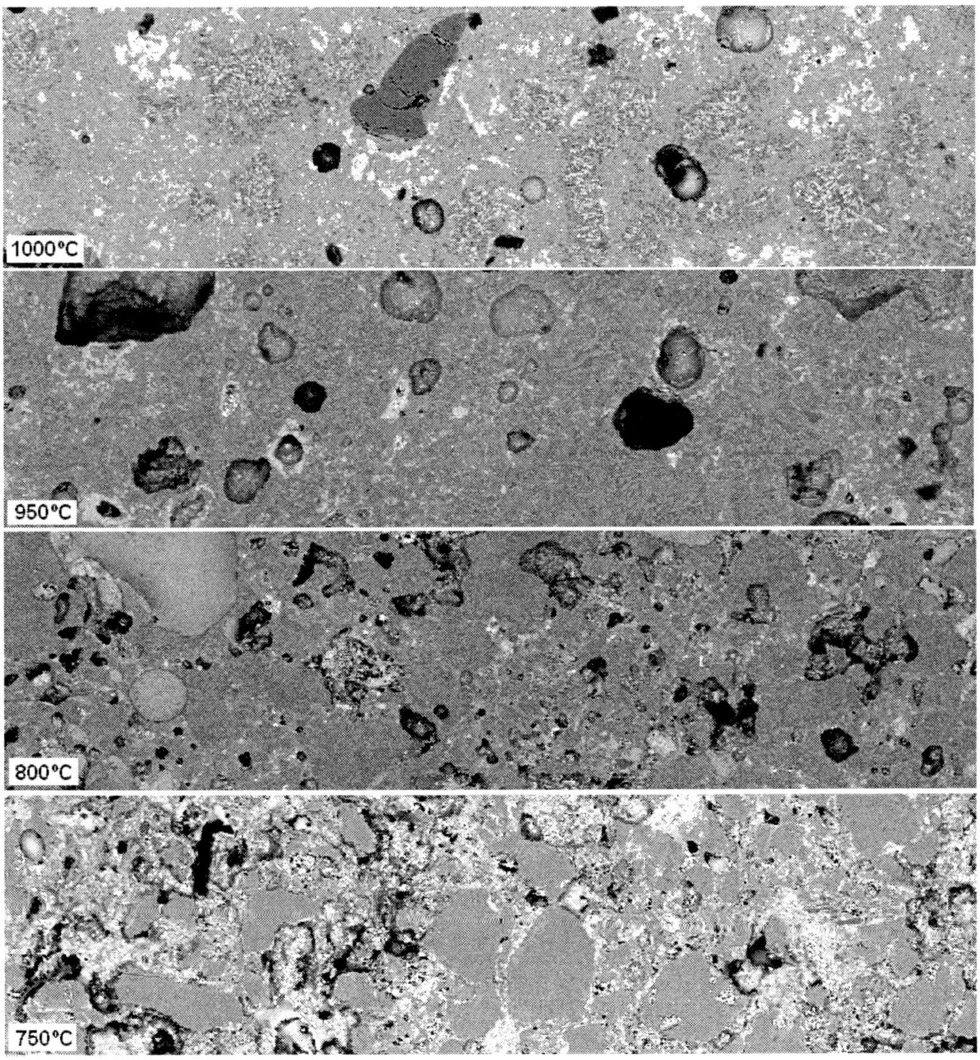

Figure 3.22. Micrograph of ceramic bodies of formulation 5Ti85G with increasing firing temperatures.

On possible explanation to such fact is the formation of soluble phases with the addition of TiO_2 to ACS, since it has in its composition other oxides such as Na_2O, which is also present in glass. So, the oxidation of Cr^{+6} is facilitated and sodium chromate is formed, making the chromium lixiviation higher.

In fact, as can be seen in Table 3.12, the phase sodium chromate is present in ceramic bodies produced from formulation 5Ti85G (10% ACS + 5% TiO_2 + 85% glass) for all firing temperatures. This phase was detected in all ceramic bodies that had low efficiency in chromium immobilization in all investigate formulations.

In the Table 3.12 the present phases in ceramic bodies produced with formulation 5Mg85G (10% ACS + %5 MgO + 85% glass) fired at 800°C are also shown. Again, the phase sodium chromate was detected, indicating the cause of the high chromium lixiviation for this firing temperature (819 mg/L), as can be seen in Figure 3.23.

Table 3.12. Present phases in ceramic bodies ACS-glass-oxide.

	Firing Temperature (°C)			
	750	**800**	**950**	**1000**
5Ti85V	Cristobalite Quartz Ca/Na silicate Rutile Corundum Eskolaite Sodium chromate	Cristobalite Quartz Ca/Na silicate Rutile Corundum Eskolaite Sodium chromate	Cristobalite Quartz Ca/Na silicate Rutile Corundum Eskolaite Sodium chromate	Cristobalite Quartz Ca/Na silicate Rutile Corundum Eskolaite Sodium chromate
5Mg85V	(did not sinter)	Quartz Corundum Aluminum phosphate Wollastonite Clinoenstatite Ompacite Eskolaite Sodium chromate	Quartz Corundum Aluminum phosphate Ompacite Eskolaite	Quartz Corundum Ompacite Eskolaite

Ompacite:$NaMgAl(Si_2O_6)_2$ Clinoenstatite: $MgSiO_3$

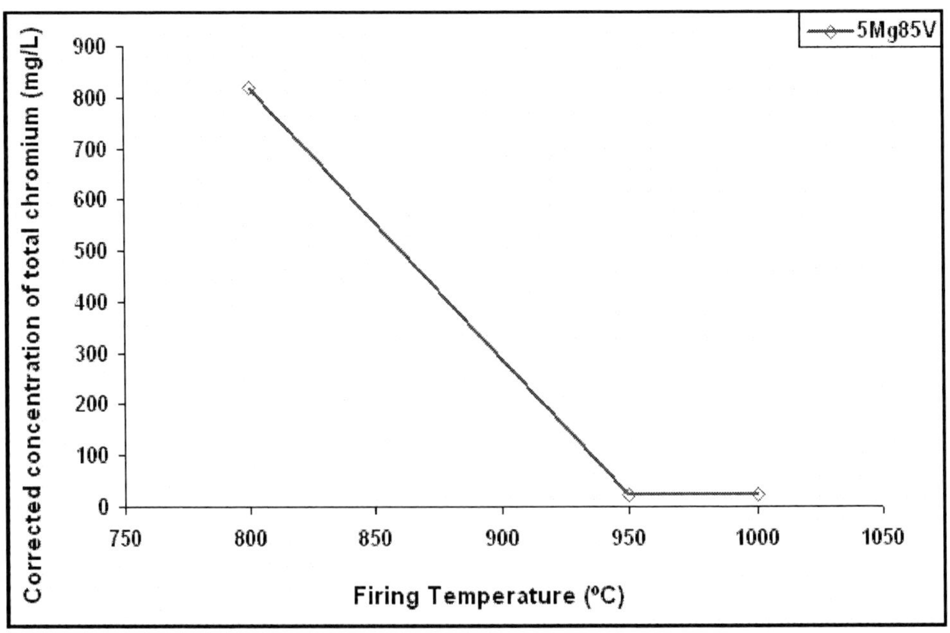

Figure 3.23. Total chromium concentration in the lixiviation of ceramic bodies produced from formulations glass-ACS with addition of 5% of MgO as a function of firing temperature.

Yet, from the same figure, a drastic reduction in lixiviation of ceramic bodies fired above 850°C can be observed. In this temperature range, according to Table 3.12, there is no more sodium chromate. The concentration of Cr^{+6} in the lixivia decreases to 23 mg/L and 2.4 mg/L for 1000°C (about ten times lower than for 950°C). This decrease must be probably associated to the higher densification of the ceramic bodies fired at this temperature, since the phase sodium chromate was not detected for both temperatures. The gresification curve of ceramic bodies 5Mg85G as a function of firing temperature is shown in Figure 3.24.

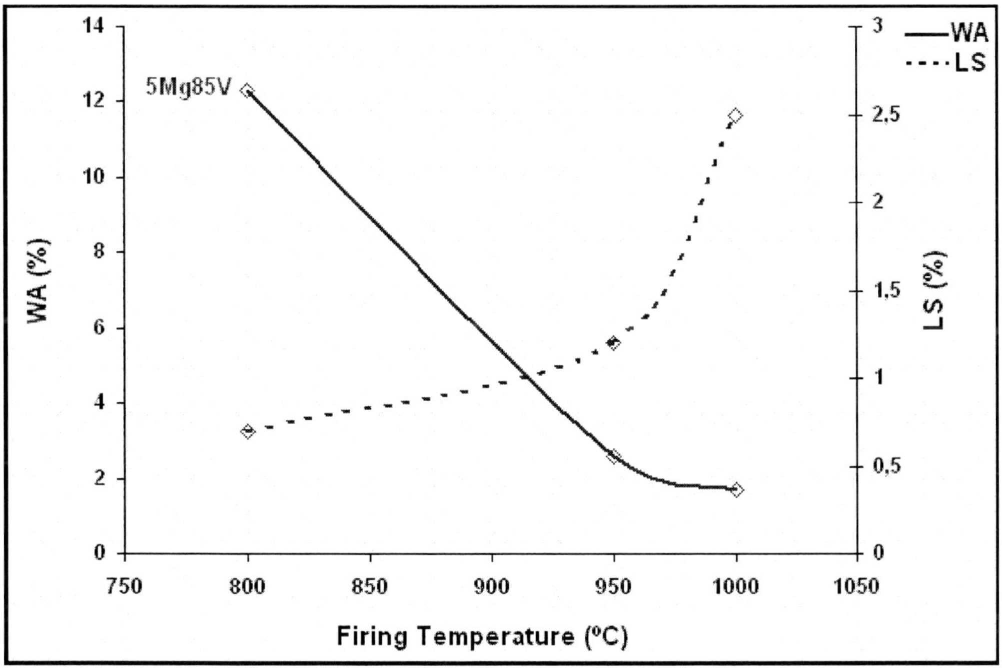

Figure 3.24. Gresification curve (water absorption – WA and linear shrinkage – LS) as a function of firing temperature of samples glass-ACS with addition of 5% of MgO.

As can be seen from Figure 3.24, the porosity was lower and linear shrinkage was higher at 1000°C. Hence, a higher densification was obtained, lowering the surface are available for chromium lixiviation, explaining the lower chromium lixiviation observed for this temperature.

Figure 3.25 shows the micrographs of ceramic bodies of formulation 5Mg85G with increasing firing temperatures. Pores get more rounded and the microstructure becomes more closed, decreasing the surface area available to promote the contact between the media and the ceramic body surface.

Figure 3.26 shows the total lixiviate chromium as a function of the percentage of MgO added to the formulation 90G, fired at 1000°C. As can be observed, addition of 5% of MgO reduces the chromium lixiviation drastically. However, addition of 10% of MgO and firing at the same temperature leads to a increase in chromium lixiviation. Water absorption and linear shrinkage for this ceramic body were 6.81% and 1.29%, respectively, indicating a significantly lower densification of this ceramic body when compared to those shown in graph of Figure 3.24. The higher solubility of ceramic body 10Mg80G would be then associated to its higher open porosity (water absorption).

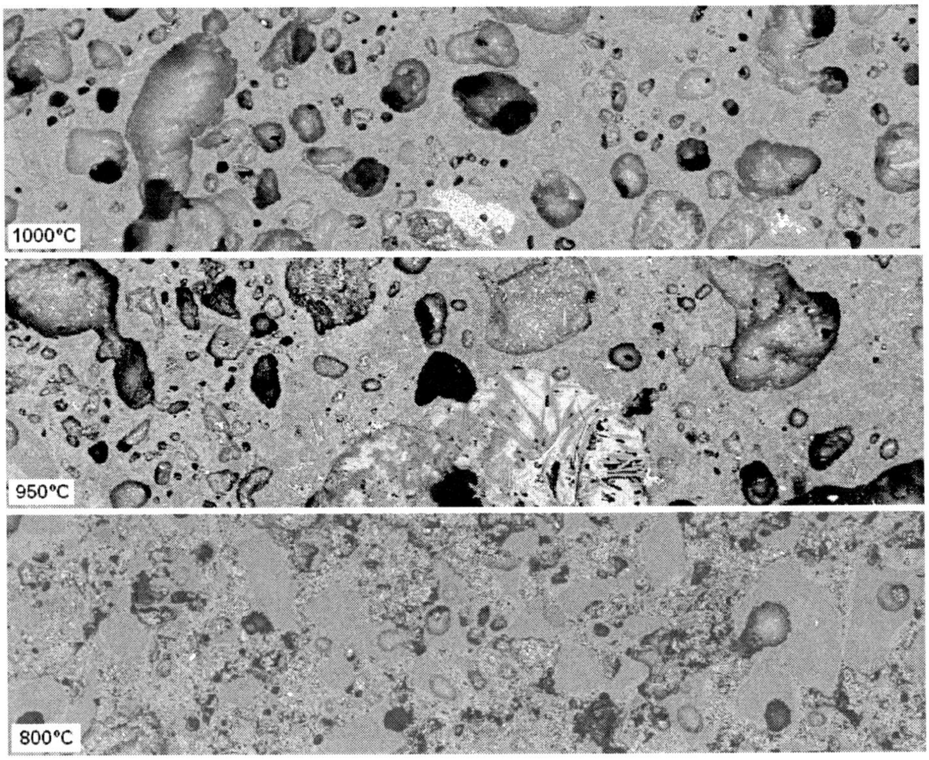

Figure 3.25. Micrograph of ceramic bodies produced from formulation 5Mg85G. Firing temperature is indicated in the figure.

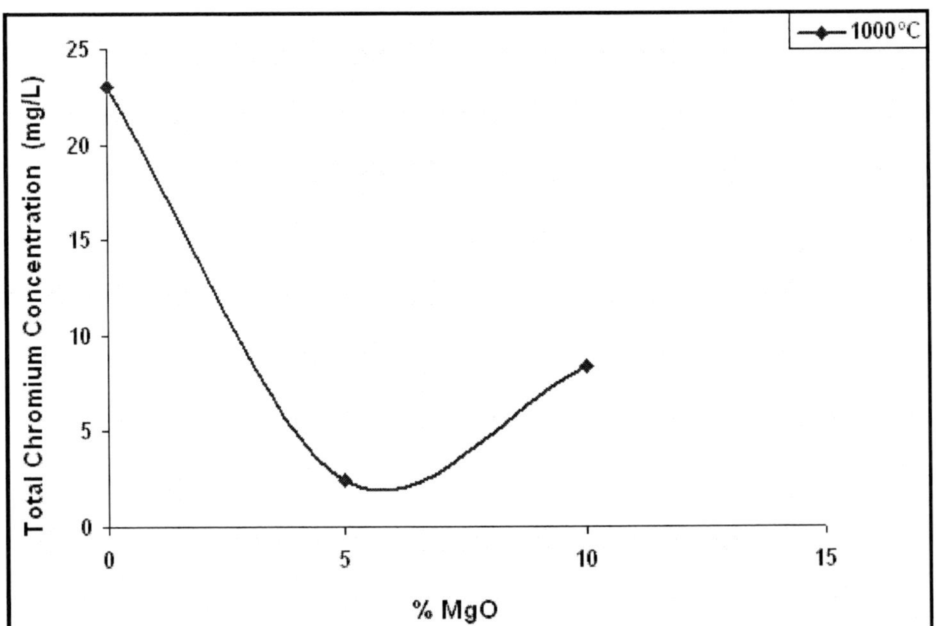

Figure 3.26. Total lixiviate chromium concentration as a function of MgO percentage added in sample containing 10% ACS and fired at 1000°C.

Table 3.13 shows the mapping of elements O, Cr, Na, and Si of ceramic bodies 5Mg85G, for firing temperatures of 800°C and 1000°C.

Table 3.13. Image mapping of sample 5Mg85G fired at 800°C.

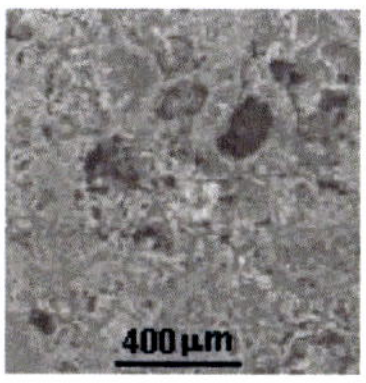		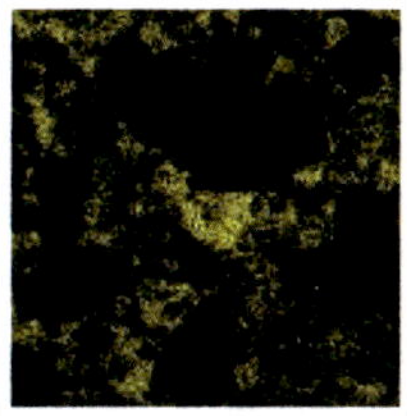	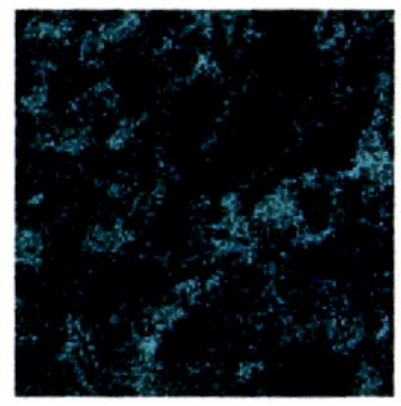	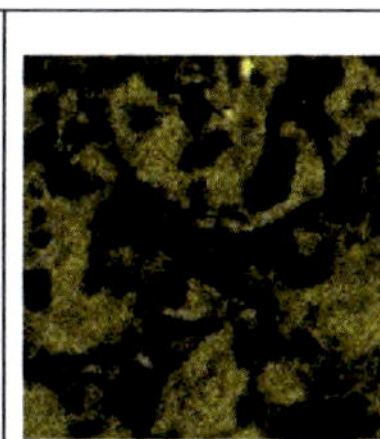
5Mg85G-800ºC	O	Cr	Na	Si
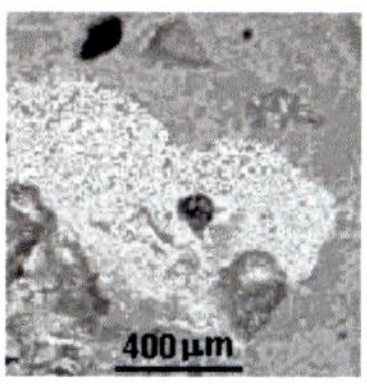	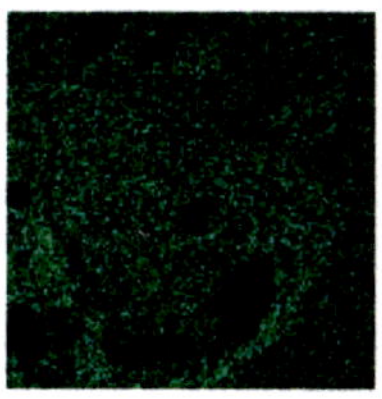	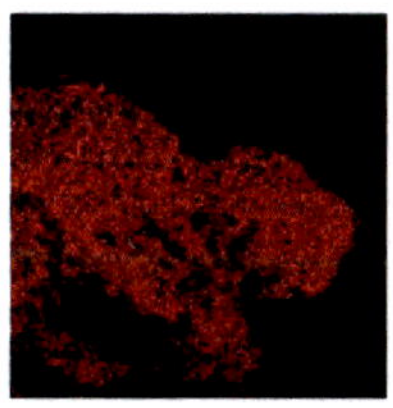	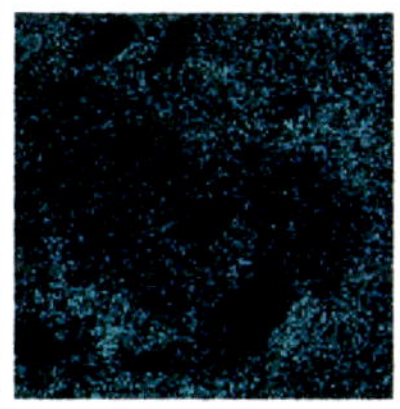	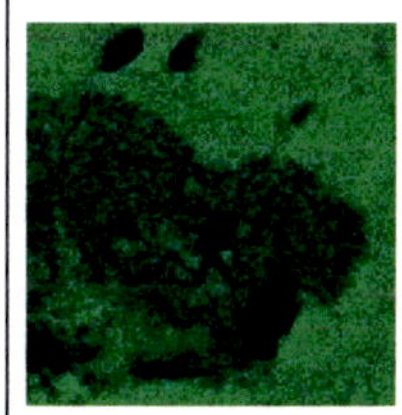
5Mg85G-1000ºC	O	Cr	Na	Si

From the shown mappings, a more homogeneous distribution of O, Na, and Si in the microstructure of the ceramic body fired at 1000°C can be observed. Cr is concentrated in one single grain (probably eskolaite). When comparing to the mapping of the sample fired at 800°C, it can be verified that the distribution of Si and Cr are almost complementary. This suggests that the higer densification of the ceramic body fired at the highest temperature is associated to the microstructure vitrification, in which both Cr and Si are not totally segregated.

Because of the immobilization of the chromium obtained for ceramic bodies of formulation 5Mg85G (10% ACS + 5%MgO + 85% glass), this formulation was submitted to gaseous emissions evaluation. Obtained results are shown in Table 3.14.

Table 3.14. Gaseous emissions evaluation of ceramic bodies 5Mg85G for firing temperatures of 750°C and 1000°C.

Samples	Ca (mg/L)	Na (mg/L)	Mg (mg/L)	Cr $_{total}$ (mg/L)
5Mg85G-750ºC	0,40	0,19	0,11	< 0,02
5Mg85G-1000ºC	0,19	0,04	0,03	< 0,02

As can be seen in Table 3.14, the chromium ion concentration was bellow the atomic absorption detection limits for all gaseous emissions during firing of ceramic bodies.

4. Conclusion

From the results obtained in this work, the following conclusions can be drawn:

It was possible to obtain the immobilization of the chromium ion from ACS according to the limit established by NBR 10.004 standard (5mg/L). This immobilization was evaluated by lixiviation tests, according to NBR 10.005, applied to ceramic bodies produced from formulations containing ACS and glass, and addition of magnesium oxide as a vitrification/densification agent.

Concerning pure oxides formulations, the one with addition of TiO_2 showed the lowest chromium lixiviation for all firing temperatures, reaching a minimum of 0.56 mg/L, followed by formulation with MgO addition fired at 1000°C, with a chromium lixiviation of 2.2 mg/L.

From gaseous emissions data, it was verified that there was no detectable chromium evolution by gas emissions during firing of ceramic bodies. Results of chromium ion analysis in washing gases were bellow the atomic absorption detection limits.

The chromium immobilization in ceramic bodies was more effective for higher firing temperatures. Increasing firing temperatures decrease open porosity by vitreous phase formation, leading to a decrease in chromium lixiviation.

Addition of ACS to glass reduces water absorption and increases linear shrinkage of ceramic bodies. Hence, there is a higher densification of ceramic bodies, reducing the contact area available for lixiviation. However, sole addition of glass to ACS increased chromium lixiviation of ceramic bodies, even if it is lower for increasing firing temperatures. This is due to the increase in lixiviable phases, such as sodium chromate, from the reaction of sodium, calcium sodium glass, and chromium from ACS.

Chromium lixiviation is directly related to the presence of sodium in the glass-containing ceramic bodies. The presence of Na_2O leads to the formation of sodium chromate, which is highly lixiviable. Addition of CaO also increases chromium lixiviation, but this increase is associated to a lower densification of ceramic bodies.

Open porosity is not the only determining factor of lixiviation in ceramic bodies. However, a relationship between chromium lixiviation decrease for 950°C and densification increase was evidenced. Thus, higher lixiviation not only by lower densification, but also by the sodium presence.

The chromium ion immobilization was poor for all investigated ceramic bodies that had the presence of sodium chromate in their constitution.

Ceramic bodies obtained only from ACS (100C) with no glass additions had the phases quartz (SiO_2), eskolaite (Cr_2O_3), and magnesiochromite ($MgCr_3O_4$) in their composition for all firing temperatures. These phases are already present in ACS before firing, except magnesiochromite. Appearing of hexavalent chromium as magnesium chromate in ACS, could explain the chromium lixiviation for the ceramic bodies 100C.

The decrease in chromium lixiviation in ceramic bodies for the investigated firing temperatures is associated to a lower amount of phases with hexavalent chromium, and to a more closed microstructure and the presence of vitreous phase in the ceramic body.

References

[1] Basegio, T.; Berutti, F.; Bernardes, A.; Bergmann, C.P. *Journal of the European Ceramic Society*. 2002, vol. 22, 2251-2259.

[2] European Commission, Directorate-General JRC, Joint Research Centre, Institute for Prospective Technological Studies (Seville),Technologies for Sustainable Development, European IPPCBureau, Integrated Pollution Prevention and Control (IPPC),Reference Document on Best Available Techniques for the Tanning of Hides and Skins, Final Draft/February 2003, http://eippcb.jrc.es.

[3] Basegio, T.M. Imobilizacão do Ion Cromo de Cinzas da Incineração de Serragem de Couro em Corpos Cerâmicos Vitrificados, Porto Alegre, Ph.D. Thesis, UFRGS, Brazil, 2004, 118 pp.

[4] Cassano, A.; Drioli, E.; Molinari, R. *Desalination,* 1997, vol.113, 251–261.

[5] Cassano, A., Drioli, E. ; Molinari, R., *Water Resarch*, 1999,vol. 4, 443–450.

[6] Cassano, A., Drioli, E., Molinari, R. ;Bertolutti, C. *Desalination* , 1996, vol.108, 193–203.

[7] Environmental Protection Agency. USA. 40 CFR. Chapter 1, Part 261, *Identification and Listing of Hazardous Waste*. App VIII. 1998.

[8] Associação Brasileira de Normas Técnicas: NBR 10004. Resíduos:Classificação. 1987.

[9] Bundesrat: Zweite allgemeine Verwaltungsvorschrift zum Abfallgesetz (TA-Abfall). GMDI 170, 1990, Germany.

[10] Castilhos, D.D. Alterações Químicas e Biológicas Devidas à Adição de Resíduos de Curtume e de Cromo Hexavalente no Solo, Master Degree Thesis, UFRGS, Brazil, 1998.

[11] Ferreira, S.A. Efeitos da Adição de Resíduos de Curtume e Carbonífero nas Plantas e no Solo, Master Degree Thesis, UFRGS, Brazil, 1998.

[12] Rivela, B.; Bornhardt, C.; Vidal, G.; Mendez, R. *Waste Management & Resarch*. 2004, vol. 22, 131–141.

[13] Albino, V. *Environmental. Technology*.1996, vol. 17, 309–315.

[14] Singh, D.; Wagh, A.S; Tlustochowicz, M. Jeong, S.Y. *Waste Management*. 1998, vol.18, 135–143.

[15] Lin, J.S.; Shen, P. *J. Non-Cryst. Solids*.1996, vol. 204, 135–140.

[16] Yalcin, N.; Sevinc, V. *Ceram. Int*.2000, vol. 26, 485–493.

[17] Pelino, M. *Waste Management*.2000, vol. 20, 561–568.

[18] I. Kim, T.; Kim, J.H.;Lee, K.S.; Seo,Y.C.; Koo, J.K.Waste Management. 2000, vol. 20, 409–416.

[19] Romero, M.; Rawlings, R.D.; Rincón, J. Ma. *J. Eur. Ceram. Soc*. 1999, vol.19, 2049–2058.

[20] Barbieri, L.; Bonamartini, A.C. ; Lancellotti, I. *J. Eur. Ceram. Soc*. 2000, vol. 20, 2477–2483.

[21] Dominguez, E.A. ; Ullmann, R., *Appl. Clay Sci*. 1996, vol.11, 237–249.

[22] Pisciella, P.; Crisucci, S.; Karamonov, A. ;Pelino, M. *Waste Manage*. 2001, vol. 21, 1–9.

[23] Cimdins, R.; Rozenstrauha, A.; Berzina, L.; Bossert, J.; Bucker, M. Glass
Resour. Conserv. Recycl., 2000, vol. 29, 285–290.

[24] Herman, D.; *Ceram. Int*. 1998, vol. 24, 515–520.

[25] Dweck, J.; Buchler, P.M.; Coelho, A.C.V.; Cartledge, F.K. *J. Environ. Sci. Health*, Part A 35, 2000, 715–740.
[26] Wang, S.; Vipulanandan, C. *Cem. Concr. Res.*, 2000, vol. 30, 385–389.
[27] Filibeli, A.; Buyukkamaci, N. ; Senol, H. *Resour. Conserv. Recycl.*, 2000, vol. 29, 251 –261.
[28] Kindness, A.; Macias, A.; Glasser, F.P. *Waste Manage*. 1994, vol.14, 3–11.
[29] Wu, C.-H.; Lin, C.-F.; Horg, P.-Y. *J. Environ. Sci. and Health*, 2004, vol.A39, n.1, 237-252.
[30] Swarnalatha, S.; Ramani, K.; Geettha Karthi, A.; Seakaran *Journal of Hazardous Materials*, 2006, vol. B137, 304-313.
[31] Basegio, T.; Berutti, F.;Bernardes, A.; Bergmann, C.P. *Journal of European Ceramic Society*, 2002, vol. 22, 2251-2259.
[32] Basegio, T.; Hass, C.; Pokorny, A.;Bernardes, A.; Bergmann, C.P. *Journal of Hazardous Materials*,2006, vol.137, issue 2,1156-1164.
[33] Basegio,T.;Machado,A.;Bernardes, A.; Bergmann, C.P. *Waste Management & Resarch*, 975-9, 2006, vol.24, n.6,537-544.
[34] Campos,M.;Velasco,F.; Martinez, M.A.;Torralba,J.M. *Journal of European Ceramic Society*, 2004, vol. 24, 811-819.
[35] Bernstein, A.G.; Bonsembiante, E.;Brusatin, G.;Calzolari, G.; Colombo, P.; Dall'Igna, R.; Hreglich, S.; Scarinci, G. Waste Management, 2002, vol. 22, 856-869.
[36] Barbieri, L. *Journal of European Ceramic Society*, 2000, vol.20, 2477-2483.
[37] Karamanov, A.; Gutzov, I.;Chomakov, I.; Christov, J.; Kostov, L. *Glastech. Ber. Glass Sci. Technol.*,1994, vol.67, n.8, 227-230.
[38] Romero,M.;Rawlings, R.D.;Rincón, J. Ma. *Journal of European Ceramic Society*, 1999, vol.19, 2049-2058.
[39] Pelino,M.;Karamanov, A.;Pisciella, P.; Crisucci, S.;Zonetti, D. *Waste Manag.*, 2002, vol.22, 945-949.
[40] Ferraris, M.;Salvo, M.; Smeacetto, F.; Augier, L.;Barbieri, L.; Corradi, A.; Lancellotti, I. *Journal of European Ceramic Society*, 2001, vol.21, 453-460.
[41] Pisciella, P.; Crisucci, S.; Karamanov, A. Pelino, M. *Waste Manag.*,2001, vol.21, 1-9.
[42] MEANS, J. L. The Application of Solidification/Stabilization of Waste Materials. Lewis Publishers. 1995. Ohio.
[43] GODOLPHIM, R. R., "Incineração em Leito Fixo de Resíduos de Couro *wet-blue*" , Dissertação de Mestrado em Engenharia Química, UFRGS, 2000.
[44] CHINCÓN, J.S. et al. *Environmental 1 Geol. Water Science*,1991, vol. 18, n°1, 11-15.
[45] WEYL, W. A. Coloured Glasses, 6ª Edición, Society of Glasses Technology, Sheffield, Reino Unido, 1999.
[46] Associação Brasileira de Normas Técnicas: NBR 10005. Lixiviação de Resíduos Sólidos, set. 1987. 20 p.
[47] US EPA, 1986. Extraction procedure toxicity test. In:stabilization/solidification of CERCLA and RCRA wastes. US EPA 625/6-89/022. US EPA, Cincinatti, Ohio.
[48] Associação Brasileira de Normas Técnicas: NBR 10004. Resíduos Sólidos – Classificação, set. 1987. 63p.
[49] ASTM C373-88(2006): Standard Test Method for Water Absorption, Bulk Density, Apparent Porosity, and Apparent Specific Gravity of Fired Whiteware Products.

[50] ASTM C326-03: Standard Test Method for Drying and Firing Shrinkages of Ceramic Whiteware Clays.

[51] BRAY, D.J. Toxicity of Chromium Compounds Formed in Refractories. Ceramic Bulletin, vol. 64, n° 7, 1985.

[52] BERNARDES, A.M. Charakterisierung von Galvanikschlämmen und pyrometallurgische Verarbeitung zur Verbeserung ihrer Umweltverträglichkeit Technichen Universitat Berlin, 1997,136 p.

[53] JOHN, V.M. Cimentos de Escória Ativada com Silicato de Sódio, Tese de Doutorado, PCC, USP, São Paulo,1995. 199p.

[54] KIRK, D.W.; CHAN, C. C. Y.; MARSH, H. *Journal of Hazourdous Materials*,2002, B90, 39-49.

[55] ILLER, R. K. The Chemistry of Sílica: solubility polymerization, colloid and surface properties, and biochemistry. New York, Cap. 1, 2 e 3, 1979.

[56] KALOUSEK, l. *J. Res. National Bureau of Standards*, 1944, v. 32, 285-302.

In: Ceramic Materials Research Trends
Editor: Paul B. Lin, pp. 325-355
ISBN: 1-60021-769-9

Chapter 12

IMPEDANCE SPECTROSCOPY OF ELECTROCERAMICS

Rainer Schmidt*
University of Cambridge, Department of Materials Science & Metallurgy,
Pembroke Street, Cambridge CB2 3QZ, United Kingdom

Abstract

Impedance spectroscopy enables the many contributions to the dielectric and resistive properties of electroceramic materials to be deconvoluted and characterised separately. The charge transport mechanism and the macroscopic dielectric constant in polycrystalline materials commonly consist of several contributions such as sample-electrode interface, grain boundary and bulk resistance/capacitance. Impedance spectroscopy is shown to be a powerful tool to account for and model such behaviour. In order to gain precise understanding of the functionality of polycrystalline electroceramic materials it is essential to deconvolute these contributions.

I. Introduction

The application of Impedance Spectroscopy (IS) has been extended to new areas of research continuously during the last decades. Originally, the method was mainly used to characterise conventional liquid phase electrochemical systems, but its use to characterise dielectric and resistive properties of solid materials has become more and more popular ever since. The most recent developments enable application of solid state IS on small length scales down to the nano-scale to achieve local characterisations in nano-structured materials and devices.

The first comprehensive overview of solid state IS was presented in the book of Macdonald (Macdonald 1987). Characterisation of the dielectric and resistive properties in ionic-charge transport dominated materials, mixed ionic - electronic and purely electronic conductors is well established now. The foundations of IS applied to bulk electroceramic materials have been laid by Irvine, Sinclair and West (Irvine et al. 1990), which is also part of

* E-mail address: rainerxschmidt@googlemail.com

the focus of this work. Here, the application of IS to electroceramic bulk single- and polycrystals, but also to granular and epitaxial layers will be discussed. Particular focus will be made on film type materials since this is a relatively new field in impedance spectroscopy, in case of epitaxial layers it is novel.

First, the basic principles of solid state IS will be reviewed. A thorough discussion follows about fitting procedures to extract information from IS data sets by means of an equivalent circuit. Several common problems and experimental limitations are described. Experimental data are presented from photorefractive single crystal $Bi_{12}TiO_{20}$ and single crystal $Pb(Mg_{1/3}Nb_{2/3})O_3$-$PbTiO_3$ piezoelectrics, bulk polycrystalline $Ca_2Mn_{1.5}Nb_{0.5}O_\gamma$ and $Ca_{12}Al_{14}O_{33}$, thick polycrystalline $NiMn_2O_4$ NTC thermistor films, and epitaxial layers of $BiFeO_3$ multiferroics.

II. Basic Principles of Impedance Spectroscopy

A. Impedance on the Complex Plane

There are two possibilities to measure impedance: In the time (t) dependent approach a step function of voltage is switched on and applied to a measurement cell or electrode - sample configuration and the time dependent current response is measured. Such time-dependent results are usually Fourier transformed into the frequency domain. Therefore, it is more convenient to measure the impedance in the frequency domain directly, which does not involve major experimental difficulty nowadays by the use of state of the art broad frequency range impedance analyzers. Such standard frequency domain impedance spectroscopy consists effectively of a time dependent voltage signal with angular frequency ω and amplitude $U(\omega ,t) = U_0 \cos(\omega t)$ applied to the sample, with the amplitude and phase shift δ of the current response measured: $I(\omega ,t) = I_0 \cos(\omega t + \delta)$. One period of the applied voltage signal corresponds to a 2π rotation of the $U(\omega,t)$ arrow in Fig 1. The current responses of ideal circuit elements are (1) in phase with the applied voltage for a resistor, (2) out of phase by $\delta = -\pi/2$ for a capacitor and (3) out of phase by $\delta = +\pi/2$ for an inductor. These phase angles are all time independent, which enables the impedance to be defined as a time independent complex number. The impedance fraction in phase with $U(\omega,t)$ is defined as real, and the part +/- $\pi/2$ out of phase as the positive/negative imaginary part. The modelling of impedance data obtained from solid materials is usually based on the use of three main equivalent circuit components: ideal resistors, capacitors and inductors. The definitions of the complex impedance for these three main components Z_R*, Z_C* and Z_L* are given in Fig 1. However, in reality very few systems behave in an ideal Debye manner as represented by ideal equivalent circuit elements.

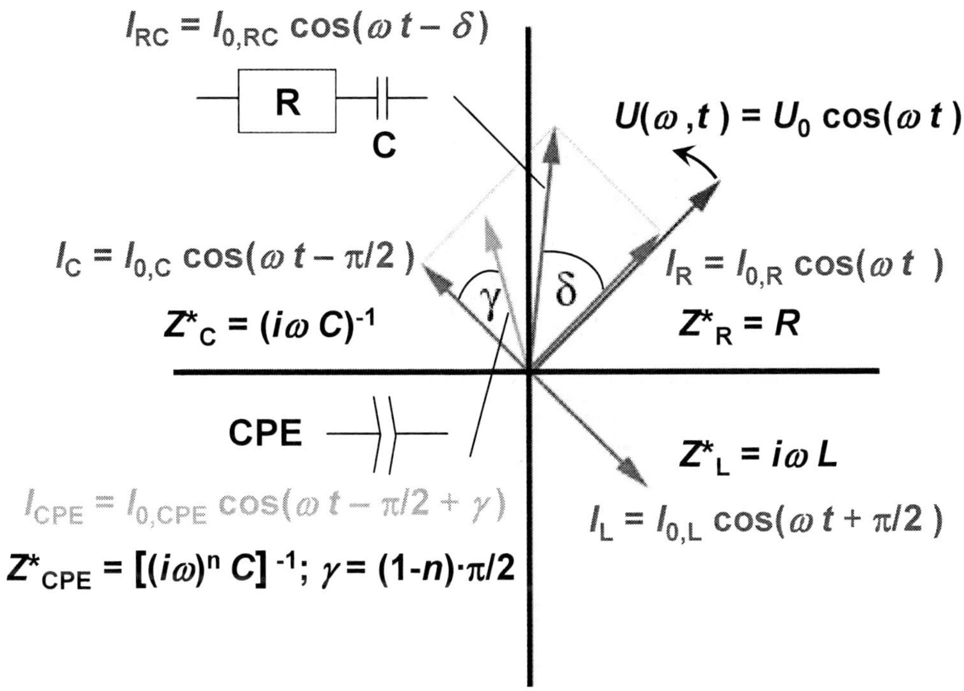

Figure 1. Impedance response of circuit elements on a phasor diagram: applied voltage $U(\omega t)$, current response I_R from an ideal resistor, I_C from an ideal capacitor, I_{RC} from an ideal series resistor-capacitor combination with phase shift δ, I_L from an inductor, and I_{CPE} from a Constant Phase Element describing a non-ideal capacitor (CPE_C) with a frequency independent phase shift γ with respect to I_C

B. Non-ideal Impedance Response

In order to describe non-ideal impedance response several empirical expressions have been proposed, which are mostly based on a specific equivalent circuit. A detailed discussion of such expressions is given in Part III. The most commonly used approach is the replacement of ideal resistors and capacitors by Constant Phase Elements (CPE), which can be done for any equivalent circuit. CPEs exhibit a frequency independent constant phase angle γ with respect to the response from ideal elements (Fig 1 shows the example of a non-ideal CPE_C capacitor). CPEs replacing ideal capacitors (CPE_C) and resistors (CPE_R) have an impedance of

$$Z^*_{CPE_C} = \frac{1}{(i\omega)^n C_m} \quad ; \quad Z^*_{CPE_R} = \frac{R_m}{(i\omega)^q} \quad ; \tag{1}$$

where R_m is the resistance and C_m the capacitance of the respective elements in modified units of $[\Omega / s^q]$ and $[\text{Farad}\cdot s^{n-1}]$. n and q are empirical parameters with ($q \geq 0$, $q \sim 0$) and ($n \leq 1$, $n \sim 1$). The relationship of n and the constant phase angle γ between an ideal capacitor and a CPE_C is displayed in Fig 1. By replacing R_m in (1) with the modified reciprocal conductance ($R_m = 1/Y_m$) it can be seen that both expressions are in fact identical. Nevertheless, both are distinguished from each other due to the fact that only CPE_R's with $q \geq 0$, $q \sim 0$ and CPE_C's with $n \leq 1$, $n \sim 1$ are physically meaningful. A transition from $q = 0$ to $q = 1$ corresponds to a transition from purely resistive to purely capacitive behaviour of the component, where the intermediate stage of $q = 0.5$ of semi resistive - capacitive behaviour has no physical equivalent.

C. Complex Relationships

Impedance spectroscopy data are often obtained in the complex impedance notation $Z^*(= Z' + iZ'')$, i.e. real and imaginary part Z' and Z'' at each frequency are considered. For data interpretation it can be very useful to transform the data into the modulus $M^*(= M' + iM'')$, capacitance $C^*(= C' - iC'')$, admittance $Y^*(= Y' + iY'')$ or dielectric $\varepsilon^*(= \varepsilon' - i\varepsilon'')$ notation, using the following complex relationships:

Table 1. Complex Relationships

	Z^*	Y^*	C^*	M^*
$Z^*=$	Z^*	$\frac{1}{Y^*}$	$\frac{1}{i\omega C^*}$	$\frac{M^*}{i\omega C_0}$
$Y^*=$	$\frac{1}{Z^*}$	Y^*	$i\omega C^*$	$\frac{i\omega C_0}{M^*}$
$C^*=$	$\frac{1}{i\omega Z^*}$	$\frac{Y^*}{i\omega}$	C^*	$\frac{C_0}{M^*}$
$M^*=$	$i\omega C_0 Z^*$	$\frac{i\omega C_0}{Y^*}$	$\frac{C_0}{C^*}$	M^*

C_0 denotes the capacitance of the measuring cell in vacuum, i.e. $C_0 = \varepsilon_0 \cdot A/d$, where A is the current cross section, d the contact distance and ε_0 the permittivity in vacuum. A/d is often referred to as the geometrical factor g. For physical interpretations it may be more appropriate to use the notation of specific impedance z^* and specific capacitance c^* or dielectric constant ε^*: $z^* = g \cdot Z^*$, $c^* = \varepsilon_0 \cdot \varepsilon^*$ with $\varepsilon^* = C^*/C_0$.

III. Equivalent Circuit Modelling

The charge transport mechanism and the macroscopic dielectric constant in polycrystalline materials commonly consist of several components such as electrode - sample interface, grain boundary and bulk contributions. In order to gain precise understanding of the functionality of a polycrystalline electroceramic it is essential to deconvolute these contributions using an appropriate equivalent circuit to describe and analyse each relaxation separately. In order to use IS in such way, several assumptions need to be made, which are summarised in the brick layer model:

A. The Brick Layer Model

In the brick layer model it is assumed that a polycrystalline sample, contacted with appropriate electrodes to perform IS, consists of several distinct regions, dominated by (a) grain boundary (GB) effects (b) electrode sample interface and (c) grain interior properties.

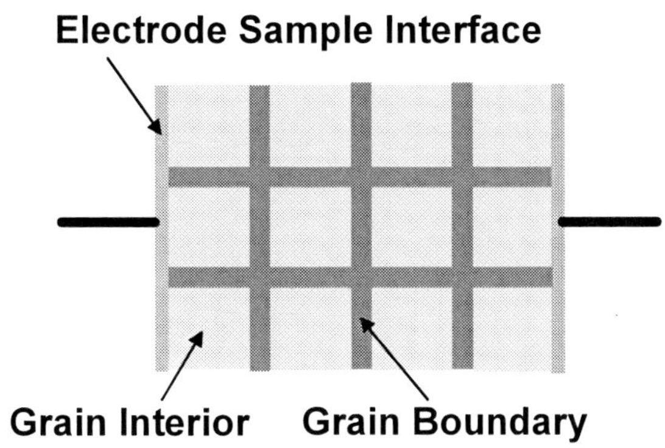

Figure 2. Brick Layer Model.

In order to account for potentially different dielectric and resistive behaviour in GBs compared to bulk regions, it is required that (a) all GB areas across the sample show similar properties, such that they can be averaged by a single equivalent circuit component, and that (b) GBs show distinctively different properties compared to grain interior regions or other contributions to allow discrimination. The same argument holds of course for any other contribution such as interface and grain interior. From these considerations it is clear that IS analysis on inhomogeneous materials has to be carried out with great care. For example, in the case of compositional variations in different grains in one single sample, current passes through the more conducting grains and bypasses highly resistive sample areas. It is also implied that grain interior contributions do not appear in the impedance spectra if they are significantly more resistive than grain boundary areas. In case of highly resistive grains, they will be bypassed and only the conducting GB contribution can be analysed in terms of resistive and dielectric behaviour. However, in carefully prepared homogeneous polycrystalline electroceramic materials the GB areas are very often more resistive than grains and all three contributions shown in Fig 2 are expected to be manifested in the impedance spectra. In (transition-) metal oxide ceramics, grain boundaries are often oxygenated compared to the grain interior, which leads to charge carrier depletion for localised electron hopping between cations.

B. RC and Debye Circuits and the Classification Scheme from Irvine, Sinclair and West

In the simplest case each dielectric relaxation process in a homogeneous sample may be described by an RC element consisting of an ideal resistor and capacitor in parallel. The macroscopic impedance is then the sum of all series RC impedances. A large variety of alternative circuits has been proposed, of which the most common one may be the Debye circuit of resistor and two capacitors in series and in parallel (Fig 3). Such circuits are difficult to justify though if the data set can be modelled also by a simpler physically meaningful circuit of series RC elements, which is almost always the case in chemically stable electroceramic bulk or film materials. An intuitive understanding of the physical meaning of an RC element may be obtained best by considering the example of an electric insulator (e.g. a dielectric or ferroelectric): The ideal capacitor describes the insulating dielectric properties of the material. The resistor in parallel describes the leakage behaviour, since ideal insulators do not exist in reality. Leakage currents can pass through the capacitor and can thus "evade" the charge storage device; in terms of the RC circuit such currents just bypass the capacitor.

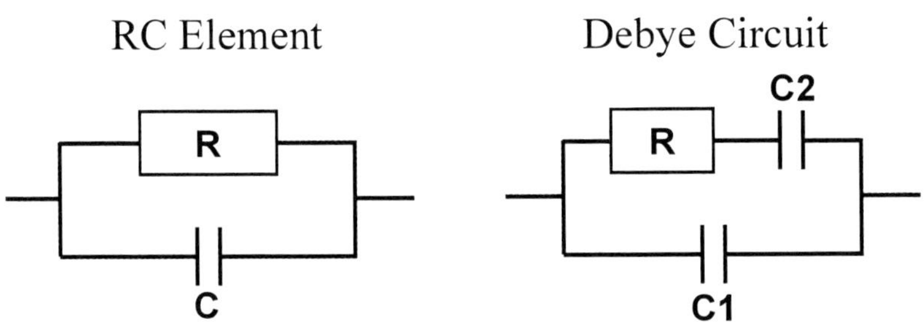

Figure 3. RC element and Debye circuit (R-C1-C2).

Strictly speaking the Debye circuit has no microscopic equivalent and is therefore physically not meaningful. Therefore, the Debye circuit may be interpreted as an empirical attempt to account for limiting low frequency and high frequency dielectric constants $\varepsilon'(0)$ and $\varepsilon'(\infty)$ within a single relaxation process, i.e. a frequency dependent dielectric constant/ capacitance.

In a regular RC element the magnitude of the specific capacitance c (= $\varepsilon_0 \cdot \varepsilon'$) allows identification of the origin of the relaxation. The following classification scheme has been proposed by Irvine, Sinclair and West (Irvine et al. 1990), containing all potentially important relaxation processes in an electroceramic material. The relaxation processes most relevant are printed in bold, the others are mainly unwanted contributions.

Table 2. The specific capacitance classification scheme according to Irvine, Sinclair and West

Specific Capacitance (F/cm)	Origin of the RC Element
10^{-12}	**bulk**
10^{-11}	minor second phase
$10^{-11} - 10^{-8}$	**grain boundary**
$10^{-10} - 10^{-9}$	**bulk ferroelectric**
$10^{-9} - 10^{-7}$	surface layer
$10^{-7} - 10^{-5}$	sample – electrode interface
10^{-4}	electrochemical reaction

An important parameter in an RC element is the relaxation time τ [s], which is defined as

$$\tau = R \cdot C = \rho_{dc} \cdot c; \tag{2}$$

where ρ_{dc} is the resistivity of the resistor. The notation in specific units may be physically more meaningful and contains material specific, size independent information on the resistive and dielectric properties of the relaxation process described by the respective RC element. In the Debye circuit τ is defined as:

$$\tau = R \cdot C_2 = \rho_{dc} \cdot c_2; \tag{3}$$

where C_2 and c_2 are the nominal and specific capacitance of C2 in Fig 3.

C. -*Z*''-*Z*' and Cole-Cole Plots

Impedance data, which can be fitted to an RC element, may be represented best on a plot of -Z'' vs Z' [or -z'' vs z'] and in case of Debye circuit fitting on a Cole-Cole ε'' vs ε' plot (Cole et al. 1941). In both cases a semicircle arc develops. For RC elements $Z'(0) = R$ [$z'(0) = \rho_{dc}$] and $\omega_{max} = \tau^{-1}$, implying that R is given by the semicircle diameter. For the Debye circuit $c1/\varepsilon_0 = \varepsilon'(\infty)$, $(c1+c2)/\varepsilon_0 = \varepsilon'(0)$; and $\omega_{max} = \tau^{-1}$, and the semicircle diameter is given by $\varepsilon'(0) - \varepsilon'(\infty)$, where $\varepsilon'(0)$ and $\varepsilon'(\infty)$ represent the low and high frequency limit of the real part of the dielectric constant ε'.

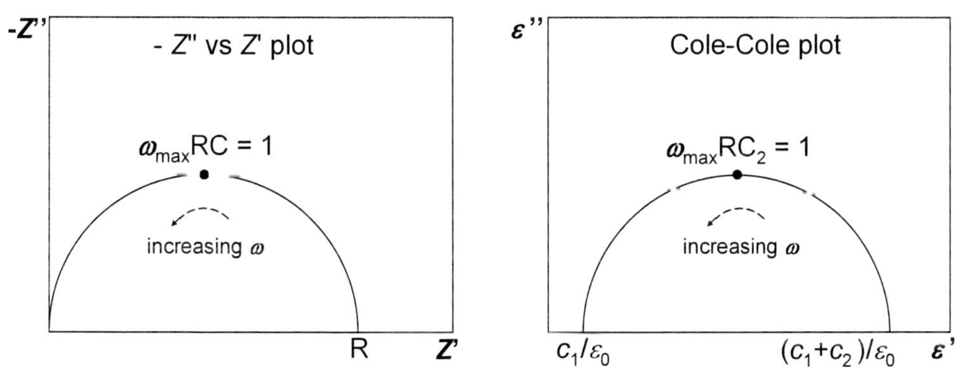

Figure 4. -Z''-Z' plots computed for an RC element (left image) and Cole-Cole plot for a Debye circuit (right image).

D. Empirical and Semi-empirical Expressions to Model Non-ideal Behaviour

1. The R-CPE Circuit

It has been mentioned in section II./B. that ideal circuit elements can be replaced by CPEs to account for non-ideal impedance response. The use of CPEs is usually explained by the presence of some distributed elements in the material - electrode system or the fact that certain parameters may be distributed across a sample (i.e. τ, ρ or ε'). The parameter q and n in eq.(1) have been claimed to be related to the width of such a distribution, i.e. $n = 1$ corresponds to a narrow distribution whereas $n < 1$ reflects an increased distribution width. Nevertheless, CPEs can be regarded only as semi-empirical elements at most.

The replacement of ideal capacitors in an RC element by a CPE_C (R-CPE circuit) is common and may be the most successful approach to model non-Debye behaviour for a single relaxation. In this case, R-CPE semicircles on a -Z''-Z' plot are suppressed with the arc centre displaced below the real axis. Alternatively, the ideal resistor may be replaced by a CPE_R, but this approach is less common and most impedance spectra obtained from electroceramic materials show semicircles suppressed to a certain degree in agreement with the CPE_C. The specific impedance of an R-CPE is:

$$z^*_{R-CPE} = \frac{\rho_{dc}}{1 + \rho_{dc}\, c_m (i\omega)^n}; \tag{4}$$

The modified specific capacitance c_m is given in $F \cdot s^{n-1} \cdot cm^{-1}$. In the ideal RC case $n = 1$, and $\rho_{dc} \cdot c_m$ can be replaced by the relaxation time τ. The impedance is then a Lorentzian type profile.

2. The Havriliak - Negami (HN) Expression

The most versatile but purely empirical expression may be the Havriliak-Negami (HN) relaxation function Φ^*_{HN} (Havriliak et al. 1966) given by:

$$\Phi^*_{HN} = \frac{1}{\left[1 + (i\omega\tau)^{\alpha}\right]^{\gamma}}; \tag{5}$$

where α and γ are empirical parameters. An alternative expression in the time-domain notation has been suggested by Kohlrausch, Williams and Watts (KWW) (Kohlrausch 1863; Williams et al. 1970). Although the Fourier transform of the KWW expression does not exactly correspond to the frequency domain HN function, it has been shown that the empirical KWW and HN parameters are related to each other in a simple way (Alvarez et al. 1991).

For an RC element the HN function describes the complex impedance by $\Phi_{H\text{-}N}^* \cdot R = Z_{RC}^*$ (or $\Phi_{H\text{-}N}^* \cdot \rho_{dc} = z_{RC}^*$ in the specific impedance notation). The case of $\gamma = 1$ is similar to an R-CPE circuit with $\alpha = n$, but c ($\tau = \rho_{dc} \cdot c$) has regular units [F] in the HN formalism. α describes the degree of suppression and γ the "skewness" of the RC semicircle on a -Z'' - Z' plot (see Fig 5b). For the Debye circuit the HN expression corresponds to [$\Phi_{H\text{-}N}^* \cdot (\varepsilon'(0)$ -

$\varepsilon'(\infty))] + \varepsilon'(\infty) = \varepsilon^*$. The computed -*Z*''-*Z*' plots for a R-CPE circuit and Havriliak - Negami RC element are shown in Fig 5 a,b.

The physically most meaningful way to describe non-Debye impedance response may be the use of a random Gaussian distribution of relaxation times. This is explained in more detail in the next section. The -*Z*''-*Z*' plots for such Voigt circuits are shown in Fig 5 c.

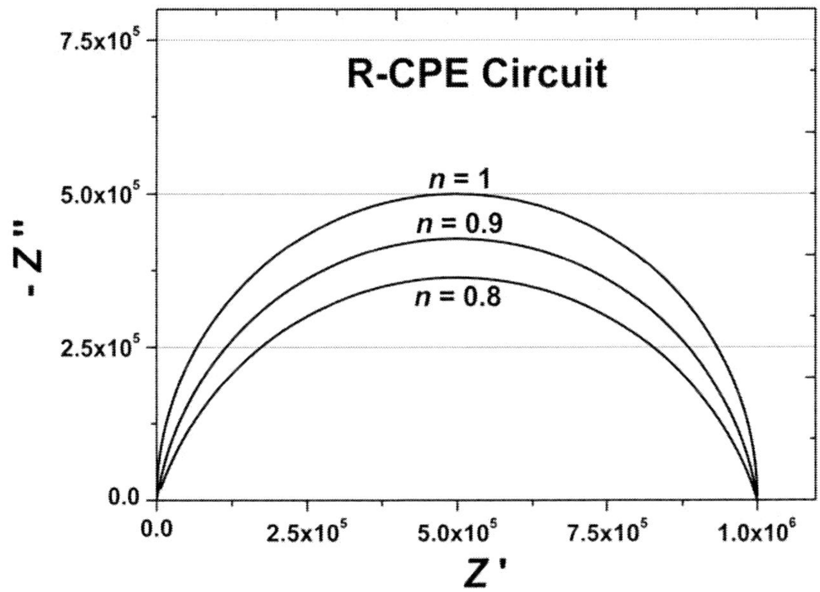

Figure 5a. -*Z*''-*Z*' plots for a R-CPE circuit with empirical exponents *n* of 1, 0.9 and 0.8.

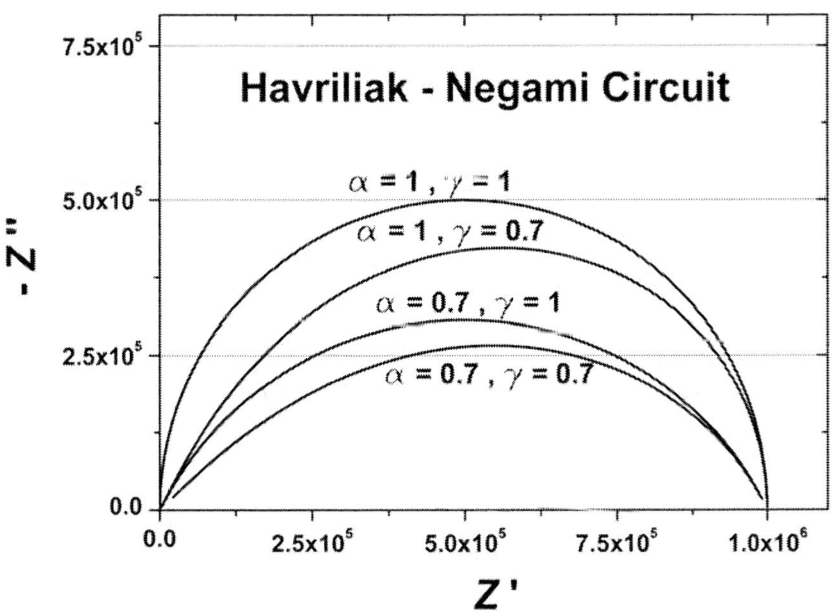

Figure 5b. -*Z*''-*Z*' plots for a Havriliak - Negami RC circuit with empirical exponents $\alpha = 1$, $\gamma = 1$ // $\alpha = 1$, $\gamma = 0.7$ // $\alpha = 0.7$, $\gamma = 1$ // $\alpha = 0.7$, $\gamma = 0.7$.

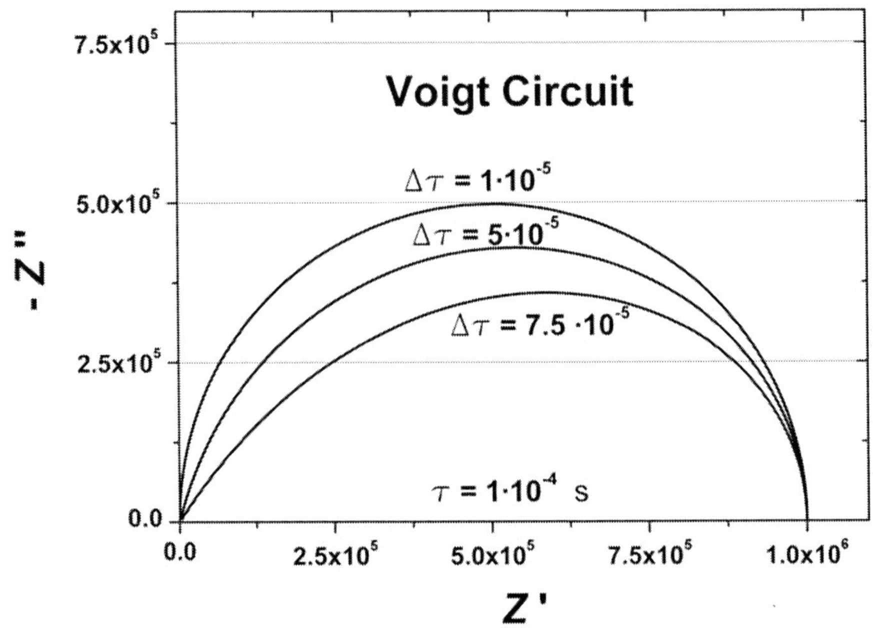

Figure 5c. -Z''-Z' plots for a Voigt circuit with a mean relaxation time $\bar{\tau}$ of $1\cdot10^{-4}$ s and standard deviations $\Delta\tau$ of $1\cdot10^{5}$ s, $5\cdot10^{5}$ s and $7.5\cdot10^{5}$ s;

It has been claimed that the empirical HN parameters α and γ can be translated directly into a physically meaningful distribution of relaxation times $g_{HN}(\tau)$ (Alvarez et al. 1991). However, the expression quoted for $g_{HN}(\tau)$ can not be normalised (and is not normalised in the state presented), and it may only represent a tool for qualitative understanding.

3. Random Gaussian Distribution of Relaxation Times

The physically most meaningful way to describe non-ideal impedance response may be the use of a normal or Gaussian distribution of relaxation times. The HN expression is still purely phenomenological, and the $g_{HN}(\tau)$ function deduced from empirical α and γ parameters is difficult to interpret in a physically meaningful way. It is most reasonable to approximate the distribution of relaxation times in a real system by a random distribution, such that infinitesimal areas i within a sample show random variations in $\tau_i = \rho_i \cdot c_i$, which then results in a Gaussian profile around the mean relaxation $\bar{\tau}$ with a standard deviation of $\Delta\tau$ (Fig 6) (Song et al. 2004).

This approach to model distributed relaxation times τ with a random Gaussian distribution is an approximation, which is justified by the fact that the constituent parameters of τ, i.e. ε' and ρ_{dc}, both possess independent distributions and it is impossible to account for each distribution separately. In fact, the situation may be much more complicated, because microscopically both ε' and ρ_{dc} may each depend on several material specific parameters again with separated distributions.

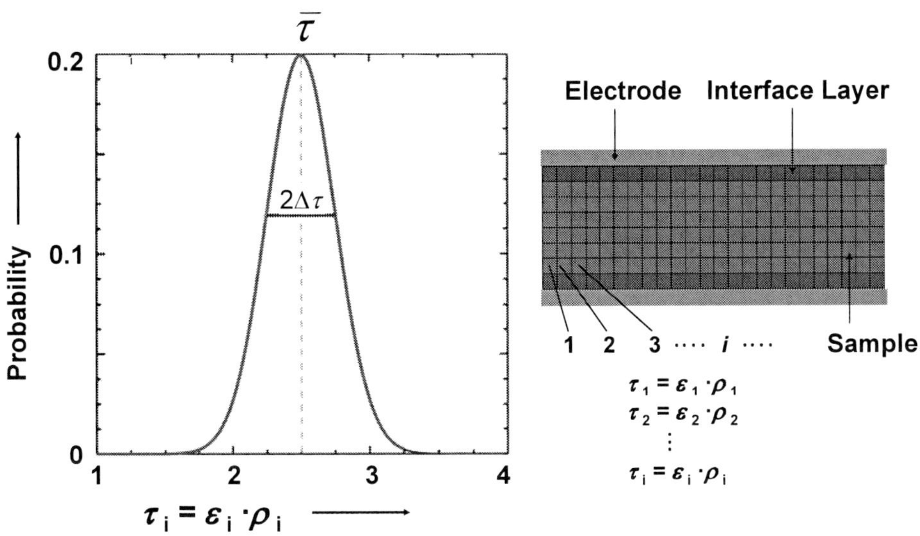

Figure 6. Electrode - sample configuration (right image) with infinitesimal sections showing a random distribution of relaxation times $\tau_i = \rho_i \cdot c_i$; the probability function vs the relaxation times τ_i are plotted in arbitrary units (left image); $2\Delta\tau$ indicates the width of the first standard deviation including 68 % of all τ values.

This will be described in more detail on the example of localised electron hopping transport:

An intrinsic sample hopping resistivity ρ_{dc} contains several independent parameters such as exponential activation energy and hopping distance, and a pre-exponential term. A random distribution of activation energies for localised charge transport has been suggested previously to fit impedance data (Macdonald 1987), but this approach may not always be valid for several reasons. In case of localised electron transport, arguing on a microscopic level requires to take into account that not only the activation energies between two localised electron states vary across a sample but also the hopping distances. Furthermore, activation energies and hopping distances are both not randomly distributed but have an upper limit determined by the percolation threshold (Shklovskii et al. 1984). The distribution of activation energies can also depend on the density of localised electron states: a parabolic density of localised electron states around Fermi level (e.g. in the presence of a Coulomb gap) may result in a different distribution of activation energies than a uniform density of states). Further arguments for the daunting complexity of the situation can be found, such as a potential influence of the pre-exponential term in the expression for the localised electron hopping transport resistivity ρ_{dc}, which can depend on several independent material specific parameters (Shklovskii et al. 1984), again each exhibiting a possible distribution. The same argument holds for the dielectric constant, which may microscopically depend again on several parameters.

Therefore, it may be impossible to develop a reliable expression, which can fit IS data and account for all distributed parameters separately. An impedance spectrum does not contain enough information to conclude on several different distributed parameters. In fact, it may not even be possible to precisely determine which parameters determine ε' and ρ_{dc} on a microscopic or atomic scale. This depends on the type of material, i.e. crystal structure, type of charge transport, etc., and it is an impossible task to work this out from IS data. Therefore, the best approximation possible is a random Gaussian distribution of relaxation times τ, because τ (= $\rho_{dc}\cdot\varepsilon_0\cdot\varepsilon$ ') contains all material specific resistive and dielectric property parameters. This argumentation is consistent with the "central limit theorem", which states that functions of a large number of independent or weakly-dependent random variables have a probability distribution close to the normal distribution [for a review see for example (Sazonov 1981)].

4. The Voigt Circuit: A Gaussian Distributed RC Element

The impedance of an ideal RC element is a Lorentzian function. Therefore, accounting for a Gaussian distribution of relaxation times in an RC circuit requires the deconvolution of the Lorentzian with a Gaussian function, which constitutes a Voigt profile. In order to determine the impedance of a Voigt circuit, the ideal RC Lorentzian may be integrated around the mean relaxation $\tau = \bar{\tau}$ and weighted by a standard Gaussian function normalised to 1. This leads to the following impedance for a Voigt circuit:

$$z^*_{Voigt} = \frac{\rho_{dc}}{\Delta\tau\ \sqrt{2\pi}} \int_{-\infty}^{+\infty} \frac{\exp\left(\dfrac{-x^2}{2\ \Delta\tau^2}\right)}{1+i\omega\ (\bar{\tau}+x)}\ dx \tag{6}$$

where $\Delta\tau$ is the standard deviation of the Gaussian distribution. The error made by accounting for physically meaningless $\bar{\tau} < 0$ values may be neglected in case of narrow distributions. In case of broad distributions, $\Delta\tau$ values may not always be reliable, whereas $\bar{\tau}$ can be expected to be valid.

Voigt profiles are commonly used in physical measurements, for example to describe the Lorentzian and Gaussian broadening of atomic or molecular spectral lines (Armstrong 1967), but they can be used also to describe impedance response of solid matter. Eq. (6) can be solved analytically as shown in Appendix A. The Voigt circuit impedance is computed in Fig 5c and a Gaussian distribution of relaxation times is shown in Fig 6 (red curve).

E. The Conductivity Notation and Jonscher's Law

In the complex conductivity notation [$\sigma^* = (z^*)^{-1}$] parallel circuit elements are additive, which leads to a simple expression for the complex conductivity $\sigma_{RC}{}^*$ of an RC element:

$$\sigma^*_{RC}(\omega,T) = \sigma_{dc}(T) + i\omega\ c \tag{7}$$

σ_{dc} is the D.C. conductivity of the resistor. In a single ideal RC element the real part of the conductivity $\sigma' = \sigma_{dc}$ is frequency independent and the imaginary part σ'' shows a linear ω dependence. CPEs (CPE_R and CPE_C) exhibit a complex specific conductance of

$$\sigma^*_{CPE_C} = (i\omega)^n \, c_m \; ; \quad \sigma^*_{CPE_R} = (i\omega)^q \, \sigma_{dc,m} \; ; \tag{8}$$

$\sigma_{dc,m}$ and c_m are the conductivity and specific capacitance in modified units of $[s^q/\Omega cm]$ and $[(Farad \cdot s^{n-1})/cm]$. This leads to the following real part of the conductivity σ' for a standard R-CPE_C circuit using the relationship $(i)^n = \cos(n\pi/2) + i\sin(n\pi/2)$:

$$\sigma' = \sigma_{dc} + c_m \, \omega^n \left(\cos \frac{n\pi}{2} \right); \tag{9}$$

which can be expressed in a modified and simplified form (Naohiro 1995):

$$\frac{\sigma'}{\sigma_{dc}} = 1 + A\,\omega^n \tag{10}$$

This corresponds to Jonscher's law (Jonscher 1972; Jonscher 1996), which has been claimed to be universally valid in disordered systems including electroceramic materials with localised charge transport. The parameters n and A can be determined from the slope of double-logarithmic plots of $\log(\sigma')$ vs $\log(\omega)$ or $\log[(\sigma'/\sigma_{dc})-1]$ vs $\log(\omega)$. This universal Jonscher behaviour can be reproduced using Voigt element equivalent circuits. Fig 7 shows a plot of the conductivity σ'/σ_{dc} vs f obtained from a single simulated Voigt element on double logarithmic axes. A constant (non-zero) slope of the graph at high frequencies can be observed, which corresponds to the parameter n. This is fully consistent with Jonscher's behaviour. This implies that the Voigt element model is not only physically meaningful but can also reproduce Jonscher's law, which is observed on a vast amount of experimental data. According to the universality of Jonscher's law, Voigt elements may thus be used for a large variety of experimental data to describe intrinsic sample contributions. For electrode interface effects, the use of Voigt elements is questionable. In this case, the fractal nature of the interface may also have a perceptible influence on or even be the dominant effect leading to non-Debye behaviour. In this case, the random distribution of relaxation times at the interface may be less influential.

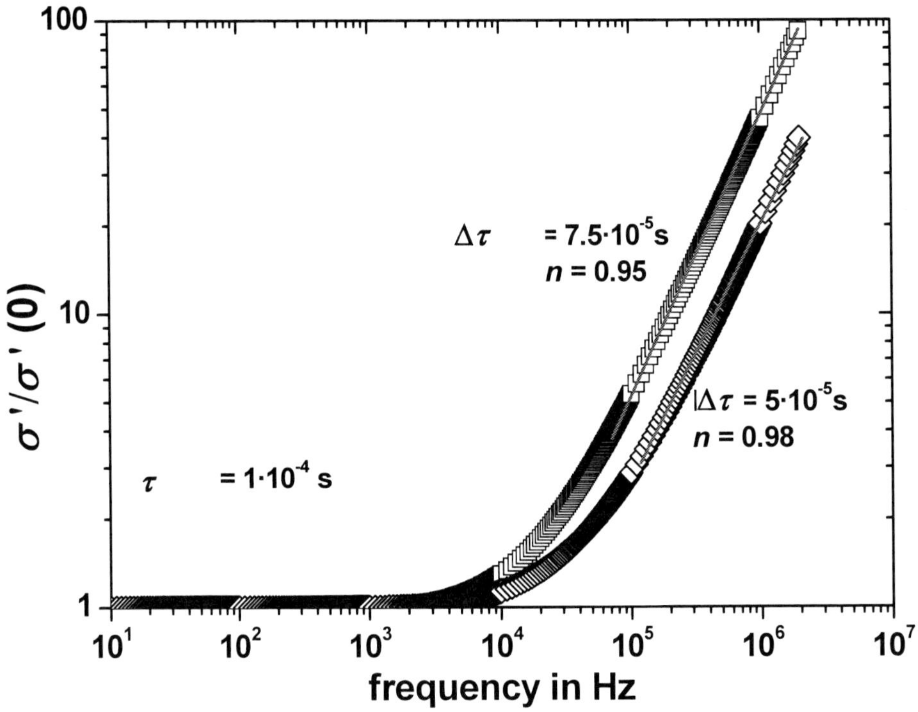

Figure 7. log-log plot of σ'/σ_{dc} vs f for simulated Voigt elements, Jonscher's exponent n is determined as the slope of the plots at high f

The standard deviation $\Delta\tau$ of $7.5\cdot10^5$ s shown in Fig 7 may represent the maximum ratio $\Delta\tau/\bar{\tau}$ = 0.75, which still leads to valid fits. At higher standard deviations the approximation of neglecting meaningless $\tau < 0$ values from the normal Gaussian distribution may no longer be valid. R-CPE$_C$ circuits with an n value of $0.95 < n < 1$ may indicate that a Voigt model can be used alternatively. For $n < 0.95$ values, non-Debye impedance response may be dominated by a wide asymmetric distribution of relaxation times, or from different phenomena such as the fractal nature of a possible electrode sample interface contribution.

IV. Experimental Limitations

Experimental limitations in IS arise mainly from unwanted contributions to the impedance spectra collected from a sample. In particular, these may be effects from cabling, leads, sample holder, contacts, apparatus and electrode sample interface contributions.

A. Contact Contributions

For reliable deconvolution of electrode - sample interface and intrinsic contributions the use of planar electrodes is favourable. Point contacts to the sample may not result in a clearly separated low frequency electrode semicircle on a -Z''-Z' plot (Hwang et al. 1997). An example of planar electrodes is depicted in Fig 8 for a cubic shaped polycrystalline bulk sample, where two faces of the cube are covered with electrode material (usually noble metals such as Pt, Au, Ag; or other metals such as In, Ni, Ga etc. are used).

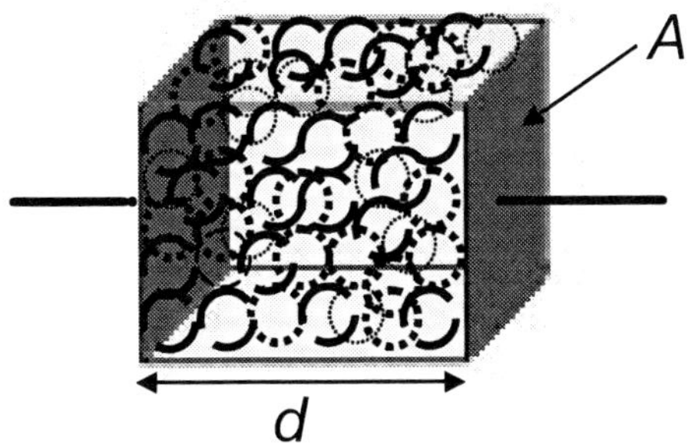

Figure 8. Cubic shaped polycrystalline bulk sample; two faces of the cube are covered with electrode material (grey); polycrystalline grains are indicated by black spheres.

In Fig 8, contact to the sample is made by attaching two cables to the planar electrodes. This approach, however, is less favourable, because it can lead to significant contact resistance. The best possibility may be the use of rigid probes, which allow exerting a mechanical load onto the electrode surfaces. Spring loaded drop down contacts may be ideal. The positive effect of such a mechanical load is demonstrated in Fig 9, for Au contacts on CeO_2 bulk polycrystalline ceramics. The Au contacts were deliberately fabricated imperfectly for demonstration purposes. The low frequency electrode interface arc clearly decreases with increasing load, indicating a decreasing contact resistance.

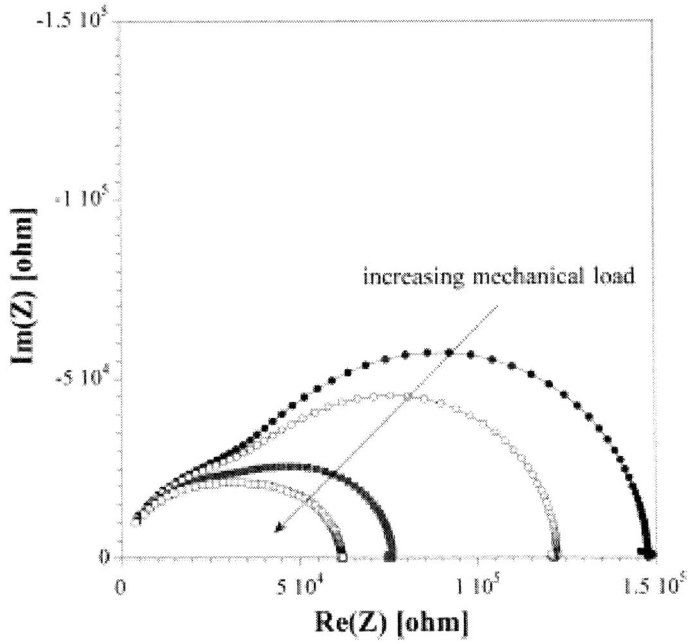

Figure 9. $-Z''$-Z' plot for CeO_2 ceramics contacted with Au and applied mechanical load to the contacts (T = 550°C in air); reproduced from (Hwang et al. 1997) with permission from Elsevier.

B. Contributions from Measurement Leads and Apparatus

Contributions from the measurement leads are the main limiting factor at high frequencies, when the inductance of the wires will dominate the impedance spectra eventually. This is indicated by a change of sign of the imaginary part of the impedance from negative (capacitive behaviour) to positive (inductive). To minimise such effects the measurement leads should be as short as possible and possess an inductance as low as possible. Fig 10 illustrates such a change in sign of Z'' and the spectrum shown is dominated by inductive effects at high frequency. Data were taken from a sample made of Portland cement paste (Edwards et al. 1997).

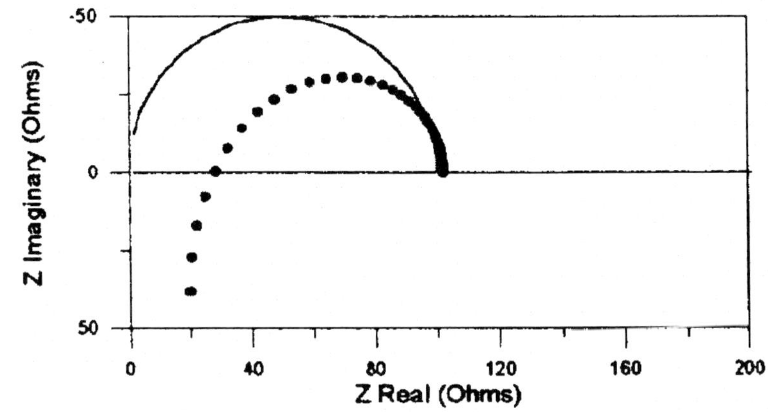

Figure 10. $-Z''$-Z' plot for impedance data obtained from Portland cement paste (black dots) and the expected pure impedance (solid line); reproduced from (Edwards et al. 1997) with permission from Elsevier.

It was claimed that such effects from measurement leads can be minimised by decreasing the geometrical factor g by a decrease in contact area A or increase in contact distance d, i.e. sample size. However, it is impossible to completely avoid such contributions, because all measurement leads will exhibit a certain inductance, which will dominate the impedance spectrum at high enough frequencies. Due to these effects, impedance spectroscopy data are usually reliable up to ~ 1MHz and usually unreliable above ~ 10MHz. Furthermore, capacitive contributions can arise from the measurement leads. If single line insulated measurement leads are placed close to each other, or are in contact with each other, a high capacitance between the leads occurs and can dominate the impedance spectra. This can be avoided by using shielded wires (e.g. coaxial cables). The shieldings of the cables should be connected to each other as near to the sample as possible. A large distance between measurement leads can also reduce the capacitance significantly.

Modern state-of-the-art impedance analysers are usually shielded effectively, such that internal contributions from the apparatus are insignificant and can be neglected. It was claimed that all capacitive contributions from wires and apparatus can be minimised by increasing the geometrical factor g by an increase in contact area or decrease in contact distance. This would result in an increase in the sample capacitance, and make parasitic capacitance contributions less significant.

A further unwanted contribution is the resistance of the measurement leads and the contact material. Such contributions are usually small (< 10Ω) and can often be neglected, except the sample under investigation contains resistances in the same order of magnitude.

C. Accounting for Unwanted Contributions in Impedance Analysis

Possible unwanted contributions from leads, electrodes and sample or probe holders are illustrated in Fig 11 (a), showing a top-top measurement of a thin insulating film deposited on a conducting (metallic) substrate. The set-up in Fig 11 (a) is favourable to characterise insulating materials, because the nominal film resistance along the thin film normal axis is small despite a high resistivity. Inductive contributions from the measuring leads are represented by the inductor L0, capacitive contributions arising between the leads or the probe holder by C0, and the resistance from leads, probes and electrodes by R0.

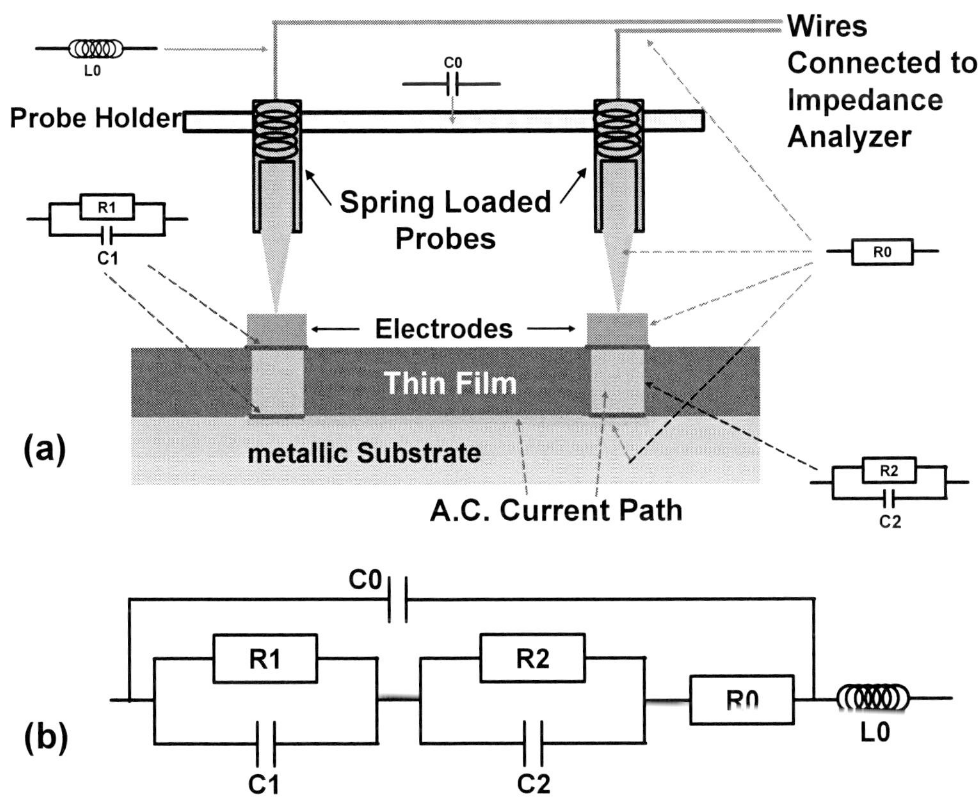

Figure 11. (a) Experimental set-up of an IS experiment, unwanted and intrinsic contributions are illustrated (b) Equivalent circuit to account for all intrinsic and unwanted contributions.

The electrode interfaces (red bars) are described by R1/C1 and the intrinsic film contribution by R2/C2. In case of an interface contribution, the use of a Voigt element may be questionable as mentioned above. The fractal nature of the interface may dominate the non-ideality of such a contribution. It has been argued that the dimensionality D of the interface is

$2 < D < 3$ and can be directly related to the CPE_C and Jonscher parameter n (Liu 1985). Grain boundary contributions were assumed to be absent, in case of highly crystalline films.

There are two possibilities to eliminate components such as R0, L0 and C0 from the data analysis.

(A) By performing an open circuit calibration measurement in advance, the capacitance C0 can be measured and later be related to the impedance measured with the sample in place. In the same way, a short circuit calibration run (for example with touching probes) allows measuring R0 and L0, and the data obtained from the sample can be stripped off the contributions from R0, L0 and C0.

(B) Parasitic contributions can be accounted for by introducing the respective circuit components into the equivalent circuit model. Fig 11 (b) shows a circuit with all major unwanted contributions.

Method (B) may be favourable, because it gives a clear indication on whether the experimental set-up is adequate, and allows analysis of the parasitic contributions in order to conclude on their origin. High R0, L0 and C0 values obtained from fits of the data to the circuit model may indicate the necessity to improve the experimental design. Especially a high contribution from C0 can be critical, because it can dominate the spectra and mask sample capacitances and hinder analysis. C0 can be reduced effectively by the correct choice of the sample holder material (high resistivity, low dielectric constant), e.g. Poly-Tetra-Fluoro-Ethylene (PTFE), and the use of shielded measurement leads. C0 may then be negligible and can be removed from the circuit model. The resistance of the probe holder or between the measurement leads was assumed to be high compared to R1, R2 and R0 and was omitted in the circuit in Fig 11 (b).

V. Impedance Spectra of Electroceramics

A. Single Crystal Studies

The investigation of impedance response from single crystal material has the distinct advantage that grain boundary (GB) contributions are absent, which may enable a more detailed insight into intrinsic bulk properties. In case of high crystalline order, the distribution of relaxation times may be narrow due to the absence of extended defects such as crystal imperfections, GBs and strain fields near GBs or near a possible film substrate interface in case of film type samples. Fig 12 shows an impedance spectrum obtained from a $Bi_{12}TiO_{20}$ single crystal (Lanfredi et al. 2000). The $-Z''$-Z' semicircle arc is almost perfectly developed, indicating a narrow distribution of relaxation times. The impedance response can be modelled effectively with one single ideal RC element. Insignificant contributions are visible at the low frequency end, which do not affect the bulk semicircle.

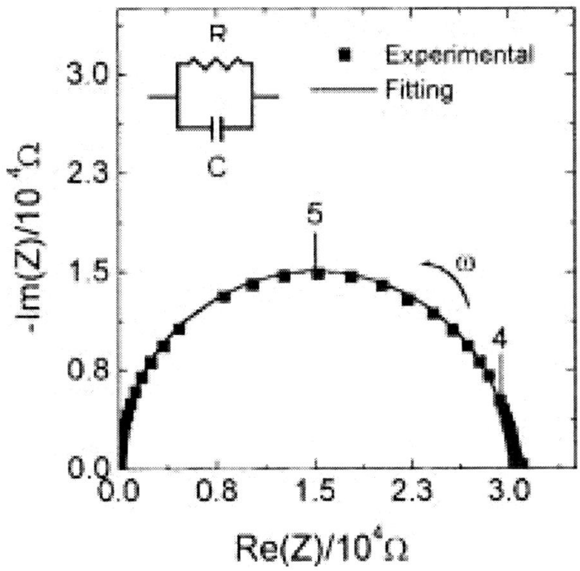

Figure 12. -Z''-Z' plot of a $Bi_{12}TiO_{20}$ single crystal, indicating almost perfect Debye behaviour; reproduced from (Lanfredi et al. 2000) with permission from The American Institute of Physics.

In order to resolve additional contributions, the low and high frequency ends of a single semicircle can be investigated in more detail. Often, a small second contribution becomes evident only by displaying the magnified high and low frequency ends separately. An alternative method is the analysis of the imaginary part of the modulus function M'' (for a definition see Table 2). Fig 13 shows an example of data taken from a $Pb(Mg_{1/3}Nb_{2/3})O_3$-$PbTiO_3$ single crystal (James et al. 2001). One main contribution at intermediate frequencies can be seen clearly, where the peak shifts to higher frequencies with increasing temperature (T). Using the RC element formalism, this indicates a decrease in relaxation time τ (possibly due to a decrease in resistance with increasing T), because the peak frequency $\omega_{max}(M'')$ is related to τ by $\omega_{max} = \tau^{-1}$ (in an RC element this is the identical relationship as for $\omega_{max}(-Z'')$ and τ). The value of M'' at the peak frequency is given by $M''(\omega_{max}) = C_0/2C$. Therefore, an approximately T independent capacitance is indicated at 200°C/ 225°C.

The intermediate frequency capacitance is in a typical range for ferroelectric bulk: ~ $1{\cdot}10^{-10}$ F/cm (see table 2). At 250°C and above the capacitance decreases (since $M''(\omega_{max})$ increases with T). Such a T dependent capacitance is commonly observed in ferroelectrics above the transition temperature, whereas below the transition and in non-ferroelectric materials the capacitance is usually expected to be T independent. Clearly, a second contribution develops at the high frequency end of the spectra in Fig 13. The capacitance of this second contribution is similar or smaller than the first contribution, which excludes the possibility of an electrode interface effect. Interface contributions commonly exhibit larger capacitance and would occur at the low frequency end (see table 2). Since the sample under investigation was a single crystal, the second contribution could potentially represent a secondary phase, surface layer or another unwanted contribution.

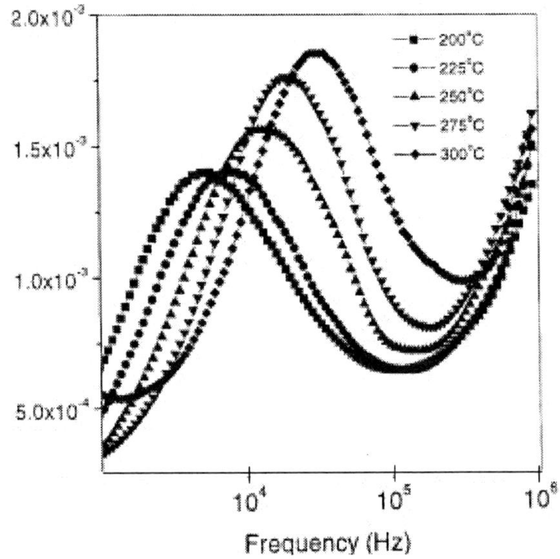

Figure 13. M'' vs f modulus spectrum of $Pb(Mg_{1/3}Nb_{2/3})O_3$-$PbTiO_3$ single crystals; reproduced from (James et al. 2001) with permission from The American Institute of Physics.

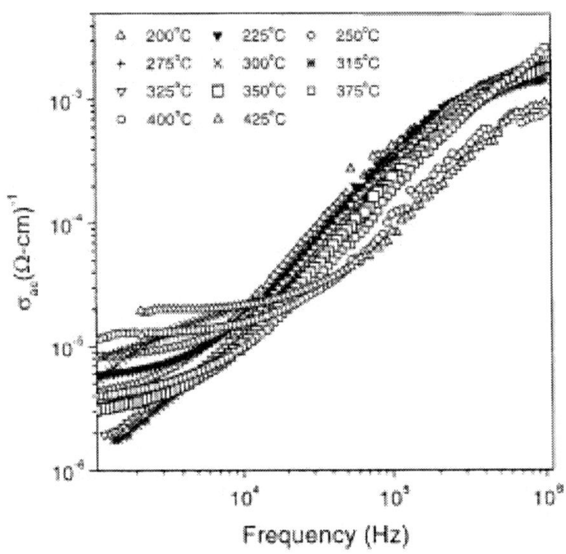

Figure 14. Double logarithmic plot of log(σ') vs log(f) from $Pb(Mg_{1/3}Nb_{2/3})O_3$-$PbTiO_3$ single crystals to determine $n \sim 0.95$; reproduced from (James et al. 2001) with permission from The American Institute of Physics.

Jonscher's universal power law (eq.10) was tested for the data set shown in Fig 13. Fig 14 shows a double logarithmic plot of log(σ') vs log(f), where Jonscher's exponential parameter n was determined from the slope of the graph in the intermediate frequency regime to be ~ 0.95.

This implies that the main relaxation at intermediate frequencies may be described best by an R-CPE_C circuit (eq. 4) with a CPE_C parameter of $n \sim 0.95$, or a Voigt element (eq. 6) with a large $\Delta\tau/\bar{\tau} = 0.75$ ratio. This is a perceptible deviation from the ideal $n = 1$ behaviour.

B. Polycrystalline Bulk Electroceramics

Fig 15 (a) shows the example of impedance spectra obtained from $Ca_{12}Al_{14}O_{33}$ polycrystalline ceramics, presented on a -Z''-Z' plot (Irvine et al. 1990).

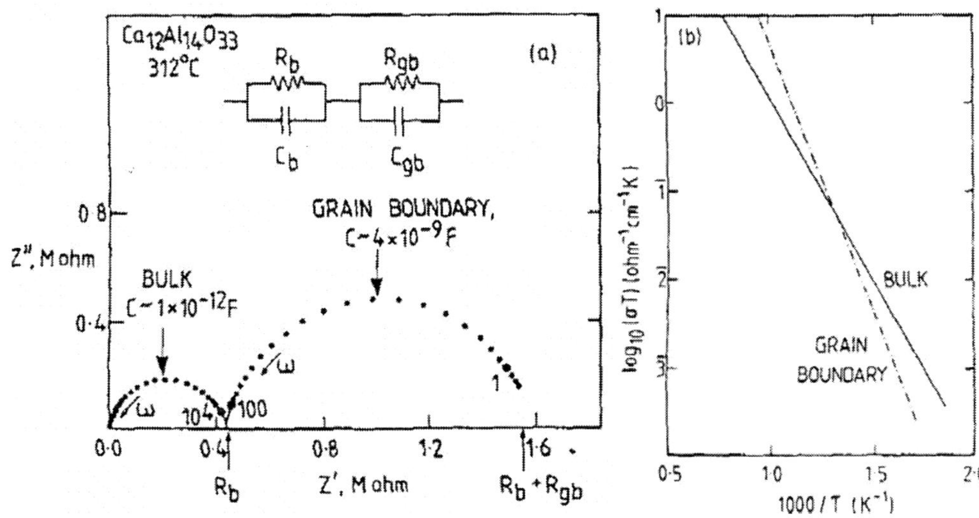

Figure 15 (a,b). -Z''-Z' impedance plots from $Ca_{12}Al_{14}O_{33}$ polycrystalline ceramics; reproduced from (Irvine et al. 1990) with permission from Wiley-VCH.

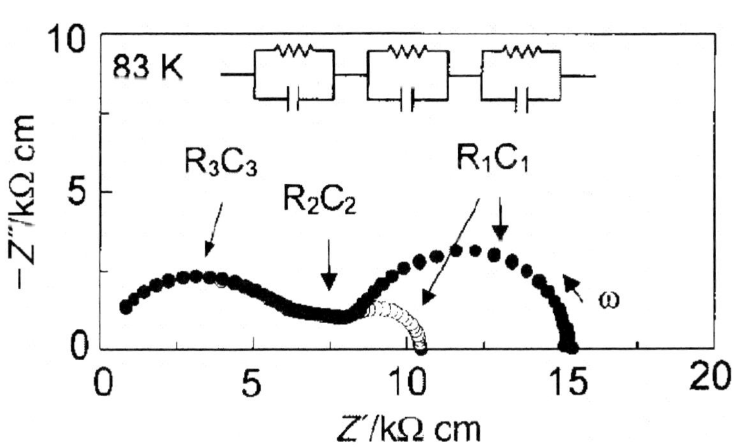

Figure 16. -Z''-Z' impedance plots from $Ca_2Mn_{1.5}Nb_{0.5}O_y$ polycrystalline ceramics; full circles correspond to an ac amplitude of 100 mV and open circles to 1V; this indicates non-ohmic behaviour of the electrode sample interface; reproduced from (Kruth et al. 2001) with permission from The Royal Society of Chemistry.

The spectra are ideal in a sense that bulk and grain boundary semicircles can be resolved clearly and analysed separately. The grain boundary and bulk resistivity display both thermally activated charge transport with different activation energies (as can be seen from the slopes of the $\log(\sigma \cdot T)$ vs $1/T$ plots in Fig 15 (b).

Very often, different contributions can not be resolved as clear as in Fig 15 (a), but show considerable overlap. Such a case is demonstrated on the example of an impedance spectrum on a -*Z*"-*Z*' plot from polycrystalline $Ca_2Mn_{2-x}Nb_xO_Y$ ceramics shown in Fig 16 (Kruth et al. 2001). In case of a strong overlap of different contributions, it may often not be reliable to determine R and C values from the semicircle dimensions and the peak frequencies $\omega_{max}(Z")$. Alternatively, fitting software should be used, where real and imaginary parts of each data point are fitted to the relevant equivalent circuit model.

Most commercial software enables the use of CPEs in the equivalent circuit, which usually improves the fit significantly since perfectly ideal Debye behaviour is hardly ever observed in real systems.

C. Polycrystalline Thick Films

The use of commercial fitting software is demonstrated on the example of NTC thermistor $NiMn_2O_4$ polycrystalline thick films on Al_2O_3 substrates. Fig 17 shows the -*z*"-*z*' plots on logarithmic axes, which leads to a distortion of the semicircles. The decreasing intercept on the real axis and decreasing size of the semicircles with increasing temperature confirms the negative temperature coefficient (NTC) of resistance. The impedance was measured in-plane, with a top-top arrangement of spring-loaded probes on Al/Ag contacts. The capacitance of the main contribution is typical for a grain boundary effect and was found to be approximately constant with temperature (in the absence of phase or structural transitions) (Schmidt et al.). An additional contribution at the low frequency end indicated a contact interface contribution. Although the data at the low frequency contribution are very limited, the small resistivity and large specific capacitance could be extracted from the fit of the data to the model. The good agreement of data and model confirms the validity of the equivalent circuit, consisting of two RC elements of parallel ideal resistor and non-ideal capacitors.

An alternative way to present impedance data is an Abrantes-Labrincha-Frade plot of *z*' vs $-z"/\omega$ (Abrantes et al. 2000), where the transition between different contributions corresponds to a distinct change in slope of the curve. The inset in Fig 18 shows that the low frequency interface contribution can not be resolved on the linear -*z*"- *z*' plots due to the small resistance, but the alternative *z*' vs $-z"/\omega$ plots show a clear change in slope displaying the low frequency contribution clearly.

The good linearity of the *z*' vs $-z"/\omega$ curve in both, the low and high frequency regime, indicates an almost ideal behaviour of the interface and grain boundary relaxations. In case of non-Debye behaviour, the curve would be bended in the respective frequency regime, the more bended the curve the larger the deviation from ideal behaviour. Close inspection of Fig 18 reveals a slight curve bending at the low frequency interface regime. No additional contribution can be seen at the high frequency end. The bulk contribution may be (a) non-existent, in case the grain boundary conductivity is higher and bypasses the bulk, or (b) the grain resistivity and/or the grain capacitance are relatively small and would appear at much

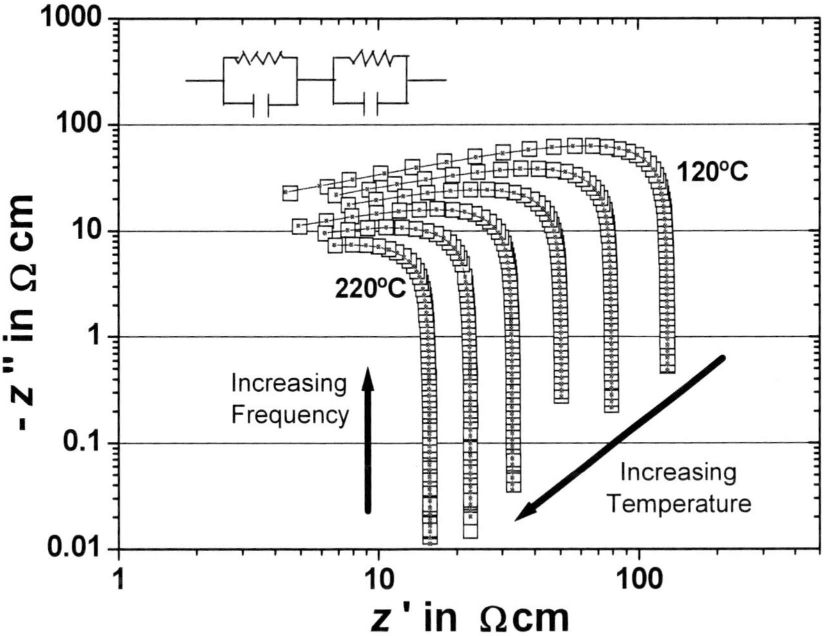

Figure 17. Double logarithmic -z''-z' impedance plots from $NiMn_2O_4$ polycrystalline ~ 25 μm thick films on Al_2O_3 substrates; □ = data and ■–■ = model; the model consists of two R-CPE elements; data taken at 120°C -220°C in 20°C increments

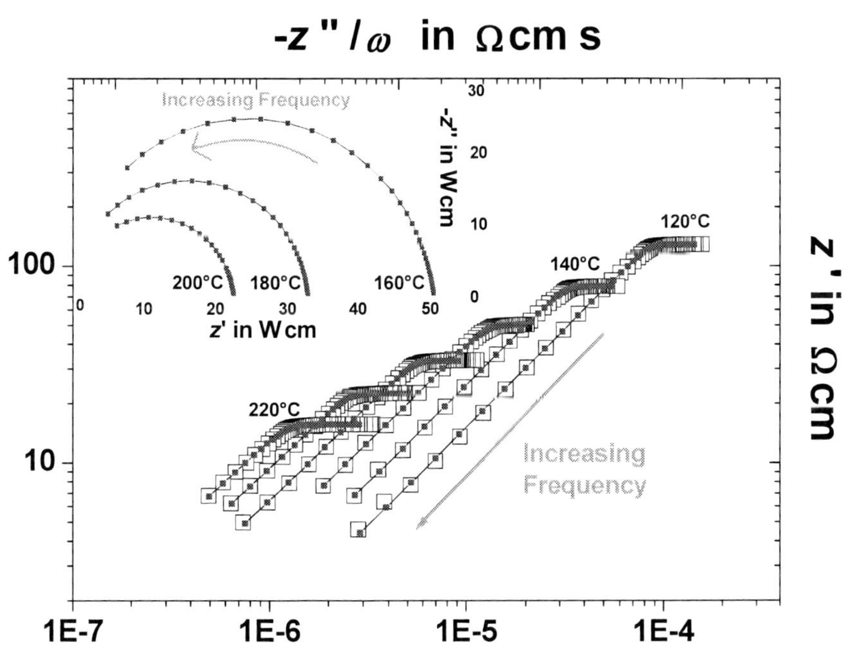

Figure 18. Abrantes-Labrincha-Frade plot of z' vs $-z''/\omega$ for the same data set as shown in Fig 17; inset: linear -z''-z' plots, □ = data and ■–■ = model;

higher frequencies. However, the detection of a small capacitance in the $NiMn_2O_4$ thick film may be difficult, because the small substrate capacitance (Al_2O_3 has a dielectric constant of ~ 10) is measured in parallel to the thick film and may show a perceptible contribution in case it is of the same order of magnitude as the film.

D. Thin Films

In case of ceramic thin films with considerable resistivity (non-metallic), it may difficult to measure the impedance in the in-plane direction using insulating substrates, because of contributions from the substrate, high nominal sample resistance and possible thin film size effects leading to non-representative resistivity and dielectric constants. In this case, the impedance may be better analysed in the out-of-plane direction by using a conducting substrate. The impedance can then be measured between one top-electrode and the substrate, or between two top-top electrodes. The ac current will pass through the metallic substrate, the case of top-top electrodes is shown in Fig 11(a). A distinct advantage of out-of-plane measurements is the possible absence of domain walls in case of ferroelectric or (anti-) ferromagnetic films. If the film thickness is small (< ~ 100 nm), single domain structure can usually be assumed in the out-of-plane direction. Problems can arise though due to a geometric effect of nano-sized films: the absolute nominal capacitance can be significantly different to the specific capacitance, and different relaxations can display similar capacitance. Deconvolution based simply on the magnitude of capacitance (table 2) is often not possible in this case.

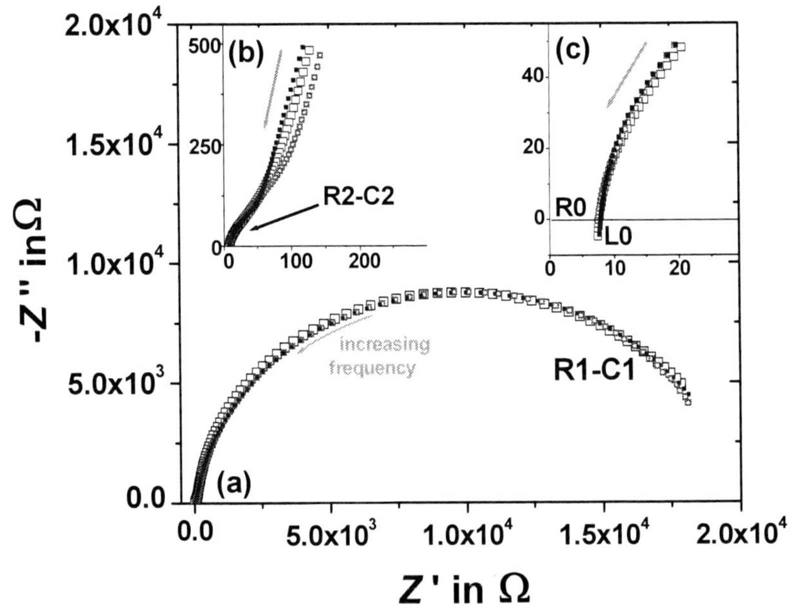

Figure 19. (a,b,c) -Z''-Z' impedance plots from 50 nm $BiFeO_3$ epitaxial films; the half-developed semicircle of the main contribution R1-C1 corresponds to the interface, the R2-C2 contribution shows strong overlap with the main contribution and represents the intrinsic film response; R0 and L0 are parasitic contributions, data taken at 150°C, □ = data and ■ = model using R-CPE elements; □ = model using Voigt elements.

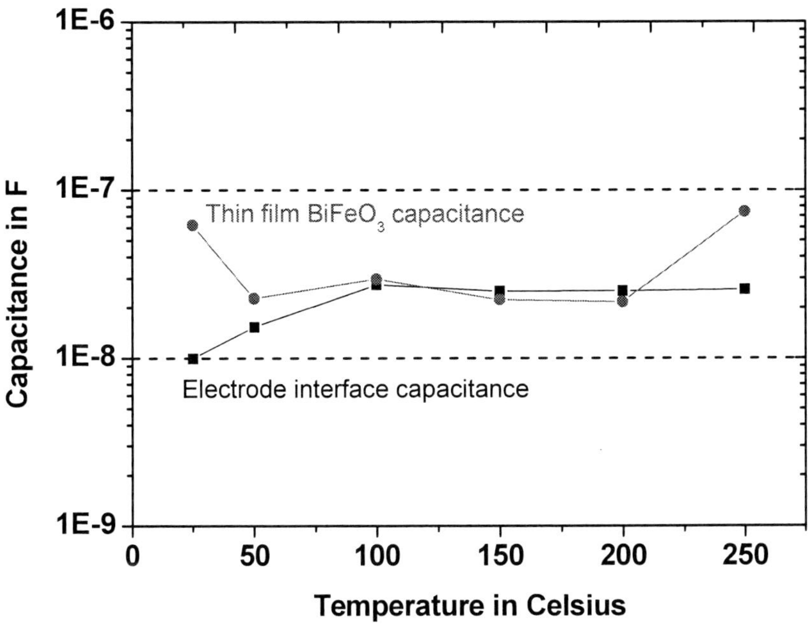

Figure 20. 50 nm epitaxial thin film $BiFeO_3$ on 1 at% Nb-doped highly conducting $SrTiO_3$ substrates; film capacitance and electrode film interface capacitance vs temperature.

The example of epitaxial thin films based on multiferroic (ferroelectric and anti-ferromagnetic) $BiFeO_3$ will be discussed in the following. An impedance spectrum obtained from a 50 nm $BiFeO_3$ epitaxial layer at 150°C is shown on a -*Z*''-*Z*' plot in Fig 19 (Schmidt et al. 2007). Two distinct contributions are visible. Grain boundary or domain wall contributions may be absent in highly crystalline single domain epitaxial films measured in the out-of plane direction, and the two contributions can be assumed to arise from film and interface relaxations. Fig 19 displays two semicircles of distinctively different size. The large semicircle dominates the spectrum in Fig 19 (a), whereas the small semicircle is evident only by zooming into the intermediate frequency regime (Fig 19 (b)). At high frequencies (Fig 19 (c)) a parasitic resistor R0 is evident as the intersect of the data curve on the real axis, and a change of sign in -*Z*'' indicates the presence of a residual inductor L0 (for a discussion of such unwanted contributions see section IV./C).

Capacitance values for both sample contributions were obtained from data fitting procedures using the equivalent circuit model shown in Fig 11 (b), using R-CPE_C elements. The capacitances C1 and C2 are of the same order of magnitude as shown in Fig 20. In order to achieve deconvolution of the two contributions it is necessary to analyse both capacitances in dependence on film thickness. Fig 21 shows that C1 is approximately constant with film thickness and in the range of a typical interface contribution by normalising (multiplying) the capacitance with the contact area *A*. C2 is approximately constant and in the range of a typical ferroelectric material by normalising with the geometrical factor *g* (see table 2). Therefore, C1 represents an interface capacitance and C2 the intrinsic film capacitance. The associated resistors and capacitors can then be analysed in dependence on temperature and associated with interface or intrinsic film behaviour.

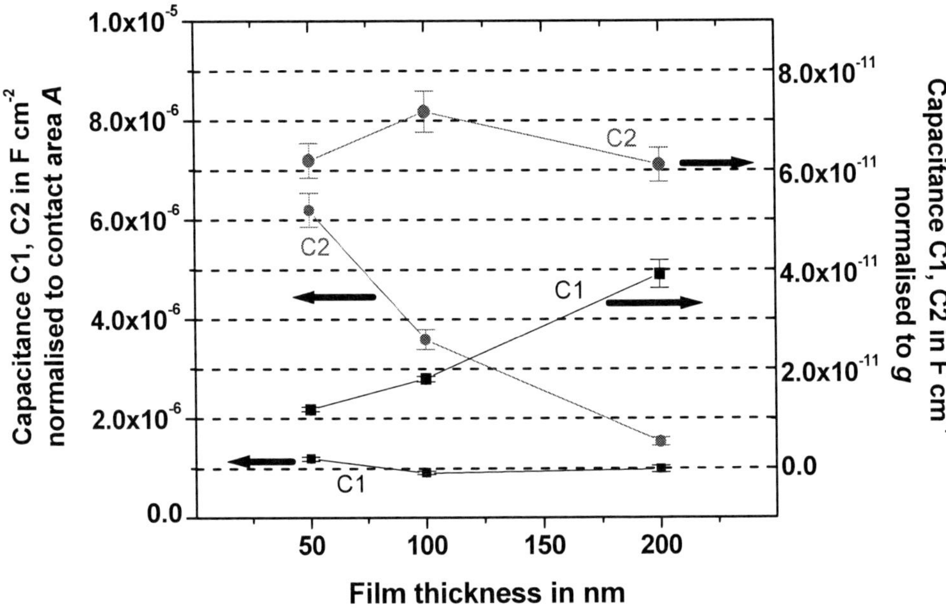

Figure 21. $BiFeO_3$ thin film samples: Capacitance C1 (■) and C2 (♦) vs film thickness; C1 and C2 normalised to contact area *A* and geometrical factor *g*.

Deviations from linearity in *C* vs *T* in Fig 20 indicate problems with the fits due to strong overlap of several contributions, and the fitted parameters may only be reliable in the approximately linear regime.

In the frequency range where the film contribution R2-C2 is dominant, both R-CPE and Voigt models resulted in a reasonable fit (Fig 19 (b)). In the intermediate regime of overlapping intrinsic film and electrode interface contributions, both models showed slight deviations from the data. This is because the Voigt model is probably not suitable for interface contributions as mentioned before, and the R-CPE is an empirical model and can obviously not perfectly model the intermediate regime of overlapping non-ideal contributions, where the origin of the non-ideality is different.

The large resistance of the interface compared to the intrinsic film contribution indicates a blocking effect of the electrodes. From temperature dependent analysis of the film resistance in 50 nm epitaxial $BiFeO_3$, the charge transport activation energy was found to be ~ 0.6 eV (Fig 22).

This is in a typical range for ionic type transport, and can explain the blocking effect of the Pt electrodes. Ionic oxygen vacancy charge carriers with comparatively high mobility in the $BiFeO_3$ layer can not cross the interface easily to transfer into the electronic charge carrier material Pt with a low oxygen ionic mobility. On the other hand, the electronic charge carriers in the Pt electrodes have a low mobility in an insulating ferroelectric material. Such electrode blocking effects are commonly observed at metal - ferroelectric insulator interfaces.

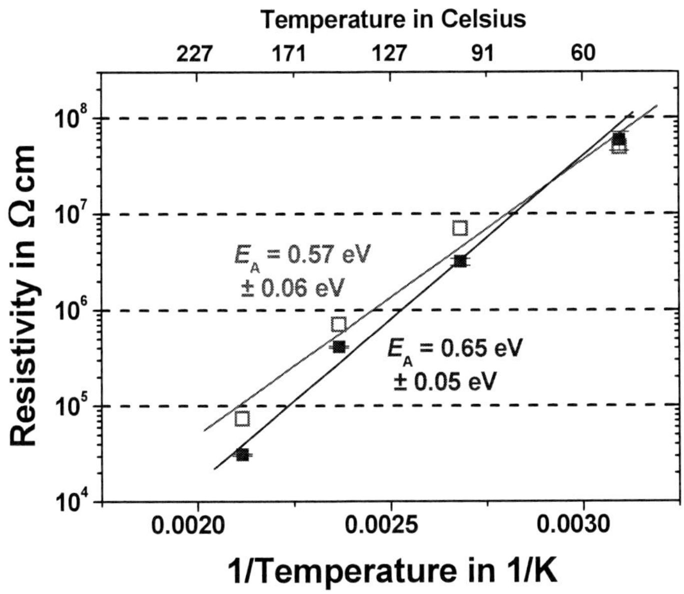

Figure 22. Intrinsic 50 nm $BiFeO_3$ thin film resistivity ρ vs T from R2 on a log(ρ) vs $1/T$ plot, the slopes correspond to the activation energies; ■ = R-CPE model; □ = Voigt model.

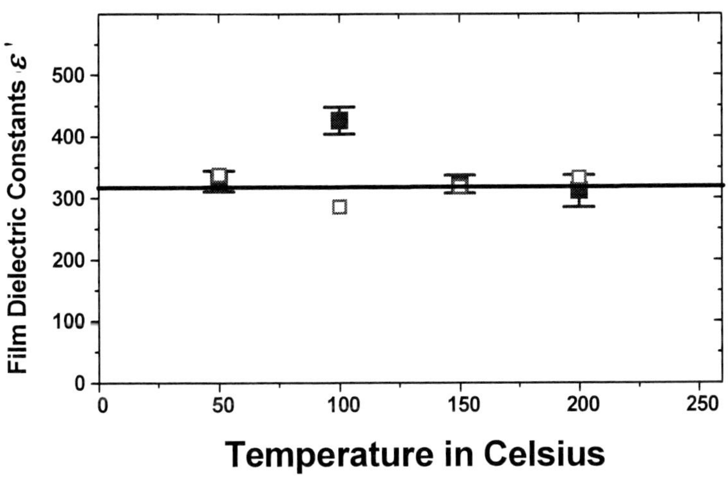

Figure 23. Intrinsic 50 nm $BiFeO_3$ thin film dielectric constant ε' vs T from C2; ■ = R-CPE model; □ = Voigt model

The intrinsic thin film $BiFeO_3$ dielectric constant was determined from C2 to be $\sim 320 \pm 75$ as shown in Fig 23.

Using the Voigt model, the capacitance C2 was determined from the mean relaxation time $\overline{\tau}$, with an approximately temperature independent standard deviation $\Delta\tau$ with a relatively small $\Delta\tau/\overline{\tau}$ ratio of 0.41 - 0.47. This indicates a narrow distribution of relaxation times and the validity of the Voigt model. Using the R-CPE circuit, the capacitance is obtained from the model in specific units of [Farad·s^{n-1}], but can be transformed in SI units

[Farad] using a standard method (Hsu et al. 2001). In Fig 23, the Voigt model shows slightly better consistency in the dielectric constant compared to the R-CPE model.

It is believed that the value of ~ 320 ± 75 for the intrinsic $BiFeO_3$ epitaxial thin film dielectric constant is a reliable value, because all unwanted contributions such as electrode sample interface (R1-C1), residual inductance (L0) and resistance (R0) have all been deconvoluted.

VI. Conclusions

Impedance spectroscopy can be regarded to be a powerful tool to reveal the composite nature of the resistive and dielectric properties of a wide variety of single crystal, polycrystalline bulk and thick film, and epitaxial thin film electroceramic materials. The analysis of bulk material is well established, whereas thick and thin film impedance analysis is relatively new.

In case of film samples, a possible contribution from the substrate has to be taken into account. In case of insulating substrates and the measurement of in-plane impedance, the substrate contribution is in parallel to the film in terms of the equivalent circuit. At low frequency, current may pass through the more conducting films and the insulating substrate contribution may be neglected. However, at high enough frequencies the substrate can show a significant contribution due to the appearance of dielectric relaxation. This problem can be circumvented using conducting (metallic) substrates, where film and substrate are in series, which allows better discrimination. In case of relatively resistive films compared to the metallic substrates the impedance response can be expected to be distinctively different.

Non-Debye impedance response can be modelled using R-CPEs, Voigt elements or the Havriliak-Negami formalism. The use of the R-CPE circuit can be justified by the fact that its impedance is consistent with Jonscher's law, which holds for dielectric relaxations in almost all kinds of solid matter. Still, no microscopic interpretation of the R-CPE circuit is available to date. The Voigt element constitutes an improvement in this context, because it can be associated with a random distribution of microscopic relaxation times across the sample. Furthermore, Voigt elements can reproduce Jonscher's behaviour reasonably for small deviations from the ideal case ($0.95 < n < 1$). Contrarily, the Havriliak-Negami expression may be regarded as purely empirical, although it can fit almost all impedance response data precisely.

A. Appendix: The Voigt Equivalent Circuit

The specific complex impedance of an ideal RC element of parallel ideal resistor and capacitor is a Lorentzian profile:

$$z^*_{RC} = \frac{\rho_{dc}}{1 + i\omega\tau} \tag{11}$$

where τ is the relaxation time ($= \rho_{dc} \cdot c$), and ρ_{dc} is the resistivity of the resistor and c the specific capacitance of the capacitor. In order to determine the macroscopic impedance of a

RC element with a random Gaussian distribution of relaxation times, the ideal RC Lorentzian function (eq. 11) may be integrated around the mean relaxation time $\tau = \bar{\tau}$ and weighted by a standard Gaussian function normalised to 1. This convolution of Lorentzian and Gaussian functions constitutes a Voigt profile, describing the impedance of the distributed RC element by:

$$z^*_{Voigt} = \frac{\rho_{dc}}{\Delta\tau\ \sqrt{2\pi}} \int_{-\infty}^{+\infty} \frac{\exp\left(\frac{-x^2}{2\ \Delta\tau^2}\right)}{1 + i\omega\ (\bar{\tau} + x)}\ dx \tag{12}$$

The small error made by accounting for physically meaningless $\bar{\tau} < 0$ values was neglected. The above integral can be evaluated using the complex scaled error function (Armstrong 1967):

$$z^*_{Voigt} = \frac{\rho_{dc}\ \sqrt{\pi}}{\Delta\tau\ \sqrt{2}\ \omega}\ W\left(\frac{i/\omega - \bar{\tau}}{\Delta\tau\ \sqrt{2}}\right) \tag{13}$$

The function W(i·*b*) denotes the complex scaled complementary error function *erfcx*(*b*):

$$W(i \cdot b) = erfcx(b) = \exp(b^2) \cdot (1 - erf(b)) \tag{14}$$

where the error function *erf*(*b*) is defined as:

$$erf(b) = \frac{2}{\sqrt{\pi}} \int_0^b e^{-x^2} dx \tag{15}$$

The analytical solution of eq.(12) was confirmed numerically, and a data fitting tool was developed based on eq.(13) using MatLab® software, which minimized the least-squares error using a statistical weighting of the squared model – data distances. The basic MatLab package does not provide the complex error function, but a routine that calculates this special function for any complex argument has been developed for this purpose (Winiecki 2006). The algorithm was published first for a Fortran program (Poppe et al. 1990).

Acknowledgments

The author wishes to thank Thomas Winiecki for invaluable help with solving integrals and equations. The author is also grateful to Finlay Morrison for useful discussions. Furthermore, the author would like to thank Juan Aguilar for the kind invitation for a research visit to FIME at UANL Monterrey (Mexico). This work was supported by the Leverhulme Trust.

References

Abrantes, J. C. C., Labrincha, J. A. and Frade, J. R. An alternative representation of impedance spectra of ceramics. *Materials Research Bulletin* 35 (2000) p.727.

Alvarez, F., Alegria, A. and Colmenero, J. Relationship between the time-domain Kohlrausch-Williams-Watts and frequency-domain Havriliak-Negami relaxation functions. *Physical Review B* 44 No.14 (1991) p.7306.

Armstrong, B. H. Spectrum Line Profiles: The Voigt Function. *Journal of Quantitative Spectroscopy and Radiation Transfer* 7 (1967) p.61.

Cole, K. S. and Cole, R. H. Dispersion and Absorption in Dielectrics. I. Alternating Current Characteristics. *Journal of Chemical Physics* 9 (1941) p.341.

Edwards, D. D., Hwang, J.-H., Ford, S. J. and Mason, T. O. Experimental limitations in impedance spectroscopy: Part V. Apparatus contributions and corrections. *Solid State Ionics* 99 (1997) p.85.

Havriliak, S. and Negami, S. *Journal of Polymer Science Part C* 14 (1966) p.99.

Hsu, C. H. and Mansfeld, F. Technical Note: Concerning the Conversion of the Constant Phase Element Parameter Y_0 into a Capacitance. *Corrosion* 57 No.9 (2001) p.747.

Hwang, J.-H., Kirkpatrick, K. S., Mason, T. O. and Garboczi, E. J. Experimental limitations in impedance spectroscopy: part IV. electrode contact effects. *Solid State Ionics* 98 (1997) p.93.

Irvine, J. T. S., Sinclair, D. C. and West, A. R. Electroceramics: Characterization by Impedance Spectroscopy. *Advanced Materials* 2 No.3 (1990) p.132.

James, A. R., Priya, S., Uchino, K. and Srinivas, K. Dielectric spectroscopy of $Pb(Mg_{1/3}Nb_{2/3})O_3$ - $PbTiO_3$ single crystals. *Journal of Applied Physics* 90 No.7 (2001) p.3504.

Jonscher, A. K. Frequency-dependence of conductivity in hopping systems. *Journal of Non-crystalline Solids* 8-10 (1972) p.293.

Jonscher, A. K. Universal Relaxation Law. Chelsea Dielectric Press London 1996.

Kohlrausch, F. *Pogg.Ann.Phys.* 119 (1863) p.352.

Kruth, A. and West, A. R. Electrical properties of the oxygen-deficient perovskites, $Ca_2Mn_{2-x}Nb_xO_\gamma$: $0<x<1.2$, with Mn valence varying from +2.0 to +4.0. *Journal of Materials Chemistry* 11 (2001) p.153.

Lanfredi, S., F, C. J. and Hernandes, A. C. Electric and dielectric properties of $Bi_{12}TiO_{20}$ single crystals. *Journal of Applied Physics* 88 No.1 (2000) p.283.

Liu, S. H. Fractal Model for the ac Response of a Rough Interface. *Physical Review Letters* 55 No.5 (1985) p.529.

Macdonald, Ed. (1987). Impedance spectroscopy. New York, John Wiley & Sons.

Macdonald, J. R. *Journal of Applied Physics* 61 No.2 (1987) p.700.

Naohiro, H. (1995). Impedance Characterisation of Conducting Materials. Department of Chemistry. Aberdeen, University of Aberdeen.

Poppe, G. P. M. and Wijers, C. M. J. Algorithm 680: Evaluation of the Complex Error Function. *Transactions on Mathematical Software* 16 No.1 (1990) p.47.

Sazonov, V. V. Normal approximation: some recent advances. Springer - Verlag Berlin 1981.

Schmidt, R. and Brinkman, A. W. Studies of the temperature and frequency dependent impedance of an electroceramic functional oxide thermistor. *Advanced Functional Materials* to be published.

Schmidt, R., Eerenstein, W., Winiecki, T., Morrison, F. D. and Midgley, P. A. The impedance of epitaxial multiferroic thin films. *Physical Review B* 75 No.24 (2007) p.245111.

Shklovskii, B. I. and Efros, A. L. Electronic properties of doped semiconductors. Springer - Verlag Berlin 1984.

Song, S. and Placido, F. The influence of phase probability distributions on impedance spectroscopy. *Journal of Statistical Mechanics: Theory and Experiment* OCT 2004 (2004) Art No P10018.

Williams, G. and Watts, D. C. *Transactions of the Faraday Society* 66 (1970) p.80.

Winiecki, T. (2006). Complex scaled complementary error function, http://www.mathworks.com/matlabcentral/fileexchange/loadFile.do?objectId=12091.

In: Ceramic Materials Research Trends
Editor: Paul B. Lin, pp. 357-378
ISBN 1-60021-769-9

Chapter 13

EXPERIMENTS AND SIMULATIONS OF THIN SOLID FILMS ISLANDING

F. Lallet, N. Olivi-Tran and A.Dauger
Laboratoire de Sciences des Procédés Céramiques et Traitements de Surface, UMR-CNRS 6638, Ecole Nationale Superieure de Céramiques Industrielles, 47 avenue Albert Thomas, 87065 Limoges cedex, France

Abstract

Basically, the formation of nanoislands can be achieved through two main mechanisms:

- Nucleation of islands, growing and sometimes coarsening; this scheme is often observed for PVD-CVD processes (refs).
- Surface roughening, islanding and growing; this scheme is observed both for sol-gel and PVD-CVD processes (refs).

In the following, we focus on the second idea and the main objective of this study is to highlight the main physical parameters responsible for the islanding of thin solid films through surface roughening synthesized either by sol-gel or PVD-CVD processes. Therefore we present two physical models dedicated to simulate the formation of nanoislands without nucleation, either during thermal treatments of thin solid films synthesized by sol-gel dip-coating or during the deposition of matter on a substrate by PVD-CVD processes. We highlight our numerical results by experimental data concerning either the islanding process of yttria-stabilized zirconia thin films deposited on corundum substrates or the formation of germanium QD's on silicon substrates.
The models describe thin solid films surface rougheness evolution from continuous layers to nanoislands. We chose classical Monte Carlo approaches with Metropolis scheme to describe the evolution of the systems from mesoscopic length scale. These approach allows one to compare, at least qualitatively, the main physical parameters responsible for the formation of nanoislands between the two synthesis process described previously through simple physical concepts.
In section 2 we present experimental details and morphological analysis of yttria-stabilized zirconia nanoislands on sapphire substrates. The section 3 gives a description of the physical models used to simulate the islanding mechanism for thin solid films synthesized either by sol-gel or PVD-CVD processes and section 4 is the general conclusion.

1. Introduction

The fabrication of epitaxial crystalline nanoislands distributed over the surface of a crystalline substrate has been an active field of experimental and theoretical researches since the 1980's. Several synthesis processes have been developed to achieve this task. In our view, two drastically different approaches can be highlighted:

On one hand, a thin liquid film of nanometric thickness (10-50 nm) is deposited on a substrate by a sol-gel dip-coating or a sol-gel spin-coating process [1]. In both cases the thin liquid film (sol) solidifies and dries at room temperature to form an amorphous layer (gel) of nanometric thickness (1-20 nm) on the substrate. Further thermal treatments at medium temperature (600°C-1200°C) induce the crystallisation of the film and normal grain growth of crystallites. Finally the formation of epitaxial nanoislands on the surface of the substrate is achieved during thermal treatments at high temperature (1300 °C-1500°C) through abnormal grain growth. In that case, the evolution from a continuous film to a collection of discrete nanoislands is achieved after the deposition of matter on the surface of the substrate. This process has been successfully applied to synthesize epitaxial oxide thin films on oxide substrates [2, 3, 4].

On the other hand, atomic species are either deposited by a Physical Vapor Deposition (PVD) or a Chemical Vapor Deposition (CVD) process. The thin solid film grows during the deposition of atomic species from a gaseous phase on the surface of the substrate which usually involve a low deposition pressure and medium temperature (600 °C-700°C). The atomic species can stick with each other to form clusters, diffuse on the substrate etc. Discrete nanoislands can appear at the top of the substrate at the beginning of the deposition process which is the Volmer-Weber (V-W) growth mode or at the top of a thin layer called wetting layer when its thickness overcome a critical value which is the Stranski-Krastanov (S-K) one [5]. In that case the formation of islands is achieved during the deposition of matter which is a process drastically different than the previous one. The nanoislands which are formed in those processes are usually smaller than in the previous case; therefore, they are called Quantum Dots (QD's). Indeed, those processes have been developed to create distributions of very small crystals in order to understand and apply new interesting physical properties associated to quantum confinement effects. Those processes have already been studied both theoretically [6, 7, 8, 9, 10, 11] experimentally [12, 13, 14, 15] and in particular widely used to create epitaxial germanium QD's on silicon substrates [16, 17, 18, 19, 20, 21]. Whatever the kind of process which is considered, the objectives are still the control of the shape, the size and the distribution function of the islands over the whole surface of the substrate.
Basically, the formation of nanoislands can be achieved through two main mechanisms:

- Nucleation of islands, growing and sometimes coarsening; this scheme is often observed for PVD-CVD processes (refs).

- Surface roughening, islanding and growing; this scheme is observed both for sol-gel and PVD-CVD processes (refs).

In the following, we focus on the second idea and the main objective of this study is to highlight the main physical parameters responsible for the islanding of thin solid films through surface roughening synthesized either by sol-gel or PVD-CVD processes. Therefore we present two physical models dedicated to simulate the formation of nanoislands without nucleation, either during thermal treatments of thin solid films synthesized by sol-gel dip-coating or during the deposition of matter on a substrate by PVD-CVD processes. We highlight our numerical results by experimental data concerning either the islanding process of yttria-stabilized zirconia thin films deposited on corundum substrates or the formation of germanium QD's on silicon substrates.

The models describe thin solid films surface rougheness evolution from continuous layers to nanoislands. We chose classical Monte Carlo approaches with Metropolis scheme [22] to describe the evolution of the systems from mesoscopic length scale. These approach allows one to compare, at least qualitatively, the main physical parameters responsible for the formation of nanoislands between the two synthesis process described previously through simple physical concepts.

In section 2 we present experimental details and morphological analysis of yttria-stabilized zirconia nanoislands on sapphire substrates. The section 3 gives a description of the physical models used to simulate the islanding mechanism for thin solid films synthesized either by sol-gel or PVD-CVD processes and section 4 is the general conclusion.

2. Yttria-Stabilized Zirconia Nanoislands on Sapphire Substrates by Sol-Gel Dip-Coating

In this section we present the experimental procedure followed to synthesize yttria-stabilized zirconia nanoislands on sapphire substrates with random crystallographic orientation at the solid-air interface. Then we present an Atomic Force Microscopy (AFM) characterisation of thin solid films surface rougheness after thermal treatments.

2.1. Experimental Procedure

2.1.1. Synthesis of Thin Zirconia-Precursor Xerogel Films by Sol-Gel Dip-Coating

We synthesized a clear and homogeneous sol of zirconia precursor from zirconium n-propoxyde, acetylacetone and n-propanol [23]. We used hydrated yttrium nitrate ($Y(NO_3)_3(H_2O)_5$) disolved in n-propanol (yttria solution: 2g of $Y(NO_3)_3(H_2O)_5$ in 20mL of n-propanol) as an yttria precursor, and the molar ratio Y_2O_3/ZrO_2 was fixed to 10% so that zirconia is expected to crystallize in its cubic phase [24].

The quantities of matter used in the process are listed in the table below; the total volume of a sol is fixed to 15mL:

The chemicals are mixed under mechanical stirring in dried air to avoid the precipitation of the zirconia precursor. The films were deposited on corundum substrates by sol-gel dip-coating [25]. The α-alumina substrates are mirror polished with a random crystallographic orientation at the solid-air interface. The dipping speed was fixed to 2mm.s^{-1}. When the substrate is removed from the sol in air at ambient temperature, a thin liquid film is deposited on the surface of the substrate and dries because of solvent evaporation. At the

Table 1. Molar quantities and volume of chemicals used in the fabrication of the sol

Chemicals	Molar quantity (mol)	Volume (mL)
zirconium n-propoxyde	0.008	3.178
acetylacetone	0.005	0.541
n-propanol	-	8.529
hydrated yttrium nitrate	0.0016	-
yttria solution	-	2.752
	Total	15

same time, the zirconium n-propoxyde is submitted to hydrolysis and polycondensation reactions, because of humidity in ambient air, which leads to the formation of a thin solid polymeric lattice at the surface of the substrate. After drying at 60 °C for 20min in air a thin xerogel film of zirconia precursor is formed.

2.1.2. AFM Study of Surface Roughening during Thermal Treatment of Yttria-Stabilized Zirconia Thin Solid Films

In this section we present the influence of thermal treatments on surface roughness evolution of the thin solid films previously synthesized. The evolution of surface roughness has been followed by AFM measurements after thermal treatments in air.

In previous experiments, the microstructural evolution of thin zirconia films from amorphous thin solid film (xerogel) to collections of epitaxial nanoislands more or less interconnected has been studied by Mary [26, 27] for tetragonal zirconia thin films and Bachelet [24] for cubic zirconia thin films deposited on c-cut α-alumina substrates. From those previous experimental investigations, it has been demonstrated that the crystallisation of zirconia precursors starts around 600 °C. After thermal treatments at higher temperature, the grains which compose the film grow in the three directions of space until their height reach the average initial thickness of the amorphous thin solid film; the islanding process starts from this point. Phenomenologically, this result is described by the transition from a normal grain growth of zirconia crystals in the three directions of space when their height is lower than the average thickness of the film to an abnormal grain growth when their height reaches the average initial thickness of the film. In the case of cubic zirconia thin films, the microstructural characterization was achieved by combining High Resolution X-Ray Diffraction (HR-XRD) [28, 29] through Reciprocal Space Mapping (RSM) [30, 31] and AFM analysis of the surface of the films which have demonstrated that the shape and size of the epitaxial zirconia islands are linked to specific out of plane and in plane crystallographic orientations with regard to the c-cut surface of the α-alumina substrate [24].

In our experiments, we studied zirconia thin solid films surface morphology evolution by AFM in tapping mode in air. We used a silicon tip with 2.2 N/m of spring constant and 76kHz of resonance frequency.

The left part of figure (1.1) presents the AFM analysis of the surface of a thin film synthesized by the previous process after a first thermal treatment at 1350°C in air for 1h. The first picture on the top is a top view of the surface of the thin solid film. The two other pictures correspond to the same surface but the middle one is a 3 dimensional (3D) view of the surface and the bottom one is a 3D zoom of the square area delimited by the white dashed lines on the top picture. One can note that, after this thermal treatment, the surface of the thin solid film is rough. The two 3D pictures highlight the beginning of the islanding process. Indeed, from the picture on the center it is clear that the substrate is partially unwetted. At this stage, the thin film is composed of several crystalline islands more or less interconnected. The zoom of the right picture point out the presence of grain boundaries and confirm this description. Although the islanding process has just begun, two kinds of cristalline islands are clearly identified:

- Flat top crystalline domains.
- Rounded crystalline domains.

The first ones are much more larger than the other ones. This kind of microstructures has already been observed for yttria stabilized zirconia (YSZ) thin films deposited on c-cut α-alumina substrates [32]. From those works, we know that the two types of crystalline islands previously reported are linked to a specific out of plane crystallographic orientation with regard to the substrate deduced from RSM analysis. Indeed, for a c-cut α-alumina substrate the flat-top islands demonstrate:

- $(100)_{YSZ}//(001)_{Al_2O_3}$ for the flat-top ones
- $(111)_{YSZ}//(001)_{Al_2O_3}$ for the rounded ones

which correspond to a fundamental result: the family planes parallel to the surface of the substrate are always dense family planes of zirconia. Theoretically, in the case of cubic YSZ, three dense and inequivalent planes should appear (100), (110) and (111). One can already note that for cubic YSZ islands on c-cut α-alumina substrates, the (110) family planes is not observed. From those results we can deduce that the nanoislands presented on the left of figure (1.1) demonstrate the following out of plane crystallographic orientations with regard to the substrate:

- $(100)_{YSZ}//(hkl)_{Al_2O_3}$ for the flat-top ones
- $(111)_{YSZ}//(hkl)_{Al_2O_3}$ for the rounded ones

The right part of figure (1.1) corresponds to the AFM analysis of the same surface but after another thermal treatment at 1500°C in air for 15min. The islanding process is drastically increased and the substrate is much more unwetted. One can see that the microstructure is still composed of flat-top and rounded islands. However, the flat-top islands are much more larger than in the previous case whereas the size of rounded ones is still the same. Moreover, the number of rounded islands decreases. From those observations, we can deduce that the increase of the temperature lead to the growing of the flat-top islands through local densification and coalescense of the rounded islands. Indeed, from the zoom on the right picture, one can see that the number of grain boundaries decreases which is linked to the increase of the size of the flat-top islands. Consequently, the growing of flat-top islands seems to be energetically more favorable than the formation of rounded ones. There is a competitive growing mechanism between the two families of islands of different out of plane crystallographic orientations.

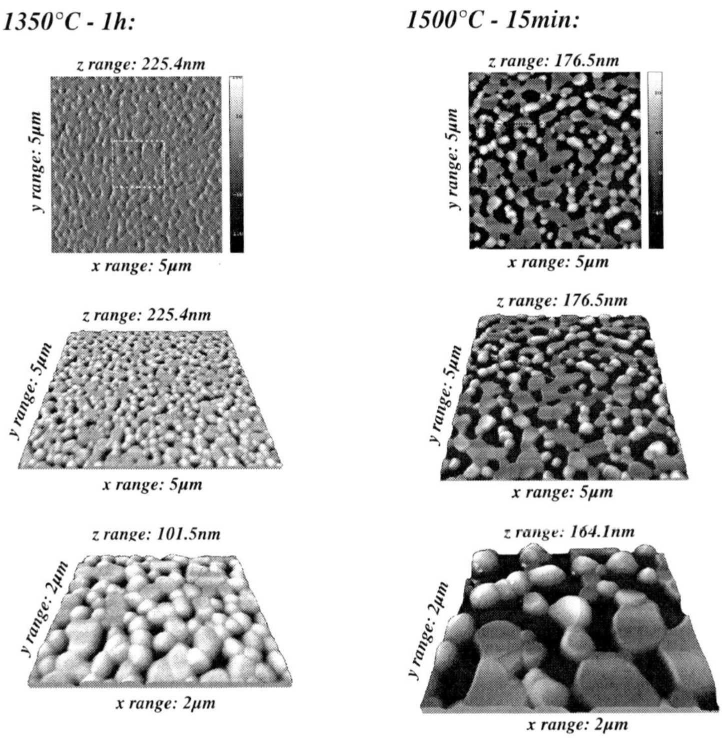

Figure 1. AFM characterization of the surface of an yttria-stabilized zirconia thin solid film after thermal treatment, 1350°C for 1h top and 1500°C for 15min bottom.

3. Numerical Simulations of Thin Solid Film Islanding

In this section we present the physical models dedicated to simulate the islanding process of thin solid films synthesized either by sol-gel or PVD-CVD processes.

3.1. Sol-Gel Process: Islanding after Matter Deposition

The numerical model dedicated to simulate the islanding process of a thin solid film synthesized by sol-gel dip-coating is inspired from the experimental results of Bachelet [24, 32] and derived from the previous theoretical approaches of Olivi-Tran [33, 34, 35]. The aim of the numerical simulation is to mimic the islanding process of thin films without matter evolution through surface roughness evolution.

3.1.1. Model of Thin Solid Film and Substrate

As previously reported for the ZrO_2/Al_2O_3 system, the islanding process of the thin film starts when the height of the crystals which compose the film reaches its average initial thickness. Therefore, we chose to model a thin polycrystalline film by a square lattice, composed of 10^4 elementary domains i of height $h_i = H = 1nm$, width $\ell = 1nm$ and random crystallographic orientation with regard to the substrate and to their next nearest neighbors. The figure (1.2) gives a schematic description of this virtual system; we applied periodic boundary conditions to overcome boundaries effects. The out of plane crystallographic orientation of each domain i is associated during the whole computation process to the inter-plane spacing c_{ih} of the family plane parallel to the surface of the substrate. Each domain has a square horizontal section l^2 and four next nearest neighbours. This virtual system is dedicated to simulate a continuous and polycrystalline thin solid film deposited on a perfect single crystalline substrate before the beginning of the islanding process. In our computations, we studied both the evolution of the height and the crystallographic orientation of the domains across time. Indeed, the evolution of the height of virtual domains mimic the evolution of the roughness of real thin solid film until discrete island are formed, whereas the modification of crystallographic orientations of the domains simulates the out of plane epitaxial relationships between the film and the substrate.

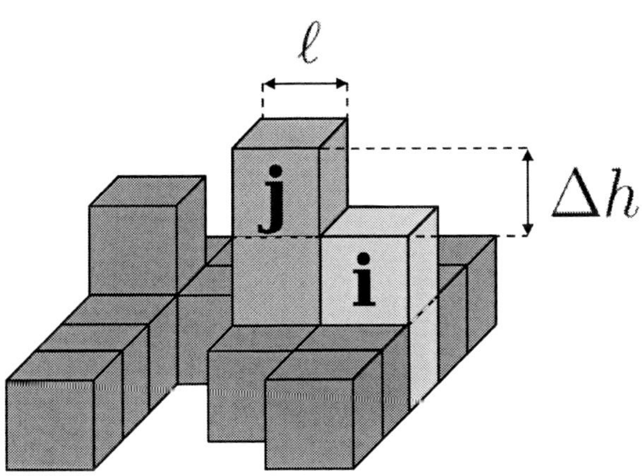

Figure 2. Discretization of a virtual thin solid film deposited by sol-gel dip-coating.

3.1.2. Physical Model

The physical model is derived from previous works on thin solid films islanding without matter evolution [35]. We consider that the matter is reorganized by surface diffusion [36]. We consider three aspects which contribute to the energy of our thin film:

- The free surface energy γ_{si} which is related to the height of each elementary domain i.

- The lateral grain boundary energy, which simulates the interfacial energy between two neighbouring domains.
- The interfacial grain boundary energy which corresponds to the interfacial energy between one elementary domain and the substrate.

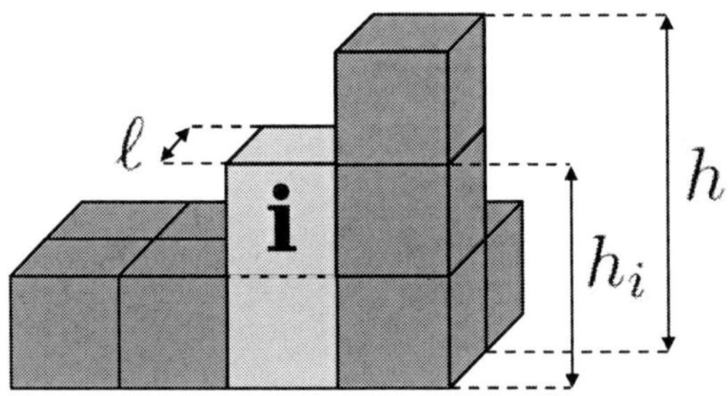

Figure 3. Discretization of a virtual thin solid film deposited by PVD-CVD.

The figure (1.3) presents two elementary domains i and j and the position of the interfaces. The variation of energy related to the evolution of the height of a domain i with regard to one of its four next nearest neighbors using the work of elastic forces, the Young modulus Y and the Poisson ratio ν is:

$$\Delta Eh = Y(1+\nu)\Delta h\ell(h_i - h_j) \tag{1}$$

where Δh is related to the physical parameters of the system through:

$$\Delta h = \sqrt{\frac{\ell^2 D_{si}\gamma_{si}\Delta t}{k_B T}} \tag{2}$$

where D_{si} is the surface diffusion coefficient, Δt the diffusion time, $k_B = 1.38 \times 10^{-23} J.K^{-1}$ the Boltzmann's constant, T the absolute temperature and $(h_i - h_j)$ is the variation of the height of a domain i between two Monte Carlo Steps (MCS). Finally, the previous equation is reformulated as:

$$\Delta E_h = A(h_i - h_j) \tag{3}$$

$$A = Y(1+\nu)\sqrt{\frac{D_{si}\gamma_{si}\Delta t}{k_B T}}\ell^2 \tag{4}$$

The variation of energy related to the evolution of the crystallographic orientation of a domain i with regard to one of its four next nearest neighbors or to the substrate is expressed as [35]:

$$\Delta E_b = \gamma_b \left(\frac{\ell^2}{h_i} + \ell\right)(c_i - c_j) \tag{5}$$

where γ_b is the grain boundary tension induced by the interfaces of domain i with regard to one of its four next nearest neighbors in the vertical plane or with regard to the substrate in the horizontal plane. The last term, $(c_i - c_j)$, is the evolution of the inter-plane spacing of domain i in the vertical plane or in the horizontal plane during one MCS. The figure (1.3.a) presents a schematic description of those interfaces in the general case. We denote the inter plane spacing of the family planes of domain i by c_{iv} in the vertical plane (family planes orthogonal to the surface of the substrate) and c_{ih} in the horizontal plane (family planes parallel to the surface of the substrate). The grain boundary energy γ_b which simulates the lateral domain-domain interfacial energy induced by the lattice mismatch between the family planes of a domain i and the one of its neighbouring domains j in the vertical plane is denoted γ_{bv}. When two crystalline domains i and j are in contact through a grain boundary, the lateral interfacial energy is expressed by the difference of the lateral free surface tensions γ_{svi} and γ_{svj} of domains i and j taken separately:

$$\gamma_{bv} = |\gamma_{svi} - \gamma_{svj}| \tag{6}$$

This is equivalent to say that when the grain boundary is created, the lateral free surface energy of each domain which form the grain is relaxed. This simple idea is consistent with the fact that generally talking a solid-solid interface is more stable than two separated solid-air interfaces. In the same way, the grain boundary energy γ_{bh} which simulates the domain-substrate interfacial energy in the horizontal plane is written:

$$\gamma_{bh} = |\gamma_{si} - \gamma_{ss}| \tag{7}$$

where γ_{ss} is the free surface tension of the substrate.
Finally, whatever the kind of interface which is considered, the physical model describes the evolution of the interfacial energies in the same way:

$$\Delta E_{bv} = |\gamma_{svi} - \gamma_{svj}| B(c_{ih} - c_{jh}) \tag{8}$$

$$\Delta E_{bh} = |\gamma_{si} - \gamma_{ss}| B(c_{iv} - c_{ss}) \tag{9}$$

$$B = \left(\frac{\ell^2}{h_i} + \ell\right) \tag{10}$$

where c_{ss} is the lattice constant of the family plane of the substrate orthogonal to its surface. In order to be consistent with the experimental systems, we computed the evolution of the crystallographic orientations with regard to the three dense and inequivalent family planes of YSZ in its cubic form. Therefore, c_{ih} owns three states which correspond to the inter-plane spacings of those family planes, $c_1 = c_{100} = a$, $c_2 = c_{110} = \frac{\sqrt{2}}{2}a$ and $c_3 = c_{111} = \frac{\sqrt{3}}{3}a$ with a the lattice parameter of cubic zirconia which is approximated to $a = 5\mathring{A}$ [37]. One can note that rigorously we should deduce the inter-plane spacing in the horizontal plane, c_{ih} from c_{iv}. However, for the sake of simplicity, we chose to approximate the horizontal crystallographic orientation of a domain i by $c_{ih} = c_{iv} = c_i$. Although this assumption is not totally satisfactory from a crystallographic point of view it is sufficient for the physical phenomena that we want to highlight. The subsequent modifications are represented in figure (1.3.b.). The free surface tension of each domain i, γ_{si}, is a function of its crystallographic orientation. Indeed, if we consider the three inequivalent planes of cubic zirconia (100), (110) and (111), the free surface tensions are [38]:

$$\gamma_s(100) = 3J.m^{-2} \quad (11)$$
$$\gamma_s(110) = 2.3J.m^{-2} \quad (12)$$
$$\gamma_s(111) = 1.2J.m^{-2} \quad (13)$$

Therefore, if we choose to define the crystallographic orientation of a domain i by the family planes parallel to the surface of the substrate, the inter-plane spacing c_i is fixed so is the free surface energy γ_{si}. Consequently,the previous equations (1.8) and (1.9) can de reformulated as:

$$\Delta E_{bv} = |\gamma_{si} - \gamma_{sj}|B(c_i - c_j) \quad (14)$$
$$\Delta E_{bh} = |\gamma_{si} - \gamma_{ss}|B(c_i - c_{ss}) \quad (15)$$

Moreover, the surface diffusion coefficient D_{si} is also a function of the crystallographic orientation. Indeed, the surface diffusion coefficient is a function of the free surface tension γ_{si} at the top of each domain i. We used a dimensional analysis to express D_{si} as a function of γ_{si}, Δt, the volumic density of cubic zirconia ρ and the characteristic length of a domain ℓ.
We denote the dimension of a physical parameter X by $[X]$.
The dimension of D_{si} is $[D_{si}] = m^2.s^{-1}$.
γ_{si} is the surface tension and $[\gamma_{si}] = kg.s^{-2}$, therefore $[\gamma_{si}\Delta t] = kg.s^{-1}$.
In order to express a physical parameter homogeneous to $m^2.s^{-1}$, we have to multiply the previous term by ℓ^2 $[\gamma_{si}\Delta t\ell^2] = kg.m^2.s^{-1}$ and divide the new expression by a mass in kg, in that case the mass which diffuses during one MCS $m = \rho\ell^3$. Consequently, the dimension of the expression $(\gamma_{si}\Delta t\ell^2)/(\rho\ell^3)$ is: $[(\gamma_{si}\Delta t\ell^2)/(\rho\ell^3)] = [(\gamma_{si}\Delta t)/(\rho\ell)] = (kg.m^2.s^{-1})/kg = m^2.s^{-1}$. Therefore we express D_{si} as follows:

$$D_{si} = \frac{\gamma_{si}\Delta t}{\rho\ell} \quad (16)$$

Consequently, we can express the total energy variation related to the evolution of the height and the crystallographic orientation of a domain i with regard to its four next nearest neighbors and the substrate:

$$\Delta E_{ij} = A\sum_{j=1}^{4}(h_i - h_j) + B\left[\left(\sum_{j=1}^{4}|\gamma_{si} - \gamma_{sj}|(c_i - c_j)\right) + |\gamma_{si} - \gamma_{ss}|(c_i - c_{ss})\right] \quad (17)$$

The first term corresponds to the evolution of local surface roughness around the domain i and the last two terms of right hand side of the equality correspond to the evolution of crystallographic orientation of the domain i with regard to its neighbors and to the substrate. With the previous expression of D_{si} the parameter A can be reformulated as follows:

$$A = Y(1+\nu)\gamma_{si}\Delta t\sqrt{\frac{1}{\rho k_B T}} \quad (18)$$

Table 2. Physical parameters used in the computations

Parameter	Value	Dimension	Reference
Temperature T	1800	K	-
Lattice parameter of YSZ a	5×10^{-10}	m	[37]
Inter plane spacing $c_1 = c_{100}$	5×10^{-10}	m	-
Inter plane spacing $c_2 = c_{110}$	3.5×10^{-10}	m	-
Inter plane spacing $c_3 = c_{111}$	2.9×10^{-10}	m	-
Free surface energy $\gamma_s(100)$	3	$J.m^{-2}$	[38]
Free surface energy $\gamma_s(110)$	2.3	$J.m^{-2}$	[38]
Free surface energy $\gamma_s(111)$	1.2	$J.m^{-2}$	[38]
Density ρ	6	$g.cm^{-3}$	[39]
Young modulus Y	300×10^9	$N.m^{-2}$	[39]
Poisson's ratio ν	0.3	-	[39]

The parameters used in our calculations are summarized in the table (1.2) below:

One can note that if the crystallographic orientation of the substrate is controlled then the surface of the substrate is perfectly defined and so are the lattice constant c_{ss} and the free surface tension γ_{ss}. For example if the substrate is a c-cut α-sapphire one, like in the experiments of Bachelet, the z axis in figure (1.3) is colinear to the $[hkl] = [001]$ crystallographic direction of the substrate. In our experiments we used α-sapphire substrates with random crystallographic orientations. Then it is not possible to define c_{ss} or γ_{ss} in that case. Obviously, this aspect is not critical. Indeed our calculations are based on a Monte Carlo approach which involves a statistical evolution of the system. Therefore, the energetic model allows us to simulate a general tendency of evolution of the system from one state to another. Consequently, the calculations are not sensitive to small variations of the lattice parameter of the substrate. Therefore in the following we consider that the α-sapphire substrate is a c-cut one which leads to $c_{ss} = 4.8 \times 10^{-10}m$ [40] and $\gamma_{ss} = 2.8J.m^{-2}$ [41].

3.2. PVD-CVD Processes: Islanding during Matter Deposition

In order to be consistent with experimental data, we compare our numerical results to experimental data of the Ge/Si model system.

3.2.1. Model of Thin Solid Film and Substrate

In that case, we also chose to model a thin polycrystalline film by a square lattice, composed of 10^4 elementary domains i. At the beginning of the computation process, the height of each domain i is $h_i = H$, with H the average thickness of the thin solid film. A domain i demonstrate a width ℓ and an unique crystallographic orientation with regard to the substrate. Indeed we consider that thin films synthesized by PVD-CVD processes are monocrystalline and epitaxially deposited on the substrate. Therefore, there is no grain boundaries between the domains which compose the virtual thin solid film. This virtual system is used to model a continuous and monocrystalline thin solid film deposited on a perfect single crystalline substrate. In our computations, we study the evolution of the height of each mesoscopic domain across time.

3.2.2. Physical Model

The numerical model is inspired from the previous works of Kawamura [42] and Russo [43] who established MC algorithms at the atomic scale to model QD's formation modes during deposition processes. Each domain i is submitted both to its surface tension γ_s and to an elastic stress field induced by the lattice mismatch between the film and the substrate, ϵ [44]. The figure (1.4) gives a schematic description of this virtual system.
The volume energy density of a thin film of initial thickness H, due to its free surface energy, is expressed as γ_s/H with H the average thickness of the film. Thus, the surface tension of a domain i can be written as $(\gamma_s h_i)/H$ and therefore the surface energy E_s of domain i induced by its free surface energy is:

$$E_s = \frac{\ell^2 \gamma_s h_i}{H} \tag{19}$$

where ℓ^2 is the free surface area at the top of each domain.
The elastic stress field leads to an elastic energy inside each domain of the film. This elastic energy can be expressed through linear elasticity theory. We suppose for the sake of simplicity that the elastic stress tensor is diagonal (for example for cubic phased materials). With the same assumption, ϵ, which is the lattice mismatch between the film and the substrate, is constant in the horizontal plane. Thus the resulting force in the horizontal plane $F_{//}$ associated to the displacement h_i in the vertical plane reduces to:

$$F_{//} = 2Y\epsilon\ell h_i \tag{20}$$

with ℓh_i the area of surfaces of domain i perpendicular to the x and y axis. Furthermore, the resulting force supported by the domain i on the vertical axis z, $F_\perp$, is equal to:

$$F_\perp = \nu Y\epsilon\ell^2 \tag{21}$$

Therefore, the resulting elastic energy E_e related to domain i and induced by the lattice mismatch ϵ is given by the sum of the works of those elastic forces, for a characteristic displacement $\epsilon\ell$ in the horizontal plane:

$$E_e = (F_{//} + F_\perp)\epsilon\ell \tag{22}$$

$$E_e = Y\epsilon(2\ell h_i + \nu\ell^2)\epsilon\ell \tag{23}$$

$$E_e = 2Y\epsilon^2\ell^2 h_i + Y\epsilon^2\ell^3\nu \tag{24}$$

The total energy, $E_t = E_s - E_e$ [44], associated to the evolution of the height of a domain i is:

$$E_t = \left(\frac{\ell^2\gamma_s}{H} - 2Y\epsilon^2\ell^2\right) h_i - Y\epsilon^2\ell^3\nu \tag{25}$$

$$E_t = \ell^2\left(\frac{\gamma_s}{H} - 2Y\epsilon^2\right) h_i - Y\epsilon^2\ell^3\nu \tag{26}$$

Similarly to the previous model, we express the variation of energy of a domain i with regard to its four next nearest neighbors j by the following expression:

$$\Delta E_{ij} = \ell^2 \left(\frac{\gamma_s}{H} - 2Y\epsilon^2 \right) \sum_{j=1}^{4} (h_i - h_j) \tag{27}$$

As previously reported for sol-gel process, the parameters used in the calculations related to PVD-CVD processes are summarized in the table below:

Table 3. Physical parameters used in the computations

Parameter	Value	Dimension	Reference
Temperature T	800	K	-
Free surface energy γ_s	1.1	$J.m^{-2}$	[45]
Young modulus Y	103×10^9	$N.m^{-2}$	[46]

3.3. Computation Process

In our computations, we used a classical Metropolis scheme [22]. At each MCS, a domain i is chosen at random. The algorithm determines which one of its four next nearest neighbors or the substrate is more likely to react. Therefore the probability of evolution of the height or the cristallographic orientation of a domain i with regard to one of its four next nearest neighbors or the substrate j is expressed as:

$$P_{ij} = \frac{e^{-\Delta E_{ij}/(k_B T)}}{\sum_j P_{ij}} \tag{28}$$

which involves that the lower ΔE_{ij}, the higher the probability of the transformation. This scheme is applied both for the first and the second model.

3.4. Results and Discussion

3.4.1. Surface Roughening: Comparison between Sol-Gel and PVD-CVD Processes

The formation of islands (sol-gel process) or QD's (PVD-CVD process) without nucleation is achieved through the evolution of the roughness of a thin film until the free surface of the substrate appears. In our numerical models, the evolution of the roughness of a virtual film is simulated by the evolution of the heights of the discrete mesoscopic domains which compose the film. The energy of a mesoscopic domain i of a thin film synthesized via sol-gel route is calculated with equation (1.17) which involves that the evolution of the roughness of the surface of the film is driven by surface diffusion and surface tension. The energy of a mesoscopic domain i of a thin film synthesized via a PVD-CVD process is calculated through equation (1.27). This equation implies that surface roughness evolution of the film is driven by a competition effect between free surface and interfacial stress energies.

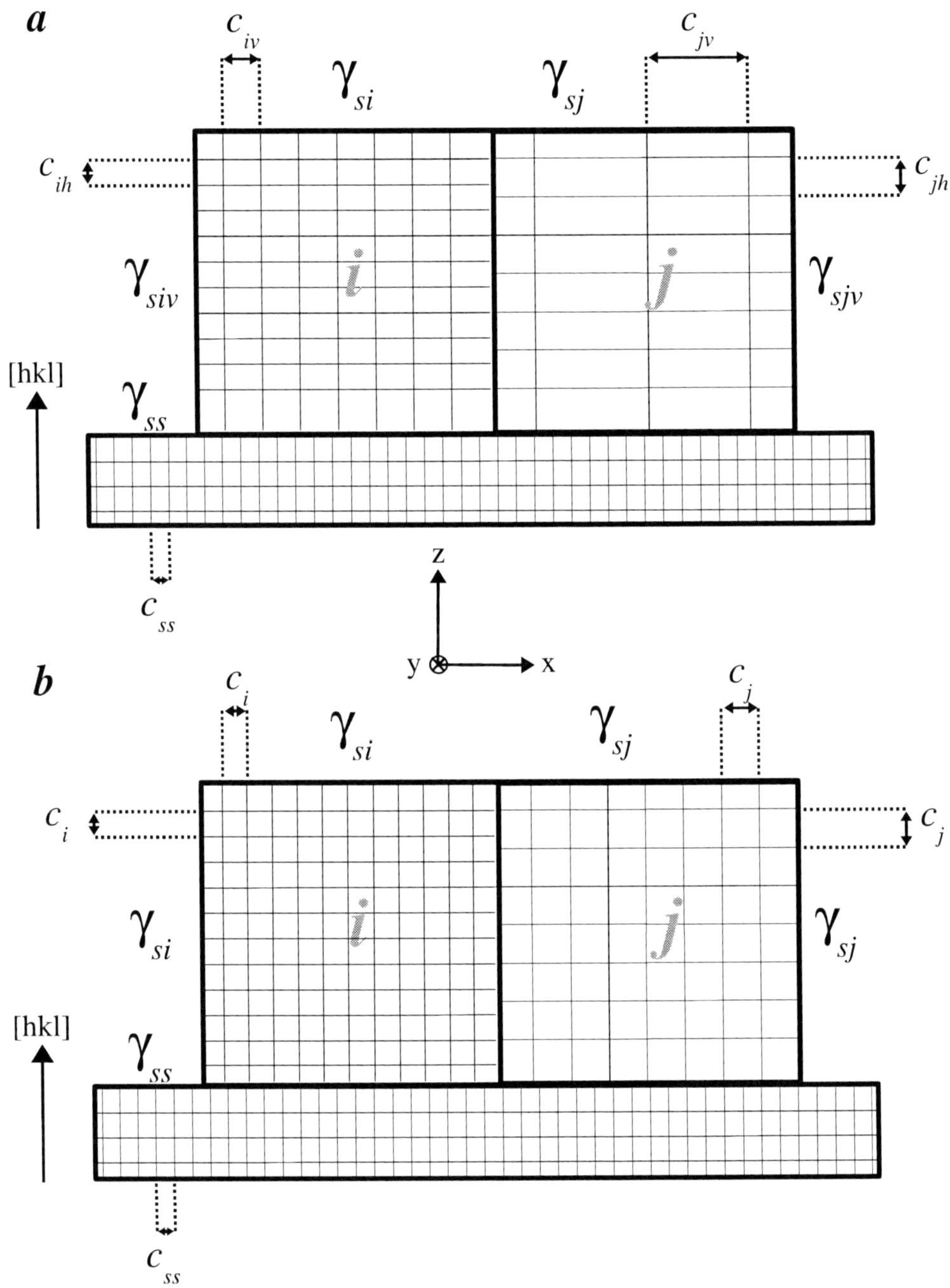

Figure 4. Schematic view of two elementary domains of a virtual thin solid film on a crystalline substrate which mimic a thin solid film deposited on a substrate by sol-gel dip-coating. This picture highlights the interfaces and inter-plane spacings taken into account in our model before a and after simplifications c.

The probability of changing the height of a domain with regard to equation (1.17) follows a monotonic tendency with regard to the intrinsic parameters of the thin film. The figure (1.5) presents top views of the evolution of the islanding of numerical thin films deposited on a substrate by sol-gel dip-coating after 10^6 MCS as a function of the initial thickness of the film H. The annealing temperature is fixed to $1800K$, so all the parameters of the equation are fixed. One can see that as the initial thickness of the film increases, the substrate is less dewetted. Indeed, as the initial thickness increases ($H = 27nm$), the amplitude of the roughness which appears at the top of the layer can not reach the substrate as fast as for thinner films ($H = 1nm$).

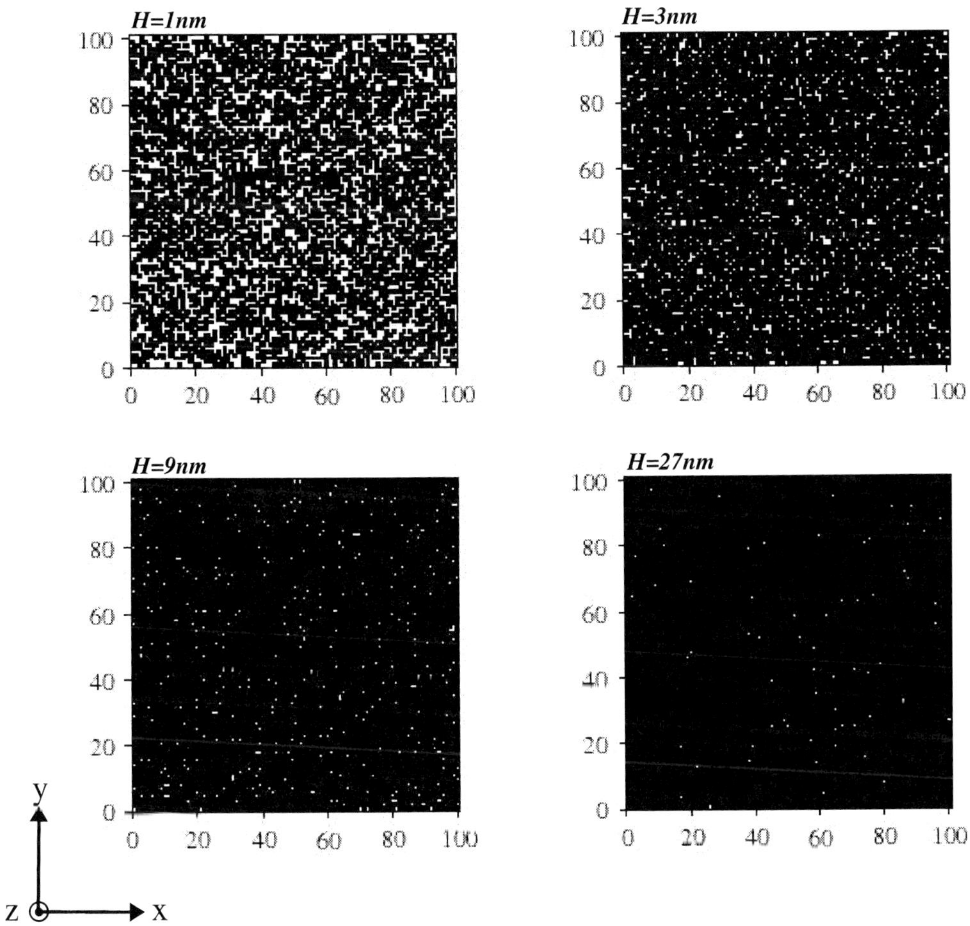

Figure 5. Top view of virtual thin solid films synthesized by sol-gel process after 10^6 MCS; effect of the initial thickness of the film on the islanding process.

The probability of changing the height of a domain with regard to equation (1.27) depends on the relative values of the parameters γ_s/H and $2Y\epsilon^2$. Indeed, if $\gamma_s/H > 2Y\epsilon^2$ then the energy of the domain i is expressed as $\Delta E_{ij} = K\sum_{j=1}^{NN}(h_i - h_j)$ (where $K = \gamma_s/H - 2Y\epsilon^2$) with $K > 0$, whereas $K < 0$ otherwise. On the one hand, if the lattice mismatch ϵ is low and/or with a low film thickness ($K > 0$), then the effect of surface energy is predominant. Therefore the probability of roughening is low and the film remains almost flat during the growing of the film which is described through the Frank-van-der-Merke (F-M) deposition process (or ALD Atomic Layer Deposition process). On the other hand, if the numerical values of ϵ and/or film thickness are high ($K < 0$), then the probability of roughening increases; therefore the effect of stress field inside the film is predominant which leads to the formation of islands. The surface roughening stage can either initiate from the top of a wetting layer, which corresponds to Stranski-Krastanov (S-K) growth mode, or directly from the top of the substrate without previous deposition of a wetting layer (ϵ very high) which describes the Volmer-Weber (V-W) growth mode. The Figure (1.6) illustrates the numerical calculations of the islanding process through the effect of the initial film thickness (ϵ=0.04, which is a typical lattice mismatch between Ge and Si [16, 17]) after 10^6 MCS while the figure (1.7) presents this evolution with regard to the lattice mismatch (H=1nm) after 10^5 MCS. We focus on the transition of K from positive to negative values as the islands are formed when $K < 0$. One can note that the increase of H or ϵ leads to a higher number of QD's, which is contrary to the evolution of a thin film synthesized via sol-gel processing where the higher thickness of the film leads to a lower number of islands. Moreover, it is clear that QD's formation is more sensitive to the variation of the lattice mismatch than to the variation of the initial thickness of the film as ΔE_{ij} is a function of ϵ^2 and $1/H$. This model describes qualitatively the evolution from F-M or S-K growth modes ($K > 0$ and $K \approx 0$ respectively) to V-W growth mode $K < 0$ as the number of QD's synthesized grows with the increase of ϵ and/or H. Those tendencies have already been reported for the Ge/Si system [21].

This section highlighted the main physical differences responsible for the islanding process of thin solid films synthesized either by sol-gel or PVD-CVD processes through surface rouhgening. In the case of thin solid films synthesized via sol-gel processes, the thinner the initial thickness of the films H, the greater the number of islands after thermal treatment whereas in the case of thin solid films synthesized by PVD-CVD processes, our computations demonstrate the opposite behaviour. This result is explained through equations (1.17) and (1.27) which point out that surface roughening of thin films synthesized via sol-gel processes is driven by surface diffusion and surface tension, whereas it is the consequence of a competitive mechanism between elastic stress energy and surface tension for thin films synthesized by PVD or CVD processes.

In the following we focus on the evolution of the out of plane crystallographic orientations of solid nanoislands during thermal treatment of thin solid films synthesized by sol-gel processes. Indeed, thin solid films synthesized by PVD-CVD processes demonstrate stable epitaxial relationships with regard to the substrate since the beginning of the deposition process. However, in the case of sol-gel processes, the formation of islands is linked to specific out of plane crystallographic orientations which are likely to evolute during thermal treatments.

3.4.2. Out of Plane Crystallographic Orientation: ZrO_2/Al_2O_3

In this section, we focus on the evolution of the out of plane crystallographic orientations of YSZ islands which appear on c-cut $alpha$-alumina substrates. In other words the ques-

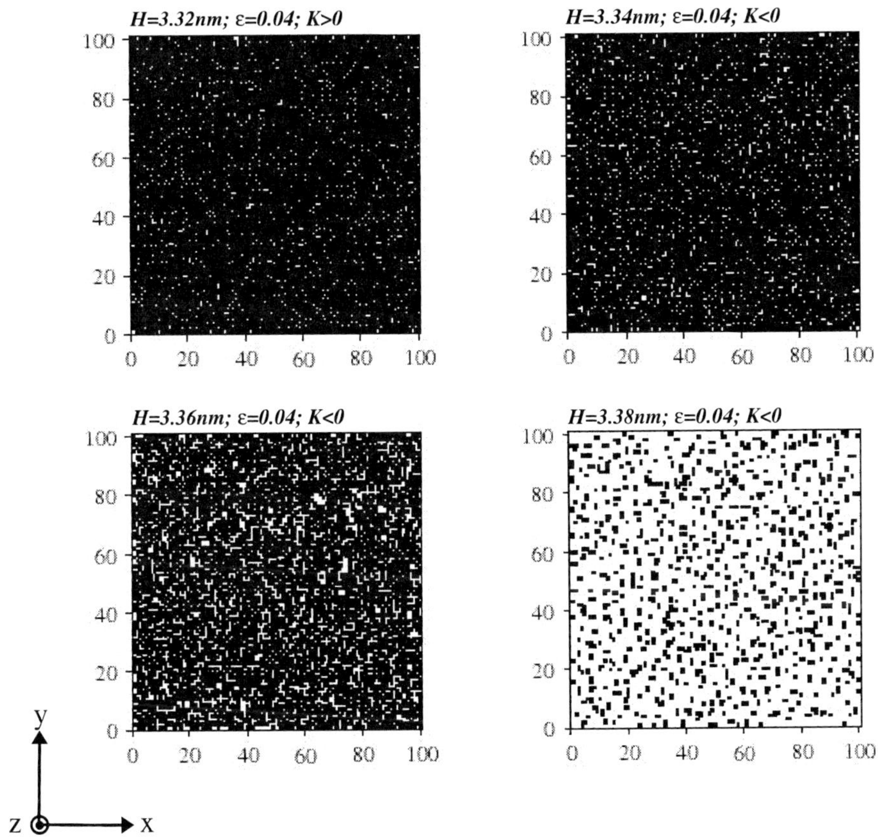

Figure 6. Top view of virtual thin solid films synthesized by PVD-CVD process after 10^6 MCS; effect of the initial thickness of the film on the islanding process.

tion is: which family planes of YSZ is parallel to the surface of the substrate after thermal treatment? In order to be consistent with the experimental system, we studied the evolution of the three dense and inequivalent family planes of yttria-stabilized cubic zirconia (100), (110) and (111) with regard to the (001) family planes of c-cut $alpha$-alumina substrate. The results are summarized on figure (1.8). The two pictures at the top of this figure represent top views of virtual thin solid films after two different computation times, 10^5 MCS on the left and 10^6 MCS on the right. The virtual crystalline domains are colored as a function of their out of plane crystallographic orientation with regard to the substrate; $c_1 = c_{100}$, $c_2 = c_{110}$ and $c_3 = c_{111}$. From the left picture one can note that the number of crystalline domains which demonstrate the (110) family planes parallel to the surface of the substrate is drastically lower than the domains which demonstrate the (100) or the (111) ones. This result is quantified on the picture at the bottom of figure (1.8). Indeed this picture represents the evolution of the number of virtual crystalline domains with one of the three crystallographic orientation as a function of the computation time. One can clearly see that the number of domains with the (110) crystallographic orientation is about 250 whereas the number of domains with the (100) and (111) crystallographic orientation is about 2800 and 2200 respectively. After 10^6 MCS the number of crystalline domains with the (110)

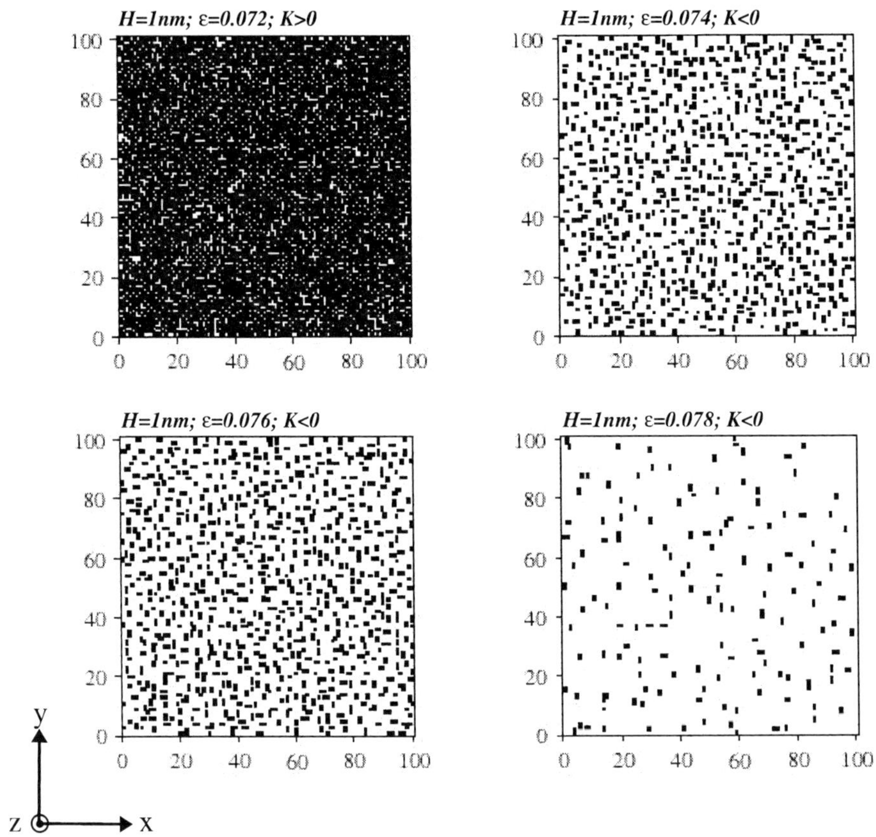

Figure 7. Top view of virtual thin solid films synthesized by PVD-CVD process after 10^5 MCS; effect of the lattice mismatch on the islanding process.

out of plane crystallographic orientation is equal to zero and the number of domains which demonstrate the (100) or (111) out of plane crystallographic orientations is constant.

The analysis of those pictures demonstrate that the system tends to eliminate the (110) out of plane crystallographic orientation. Finally, the virtual thin solid film is textured with to different out of plane crystallographic orientation distributed over the surface of the substrate. Those results are perfectly consistent with the experiments of Bachelet [32] who demonstrated a competitive growing mechanism between flat-top islands with the (100) out of plane crystallographic orientation and rounded islands with the (111) one on c-cut α-alumina substrates.

This result can be understood simply with our model. On one hand, one can note from equation (1.8) that the crystallographic orientation which is the more likely to be chosen is (111). Indeed this family planes demonstrate the lowest inter plane spacing $c_{111} = 2.9 \times 10^{-10} m$ and the lowest free surface tension $\gamma_s(111) = 1.2 J.m^{-2}$; therefore the probability to be chosen for the (111) crystallographic orientation is higher than the others. On the other hand, one can note from equation (1.9) that the crystallographic orientation which is the more likely to be chosen is (100). Indeed the crystallographic orientation which leads to the more stable state with regard to the surface of the substrate ($c_{ss} = 4.8 \times$

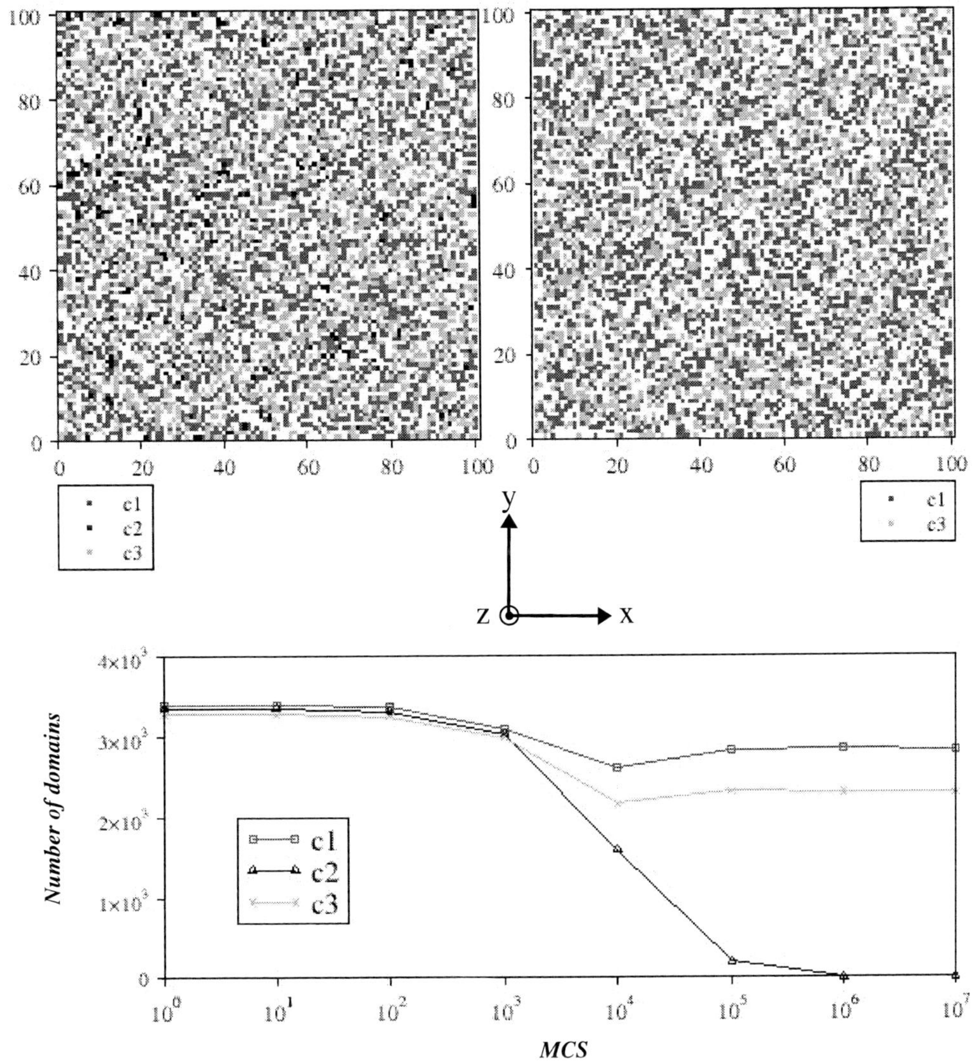

Figure 8. Evolution of the out of plane crystallographic orientations of virtual cubic zirconia islands on a c-cut *alpha*-alumina substrate. The top pictures are top views of the virtual crystalline domains colorized as a function of their out of plane crystallographic orientations: c1, c2 or c3. The bottom picture illustrates the evolution of the number of crystalline domains of specific crystallographic orientation as a function of the computation time (MCS).

$10^{-10}m$, $\gamma_{ss} = 2.8J.m^{-2}$) is (100) ($c_{100} = 5 \times 10^{-10}m$ and $\gamma_s(100) = 3J.m^{-2}$). In conclusion the formation of the (111) crystallographic orientation is favorized with regard to the neighbouring domains whereas the formation of the (100) one is induced and driven by the interface between the film and the substrate which explains the competitive growing mechanism observed experimentally.

4. Conclusion

In this paper we presented two physical models dedicated to simulate the islanding process of crystalline thin solid films of nanometric thickness, deposited on perfect single crystalline substrates either by sol-gel or PVD-CVD processes.
This approach allows one to highlight the physical parameters involved in the transition from continuous thin solid films to islands and to understand the differences which are observed experimentally between those processes.
In particular, the model associated to sol-gel process is releavent to simulate the evolution of the out of plane crystallographic orientation of the crystals which compose the thin solid film with regard to the substrate. However, we used some approximations to edify our physical models and a classical MC approach is not releavant to simulate the effects of real physical time sepcifically for diffusion processes.

References

[1] Brinker J.F., Scherer G.W., *Sol Gel Science: the physics and chemistry of sol gel processing* Academic Press Inc. (1990)

[2] C. Legrand, J.H. Yi, P. Thomas, R. Guinebretière, J.P. Mercurio *J. Eur. Ceram. Soc.* 19, 1379-1381 (1999).

[3] R. Guinebretiere, A. Dauger, B. Soulestin *Phyl. Mag. A* 79, 1517-1531 (1999).

[4] J.P. Mercurio, P. Marchet *Integrated Ferroelectrics* 61, 163-165 (2004).

[5] Y. Kajikawa, S. Noda *Appl. Surf. Sci.* 245, 281-289 (2005).

[6] L. Nurminen, A. Kuronen and K. Kashi *Phys. Rev. B* 63, 035407 (2000).

[7] M. Kalke, D.V. Baxter, *Surf. Sci.* 477, 95-101 (2001).

[8] P. Liu, Y.W. Zhang, C. Lu *Phys. Rev. B* 68, 035402 (2003).

[9] R.J. Wagner and E. Gulari *Phys. Rev. B* 9, 195312 (2004).

[10] R.J. Wagner, E. Gulari *Surf. Sci.* 590, 1-8 (2005).

[11] J. Tersoff *Phys. Rev. B* 43, 9377 (1991).

[12] K. Alchalabi, D. Zimin, G. Kostorz, H. Zogg *Phys. Rev. Lett.* 90, 026104 (2003).

[13] J.C. Nie, H. Yamasaki, Y. Mawatari *Phys. Rev. B* 70, 195421 (2004).

[14] D. Barreca, A. Gasparotto, E. Tondello, C. Sada, S. Polizzi and A. Benedetti *Chem. Vap. Deposition* 9, 199-206 (2003).

[15] T. Baron, F. Mazen, J.M. Hartmann, P. Mur, R.A. Puglisi, S. Lombardo, G. Ammendola, C. Gerardi, *Solid-State Electronics* 48, 1503-1509 (2004).

[16] P. Sutter, M.C. Lagally *Phys. Rev. Lett.* 84, 4637 (2000).

[17] R. M. Tromp, F.M. Ross and M.C. Reuter *Phys. Rev. Lett.* 84, 4641 (2000).

[18] G. Capellini, M. De Seta, F. Evangelisti *Mater. Sci. Eng. B* 89, 184-187 (2001).

[19] A. Portavoce, M. Kammler and R. Hull *Phys. Rev. B* 70, 195306 (2004).

[20] P.S. Chen, Z. Pei, Y.H. Peng, S.W. Lee, M.-J. Tsai *Mater. Sci. Eng. B* 108, 213-218 (2004).

[21] B. Voigtländer, *Surf. Sci. Rep.* 43, 127-254 (2001).

[22] N. Metropolis, A.W. Rosenbluth, M.N. Rosenbluth, A.T. Teller, E.J. Teller, *Chem. Phys.* 21, 1087 (1953).

[23] R. Guinebretière, A. Dauger, A. Lecomte, H. Vesteghem, *J. Non. Cryst. Sol.* 147, 542 (1992).

[24] R. Bachelet, A. Boulle, B. Soulestin, F. Rossignol, R. Guinebretière and A. Dauger, *Thin Solid Films* in press.

[25] J. Livage, C. Sanchez, F. Babonneau, *Chem. Ad. Mater* 12, 389 (1998).

[26] C. Mary, P. Lenormand, R. Guinebretière, A. Lecomte, A. Dauger, *J. Phys. IV* 10, 377-386 (2000).

[27] C. Mary, *PhD. Thesis: "Couches de zircone par voie sol-gel sur substrat de saphir: croissance granulaire et orientation cristalline."* (2001).

[28] A. Boulle, L. Pradier, O. Masson, R. Guinebretière, A. Dauger, *Appl. Surf. Sci.* 188, 80-84 (2002).

[29] A. Boulle, O. Masson, R. Guinebretière, A. Dauger, *Thin Solid Films* 424, 1-6 (2003).

[30] R. Guinebretière, R. Bachelet, A. Boulle, O. Masson, A. Lecomte, A. Dauger, *Mater. Sci. Eng. B* **109**, 42-46 (2004).

[31] R. Guinebretière, A. Boulle, O. Masson, A. Dauger, *J. Phys. IV* 12, 273-281 (2002).

[32] R. Bachelet, *PhD. Thesis: "Couches minces d'oxydes labores par voie sol-gel, pitaxies et nanostructures par traitements thermiques post-dpt."* (2006).

[33] N. Olivi-Tran, A. Boulle, A. Gaudon and A.Dauger, *Phys. Lett. A* 351, 426-430 (2006).

[34] X. Zhou, N.Olivi-Tran, A.Gaudon and A.Dauger, *J. of Alloys and Comp.* in press.

[35] F. Lallet, R. Bachelet, A. Dauger and N. Olivi-Tran *Phys. Rev. B* 74, 075411 (2006), and references within.

[36] R. Thouy, N. Olivi-Tran, R. Jullien, *Phys. Rev. B* 56, 5321 (1997).

[37] E.H. Kisi, C.J. Howard, Crystal structures of of zirconia phases and their inter-relation, *Key Eng. Mater.* Vols 153-154 (1998), Zirconia Eng. Ceramics: Old challenges-new ideas, p1, Trans. Tech. Publications Inc.

[38] A. Christensen, Emily A. Carter, *Phys. Rev. B* 58, 8050 (1998).

[39] P.Pascal, *Nouveau Traité de Chimie Minérale, Tome IX, Titane, Zirconium, Hafnium, Thorium* Masson Eds., Paris, (1963).

[40] C.J. Ball, *Powder Diffraction*, 21, 19-24 (2006).

[41] P.D. Tepesch, A.A. Quong, *Physica Statu Solidi b*, 217, 377-387 (2000).

[42] T. Kawamura, T. Natori, *Surf. Sci.* 438, 148-154 (1999).

[43] G. Russo, P. Smereka, *J. Comp. Phy.* 214, 809-828 (2006).

[44] F. Lallet, A. Dauger and N. Olivi-Tran *Physica Statu Solidi c* in press.

[45] H.J. Zhang, B. Lu, X.-S. Wang, F. Hu, H.Y. Li, S.N. Bao and P. He, *Phys. Rev. B* 70, 235415 (2004).

[46] R. Komanduri, N. Chandrasekaran and L. M. Raff, *Mater. Sci. and Eng. A*, 340, 58-67 (2003).

INDEX

A

B

C

F

G

H

M

Q

R

S

T

U

V